AF616320

Contemporary Topics in Immunobiology

VOLUME 12

Immunobiology of Parasites and Parasitic Infections

Contemporary Topics in Immunobiology

Contemporary Topics in Immunobiology

VOLUME 12

Immunobiology of Parasites and Parasitic Infections

Edited by
John J. Marchalonis
Medical University of South Carolina
Charleston, South Carolina

PLENUM PRESS • NEW YORK AND LONDON

Library of Congress cataloged the first volume in this series as follows:

Contemporary topics in immunobiology, v. 1-
1972–
New York, Plenum Press.
v. illus. 24 cm. annual.

1. Immunology—Periodicals.
QR180.C632 574.2′9′05 79-179761
ISSN 0093-4054 rev
Library of Congress 72 [r74c2]

Library of Congress Card Catalog Number 79-179761
ISBN 0-306-41418-X

A Division of Plenum Publishing Corporation
233 Spring Street, New York, N.Y. 10013

Printed in the United States of America

Contributors

Joseph F. Albright	*Department of Life Sciences* *Indiana State University* *Terre Haute, Indiana 47809 USA*
Julia W. Albright	*Department of Life Sciences* *Indiana State University* *Terre Haute, Indiana 47809 USA*
Anthony C. Allison	*Institute of Biological Sciences* *Syntex Research* *Palo Alto, California 94304 USA*
Robin F. Anders	*Laboratory of Immunoparasitology* *The Walter and Eliza Hall Institute of Medical Research* *Melbourne, Victoria 3050, Australia*
John W. Barnwell	*Malaria Section* *Laboratory of Parasitic Diseases* *National Institute of Allergies and Infectious Diseases* *National Institutes of Health* *Bethesda, Maryland 20205 USA*
Dean Befus	*Department of Pathology* *Host Resistance Programme* *McMaster University Health Sciences Centre* *Hamiltion, Ontario, Canada L8N 3Z5*
John Bienenstock	*Department of Pathology* *Host Resistance Programme* *McMaster University Health Sciences Centre* *Hamiltion, Ontario, Canada L8N 3Z5*
Colin B. Chapman	*Laboratory of Immunoparasitology* *The Walter and Eliza Hall Institute of Medical Research* *Melbourne, Victoria 3050, Australia*
Kathy M. Cruise	*Laboratory of Immunoparasitology* *The Walter and Eliza Hall Institute of Medical Research* *Melbourne, Victoria 3050, Australia*
Raymond T. Damian	*Department of Zoology* *University of Georgia* *Athens, Georgia 30602 USA*

R. Dayal *WHO Research and Training Centre*
Geneva Blood Centre and
Department of Medicine
Geneva University Hospital
1211 Geneva 4, Switzerland

H. D. Engers *Department of Immunology*
Swiss Institute for Experimental Cancer Research,
CH-1066 Epalinges, Switzerland

Edward G. Fox *Allergy and Immunology Section*
University of Pennsylvania School of Medicine
Philadelphia, Pennsylvania 19104 USA

Edito G. Garcia *Department of Parisitology*
Institute of Public Health
University of the Philippines
Ermita, Manila 2801, Philippines

Emaluela Handman *Laboratory of Immunoparasitology*
The Walter and Eliza Hall Institute of Medical Research
Melbourne, Victoria 3050, Australia

Russell J. Howard *Malaria Section*
Laboratory of Parasitic Diseases
National Institute of Allergies and Infectious Diseases
National Institutes of Health
Betheda, Maryland 20205 USA

Marshall W. Lightowlers *Department of Paraclinical Sciences*
University of Melbourne Veterinary Centre
Werribee, Victoria 3030, Australia

G. Lima *WHO Immunology and Research Training Centre*
Institute of Biochemistry
University of Lausanne
CH-1066 Epalinges, Switzerland

J. A. Louis *WHO Immunology and Research Training Centre*
Institute of Biochemistry
University of Lausanne
CH-1066 Epalinges, Switzerland

R. M. Maizels *Division of Parasitology*
National Institute for Medical Research
London NW7 1AA, England

Graham F. Mitchell *Laboratory of Immunoparasitology*
The Walter and Eliza Hall Institute of Medical Research
Melbourne, Victoria 3050, Australia

Nadia Nogueira *Department of Cellular Physiology and Immunology*
The Rockefeller University
New York, New York 10021 USA

Bridget M. Ogilvie *The Wellcome Trust*
London NW1 4LJ, England

Luc H. Perrin — *WHO Research and Training Centre*
Geneva Blood Centre and
Department of Medicine
Geneva University Hospital
1211 Geneva 4, Switzerland

J. Pestel — *WHO Immunology and Research and Training Centre*
Institute of Biochemistry
University of Lausanne
CH-1066 Epalinges, Switzerland

Mario Philipp — *Division of Parasitology*
National Institute for Medical Research
London NW7 1AA, England
Present address: New England Bio-Labs
Beverly, Massachusetts 01915 USA

S. Michael Phillips — *Allergy and Immunology Section*
University of Pennsylvania School of Medicine
Philadelphia, Pennsylvania 19104

Margaret Pinder — *Centre de Recherches sur les Trypanosomoses Animales (C.R.T.A.)*
B.P. 454
Bobo-Dioulasso, Upper Volta, West Africa

Michael D. Rickard — *Department of Paraclinical Sciences*
University of Melbourne Veterinary Centre
Werribee, Victoria 3030, Australia

Ian C. Roberts-Thomson — *Laboratory of Immunoparasitology*
The Walter and Eliza Hall Institute of Medical Research
Melbourne, Victoria 3050, Australia

L. Rodriguez da Silva — *WHO Research and Training Centre*
Geneva Blood Centre and
Department of Medicine
Geneva University Hospital
1121 Geneva 4, Switzerland

Georges E. Roelants — *Centre de Recherches sur les Trypanosomoses Animales (C.R.T.A.)*
B.P. 454
Bobo-Dioulasso, Upper Volta, West Africa

R. Titus — *WHO Immunology and Research Training Centre*
Institute of Biochemistry
University of Lausanne
CH-1066 Epalinges, Switzerland

M. J. Worms — *Division of Parasitology*
National Institute for Medical Research
London NW7 1AA, England

Preface

The phenomena involved in infections of man and domestic animals with metazoan or protozoan parasites present formidable practical problems as well as a theoretical challenge to immunologists, molecular biologists, and evolutionary biologists. With respect to the public health and economic problems, malaria, for example, remains a major health problem with approximately 200 million people being infected yearly and, on the basis of World Health Organization estimates, more than 1 million children die each year of malaria infections (Chapter 4). This volume addresses state-of-the-art immunologic approaches to the development of vaccines for parasitic diseases (Chapter 9) and analyses of studies bearing on the antigenic characterization of protozoan and metazoan parasites (Chapters 4, 5, and 7), on investigations of the role of precise mechanisms underlying natural resistance or nonpermissiveness of the host to parasitic infections (Chapters 1, 2, and 12), on induced mechanisms including the generation of parasite-specific T-cell lines and clones (Chapter 6), and on the generation of monoclonal antibodies (Chapters 4 and 5) to parasite antigens of distinct developmental stages.

Great progress has been made in characterizing parasite antigens capable of inducing a protective response in the vaccinated host; further progress in this area strongly depends on biochemistry and molecular biology with the long-term goal of synthesizing such antigens chemically or producing them by means of recombinant DNA technology (Chapter 4). Antigens have now been isolated from a variety of parasite species; these have been characterized to some degree by polyacrylamide gel electrophoresis and by partial sequence analysis (Chapter 7). Many of the contributors to this volume have expressed optimism that it will be possible to develop vaccines conferring protection against parasitic infections. The current drawback is that these antigens are not available in sufficiently large quantity to be used routinely for vaccines. It will be necessary to employ means such as biochemical synthesis of antigenic determinants or recombinant DNA technology to prepare sufficient quantities of antigen to use as vaccines.

Although parasites tend to be highly antigenic, well-adapted parasites are quite adept at evading the hosts immune response. They apparently employ a variety of strategies, such as coating themselves with host serum components, including albumin (Chapter 8) or host serum antibody of a type that is relatively ineffective in activating macrophages or in mediating complement-dependent lyses (Chapter 9). Parasites have been observed to produce or alter immunomodulatory factors and lead to a state of host-immune suppression (Chapter 10).

Another parasite strategy is the production of factors that bring about a polyclonal B-cell activation that may interfere with the specific antimalarial immune response of the host (Chapter 4). In many ways, the capacity of parasites to evade immune responses of hosts is reminiscent of that of tumors. However, the analogy breaks down because tumors grow essentially without bounds and bring about the dealth of the host; whereas parasites and certain selected hosts have undergone many millions of years of coevolution leading to the result that in a well-adapted parasite, the numbers of parasite within the host reach an equilibrium level that is tolerable to the host.

It was previously predicted (Marchalonis, *Immunity in Evolution*, Harvard University Press, 1977) that a kind of arms race can occur between parasites and their hosts, in which the parasites could express a succession of varying antigens and the host would constantly be using the variability of its immune system to try to catch up and eliminate the parasite variants. As is well established, immunoglobulin variability derives from the presence of many variable region genes and involves a process of recombination and rearrangement with genes encoding constant regions and diversity (D) segments. Current data indicate that surface glycoproteins of trypanosomes have homologous C-terminal regions and also possess variable N-terminal regions that can be classified into subgroups in a manner parallel to the description of immunoglobulin isotypes. There also appears to be an intense rearrangement of genes specifying for these glycoproteins in the trypanosome genome. The capacity of parasites to alter their surface antigens in response to host-immune mechanisms is a fascinating problem in biology and may lead to new insights into mechanisms of gene rearrangement.

Parasites are, in general, susceptible to a variety of mechanisms of natural resistance such as complement mediated lyses, macrophages and natural killer cells (Chapter 1) and also are recognized by T cells, antibodies, and activated macrophages (Chapter 6). Yet they have devised many strategies for coopting or evading these host-immune responses (Chapter 10). Damian develops an integrated, or holistic, explanation of schistosome immunity, and Phillips and Fox use two apparently dissimilar parasitic diseases, filariasis and schistosomiasis, to generate a mechanistic immunologic model accounting for the clinical spectrum of parasitic diseases. In addition to the enormous practical problems currently presented by parasites from the standpoint of public health and agriculture, they offer many fascinating systems for the study of immunobiology and of the molecular biology of the concerted interactions between host recognition and parasite response. I thank all the contributors for their stimulating chapters, which illustrate how the concepts and technology of modern immunobiology and molecular immunology can be applied to an ancient problem.

John J. Marchalonis

Charleston, South Carolina

Contents

Chapter 1

Natural Resistance to Animal Parasites

Joseph F. Albright and Julia W. Albright

I. Introduction 1
II. Natural Resistance in Vertebrate Hosts 4
A. Strains of Inbred Mice 4
B. Biozzi Strains of Mice 9
C. Variations in Resistance Related to Sex and Age 10
III. Mechanisms of Natural Resistance 14
A. Humoral Substances 14
B. Immunologically Deficient Mice 17
C. Macrophages 24
D. Naturally Cytotoxic Cells 30
E. Resistance in Nonpermissive Prospective Hosts 34
IV. Conclusions 37
V. References 40

Chapter 2

Intracellular Mechanisms of Killing

Nadia Nogueira

I. Introduction 53
II. Parasite Interiorization by Mononuclear Phagocytes 54
A. Subcellular Localization 54
B. Biochemical Events 54
III. *Trypanosoma cruzi:* Macrophage Activation and Intracellular Killing 56
A. Correlation with H_2O_2 Generation in a Murine System 56
B. Correlation with H_2O_2 Generation in Human Mononuclear Phagocytes 57

C. Triggering of H_2O_2 Release by Macrophages Phagocytizing *T. cruzi* 61
IV. *Toxoplasma gondii:* Macrophage Activation and Intracellular Killing 63
A. Correlation with the Generation of Reactive Oxygen Intermediates in a Murine System 63
B. Failure to Trigger the Respiratory Burst 64
V. *Leishmania* spp.: Macrophage Activation and Intracellular Killing 64
A. *Leishmania enrietti* 64
B. *Leishmania tropica* 65
C. *Leishmania donovani* 65
VI. Conclusions 66
VII. References 66

Chapter 3

Induction and Expression of Mucosal Immune Responses and Inflammation to Parasitic Infections

Dean Befus and John Bienenstock

I. Introduction 71
II. The Interface: Host Cells Meet Parasite Antigens 72
A. Antigens and Epithelial Cells 73
B. Induction Events in Mucosal Lymphoid Aggregates 74
C. Comment 79
III. Immune Responses and Mucosal Inflammation in Parasitic Infections 79
A. Epithelium 79
B. Lamina Propria 82
C. Intestinal Lumen 85
IV. Mucosal Effectors of Host Resistance 86
A. Parasite Elimination 86
B. Host Pathology 95
V. Conclusions 96
VI. References 96

Chapter 4

Antigenic Characterization of Plasmodia

Luc H. Perrin, L. Rodriguez da Silva, and R. Dayal

I. Introduction 109
II. Plasmodial Life Cycle 110

III. Sporozoites 112
IV. Exoerythrocytic Stages 112
V. Asexual Erythrocytic Stages 113
A. Antigenic Characterization of the Various Asexual Erythrocytic Stages 113
B. Identification of Plasmodial Antigens Relevant to Protection Using Polyclonal and Monoclonal Antibodies 116
C. Circulating *P. falciparum* Antigens and Polypeptides Released in Culture Medium 120
VI. Gametes 120
VII. Conclusions and Prospects 121
VIII. References 122

Chapter 5

Roles of Surface Antigens on Malaria-Infected Red Blood Cells in Evasion of Immunity

Russell J. Howard and John W. Barnwell

I. Introduction 127
II. Immunity to Asexual Malaria Parasites 128
A. Natural Infections in Humans 128
B. Experimentally Induced Infections in Human Subjects and Animals 128
C. Active Immunization 130
D. Nature of Immune Responses 130
III. Antigens on Malaria-Infected Erythrocytes—Targets of Parasiticidal Immunity? 136
A. New Antigens Expressed on Malaria-Infected Erythrocytes . . 137
B. Parasiticidal Mechanisms—Activated *in Vivo* by Antibody Binding to Infected Cells? 139
IV. Nature of New Antigens on Infected Cells 142
A. New Antigens Associated with Morphologic Alterations of the Membrane 142
B. New Antigens Associated with Antigenic Variation: SICA Antigens 145
C. Other New Antigens on Infected Cells 160
V. Parasitic Evasion of Immunity—General Concepts 160
VI. SICA Antigen-Immune Evasion 165
A. Immune Evasion by Antigenic Variation 166
B. Comparison of SICA[+] and SICA[-] Parasites: Evidence for Immunosuppression 168
VII. Knobs and Immune Evasion by Sequestration 174
A. Sequestration *in Vivo* 174

B. Binding to Endothelial Cells and Melanoma Cells *in Vitro* . . . 179
C. Sequestration and Parasite Survival 180
D. Role of Antibody to Knob Components in Immunity 184
E. Summary of Knob Function 187
VIII. Concluding Remarks 188
IX. References 191

Chapter 6

Murine T-Cell Responses to Protozoan and Metazoan Parasites: Functional Analysis of T-Cell Lines and Clones Specific for *Leishmania tropica* and *Schistosoma mansoni*

J. A. Louis, G. Lima, J. Pestel, R. Titus, and H. D. Engers

I. *Leishmania tropica* Major 201
A. Description of Methods Used for the Functional Analysis of Specific T-Cell Responses in Mice Primed with *L. tropica* Antigens 204
B. Influence of Infection with *L. tropica* on the Development of Specific T-Cell Responses in Mice 205
C. *Leishmania*-Specific T-Cell Lines 208
D. *Leishmania*-Specific T-Cell Clones 214
E. Concluding Remarks 216
II. *Schistosoma mansoni* 216
A. Murine T-Lymphocyte-Dependent Proliferative Responses to Schistosoma Antigens: Characteristics and Specificity 217
B. Helper Activity of LN Cells from Mice Primed with Schistosoma Antigens 218
C. Schistosome-Specific T-Cell Lines 219
D. Cloned T Cells Specific for Schistosoma Antigens 219
E. Concluding Remarks 220
III. References 221

Chapter 7

Immunobiology of African Trypanosomiasis

Georges E. Roelants and Margaret Pinder

I. Introduction 225
II. Trypanosome Antigens 226
A. Variant-Surface Glycoprotein 226
B. Internal Antigenic Determinants 230

III. Protective Immune Response 231
A. Antibody 231
B. Cell-Mediated Immunity 235
IV. Deleterious Effects of the Immune Response 236
V. Effects of Trypanosome Infection on the Lymphoid System 238
A. Hypergammaglobulinemia and Polyclonal Activation 238
B. Immunodepression 238
VI. Susceptibility to Trypanosomiasis 245
A. Nonpermissive Hosts 245
B. Variations of Susceptibility in Permissive Hosts 248
VII. Conclusions 260
VIII. References 261

Chapter 8

Rodent Models of Filariasis

Mario Philipp, M. J. Worms, R. M. Maizels, and Bridget M. Ogilvie

I. Introduction 275
II. Responses of the Host to Developing Larvae and Adult Worms . . 276
A. Studies in Susceptible Hosts Infected with Filariae Naturally Parasitic in Rodents 276
B. Studies in Rodents Susceptible to Infections with Nonrodent Filariae, Including Parasites of Humans 283
C. Experimental Filariasis in Resistant Rodents 291
III. Responses of the Host to Microfilariae 294
A. Models of Latent Filariasis 295
B. Responses to Microfilariae in Rodents Resistant to Normal Infections 301
IV. Conclusions 308
V. References 311

Chapter 9

Examination of Strategies for Vaccination against Parasitic Infection or Disease Using Mouse Models

Graham F. Mitchell, Robin F. Anders, Michael D. Rickard, Colin B. Chapman, Marshall W. Lightowlers, Ian C. Roberts-Thomson, Edito G. Garcia, Emanuela Handman, and Kathy M. Cruise

I. Introduction 323

II. *Taenia taeniaeformis* (Murine Cysticercosis): Immunoprophylaxis and Immunotherapy . . . 328
III. *Leishmania tropica* (Murine Cutaneous Leishmaniasis): Vaccination against Parasite Establishment or Chronic Disease . . . 329
IV. *Schistosoma japonicum* (Murine Schistosomiasis Japonica): Vaccination against Immunopathologic Disease . . . 336
V. *Giardia muris* (Murine Giardiasis): Vaccination for Accelerated Rejection of Intestinal Parasites . . . 342
VI. *Nematospiroides dubius* (Murine Nematodiasis): Vaccination for Accelerated Rejection of Intestinal Parasites . . . 344
VII. *Fasciola hepatica* (Murine Fascioliasis): Vaccination for Neutralization of Parasite-Protective Mechanisms . . . 347
VIII. Concluding Comments . . . 348
IX. References . . . 349

Chapter 10

Immunity in Schistosomiasis: A Holistic View

Raymond T. Damian

I. Introduction . . . 359
II. Host Quality and Immunity . . . 364
III. Immunogenic Stages . . . 367
A. Concomitant Immunity and the Immunogenic Role of Adult Worms . . . 367
B. Immunogenic Role of Larval Stages . . . 369
C. Immunogenic Role of Eggs . . . 370
D. Stage-Specific or Stage-Common Antigens? . . . 372
IV. Vulnerable Stages . . . 374
A. Concomitant Immunity and the Vulnerability of Larval Stages . . . 374
B. Immune Evasion . . . 377
C. Adult Worm Vulnerability . . . 381
V. Dose and Time Effects in Induction of Immunity . . . 386
A. Effect of Cercarial Dose Size on Development of Primary Infections in Nonpermissive Hosts . . . 386
B. Effect of Cercarial Dose Size on Reinfection Immunity . . . 387
C. Effect of Adult Worm Dose Size in Surgical Transfer Experiments on Reinfection Immunity . . . 388
D. Dose Effects Using Irradiated Larvae and Dead Vaccines . . . 389

E. Interaction of Worm Burden and Residence Time in the Development of Reinfection Immunity 390
F. Conclusions 390
VI. Influences of Host and Parasite Quality on Effective Antigenic Load 391
A. Host Quality 391
B. Parasite Quality 392
VII. Interaction of Effective Antigenic Load and Host Genotype in Induction of Immunity 392
A. Influence of Host Genotype 392
B. Effector–Suppressor Switches 394
VIII. Dual-Immunity Theories in Schistosomiasis 396
IX. Connections Between Acquired Resistance and Egg-Induced Hypersensitivity 397
A. Immunopathologic Host Changes 397
B. Cercarial–Egg Antigenic Cross-Reactions and Their Possible Significance 398
C. Granuloma-Associated Nonspecific Modulatory Phenomena 399
D. Conclusions 399
X. Concluding Remarks 400
XI. References 402

Chapter 11

Immunopathology of Parasitic Diseases: A Conceptual Approach

S. Michael Phillips and Edward G. Fox

I. Introduction 421
II. Filariasis 422
A. Clinical and Histologic Pathology 425
B. Immunopathology of Filariasis 428
III. Schistosomiasis 433
A. Life Cycle 433
B. Disease Spectrum 434
C. Acute Pulmonary Schistosomiasis 438
D. Early Systemic Schistosomiasis 439
E. Chronic Schistosomiasis 440
F. *Schistosoma mansoni:* Immunopathogenesis—Granuloma Formation and Modulation 443
G. *Schistosoma japonicum:* Immunopathogenesis 448

IV. Summary 449
V. References 451

Chapter 12

Cellular Immunity to Malaria and Babesia Parasites: A Personal Viewpoint

Anthony C. Allison

I. Introduction 463
II. Relationship of the Sickle Cell Trait to Malaria 464
III. Malaria and Glucose 6-Phosphate Dehydrogenase Deficiency 465
IV. Synergistic Effects of Inherited and Acquired Immunity to Malaria 467
V. Role of Cell-Mediated Immunity in Malaria and Babesiosis 467
VI. Nonspecific Immunity to Hemoprotozoan Infections 469
VII. Antibody Independence of Immunity to Hemoprotozoa 469
VIII. Multiplication of *P. falciparum* in Cultures of Erythrocytes with Abnormal Hbs and G6PD Deficiency 470
IX. Susceptibility of Erythrocytes to Oxidant Stress 471
X. Factors That Increase and Decrease Susceptibility of Erythrocytes to Oxidant Stress 472
XI. Production of Superoxide and Hydrogen Peroxide by Leukocytes 475
XII. Oxidant Stress Exerted on Erythrocytes by Malaria Parasites 477
XIII. Effects of Oxidants on Malaria Parasites in Erythrocytes 478
XIV. Tumor-Necrosis Factor, Lipoproteins, and Oxidized Lipids 480
XV. Cell-Mediated Immunity and Premunition 482
XVI. Synergism of Cell-Mediated Immunity and Inherited Erythrocytic Resistance 483
XVII. Controversy and the Development of Science 483
XVIII. References 484

Index **491**

Chapter 1

Natural Resistance to Animal Parasites

Joseph F. Albright and Julia W. Albright

Department of Life Sciences
Indiana State University
Terre Haute, Indiana 47809

I. INTRODUCTION

There exists great variation in the degree to which parasites are fastidious with respect to their hosts. This is exemplified by the expression "host range," a descriptive feature of a given parasite used to characterize the variety of species that can be infected by that parasite. Many parasites are highly selective (i.e., specific), their range of hosts being limited to one, or a few related, species. These are believed to be well-adapted parasites having settled on a host that is optimum for their welfare. Other parasites have a broad host range and will survive, if not thrive, in a variety of hosts.

The contrast between monospecific (or oligospecific) and polyspecific parasites points to an important fact that has long been recognized but that has received relatively little attention, viz., that there exist powerful mechanisms for resisting or promptly eliminating parasitic invaders among prospective hosts. There are numerous examples of this nonpermissiveness, commonly referred to as "natural resistance," of prospective vertebrate and invertebrate hosts *vis-à-vis* all types of parasites including eukaryotic, prokaryotic, and viral. Until quite recently virtually nothing was known about the mechanisms responsible for host nonpermissiveness; indeed, until recently interest in the subject was limited to a few, lonely investigators. Currently, however, there is rapidly growing interest in the subject as revealed by the publication of the proceedings of several recent symposia (Cudkowicz *et al.*, 1978; Skamene *et al.*, 1980; Smith *et al.*, 1980) and a plethora of original research papers.

In preparing this chapter our objective has been heuristic rather than compilative; i.e., we have tried to assess the current status of research concerned with natural resistance of prospective hosts toward animal parasites and to indicate profitable directions of future research. For this reason, our use of the available literature has been selective rather than all-inclusive. We realize, of course, that to assume, even for a moment, the role of scientific soothsayer is risky. Not only do we run the risk of being ridiculed for our dotage, we also run the risk of estranging esteemed friends whose work we may have treated unjustly. It is hoped that neither of these will befall us.

The definition of the expression "natural resistance" was stated clearly by Frenkel and Caldwell (1975), who wrote

> Natural resistance is defined as the combined protective effects of anatomic barriers, baseline phagocytosis, digestion by polymorphonuclear neutrophils, and expressor mechanisms (e.g., complement), all modulated by nutritional and hormonal states and by genetic makeup.

Our discussion of natural resistance is focused on animal parasites but draws occasionally on illuminating studies of resistance to bacteria. Virtually nothing is included about resistance to other microorganisms or viruses.

This chapter does not deal at any length with anatomic barriers, despite their intrinsic interest. Recent work concerned with variation in susceptibility of human erythrocytes to invasion by malaria organisms illustrates the fascination of studying anatomic barriers. Erythrocytes appear to have surface receptors for plasmodia; e.g., erythrocytes lacking Duffy blood group substances can bind to *Plasmodium knowlesi*, although they are unable to internalize the parasite (Miller *et al.*, 1977). The glycoprotein, glycophorin, appears to be at least a component of the erythrocyte receptor for *P. falciparum* (Perkins, 1981). The process of internalization is complex and involves the redistribution of membrane components in the process of formation of a parasitophorous vacuole (McLaren *et al.*, 1979; Aikawa *et al.*, 1981; Dluzewski *et al.*, 1981). It is well known that the stage of erythrocyte maturation affects invasiveness by malaria organisms; *P. falciparum* invades preferentially recently matured normocytes, whereas *P. vivax* and *P. yoelii* seem to prefer reticulocytes (Pasvol *et al.*, 1980; Zuckerman, 1957).

Similarly, in this chapter we deal with the subjects of nutritional and hormonal influences on parasite–host relationships in a superficial fashion and only when consideration of such influences sheds light on a seemingly unrelated topic. Both nutritional and hormonal effects on host–parasite relationships are worthy subjects for investigation, and far too little is known about them.

To set the stage for the discussion that follows, and to introduce the phenomenon of natural resistance to the reader less experienced in the subject, we refer, in brief, to some of our recent studies with the rodent trypanosomes (Albright and Albright, 1981*a,b*). *Trypanosoma musculi* is rigorously host specific; it thrives in mice but fails to establish an infection even in rats. Similarly, the rat-specific *Trypanosoma lewisi* fails to become established in normal mice. The mechanisms underlying the resistance of rats and mice to the excluded trypanosome are unknown, although some progress has been made (see Section III.E). Also, there is considerable intraspecies variation among individuals and inbred strains in the degree of resistance to the trypanosomes. This has been studied among strains of mice with respect to *T. musculi* infections (Albright and Albright, 1981*a*). Highly susceptible strains such as A/J develop a 10- to 100-fold higher level of parasitemia, as well as a more prolonged course of infection, than does the relatively resistant C57BL/6 strain. This sort of variation in resistance among strains of inbred mice is typical of that reported during the last few years for a variety of infectious agents, including bacteria and protozoan and metazoan parasites (Skamene *et al.*, 1980). Genetic analysis has provided a surprise concerning the genetic control of resistance to infectious organisms. It appears that in many cases, at least, major control of resistance resides in a relatively few genes; in some well-studied cases, control has been assigned to a single locus. This relative simplicity of genetic control was unexpected. It was assumed that many genes would be involved corresponding to the assumed complexity of the combative and reactive mechanisms evoked by an infectious agent. Indeed, mathematical simulations have led to the conclusion that genetic complexity favors stability of a parasite–host relationship (see Yu, 1972).

One key to understanding the significance of variation in resistance among individuals of a given species as related to the control of host–parasite relationships comes from epidemiologic analysis. Bradley (1972) has provided a lucid discussion. He points out that the most stable relationship between host and parasite exists when parasite transmission is well above the minimum needed for persistence of the parasite and, concurrently, parasite numbers are regulated by individual hosts. Regulation of parasite number by individual hosts may involve a premunitive mechanism that prevents superinfection [e.g., the "concomitant immunity" seen in experimental primate infections with *Schistosoma mansoni* (Smithers *et al.*, 1969)] or mechanisms that limit parasite reproduction [e.g., elaboration of ablastin, characteristic of rodents infected with trypanosomes (D'Alesandro, 1962)] or both. Short of this ideal situation, and given variations in transmission efficiency and communities of hosts, Bradley's analysis clearly shows that variation in host resistance is the expected situation.

At the extremes of variation in resistance are those individuals, or species, that are totally resistant at one end and those that are uniformly susceptible at the other. Some questions that arise, then, are the following: Are the mechanisms of natural resistance present in resistant individuals of a generally permissive host species essentially the same as those employed by related but totally nonpermissive species? Or are there different mechanisms at work in the nonpermissive species of hosts? In either case, what are the phylogenetic distributions of these mechanisms? Has natural resistance to parasites been observed in invertebrates and, if so, what are the mechanisms and can they be readily analyzed?

With respect to the last question concerning natural resistance in invertebrates, some provocative initial studies have been conducted. Resistance of inbred lines of snails (*Biomphalaria glabrata*) to sporocysts of *Schistosoma mansoni* has been investigated (Richards, 1975). Of 10 inbred lines 8 were susceptible to a given strain of *S. mansoni* (PR-1), but with respect to a different strain of *S. mansoni* (PR-2) 7 of the 10 lines were nonsusceptible (Sullivan and Richards, 1981). The latter nonsusceptible lines provoked varying intensities of resistance as judged by the vigor of encapsulation of parasite sporocysts by snail hemocytes and the speed with which sporocysts were destroyed (Sullivan and Richards, 1981). Resistance to sporocysts could result from active attack by snail hemocytes, resembling certain cell-mediated immune reactions in vertebrates (Cheng, 1968), from an inadequate supply of stimulants of growth and development or from an unsuitable host snail environment (Lie *et al.*, 1977). The former cause is of primary interest in the present discussion. In other studies, it was demonstrated that the hemocytes of snails susceptible to PR-1 sporocysts were able to damage the sporocysts severely if the plasma phase of hemolymph from other resistant snails was present in the reaction mixture (Bayne *et al.*, 1980). Cytotoxicity of hemocytes for sporocysts involves close proximity of the effector and target cells in a manner resembling antibody-dependent cell-mediated cytotoxicity (ADCC) reactions in vertebrate systems. The nature of the required humoral substances present in hemolymph is of considerable interest. It appears that they have an affinity for binding to carbohydrate moieties on the surface of sporocysts, thus resembling lectins (Yoshino *et al.*, 1976).

In insects, too, there is evidence of cell-mediated resistance to parasites. A particularly fascinating system, in the special sense that a detailed genetic analysis should be possible, is the cell-mediated resistance displayed by *Drosophila melanogaster* toward the parasite *Asobara tabida* (Nappi, 1981). In these invertebrate systems it is, as yet, impossible to judge when a response should be considered to reflect natural resistance as contrasted to acquired immunity. Perhaps at this rudimentary level the distinction fades.

II. NATURAL RESISTANCE IN VERTEBRATE HOSTS

A. Strains of Inbred Mice

Scholarly reviews of this subject have been written by Mitchell (1979*a*) and Wakelin (1978). Most investigations of genetic control over parasitic infections have involved the use of strains of inbred and recombinant inbred mice (Taylor, 1980). Numerous studies have been performed dealing with a variety of parasites. The information gained from these studies has provided some surprises: (1)

meaningful genetic analysis is possible because the number of genes involved in major control mechanisms is small, and (2) in general, genes associated with the "immune response" (Ir) region, or any other region, of the major histocompatibility complex (MHC) are not primarily responsible for regulating susceptibility to parasites. The second of these conclusions comes from analyses of murine infections with each of the following parasites, which include some bacteria and a rickettsia: *Babesia microti* (Eugui and Allison, 1980), *Leishmania donovani* (Bradley, 1977; Bradley *et al.*, 1979; Bradley, 1980), *Leishmania tropica* (De Tolla *et al.*, 1980); *Listeria monocytogenes* (Cheers and McKenzie, 1978; Skamene *et al.*, 1979; Skamene and Kongshavn, 1979), *Mycobacterium bovis* (Gros *et al.*, 1981), *Plasmodium chabaudi* (Eugui and Allison, 1980), *Rickettsia tsutsugamushi* (Groves *et al.*, 1980), *Salmonella typhimurium* (Hormaeche, 1979; Plant and Glynn, 1976; O'Brien *et al.*, 1980*b*), *Taenia taeniaeformis* (Mitchell *et al.*, 1977), *Trichuris muris* (Wakelin, 1975), *Trypanosoma congolense* (Morrison and Murray, 1979), *Trypanosoma cruzi* (Trischmann *et al.*, 1978; Trischmann and Bloom, 1982), *Trypanosoma musculi* (Albright and Albright, 1981*a*) and *Trypanosoma rhodesiense* (Greenblatt *et al.*, 1980). The reason for including bacteria and a rickettsia in this list is that some of the most illuminating genetic analyses of resistance have been performed with bacterial parasites (see Section III.C). Not every analysis has provided evidence of relative independence of genetic control from the MHC, however. In the case of both *Trichinella spiralis* (Wassom *et al.*, 1979) and *Toxoplasma gondii* (Williams *et al.*, 1978) evidence has been presented for a major role of genes associated with the MHC. In several other cases (Morrison and Murray, 1979; Trischmann and Bloom, 1982; Blackwell *et al.*, 1980) secondary roles of genes linked to the MHC have been ascertained.

Table I shows the patterns of relative resistance/susceptibility among various strains of mice toward a variety of parasites. Several points are suggested by this compilation:

1. The same pattern of resistance/susceptibility of the strains is apparent for *M. bovis*, *S. typhimurium*, and *L. donovani*, all of which are intracellular parasites.
2. A different pattern may be seen with respect to *L. monocytogenes* and *T. cruzi*, which also are intracellular parasites, and the patterns displayed by these two are quite similar.
3. The strain distribution of resistance/susceptibility is similar for all the trypanosomes, with the possible exception of *T. cruzi*, the only intracellular parasite in this group.
4. *P. berghei* and *B. microti* appear to display the same pattern, which is different from that of all other parasites listed in Table I.
5. Insufficient strains are represented to draw a conclusion regarding the pattern presented by the two nematodes.

Table I. Relative Resistance of Strains of Inbred Mice to Some Infectious Organisms[a,b]

Parasite	Strain (H-2 haplotype)															
	A/J (a)	A/He (a)	AKR (k)	Balb/c (d)	CBA (k)	C3H (k)	C57BL/6 (b)	C57BL/10 (b)	C57BR (k)	C57L (b)	DBA/1 (q)	DBA/2 (d)	NZB (d)	SJL (s)	SWR/J (q)	129/J (b)
Listeria Monocytogenes (1)	s	–	–	s	s	s	r	r	–	–	s	s	r	r	–	–
Mycobacterium bovis (2)	r	–	r	s	r	r	s	s	r	–	s	r	–	–	–	–
Salmonella typhimurium (3)	r	–	–	s	r	r	s	–	–	–	s	r	–	–	–	–
Leishmania donovani (4)	r	–	r	s	r	r	–	s	r	r	s	r	r	–	–	r
Leishmania tropica (5)	r	–	–	s	r	r	r/s	–	–	–	–	s	–	–	–	–
Rickettsia tsutsugamushi (6)	–	–	r	r	s	s	r	r	–	r	s	s	–	s	–	–
Babesia microti (7)	s	–	–	r	r/s	–	–	r	–	–	–	–	–	–	–	–
Trypanosoma rhodesiense (8)	r/s	–	r	r/s	s	s	r	–	–	–	r/s	r/s	–	–	–	–
Trypanosoma congolense (9)	s	–	r	r/s	–	r/s	r	–	–	–	r/s	–	–	–	s	s

Trypanosoma musculi (10)	–	s	–	r/s	s	s	r	–	–	–	r/s	r/s	–	–	–	–
Trypanosoma cruzi (11)	s	–	s	r/s	r/s	s	–	r	–	–	r	s	–	r	–	–
Plasmodium berghei (12)	s	–	s	–	r/s	–	–	–	–	s	r	r/s	–	–	–	–
Plasmodium chabaudi (13)	s	–	–	r	r/s	–	s	r	–	–	–	–	–	–	–	–
Trichinella spiralis (14)	–	r/s	r/s	–	r	r	–	s	–	–	r	r/s	–	–	–	–
Trichuris muris (15)	s	–	–	r	s	r	r	–	–	–	–	r/s	–	–	–	–

[a]References: (1) Cheers *et al.* (1980)
(2) Gros *et al.* (1981)
(3) Plant and Glyn (1976); Hormaeche (1979)
(4) Bradley (1977)
(5) Behin *et al.* (1979); Handman *et al.* (1979)
(6) Groves *et al.* (1980)
(7) Eugui and Allison (1980)
(8) Greenblatt *et al.* (1980)
(9) Morrison and Murray (1979)
(10) Albright and Albright (1980)
(11) Trischmann *et al.* (1982)
(12) Most *et al.* (1966)
(13) Eugui and Allison (1980)
(14) Rivera-Ortiz and Nussenzweig (1976); Wakelin (1980); Wassom *et al.* (1979)
(15) Wakelin (1975)

[b]Abbreviations: s, susceptible; r, resistant; r/s, intermediate (moderate) resistance.

It must be stressed that the evaluations of relative resistance reported in Table I are useful only as a general guide to strain distributions of resistance to parasites; in most cases the tabulation should not be considered rigorous. Many variables can influence attempts to determine relative resistance, including age, sex, and health of the host; other, prior infections of the host; number of parasites inoculated into the host, and route of inoculation. Special concern should be given to variation in the parasite under investigation, i.e., Is it representative of the natural parasite, or is it perhaps modified as a result of having been maintained too long under artificial laboratory conditions? These variables can explain why identical results have not always been obtained in different laboratories.

In the case of several parasites listed in Table I, viz. *L. monocytogenes*, *M. bovis*, *S. typhimurium*, *R. tsutsugamushi*, and *L. donovani*, there are very sharp differences in the magnitude of infection in different strains of mice, and the distinction between resistant and susceptible is sharp. In these cases it has been possible to conduct precise genetic analyses and to ascertain the number and location of the genes controlling resistance. It has been shown for *L. donovani* that in the early course of infection (first 2–3 weeks after inoculation) resistant and susceptible strains differ markedly in parasite burden; this difference is controlled primarily by a single gene, designated *Lsh* located on murine chromosome 1 (Bradley, 1977, 1980; Bradley *et al.*, 1979). In the case of *M. bovis* infections, the difference in the number of viable bacteria in the spleens of resistant and susceptible mice is apparent as early as 24 hr after inoculation; this difference in relative resistance is primarily regulated by a single, autosomal gene designated *Bcg* (Gros *et al.*, 1981). The *Bcg* gene is not linked to the MHC. Similarly, a single gene designated *Ity* that maps to murine chromosome 1 controls resistance/sensitivity to infections with *S. typhimurium* (O'Brien *et al.*, 1980*b*; Plant and Glynn, 1979, 1980). In the case of murine infections with *L. monocytogenes*, relative susceptibility in the early course of infection is determined by a single, autosomal gene designated *Lr* that is not linked to the MHC (Cheers and McKenzie, 1978; Skamene and Kongshavn, 1979; Cheers *et al.*, 1980; Kongshavn *et al.*, 1980). In the case of the rickettsia, *R. tsutsugamushi*, the causative agent of scrub typhus, resistance is a dominant trait controlled by a single gene designated *Ric* that has been mapped to chromosome 5 of the mouse (Groves *et al.*, 1980). In all five of these examples, resistance is apparent relatively early during the course of infection, is not associated with acquired immune responses, and therefore probably falls within the province of natural resistance. The mechanisms through which this resistance is manifested are considered in Section III.C.

Very little work concerned with the genetics of resistance of parasites has involved host species other than mice. The reason, of course, is the relative paucity, until recently, of inbred lines of other species. One interesting series of investigations (Weiss, 1978; Neilson, 1978; Haque *et al.*, 1978; Ogilvie and MacKenzie, 1981) demonstrated that different strains of hamsters are differentially susceptible to the filarial nematode, *Dipetalonema viteae*. In most strains the persistence of microfilaria in the blood extended through a peak at around 75–85 days and then fell to low or undetectable levels by around 120 days after initiation of infection. In some strains, however, the microfilaremia lasted considerably longer. There was also considerable interstrain variation in the survival of the adult worms. Another investigation found that the severity of infection with *S. mansoni* differed greatly in two inbred lines of hamsters (Smith and Clegg, 1976).

One of the classic cases of interspecies variation in resistance is the difference displayed by zebu and ndama cattle to infection with the African trypanosomes, *T. brucei* and *T. congolense*. A recent systematic investigation found that 30 of 40 cattle of the zebu breed died as a result of infection, whereas none of 37 similarly exposed ndama cattle succumbed (Murray and Morrison, 1979*b*). After a very similar prepatent period in both breeds a considerable difference was observed in the magnitude of the first wave of parasitemia, the infection being much lower in the ndama breed. This difference in cattle was duplicated by the relative resistance of inbred strains of mice to *T. congolense* (e.g., A/J mice are susceptible, while C57BL/6 mice are resistant), except that mice of both strains ultimately died of the infections. In cattle, also, some well-established differences among breeds in the resistance to infections with *Babesia bovis* and *Babesia bigemina* have been reported (Johnston, 1967; Ranatunga and Wanduragala, 1974; reviewed in Zwart and Brocklesby, 1979). These differences were not attributable to differences in hemoglobin type, blood groups, or enzymes present in erythrocytes. Significantly, splenectomy was found to increase the susceptibility of the more resistant *Bos indicus* cattle substantially (Zwart and Brocklesby, 1979).

From an early investigation of five breeds of chickens it was reported that there were substantial differences among the breeds in the burden of the nematode, *Ascaridia lineata* (Ackert *et al.*, 1935). Highly inbred lines of chickens are available for more extensive investigations in this species. A recent report of particular interest to students of natural resistance demonstrated considerable differences in the magnitude of *Entamoeba histolytica* infections in the eggs derived from different inbred strains of chickens (Jaouni, 1979). The underlying cause of resistance was not discovered, but possible causes such as maternal antibodies, yolk or serum toxic substances, and interferon were ruled out. This system merits considerable attention to determine the nature of the phenotypic differences between the eggs of the different strains.

B. Biozzi Strains of Mice

Mice from a randomly breeding population have been selected for their ability to respond to sheep erythrocytes; two inbred lines have been developed, a low-responder line (L), that does not respond well and a high-responder line (H) (Biozzi *et al.*, 1975). The L and H lines have been found to differ in the state of activation of their macrophages, the L line being considerably more active (Wiener and Bandieri, 1974). They do not differ significantly in capacity for cell-mediated immune responses (Biozzi *et al.*, 1979). The ability of these lines of mice to resist various parasitic infections has been investigated. When exposed to 50 muscle larvae of *Trichinella spiralis*, H mice developed only half as many larvae as did the L mice; however, L mice were completely resistant to a second infection, while H mice were only partially protected (Perrudet-Badoux *et al.*, 1978). Line L mice were somewhat less susceptible than H mice to *S. mansoni* (Blum and Cioli, 1978). Results of considerable interest have been obtained from studies of relative resistance to *T. cruzi*, *L. tropica*, and *S. typhimurium*, all of which are intracellular parasites having a particular affinity for macrophages. Line L mice were found to be more susceptible to two different strains of *T. cruzi* (Kierszenbaum and Howard, 1976). Furthermore, specific vaccination provided little protection to L mice, but was quite effective in protecting H mice. In contrast, L mice were substantially more resistant than H mice to *L. tropica* (Blum and Cioli, 1978). Increased resistance in H mice suggests a role for antibodies, while greater resistance in L mice suggests that the activated macrophages are important in resisting infection. In the case of *S. typhimurium* infection, the L mice were substantially more resistant than were H mice (Plant and Glynn, 1980). Analyses of hybrids between L and H mice and between these lines and innately susceptible (Balb/c) or resistant (CBA) strains of mice indicated the following: (1) the resistance factor associated with L mice is dominant and is able to impart resistance to hybrids possessing the Ity^s (susceptible) gene and even increase resistance in hybrids having the Ity^r (resistant) gene; and (2) H mice provide a susceptibility factor that somewhat decreases the resistance displayed by hybrids possessing the Ity^r gene. It is postulated that resistance associated with the L genotype is macrophage-mediated and supplements the resistance associated with the Ity^r gene, which is also believed to be expressed by macrophages (see Section III.C).

C. Variations in Resistance Related to Sex and Age

1. Sex

Numerous reports are concerned with differences in the magnitude, duration, and mortality of infections between male and female hosts. It would be unprofitable to consider these reports in any detail here because the underlying causes of the differences in severity of infections between the sexes are largely unknown. Possible differences in resistance between the sexes of inbred strains of mice are often overlooked and may obscure the assessment of relative resistance. It could be rewarding to investigate the causes of sex-associated differences in resistance. To illustrate, some recent reports (e.g., Simpson and Cioli, 1982) indicate that the lipid composition of parasites may be markedly influenced by the lipid composition of the medium in which they reside and that the lipid composition of their membranes may influence their susceptibility to immune lysis. Thus, sex-associated differences in plasma content of lipids and lipoproteins could affect relative resistance of hosts toward parasites.

Female rats are particularly vulnerable to *T. lewisi* at certain times during pregnancy (Shaw and Dusanic, 1973). The midterm of pregnancy was found to be the most susceptible period. Rats infected at this time developed abnormally high levels of parasitemia, and most died at about the time of parturition. In contrast, rats infected during the last week of pregnancy developed unusually low levels of parasitemia and gave birth normally. A similar situation was observed in pregnant mice infected with *T. musculi* (Krampitz, 1975). Infections were abnormally high, fetal resorption and abortion common, and maternal death frequent in mice infected at midterm, whereas in mice infected during the last week of pregnancy parturition was normal, and the course of infection was typical of nonpregnant mice. In both pregnant rats and mice it was observed that the placenta was a focus of enormous numbers of dividing parasites.

2. *Age*

It is common, although not invariable, to find that young, neonatal hosts are a good deal more susceptible to infection than are adult hosts. Attempts to demonstrate this are sometimes frustrated by maternal antibodies, the influence of which may be seen for several weeks after birth. As examples of changing susceptibility after birth, consider experimental infections of rodents with their specific trypanosomes (see D'Alesandro, 1979). As shown in earlier studies (Herrick and Cross, 1936; Duca, 1939) some 70–90% of rats <1 month of age died when infected with *T. lewisi*, whereas $\leqslant 6\%$ of rats > 1 month of age expired. Approximately the same is true in the case of young mice infected with *T. musculi* (Culbertson, 1941; J. W. Albright and J. F. Albright, unpublished observations, 1980).

It is apparent that the decrease in morbidity in adolescent, infected animals coincides with the maturation of the immune system and the ability to marshal a vigorous immune response. Nevertheless, there are clear examples of natural resistance in neonates.

The topic of ontogeny of natural resistance toward parasites would appear to merit much more attention than it has received. It would be useful to know the degree of variation in resistance displayed by fetuses and newborns of different inbred strains of mice. If, indeed, neonates are uniformly susceptible it is a wonder that host species are not literally ravaged by their parasites. There must be some device or mechanism that protects a large proportion of neonates from decimation in order to maintain a balanced relationship between parasites and their preferred species of hosts. In the case of protozoan parasites, at least, it would seem that invasion of any significant proportion of neonates could lead to appearance of a more virulent form of the parasite, which in turn would be even more devastating toward other neonates and older hosts as well. An illustration of such a process is provided in the review of D'Alesandro (1979) of some early work with rodent trypanosomes. In general, *T. lewisi* are well tolerated by their rat hosts. However, there exist naturally pathogenic variants of *T. lewisi* that produce lethal infections even in adult rats (Brown, 1914, cited in D'Alesandro, 1979). Furthermore, markedly enhanced virulence can be developed by rapid passage through rats and some of these strains may even be capable of infecting mice (Roudsky, 1910*a,b*, 1911*a,b*, cited in D'Alesandro, 1979). All these more virulent strains are more pathogenic for young rats. It would seem that in the absence of intervening forces, such as maternal antibodies and natural resistance mechanisms, these virulent strains of *T. lewisi* would devastate the population of rats.

Turning attention to the other end of the life span, there is relatively little information regarding the resistance of aged hosts toward their parasites. A brief report by Crandall (1975) indicates that aged mice experience significantly greater infections with the helminth, *Trichinella spiralis*. Another report demonstrates that aged mice develop substantially greater infections of *T. gondii* than do their young-adult relatives (Gardner and Remington, 1978*a,b*). Recently, we conducted an investigation of the relative resistance of mice of varying ages to infection with *T. musculi* (Albright and Albright, 1982). Old mice (28-34 months of age) of several strains studied consistently displayed greater levels of parasitemia (2- to 8-fold) and a more prolonged course of infection than did young adults of the same strain. An illustration of this is provided in Fig. 1, in which the course of *T. musculi* infection in young and old A/J mice (a susceptible strain) is depicted. The plateau level of parasitemia was about eightfold higher and the duration of infection was considerably longer in the aged mice. The procedure of transferring graded numbers of spleen cells from young and old infected donors to irradiated, young recipients was employed to demonstrate that the cells of aged donors were much less efficient in providing protection against *T. musculi* infection. Although serum antibodies acquired from donors that have recovered from infection are able to confer resistance to infection on normal animals, we have been unable to show that humoral antibodies play a role in the cure of an established infection. Thus, the difference in severity of infection

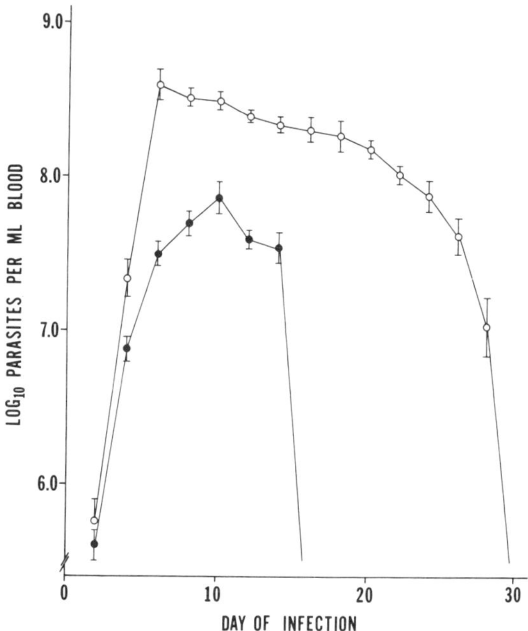

Figure 1. Course of *Trypanosoma musculi* infection in young (•) and old (○) A/He mice, aged 3 and 25 months, respectively.

between young and old mice could not be attributed to the difference in capacity for humoral antibody production. Additional experiments to test the ability of young and old spleen cells to effect antibody-dependent lysis (ADCC) of target trypanosomes demonstrated no age-related decline in ADCC. Recently we have turned to an examination of the naturally cytotoxic (NC) or "natural killer" (NK) subset of lymphocytes in search of the origin of the difference between young and old mice in susceptibility to *T. musculi*. The data presented in Table II disclose that early in the course of a *T. musculi* infection in young mice there was an approximate threefold rise in the level of NK activity of spleen cells followed by a decline to a level below normal. The rise in NK activity in the spleen of old mice and the subsequent decline were comparable to that in the young adult. However, the initial NK activity in old, control mice was less than one-half that in young mice and remained at about that proportion of young NK activity during the rise and fall accompanying infection. To what extent this difference in NK capability is related to the difference in resistance to *T. musculi* is unknown. The early rise and subsequent decline of NK activity have been observed in young mice infected with *T. cruzi* (Hatcher *et al.*, 1981), *P. yoelii* (Jayawardena, 1981), and *T. gondii* (Hauser *et al.*, 1982).

Table II. Natural Killer Activity of Spleen Cells of Young and Old Mice Infected with *Trypanosoma musculi*

Day of infection	Percent of isotope released[b] by spleen cells			
	C3H mice		C57BL/6 mice	
	Young[c]	Old[c]	Young[c]	Old[c]
Normal[a]	23.4	9.0	18.1	12.3
1	28.6	1.1	31.4	6.4
2	52.4	15.3	59.0	15.1
3	56.3	21.6	46.7	25.4
4	61.4	25.2	61.9	18.7
5	59.5	18.1	29.5	17.9
6	42.2	15.4	26.0	14.9
8	24.2	9.3	17.0	–
10	10.4	6.8	16.4	–
20	11.2	8.0	12.4	–

[a]Uninfected, control mice.
[b]Isotope released by YAC-1 target cells during 4-hr incubation (effector: target cell ratio, 200:1).
[c]Young mice, 3 months of age; old mice 25 months of age.

III. MECHANISMS OF NATURAL RESISTANCE

A. Humoral Substances

In a number of cases in nonpermissive, prospective hosts, natural resistance has been found to involve noxious, humoral substances. Complement (C), long known to function as a cytolysin or opsonin *vis-à-vis* many parasites when activated by antibodies against the parasites, may act directly on certain parasites in the absence of specific antibodies. For example, it was shown that the C system present in chicken serum can be activated via the alternative pathway (C3 shunt) by *T. cruzi* (Kierszenbaum *et al.*, 1976, 1981). This finding explained the rapid destruction of *T. cruzi* that occurred when inoculated into chickens. Both the reticulotropic (Tulahuen) strain and the myotropic (Y) strain of the parasite were lysed by chicken C. The possibility that trypanosome-bound, host immunoglobulin might have been instrumental in C activation was eliminated. In chickens with normal C levels, the inoculated *T. cruzi* were completely cleared from the bloodstream within a few minutes.

In birds treated with cobra venom factor (CVF) to destroy the C system, and in bursectomized, cyclophosphamide-treated birds similarly treated with cobra venom, clearance required > 21 hr. This latter observation is of additional interest because it indicates the presence in chickens of a mechanism for natural resistance independent of C. It is worth noting that chicken embryos appear to be susceptible to *T. cruzi* (Ganapati, 1948), but that they develop resistance soon after hatching (Nery-Guimaraes and Lage, 1972), probably initially as a result of the appearance of a functioning C system (Gabrielsen *et al.*, 1973).

In the search for natural C-activated resistance, great care is required to rule out naturally occurring antibodies. A good example was provided by Pearson and Steigbigel (1980) who showed, as had many previous workers, that serum from normal humans who had no contact with *L. donovani* antigens was able to lyse promastigotes but not amastigotes of *L. donovani*. Lysis clearly involved C activation through the classic pathway. Careful investigation demonstrated, however, that the promastigotes had both IgM and IgG bound at their surfaces. This immunoglobulin undoubtedly represented cross-reacting antibodies elicited by similar determinants present on some otherwise-unrelated immunogen.

There are a few well-studied examples of resistance to parasites involving direct activation of the C system independent of antibodies. One fascinating study was concerned with murine resistance to *Salmonella* of the Ra chemotype (Ihara *et al.*, 1982). There exists in mouse serum a polypeptide (~28,000 Mr) capable of selective binding to lipopolysaccharide of the Ra, but not of the S or Re, chemotype. The resulting complex is able to activate the C system via the classic pathway. The relevant polypeptide displays binding affinity for a side chain of LPS terminating in L-glycero-D-mannoheptose; however, the polypeptide shows no resemblance to an immunoglobulin chain. The action of this polypeptide is similar to that of the well-known C-reactive protein, found in human and rabbit serum, which is capable of binding to certain bacterial polysaccharides producing complexes capable of activating C via the classic pathway (Kaplan and Volankis, 1974; Osmond *et al.*, 1975).

Antibody-independent binding of guinea pig complement component C3 to the sheath of infective-stage larvae of *Ascaris suum* has been demonstrated (Leventhal and Soulsby, 1977). In this instance it appeared that only C3 was involved, and not the entire C system; however, the participation of some other heat-labile serum component was required for C3 binding. Of considerable interest was the fact that parasitic larvae collected from the livers of infected guinea pigs 16 hr after inoculation bound little or no C3, while larvae collected 48 hr after inoculation once again showed considerable affinity for C3. This finding suggests the presence of a defense mechanism in the guinea pig host involving direct activation of C3 followed by the well-known, manifold effects of C3b and C3a (e.g., opsonization, chemotaxis, local anaphylaxis). Larvae that survive this initial onslaught may quickly change the composition of their cuticle thus nullifying their treacherous component.

It has been known for many years that the sera of certain species are highly toxic for certain parasites. Paul Ehrlich was aware of the toxicity for *Trypanosoma brucei brucei* of normal human serum collected from humans who had never been exposed to the parasites (Ehrlich, 1907, cited in Taliaferro and Olsen, 1943). An even earlier report, apparently the first, appeared in 1902 (Laveran, 1902, cited in Rifkin, 1978). The ability of normal sheep serum to protect mice from infection with *T. musculi* was reported by Thiroux (1909) and investigated more extensively many years later (Taliaferro and Olsen, 1943).

Only recently has the trypanocidal power of normal human serum been explained (Rifkin, 1978). It is the high-density lipoprotein (HDL) of human serum that is toxic for *T. b. brucei*. Toxicity results from the slow lysis of the trypanosomes. Lytic activity of serum is lost when the HDL is removed. Furthermore, serum from patients with Tangier disease who lack HDL has no effect on the trypanosomes. Injection of normal serum into mice having an established *T. b. brucei* infection results, after a lag phase of 6–12 hr, in rapid clearance of the parasites; however, this clearance after a single inoculation of human serum or of enriched HDL does not result in complete cure, for the parasitemia reappears a few days after treatment. Human serum HDL was shown effective against trypanosomes of the *brucei* subgroup only (Rifkin, 1978). Neither rat nor rabbit serum HDL was effective.

The toxicity of sheep serum for *T. musculi* has not been analyzed. It seems quite likely, however, that it is caused by a serum lipoprotein (J. W. Albright and J. F. Albright, unpublished observations, 1981). Clearly, the analysis of lipoprotein toxicity toward parasites is worthy of considerable investment of research effort. Moreover, a related phenomenon also deserves considerable attention, viz., the effect of parasitic infections on plasma lipid and lipoprotein concentrations and on the biosynthesis of these substances. Except for some preliminary work in our laboratory we are unaware of any investigations dealing with changes in plasma lipids and lipoproteins attending parasitic infections. However, there are good reasons for suspecting that changes may occur. One reason arises out of the well-known immunosuppression that attends many parasitic infections. It has been demonstrated that both HDL and LDL (low-density lipoprotein) interfere with the early events in humoral and cell-mediated immune response of both mice (Hsu *et al.*, 1982; Curtiss *et al.*, 1977) and human subjects (Curtiss and Edgington, 1976). Furthermore, alterations in immune functions have been found in hypercholesterolemic monkeys (Fiser *et al.*, 1973). It has been demonstrated that changes in the mole ratio of cholesterol to phospholipid in the serum is reflected by an increase in the same ratio in lymphocyte membranes; the latter change is strictly correlated with a rise in membrane microviscosity and accompanied by decreased responses of lymphocytes to concanavalin A (con A) stimulation (Rivnay *et al.*, 1980). Thus, it would appear that parasite-induced changes in lipoproteins, which transport cholesterol, among other functions, could readily account for at least some of the suppression that accompanies infections. It is also possible that the decrease in resistance of hosts as they age (Albright and Albright, 1982) may be explained, at least in part, by the rise in plasma concentration of lipids characteristic of aging animals (Weibust, 1972; Rivnay *et al.*, 1980).

B. Immunologically Deficient Mice

In this section we contemplate what may be learned about natural resistance to parasites from two types of immunodeficient mice, viz., hypothymic mice (especially congenitally athymic, nude mice) and mice carrying an X-linked, recessive mutation (*xid*) displayed as an arrest in B lymphocyte maturation, much-reduced capacity to produce IgM, and little or no ability to respond to a group of thymus-independent antigens (Scher *et al.*, 1975, 1976; Amsbaugh *et al.*, 1974).

1. Hypothymic Mice

The subject of immune responses to parasites in hypothymic animals has been thoughtfully reviewed by Targett (1973) and Mitchell (1978, 1979*a*). Mitchell has carefully analyzed infections of congenitally athymic mice with a variety of selected parasites. One objective, among several, has been to compare relative resistance to a given parasite in mice of different strains carrying the nude gene. A second objective has been to ascertain the T-cell dependence of various features of parasitic infections. The importance of distinguishing clearly between infections with host-specific, or natural, parasites and those involving artificial or highly abnormal parasites has been stressed by Mitchell (1979*b*), who wrote:

> In reviewing available information on parasitic infections in hypothymic mice, the gross consequences of exposure to parasites are of two types: (1) increased susceptibility (i.e., increased parasite burdens, increased proliferative rate of parasites, failure to develop resistance, death of the host, or prolonged infection); or (2) comparable susceptibility (or even marginally greater resistance). *All* natural murine parasites are in the first category whereas the second category contains several abnormal parasites, abnormal because the parasite is not usually found in mice or because of laboratory modification of what was once a natural murine parasite. As mentioned above, any parasite which kills the majority of individuals of the host species used in the laboratory can be considered abnormal. The findings using natural parasite systems in nude mice strongly support the notion that T-cell dependent, host-protective immunological responses have been a key ingredient in the evolution of balanced host-parasite relationships.

A summary of the infections of nude mice with representatives of the major phyla, which include animal parasites, can be found in Mitchell's articles (1979*a*, *b*). In contemplating what may be learned about natural resistance from studies of infections in nude mice, let us consider first the case of *Ascaris suum* infections (Mitchell *et al.*, 1976). Mice of the C57BL/6 strain are quite susceptible to *A. suum*, while Balb/c and CBA/H mice are comparatively resistant. However, this interstrain difference vanishes when the nude gene is homozygous; i.e., C57BL/6.nu/nu, Balb/c.nu/nu and CBA/H.nu/nu mice display essentially the same degree of susceptibility. In fact, the C57BL/6.nu/nu mice display somewhat greater resistance than do the other two nude strains during the early course of infection, as judged by the numbers of larvae present in the lungs. Thus, there appears to be a basic level of resistance to this nematode that is displayed in the C57BL/6 strain but that is normally overshadowed by thymus-dependent resistance in the other strains. Perhaps this basic resistance is a reflection of the property of C3 binding displayed by *Ascaris* larvae in guinea pigs (Leventhal and Soulsby, 1977). The role played by the thymus in the relatively resistant strains presumably is in the initiation of acquired immune responses.

The relative resistance of nude mice and their normal littermates to infection with *S. mansoni* has been investigated; it was found that nude mice were not more susceptible (Phillips *et al.*, 1977). In fact, there was no significant difference in the number of worms present in intact and nu/nu mice after 6 weeks' infection. As judged by certain indices of morbidity (e.g., spleen and liver weight, reticuloendothelial activity), the hypothymic mice remained in better condition than did their euthymic littermates. Granulomas of the liver were smaller and less cellular in the hypothymic mice, and the extent of eosinophilia was much lower. Other workers have also observed that the extent of eosinophilia normally seen in schistosomiasis was greatly reduced in nude mice (Hsu *et al.*, 1976). In fact, the eosinophilia and local tissue accumulations of eosinophils characteristic of many metazoan infections are largely absent in nude mice (Hsu *et al.*, 1976; Mitchell *et al.*, 1976; Phillips *et al.*, 1977; Johnson *et al.*, 1979). The role performed by eosinophilia in parasitic infections is not well understood, although they do function as effector cells in ADCC reactions, e.g., against schistosomes (Butterworth *et al.*, 1977). There is no evidence that eosinophils are involved in natural resistance, although the possibility that there exists a mechanism that employs C and eosinophils should be considered.

P. yoelii infections are much more severe in nude than in normal mice (Jayawardena, 1981). The number of parasites in the blood rises steadily after inoculation of CBA nu/nu mice, and death occurs within 30–35 days; intact mice, in contrast, gain control of the infection, and by days 16–18 post infection virtually all parasites have been eliminated. *P. yoelii* may be considered a natural parasite in mice (Mitchell, 1979*a*). Of some interest is the fact that for the first 10 days or so postinoculation, the parasite is controlled as well in nude as it is in intact mice (Jayawardena, 1981). This early control is probably an indication of natural resistance, as suggested subsequently for the case of *T. musculi* infections. In contrast to *P. yoelii* infections, the control of *P. berghei* infections in nude mice is equal to, or better than, that in intact mice (Jayawardena, 1981; Waki and Suzuki, 1977). Similarly, neonatally thymectomized hamsters are more resistant to *P. berghei* than are intact hamsters (Wright, 1968). The effect of splenectomy on *P. yoelii* infections is the same as the effect of the nude gene, according to Jayawardena (1981); in contrast, splenectomy prolongs the survival of mice exposed to *P. berghei.* Jayawardena interpreted these results as indicating that in cases in which immune responses are not effective in controlling infections, their contribution may be detrimental to the host. Related experiments concerned with the effects of the absence of a spleen on infections with *B. microti*, a natural parasite of mice (Mitchell, 1979*a*) and *B. rodhaini* were reported by Irvin *et al.* (1981). Both congenitally asplenic [owing to the dominant hemimelia mutation; see review by Lozzio (1976)] and surgically splenectomized mice were employed. Whether or not a spleen was present mattered little in the case of the virulent *B. rodhaini* infections. In *B. microti* infections, however, higher parasitemia was seen consistently in animals without spleens. The highest parasitemia was found in surgically splenectomized mice, which suggested that in the mutant, asplenic mice some other tissue might have partially compensated for the absence of the spleen.

The influence of the thymus on relative resistance of Balb/c mice to *B. rodhaini* was investigated by Mitchell (1977), who showed that approximately 20% of nu/nu mice survived infection, whereas none of the nu/+ survived. Adoptive transfer of lymph node cells from normal to nu/nu mice resulted in restored susceptibility to the parasite. Thus, it appeared that a set of thymus-derived cells was involved in enhancing sensitivity to the parasite and in promoting the severity of the primary infection.

From the preceding discussion it appears that the situation is as follows: (1) in the case of natural infections, elimination of either the thumus or the spleen of the host results in much more severe parasitic infections; (2) in contrast, the absence of thymus or spleen in the case of infections with unnatural parasites has little effect on the severity of the infections *per se.* These conclusions may require modification as more results from studies of infections in relatively resistant and relatively susceptible hosts become available. Furthermore, it is likely that future results will reflect the variations in quality and relative immunogenicity of the various parasite components.

It is worthwhile to point out that results very similar to those discussed concerning animal parasites have been obtained in studies of bacterial and yeast infections in hypothymic and normal mice. For example, nude mice were reported to be much more resistant to acute infection with *Listeria monocytogenes* than were euthymic mice of the same strain (Emmerling *et al.*, 1977). Similarly, nude mice were more resistant than normal mice to infections with *Candida albicans*, and it was shown that transplants of thymus cells to nude mice increased their susceptibility to the yeast (Rogers *et al.*, 1976).

Several investigators have remarked about an interesting similarity in the early course of infection, assessed by the rate of increase in numbers of parasites, in normal and hypothymic hosts infected with the same parasite. It appears that the parasite growth profiles are essentially identical in the two types of hosts up to the time when immune elimination of the parasite begins in the normal animals. A clear example of this may be seen in the growth of *P. yoelii* in normal and hypothymic mice of the relatively resistant CBA strain (Jayawardena, 1981). Another example is provided by the analysis of *T. musculi* growth in normal and hypothymic mice of the resistant Balb/c strain (Brooks and Reed, 1979). During the (approximately) first 10 days after parasite inoculation, hypothymic mice controlled the course of infection as well as normal mice. At about the time that parasite clearance began in normal mice, a dramatic rise in parasitemia occurred in the nude mice, eventually leading to their death. The early resistance to parasite growth displayed by the nude mice was eliminated by treatment of the mice with trypan blue, a substance shown to be toxic for macrophages (Hibbs, 1975; Guckian *et al.*, 1978). Treatment of normal mice also resulted in a sharp rise in parasitemia; in this case, however, the rise was terminated abruptly, followed by elimination of the parasites. Trypan blue-treated nude mice were defenseless, and the rapid increase in parasitemia soon terminated in death of the hosts.

Our studies on *T. musculi* infections of immunologically deficient mice are relevant, for they seem to indicate the existence of a mechanism for natural resistance in resistant mice that is lacking in susceptible mice (Albright and Albright, 1981*a*). Mice of the C57BL/6 strain are considerably more resistant to *T. musculi* than are a variety of other strains studied. In contrast, C3H mice are among the most susceptible. Exposure of C3H mice to a moderate level of gamma radiation (400 rad) before inoculation with *T. musculi* caused marked increase in sensitivity to the parasites; the parasitemia increased rapidly to a fatal level. In contrast, the same amount of radiation given to C57BL/6 mice did not change the level of parasitemia; it did, however, markedly prolong the duration of parasitemia. The time required for eventual immune elimination of the parasites by the irradiated mice was twice as long as in the case of normal mice. The effect of trypan blue on susceptibility to *T. musculi* was also quite different in the two strains of mice (J. W. Albright and J. F. Albright, unpublished results, 1982). Mice were given repeated injections (every third day) for a period of 3 weeks, or until they died. C3H mice developed levels of parasitemia much higher than in untreated mice, and many of them died. In contrast, the treated B/6 mice once again displayed a markedly prolonged duration of infection, but the levels of parasitemia were the same as in untreated mice.

The investigations summarized allow us to surmise that mechanisms of natural resistance are able to restrict parasite growth and other activities during the early phase of infection in either susceptible or resistant strains. Subsequently, in the case of susceptible strains, natural resistance may fail, at which point survival of the host will depend on the effectiveness of acquired immune responses. In resistant hosts, however, the more powerful mechanisms of natural resistance may be able to restrict the parasite, if it is a natural parasite, even in the virtual absence of conventional immune responses. In cases in which the parasite is unnatural, and the parasite-host combination is abnormal, neither susceptible nor relatively resistant hosts are able to cope with the infection. Much more data are required before this notion can be properly tested. The results summarized here also serve to focus attention on the role of macrophages in natural resistance. The effect of trypan blue treatment on *T. musculi* infections in nude Balb/c strongly suggests an effective role of macrophages. However, the effects of irradiation and trypan blue treatment on *T. musculi* infections in C57BL/6 mice do not clearly suggest macrophages as being central to natural resistance. We will return to the discussion of macrophages and natural resistance after considering briefly the behavior of certain parasites in mice carrying an X-linked, B-lymphocyte deficiency.

2. *X-Linked, B-Lymphocyte Defect*

A substrain of CBA/H mice, designated CBA/N, carries a mutation on the X chromosome that results in a B-lymphocyte immunodeficiency (the *xid* mutation). Mice of the CBA/N substrain display little or no humoral immune response to a group of thymus-independent antigens, especially those of the TI-2 group (Mosier *et al.*, 1977). These mice suffer primarily from inability to generate IgM and IgG3 immunoglobulins. Although they fail to respond, in particular, to carbohydrate antigens, their responses to many protein antigens and hapten-protein conjugates appear to be near-normal (Perlmutter *et al.*, 1979). These mice evince no deficiency in T-lymphocyte-mediated immunity (Scher *et al.*, 1975). CBA/N mice and male offspring of crosses between CBA/N females and males of some other strain have been used in studies of relative resistance to parasitic as well as bacterial infections.

The ability of mice bearing the *xid* mutation to contend with *A. suum* infection was studied by Brown *et al.* (1977). Ascaris infections of mice trigger an IgM response in which a high proportion of the antibody is directed against phosphorylcholine (PC) (Crandall and Crandall, 1971). CBA/N mice failed to generate a significant response to PC when infected with *A. suum*, but their resistance to the parasite was equivalent to that of normal CBA/H mice. T-cell-dependent humoral immunity in infected CBA/N mice was found to be about normal. Ascaris infection of mice represents an unnatural host–parasite combination; CBA/H is a relatively resistant strain (Mitchell, 1978).

Mice that are *xid* mutant are less resistant to *P. yoelii* and *B. microti* than are their normal counterparts (Jayawardena and Kemp, 1979). Both organisms may be considered natural parasites of the mouse (Mitchell, 1979*a*). In the case of *P. yoelii*, the duration of infection in *xid* mutant mice is significantly prolonged, but there is little change in the level of parasitemia; in time the blood is cleared of parasitized cells. There is a very interesting difference in behavior of *B. microti* in normal as compared with *xid* mutant mice. In the latter there is a marked prolongation of the initial infection and, in addition, the infection becomes chronic and characterized by repeating cycles of parasitemia.

Thus, in *xid* mutant mice, and in hypothymic and splenectomized mice of the resistant CBA genotype, there is at least a temporary loss of control of infections with *P. yoelii* (Jayawardena, 1981). However, the loss of control is not immediately apparent, for the rate of increase of parasitemia in all cases is the same as in normal CBA mice up to the point at which immune responses are able to initiate cure of the infection. It is after this time that control of the infection is lost in the immunodeficient animals. Thus, it may be argued that in all the deficient mice the mechanisms of natural resistance remain unaffected; at least the functional efficiency of the mechanisms is not diminished.

It may be asked whether the loss of control of *P. yoelii* and *B. microti* infections in the *xid* mutant mice signifies that control is normally manifested by humoral responses against TI-2 antigens of the parasites. This seems unlikely; we have already seen in the case of *Ascaris* that control is not maintained by humoral antibody against PC (Brown *et al.*, 1977). In the case of *P. yoelii* there is direct evidence against this prospect, for it has been reported that *P. yoelii* lacks TI-2 antigens (Taylor *et al.*, 1982).

In contrast to *P. yoelii* infections, which are more severe in *xid* mutant mice, the course of infection with an unnatural parasite, *T. brucei*, was found to be essentially equivalent in CBA/N and normal CBA mice (Gasbarre *et al.*, 1981). Furthermore, CBA/N mice survived two to three times longer than the normal animals. The reason for the prolonged survival of the CBA/N relative to the normal mice, when there was no difference in the course of their infections, as assessed by parasitemia, was not resolved, but might have been the significantly lower concentration of immune complexes in CBA/N animals. Whatever the reasons, it is not uncommon to find that immunodeficient animals suffering from unnatural parasite infections survive considerably longer than their fully immunocompetent colleagues, even though the course of infection may be quite similar in both. A parallel situation in intact and nude mice infected with *S. mansoni* (Phillips *et al.*, 1977) was discussed in Section III.B.1.

The recent availability of mice carrying the nude and *xid* mutations simultaneously offers an excellent possibility to examine natural resistance to parasites in the absence of many of the complications arising from concurrent acquired immune responses. So far very little information about parasitic infections in these mice has been published. A recent study by Finerty *et al.* (1982) of *T. rhodesiense* infections in mice carrying both mutations (referred to as type II nudes or nu/nu-II) is quite informative. The study was performed with mice of four types: male CBA/N, NIH-white Swiss nudes, (CBA/N $\times$ nu/nu)F_1 male, and type II nudes, as well as conventional CBA/CaJ animals. The CBA strain is relatively susceptible to *T. rhodesiense*, which is an unnatural parasite in mice. It was found that death in these four types of mice, after trypanosome inoculation, occurred in a definite order: CBA/CaJ mice died earliest (mean time to death, MTD, about 23 days); type I (nu/nu) and type II (nu/nu-II) nude mice died at later times (MTD of 38 and 39 days, respectively); and CBA/N mice survived considerably longer (MTD about 61 days of infection). With respect to the infections, the highest parasitemia was found in CBA/CaJ mice, but was substantially lower in nu/nu animals. The infection in *xid* mutant mice was more controlled and lower than in nu/nu mice. Type II nude animals developed parasitemia similar to that in the CBA/CaJ normal animals. All four types of animals were able to control the initial wave of parasitemia in similar fashion, suggesting that control was exerted by natural resistance mechanisms (our interpretation of the data presented by Finerty *et al.*, 1983). The results obtained from analyses of trypanosome-stimulated, immunoglobulin-producing cells in the spleens of these mice are fascinating. The numbers of both IgM-producing and IgG2-producing cells were determined. In brief, the results showed that at peak response to infection the nu/nu-II mice had only one-tenth as many IgM-producing cells as did the nu/nu mice and about one-sixth the number present in CBA/N mice. The number of IgG2-producing cells in nu/nu-II animals was one-half that present in nu/nu, and about one-fiftieth that present in CBA/N animals.

The results of these studies of Finerty *et al.* (1983) have prompted us to suggest, albeit with caution, the following interpretation. Reduction in the capacity for IgM production in the CBA/N mice permitted full display of the potential of thymus-derived lymphocytes, acting in concert with natural resistance mechanisms (macrophages and naturally cytotoxic cells), to combat the infection. In nude, and especially in *xid* nude, mice the capacity for acquired immune responses was virtually eliminated. Under these conditions, a reasonably uncomplicated display of the potency of natural resistance mechanisms was possible—and it proved to be a considerable potential.

Some illuminating studies of bacterial infections in *xid* mutant mice have been conducted. For example, CBA/CaHN mice that are immunologically normal are resistant to *S. typhimurium* (LD_{50} 1×10^4 bacteria); in contrast, the LD_{50} of CBA/N mice is a mere 10 organisms (O'Brien *et al.*, 1980*a*). This sensitivity of *xid* mutant mice was compared with that exhibited by mice of other strains carrying the sensitive allele at the *Ity* locus on chromosome 1. The MTD of Ity^s mice was 7 days and of *xid* mice 16 days. Furthermore, a marked difference between the two types of mice was found in the number of organisms harbored in the spleen. Seven days after initiating infections with 50 bacteria, the number of organisms harvested from the average spleen of *xid* mice was 2×10^4, whereas it was 2×10^6 in Ity^s mice. These results suggested that two different mechanism of resistance were displayed by the two lines of mice, viz., a mechanism that acted early and was defective in *Ity*^{s} mice and a mechanism that acted later and was defective in *xid* mice. The defect associated with the *Ity*^{s} gene is probably a defect of the macrophages, although this is not yet firmly established (see Plant and Glynn, 1980; Hormaeche *et al.*, 1980). The defect in the *xid* mutant mice was demonstrated to be diminished ability to produce antibody against the bacteria (O'Brien *et al.*, 1980*a*). In this case, as in others discussed earlier, the use of mutant animals defective in their ability to manifest acquired immune responses permitted clearer focus on mechanisms of natural resistance.

C. Macrophages

We should inquire next about the nature and structure of natural resistance manifested by host cells. We begin with a consideration of macrophages. The subject of macrophages and their functions is far too vast to be contemplated here. Rather, we discuss a few aspects of parasite–host macrophage relationships.

Several extensive investigations have been conducted with intracellular parasites: the protozoa (*T. gondii*, *L. donovani*, and *L. tropica*) and several bacteria (*S. typhimurium*, *L. monocytogenes*, and *M. bovis*). Work with other, extracellular protozoa and with metazoa has not progressed as far; nevertheless, it is possible to discern some similarities in the information available about extracellular and intracellular protozoa.

When parasites enter the host they encounter a variety of hostile cells of which the most immediately dangerous are the macrophages. Many protozoan parasites have adapted to residing inside macrophages. The reasons for their failure to be destroyed quickly in that location are complex, varied, and not well understood. However, the proper activation of macrophages, e.g., by their being exposed to lymphokines, does lead to destruction of intracellular parasites. Consider the demonstrations of killing by activated, but not by placid, macrophages of ingested *T. gondii* (Gardner and Remington, 1978*a,b*), *T. cruzi* (Nogueira and Cohn, 1978; Nogueira *et al.*, 1980), and *L. donovani* (Chang and Chiao, 1981). Activation may not always be necessary for macrophages to destroy ingested parasites actively. As shown by McLeod *et al.* (1980) most human monocytes readily destroy ingested *T. gondii* without evident activation.

This early interaction between parasites and host macrophages can be complex and the consequences highly influential on subsequent events, particularly on the capricious interactions involving macrophages, T lymphocytes, and parasites. To highlight some of these complexities, consider some recent reports concerned with *L. donovani* and *L. tropica* murine infections.

The manner in which macrophages interact with *Leishmania* parasites depends in a major way on the effects of a single gene. In the case of *L. donovani*, the gene *Lsh* located on chromosome 1 is expressed by both the Kupffer cells of the liver and by the macrophages of the spleen, but not by the peritoneal macrophages (Bradley, 1980). The major influence of *Lsh* seems to be on the rate of multiplication of amastigotes within the macrophages during the first 2 weeks of infection. It is clear that initial parasite uptake by macrophages of resistant and susceptible mice does not differ significantly. T-lymphocyte depletion does not alter the growth of the parasites in mice of a resistant strain during the first 2 weeks of infection.

The gene that exerts major control of infections with *L. tropica* is in a different location; it is an autosomal gene not linked to the MHC and not linked to *Lsh* (Howard *et al.*, 1980*a*). More is known about the effects of this gene than about *Lsh.* As in the case of *Lsh*, the gene controlling resistance to *L. tropica* is expressed by macrophages. Behin *et al.* (1979) classified mice as healers (relatively resistant) and nonhealers (susceptible). Macrophages of healer strains could be activated by con A-stimulated lymphocytes to destroy ingested *L. tropica.* Macrophage activation was stimulated by lymphokines elaborated by con A-activated T lymphocytes. The latter could be obtained from either healer of nonhealer strains of mice. Macrophages from nonhealer mice could be induced to destroy phagocytosed *L. tropica* if they were under continuous lymphokine stimulation. Thus, it appeared that one difference between healer and nonhealer mice was in the threshold of stimulation required to activate their macrophages. Another difference appeared to reside in the lesser efficiency of activated macrophages from nonhealer strains to destroy ingested parasites.

Evidence that a major difference between mice susceptible or resistant to *L. tropica* lies in the efficiency of T-lymphocyte–macrophage interactions was provided by Handman *et al.* (1979). They first showed, by use of an *in vitro* system, that the frequency of parasitized macrophages shortly after exposure to promastigotes was the same for both resistant and susceptible mice. Afterward the course of events was entirely different in the cultures of macrophages from the two kinds of mice. There was a progressive decline in the frequency of parasitized macrophages in cultures derived from resistant C57BL/6 and NZB mice. In contrast, there was a steady rise in the percentage of infected cells of the susceptible Balb/c and CBA/H strains. A second important finding was that amastigote-infected macrophages of susceptible strains were defective in presenting *L. tropica* antigens in a manner conducive to stimulation of delayed-type hypersensitivity (DTH) responses. The most interesting finding was that infected macrophages of susceptible strains displayed a relative deficiency in their surface density of H-2 determinants. Handman *et al.* (1979) suggest the possibility that owing to this reduced surface concentration of H-2 substances to interact with *L. tropica* components, the macrophages are delinquent in presentation of immunogen for stimulating the appropriate immune responses. If this is correct, it is a clear demonstration of one of the later consequences of a defective natural resistance mechanism.

A related conclusion was drawn from an elegant series of experiments by

Gorczynski and MacRae (1982). These workers studied the growth of *L. tropica* amastigotes in macrophages obtained from skin of resistant and susceptible strains of mice. Parasite growth was found to be faster in macrophages of susceptible Balb/c mice than in resistant CBA mice. The macrophages were partially resolved into subpopulations of different density by velocity sedimentation; parasite growth was faster in both the rapidly and more slowly sedimenting cells of Balb/c than of CBA mice. Nevertheless, a burst of parasite proliferation that occurred early after infection was controlled by macrophages of both genotypes. A comparison of the ability of skin macrophages of the two genotypes to present antigen to lymphocytes of different genotypes was performed. Skin macrophages infected with *L. tropica* were grown in monolayer. Lymphocytes were obtained from chimaeras prepared by reconstituting irradiated BCF_1 hybrids either with Balb/c or CBA bone marrow. This step was taken to ensure that the lymphocytes of different genotypes matured in the same environment, i.e., in the same milieu of histocompatibility substances. The chimaeras were sensitized to *L. tropica* antigens before being sacrificed. Sensitized lymphocytes were added to the macrophage monolayer cultures; after 72 hr of cocultivation tritiated thymidine was added in order to assess the extent of lymphocyte proliferation. Lymphocytes and macrophages were mixed in all combinations of the three genotypes (Balb/c, CBA, and BCF_1). In every combination prepared with Balb/c macrophages, lymphocyte proliferation was markedly less than in all other combinations. There was no evidence of defective ability of Balb/c lymphocytes to recognize and respond to antigen properly presented. Additional experiments disclosed that the macrophage defect pertained only to parasite antigens derived from processing their own ingested parasites, for when Balb/c macrophages were exposed to an extract of *L. tropica* before cocultivation with sensitized lymphocytes, they were able to stimulate lymphocyte proliferation efficiently. Another set of experiments reported in the paper of Gorczynski and MacRae (1982) provided strong evidence that the lymphocyte population of Balb/c genotype is particularly susceptible (in contrast to CBA or the BCF_1 hybrid genotype) to elaboration of a subset of T-suppressor cells in the response to *L. tropica.* This finding confirmed earlier investigations.

Howard *et al.* (1980) were perplexed by the paradoxical findings that hypothymic mice of the susceptible Balb/c strain were able to control *L. tropica* infection much better than their euthymic relatives, while the reverse was true of the relatively resistant CBA mice. These investigators reasoned that the sensitivity of Balb/c mice might result, in part at least, from an exaggerated elaboration of suppressor cells in mice of this strain. They proved, through a series of well-conceived experiments, that such is indeed the case (Howard *et al.*, 1980, 1981*b*):

1. Adult thymectomy diminished DTH responses of CBA mice to *L. tropica* extracts, but enhanced the responses of Balb/c mice and (Balb/c × C57BL/6)F_1 hybrids.

2. Moderate doses of cyclophosphamide markedly enhanced DTH responses to *L. tropica* extracts of both Balb/c and CBA mice when the drug was given 10 days before testing, an effect that vanished in Balb/c mice by 45 days after drug treatment coincident with suppressor T-cell repopulation.
3. The suppressor cells were specific, suppressing only the response to *L. tropica* immunogens.
4. The suppressor cells were, indeed, of the T-lymphocyte lineage.
5. Irradiation of Balb/c mice during an interval extending from a few days before until the time of inoculation of *L. tropica* markedly reduced the severity of the infection and even resulted in a proportion of cured animals.

They showed further that susceptible Balb/c mice were at least as capable as resistant CBA mice in coping with *L. tropica* infection in the absence of powerful T-cell-mediated suppression.

A conclusion that may be drawn from the various investigations on Leishmania infections is reached by way of the following reasoning:

1. Macrophages of susceptible strains of mice are defective in controlling intracellular multiplication of ingested parasites.
2. Inordinate growth of parasites results in elaboration locally of high concentrations of parasite immunogens, a condition conducive to induction of T-suppressor lymphocytes.
3. Macrophages of the susceptible genotypes are also defective in their presentation of antigens to T lymphocytes, the result being either weak DTH, inadequate elaboration of lymphokines to activate macrophages, or both.
4. Part of this defect may be reduced surface concentrations of Ia and other histocompatibility substances.

Thus, the profound and long range effects of abnormalities in a natural resistance mechanism may be clearly visualized.

The defective functions of macrophages and the resulting consequences portrayed in the case of *L. tropica* are similarly invoked to explain relative resistance and susceptibility to certain bacterial infections. Indeed, it has been demonstrated that the *Ity*, *Bcg*, and *Lsh* genes are all tightly linked on murine chromosome 1 (O'Brien *et al.*, 1980*a,b*; Plant and Glynn, 1979; Bradley, 1980; Gros *et al.*, 1981). In the case of the bacteria, *S. typhimurium* and *M. bovis*, it has been shown that a major difference between resistant and susceptible strains of mice lies in the inability of macrophages of the latter strains to control intracellular growth. A defect in macrophages that accounts for the susceptibility of C3H/HeJ mice to *S. typhimurium* is controlled by a gene different from *Ity*; this gene maps to a location close to the *Lps* gene on chromosome 4—it may, in fact, be the *Lps* gene (O'Brien *et al.*, 1980*c*).

An extrinsic, rather than intrinsic, control of macrophage function has been shown to determine relative resistance of mice to *L. monocytogenes* (Kongshavn *et al.*, 1980). The early course of infection is controlled in resistant mice (e.g., those of C57BL lineage) by macrophages. Cell-mediated, specific immune response follows, in which T-lymphocyte-mediated activation of macrophages effects the clearance of the infection. The antibacterial capability of macrophages during the early phase of infection is strongly influenced by an autosomal, dominant gene designated *Lr* (Cheers *et al.*, 1980). This gene exerts its effect on macrophages indirectly through an effect on the milieu in which the precursors of the macrophages mature (Kongshavn *et al.*, 1980). The latter conclusion emerged from studies on relative resistance of chimeras constructed of different combinations of susceptible and resistant hosts and bone marrow donors. It was found that (1) resistance obtained only in combinations that involved resistant hosts, and (2) the resistance manifested in such combinations was expressed by donor, not host, macrophages. Evidence was obtained that the superiority of the milieu in resistant mice was associated with the presence in these mice, but not in susceptible strains, of a type of radiosensitive cell. Resistance was also correlated with rapid acquisition (possibly through proliferation) of circulating monocytes (Stevenson *et al.*, 1980). Thus, the possibility was raised that more rapid proliferation, or the generation of greater numbers, of macrophages and their precursors was responsible for greater resistance in certain strains of mice. It is worth noting before leaving this extraordinarily interesting system the existence of evidence suggesting that the resistance of strains such as C57BL involves not only superior performance of macrophages early during infection, but in the later, T-cell-mediated phase of the infection as well (Skamene and Kongshavn, 1979).

While the information concerning macrophage destruction of extracellular parasites is less sophisticated, at least from the standpoint of genetic control, there is ample evidence that macrophages act effectively against both protozoan and metazoan parasites. For example, murine macrophages are capable of attacking both cercariae and schistosomules of *S. mansoni*, although their capability varies with the mouse genotype (Ellner and Mahmoud, 1979; Mahmoud *et al.*, 1979). A particularly clear example is afforded by the spontaneous cytotoxicity of macrophages from human peripheral blood toward the human intestinal parasite, *Giardia lamblia* (Smith *et al.*, 1982). Incubation of mononuclear phagocytes isolated from normal human blood with isotopically labeled target *G. lamblia* revealed a slow, progressive killing of the parasites over a time course of about 16 hr. The effector cells were demonstrated to be phagocytic and plastic adherent. Radioisotope release was strictly correlated with loss of colony-forming potential of the target parasites. Although the nature of the lethal process has not yet been reported, evidence has been presented that killing is nonspecific because an unrelated organism, *Trichomonas vaginalis*, was also found to be susceptible to the macrophages.

Natural cytotoxicity of murine (Balb/c) macrophages toward *T. vaginalis* was investigated by Landolfo *et al.* (1980). These workers studied the attack of efficitor macrophages on isotopically labeled parasites by following isotope release over a 48 hr incubation period. That the effector cells were, in fact, macrophages was concluded from their adherence to nylon–wool and plastic surfaces and their removal by carbonyl iron treatment. The activity of the effector cells was enhanced by treating mice with poly I:C, LPS, light mineral oil and was strikingly reduced by treating mice with carrageenan.

Nonspecific enhancement of resistance to parasites has often been demonstrated after treatment of hosts with preparations such as bacillus Calmette-Guerin (BCG), *Corynebacterium parvum* (CP), *Escherichia* or *Salmonella lipopolysaccharide* (LPS), complete Freund's adjuvant, and poly I: poly C (poly I:C). This subject is much too complex and confusing to consider in detail here. A few examples will suffice. Treatment of mice with BCG or CP significantly enhanced resistance to infection with schistosome cercariae; the effect varied with the strain of mouse employed (Civil *et al.*, 1978; Civil and Mahmoud, 1978) but apparently was independent of the MHC. BCG and CP were shown to be able to activate macrophages for efficient killing of schistosomules *in vitro* (Mahmoud *et al.*, 1979). BCG, CP, and *Bordetella pertussis* treatment before or at the time of inoculation of *T. congolense* afforded significantly enhanced resistance of both susceptible (A/J) and resistant (C57BL) mice. This enhanced resistance was displayed as prolonged survival, a prolonged prepatent period, and reduced parasitemia (Murray and Morrison, 1979*a*). BCG and CP treatment of CBA mice has been shown to afford protection against infections with *B. microti* and *B. rodhaini* (Clark *et al.*, 1976, 1977*a,b*); however, no protection was provided against *P. berghei* and only modest protection against *P yoelii* when the latter were initiated with parasitized erythrocytes. The BCG induced resistance against *B. microti* was not affected by exposing mice to irradiation or cyclophosphamide treatment, not was it altered by splenectomy. BCG was found to enhance the resistance of hypothymic mice to *B. microti* , but not to the extent displayed by BCG-treated intact mice. Clark *et al.* (1977*a,b*) could provide no evidence that either specific antibodies or macrophages were clearly responsible for BCG-enhanced resistance. CP-induced protection of mice to both *B. microti* and *P. chabaudi* was attributed to a nonantibody, humoral substance capable of traversing erythrocyte membranes to act on intracellular parasites (Clark *et al.*, 1977*a*). An example of the effect of *Mycobacterium* in CFA on resistance of mice to *G. muris* was provided by Roberts-Thomson and Mitchell (1979). Inoculation of resistant Balb/c mice with *G. muris* in CFA provided complete protection; protection was not afforded, however, when the parasite in incomplete Freund's adjuvant (IFA) was inoculated. Susceptible C3H mice were not protected with an inoculum composed of either version of Freund's adjuvant.

Future studies of the effects of BCG on parasite resistance in resistant/susceptible strains of mice should be designed with the known interstrain differences in the responses to BCG in mind. Strains of mice have been shown to vary considerably in their pathologic reaction, collectively known as chronic granulomatous inflammation (CGI), to intravenously administered BCG (Allen *et al.*, 1977; Moore *et al.*, 1980). For example, C57BL/6 mice respond vigorously, developing markedly enlarged lungs and spleens. CBA mice are much less responsive showing slight enlargement of these organs. Balb/c mice are intermediate in their responses. BCG treatment of C57BL/6 mice caused marked suppression of the DTH response to sheep erythrocytes, but suppression was eliminated in those that were splenectomized beforehand. Suppression was shown to be mediated by cells with the properties of macrophages, probably activated macrophages. If the normally resistant CBA mice were given cyclophosphamide 2 days before BCG, they developed severe CGI. Evidence was provided that cyclophosphamide treatment eliminated T-suppressor cells, which normally restrict the response to BCG in CBA animals.

In the light of recent information, it appears that the most likely explanation of the enhanced resistance provided by these agents, such as BCG, CP, and poly I:C, is that they boost the activity of natural resistance mechanisms in which macrophages, monocytes, and naturally cytotoxic cells are pivotal. In addition to the considerable evidence that these agents incite macrophages, there is now considerable evidence that they enhance NC activity and interferon production (Wolfe *et al.*, 1976; Ojo *et al.*, 1978; Gidlund *et al.*, 1978; Djeu *et al.*, 1979). In the next section we consider the contribution of naturally cytotoxic cells to natural resistance.

D. Naturally Cytotoxic Cells

Naturally cytotoxic cells are commonly referred to as natural killer (NK) cells. However, natural cytotoxicity is displayed by several distinguishable types of cells, including at least two subsets of NK cells (Lust *et al.*, 1981), and a closely similar subset, designated NC, with activity especially against solid tumors (Stutman *et al.*, 1980*a*). It would be inappropriate here to attempt a disquisition on NK cells, their properties and functions, as thorough reviews are available (Cudkowicz *et al.*, 1978; Herberman and Holden, 1978; Cudkowicz and Hochman, 1979; Herberman, 1980; Stutman *et al.*, 1980*a*). Rather, our discussion reviews reports—all of them recent—that deal with NK cells in parasitic infections, referring, where appropriate, to known feature of NK cells.

The close correlation between relative resistance of strains of mice to *B. microti* and *P. chabaudi* and the comparatively high NK activity of those strains was demonstrated by Eugui and Allison (1980). Mice of resistant CBA and C57BL strains, known to possess relatively high NK activity (Herberman *et al.*,

1975; Kiessling *et al.*, 1975), displayed significantly elevated NK activity toward YAC-1 tumor cells during the course of infection. In contrast, strain A mice, which are deficient in NK activity, showed little or no change during infection with *P. chabaudi*. Eugui and Allison (1980) noted that strains such as CBA and C57BL also displayed striking splenomegaly during infection, whereas splenic enlargement in strain A mice was slight. Thus, the combined increase in NK activity coupled to splenomegaly could result in formidable natural resistance in the resistant mice. Eugui and Allison (1980) assayed splenic NK activity rather late, during the peak of parasitemia and subsequently during the phase of declining parasitemia. Other investigators, working with other parasites, have discovered that the peak in NK activity appears soon after parasite inoculation.

Hatcher and Kuhn (1981) and Hatcher *et al.* (1981) investigated NK activity to *T. cruzi* in a resistant (C57BL/6) and a susceptible (C3H) strain of mice. They demonstrated two peaks of enhanced activity toward YAC-1 target cells occurring around the third day and, again, around the sixteenth day of infection, the exact time depending on the infecting dose of parasites. The first peak of trypanosome-induced lytic activity (TILA-1) was associated with typical NK cells (nonadherent, not affected by anti-Thy serum, but killed by antiserum specific for NK 1.2 and complement). The characteristics of effector cells associated with TILA-2 were somewhat different; they were not affected by antiserum to NK 1.2, but they were rather sensitive to antiserum against Thy 1. Additional work (Hatcher and Kuhn, 1982*a,b*) has revealed additional features of the TIAL-2 effectors, as follows: (1) they are not defective in beige mutant mice, as is the case of conventional NK cells [see Roder (1979)]; (2) they lack Lyt 1 and Lyt 2 determinants; and (3) their attack on target cells is significantly inhibited by the sugars *d*-mannose and *d*-galactose, a characteristic of NK cells rather than of T lymphocytes (Stutman *et al.*, 1980*b*). Thus, these cells appear to have some features of both T lymphocytes and conventional NK cells.

In the only work to date that demonstrates direct killing of parasites by NK cells, Hatcher and Kuhn (1982*c*) have shown that the NK cells responsible for the early TILA-1 phase of an infection are quite effective in killing *T. cruzi*. Both epimastigotes and blood-form trypomastigotes are attacked. Furthermore, there appears to be no difference between sensitive and resistant strains in the vigor of their NK activity toward the parasites. Heat-killed blood forms of *T. cruzi* are as effective as live parasites in inducing the early rise in NK activity.

Quite similar findings concerning murine NK activity toward *T. gondii* have been reported (Hauser *et al.*, 1982). By day 3 after parasite inoculation, the NK activity displayed by peritoneal exudate (PE) cells was 10 to 20 times higher than that observed at normal control levels, and spleen cell activity was elevated some two- to sixfold. This was a transient effect, however; by the seventh day of infection NK activity was much less. The A/J strain of mice was the only one of several strains studied that failed to develop a substantial increase in NK activity; this was correlated with the finding that A/J was the only strain that failed to elaborate a significant amount of interferon as one response to infection. A sonicated preparation of *T. gondii* was nearly as effective as live parasites for inducing an NK response. When silica particles were injected into mice preceding, and at the time of, parasite inoculation, the NK response was either markedly diminished or completely inhibited, depending on the dose of silica administered. Hauser *et al.* (1982) suggested that the macrophages damaged by the silica were required to support the burst in NK activity. It is equally possible that silica induced suppressor macrophages that inhibited the NK response to infection. Carrageenan, an agent having an effect on macrophages very similar, if not identical, to that of silica, has been shown in elegant experiments of Cudkowicz and Hochman (1979) to induce the expression of macrophages that inhibit NK activity.

NK cell activity is also elevated in the case of murine infection with *P. yoelii* (Jayawardena, 1981). On the sixth day of infection, splenic NK activity against YAC-1 target was found to be twice the normal level. Jayawardena reported, in addition, that in NK-defective beige mutant mice infections with *P. yoelii* were much more severe than in normal mice.

We have investigated NK activity and interferon levels in mice infected with *T. musculi* (Albright *et al.*, 1983). Increased NK activity in the spleens of C3H and C57BL/6 mice was evident by the first day of infection and rose to a peak on days 3 and 4 (Table II). By the seventh and eighth day of infection, NK activity was declining, reaching a stable level somewhat below the normal level. The serum titers of interferon closely paralleled NK activity. Preliminary experiments have indicated that the decline in NK activity is correlated with the appearance of a population of adherent suppressor cells, but more work is required to determine the properties of these cells. Note that in Table II there is no indication of a rise in NK activity on day 20 of infection at the time when clearance of the *T. musculi* was complete.

NK activity of human peripheral blood cells from children suffering *P. falciparum* infections has been reported (Ojo-Amaize *et al.*, 1981). The target cell employed was the human erythroid leukemia line, K562. In patients ranging in age from 1 year to almost 6 years a strict correlation was found between the degree of parasitemia and the level of NK activity. Furthermore, there was a close correlation between serum interferon (IFN) titers, most of which as α-interferon, and levels of NK activity. The NK activity of peripheral blood cells obtained from patients could not be further increased by incubating the cells with exogenous IFN, suggesting that they were maximally activated, or had perhaps become refractory.

The evidence that IFN triggers the activation of NK cells is strong; it is therefore of interest to inquire about the evidence that parasite infection stimulates the elaboration of IFN. There is conflicting information. For example, Rytel and Marsden (1970) found elevated IFN titers in mice infected with

T. cruzi, whereas Hatcher and Kuhn/(1982*a*) did not. Rytel *et al.* (1973) failed to find elevated IFN titers in young adult patients suffering from either malaria (*P. vivax* and *P. falciparum*) or Chagas disease (*T. cruzi*). Recently, Sonnenfeld and Kierszenbaum (1981) and Kierszenbaum and Sonnenfeld (1982) demonstrated that *T. cruzi* infections of mice do stimulate IFN elaboration and, moreover, that administration of authentic IFN markedly reduces the severity of *T. cruzi* infections. In the case of murine infections with *P. berghei* there is agreement that IFN titers are elevated (Huang *et al.*, 1968) and that IFN inducers protect mice against *P. berghei* (Schultz *et al.*, 1968; Jahiel *et al.*, 1970).

Two other publications of immediate interest are available. One of these reports demonstrates that NK cells are elevated in *S. mansoni* infections (Attallah *et al.*, 1980). The other is a detailed study of NK activity in mice infected with the yeast *Cryptococcus neoformans* (Murphy and McDaniel, 1982). Previous work had shown that *C. neoformans* infections of nude mice were less severe than in normal Balb/c mice (Cauley and Murphy, 1979), a finding that might be attributed to the well-known elevated NK activity characteristic of nude mice (Herberman, 1978). Murphy and McDaniel (1982) demonstrated a strict correlation between elevated NK activity against YAC-1 target cells and restricted growth *in vitro* of *C. neoformans* under the following conditions: (1) higher activity in young than in old mice, (2) poly I:C stimulation of mice, and (3) treatment of mice with CP, which induced an initial rise and subsequent decline in splenic NK activity.

Our present view of the involvement of NK cells in parasitic infections is that they act early in the course of infection and contribute significantly to natural resistance. Several points concerning the distribution and characteristics of NK cells are worth noting:

1. NK activity varies considerably among individuals and strains of mice.
2. NK activity is elevated in the hypothymic mouse (Herberman, 1978) and in *xid* mutant mice (Herberman and Holden, 1978).
3. NK activity is not limited to vertebrates, but is present in invertebrates, as well (Boiledieu and Valembois, 1977).
4. Agents such as BCG, CP, and poly I:C stimulate NK activity, probably as a result of stimulating interferon production (Wolfe *et al.*, 1976; Ojo *et al.*, 1978; Gidlund *et al.*, 1978; Djeu *et al.*, 1979).
5. Agents such as silica, carrageenan, and xenogeneic antiserum against mouse cells cause marked decline in NK activity as a result of stimulating macrophage suppressors (Kiessling *et al.*, 1977; Cudkowicz and Hochman, 1979).

In our view, concerted attention to the NK (and NC) systems as they relate to parasitic infections is warranted. For example, it is essential to determine why the initially elevated NK activity so quickly declines during parasitic infection. Is the decline a reflection of a rise in suppressor macrophages or perhaps a further reflection of immunosuppression attributable to parasite-derived substances (Cunningham *et al.*, 1978; Albright and Albright, 1980, 1981*c*)? It appears that the latent potential of the NK system far exceeds its normal expression *in situ* (Cudkowicz and Hochman, 1979; Weiss *et al.*, 1982). If methods could be devised to induce sustained elevation of NK activity without undesirable consequences the benefits could be enormous.

E. Resistance in Nonpermissive Prospective Hosts

Although it has been known for at least 80 years that many parasites are highly host-specific, very little attention has been given to the nature of the solid resistance (nonpermissiveness) displayed by species that, for various reasons, might appear to be suitable hosts, but are not. For want of a better phrase, we will call them nonpermissive prospective (NP) hosts. Because more work has been done on natural resistance to trypanosomes than on any other parasite, the following discussion is devoted to trypanosomes. Natural resistance to trypanosomes was the topic of an informative and readable review by Terry (1976). For additional reading there is an excellent account of host specificity toward a number of species of *Babesia* by Zwart and Brocklesby (1979).

To facilitate their investigations of *Trypanosoma vivax* infections of sheep, Desowitz and Watson (1951, 1952) attempted to develop a strain of the parasite in laboratory rats. These workers discovered that rats were completely refractory to the parasite unless the inoculation of the parasite was accompanied by injection of sheep serum. As long as the infected rat was provided periodic injections of sheep serum, the infection would persist. To achieve successful transfer of the parasites from infected to fresh rats, the latter had to be provided with sheep serum. Finally, after more than 40 passages, a rat-adapted strain of *T. vivax* was obtained that proved to be highly virulent for rats (Desowitz and Watson, 1953). Attempts to identify the substances in sheep serum that were required for initiation of *T. vivax* infection in rats were inconclusive.

Terry (1957) discovered that the rat-adapted strain of *T. vivax* was incapable of initiating infection in the cotton rat, *Sigmodon hispidus*. Serum of the cotton rat was highly efficient in agglutinating and subsequently lysing *T. vivax*. The serum of cotton rats failed to agglutinate other species of trypanosomes. Furthermore, *T. vivax* organisms were not agglutinated by sera from a variety of other mammals. Later studies (Hudson and Terry, 1970; Hudson, 1971, 1972) were designed to (1) determine whether the agglutinating substance was present in rats reared under bacteria-free conditions, and (2) ascertain whether the substance was an immunoglobulin. It was found that the agglutinating titers of bacteria-free and conventional animals were nearly equivalent; it was therefore unlikely that the substance was a cross-reacting antibody developed originally against some other infectious organism. The question of whether the substance was an immunoglobulin was not resolved, although it appeared unlikely.

The well-known toxicity of human serum for *T. b. brucei* was mentioned in Section III. A. Since the original report (Laveran, 1902) there have been several studies on the nature of active substances in human serum. For example, evidence that antibody of the IgM isotype might be the active substance was adduced by Aaronovitch and Terry (1972), certainly a reasonable possibility at the time. Finally, an illuminating report by Rifkin (1978) demonstrated clearly that the toxicity of human serum for *T. b. brucei* is attributable to HDL. The next advance in this line of investigation should reveal how the lipoprotein effects the demise of *T. b. brucei*, which could be a promising lead toward the development of trypanolytic substances.

The strong resistance of mice to the rat trypanosome, *T. lewisi*, was investigated by Lincicome (1955, 1958). In normal, untreated mice, inoculated *T. lewisi* organisms were completely eliminated in 1 day. If the mice were injected daily with fresh rat serum the trypanosomes survived and even grew slowly for periods of about 2 weeks. Persistence and growth were best in mice starved before parasite inoculation and then maintained on a restricted diet while being injected with rat serum. Subsequent experiments (Lincicome and Francis, 1961) showed that the serum of rats was much more effective in prolonging parasite survival than was the serum of any other species tested; serum of hamster, guinea pig, and rabbit had some effect,while serum of horse, human, cow, dog, pig, and chicken was ineffective. Greenblatt and Lincicome (1966) and Greenblatt *et al.* (1969) attempted to identify the components of rat serum responsible for prolonging the survival of *T. lewisi* in mice. At first, it appeared that IgM and IgG γ_2 immunoglobulins were the active components, but subsequent work appeared to exclude immunoglobulins.

The possibility that rat serum might provide a nutrient or stimulant required by *T. lewisi* that was lacking in the mouse was largely excluded by Greenblatt and Shelton (1968). They showed that *T. lewisi* organisms grew reasonably well when confined to diffusion chambers implanted in the peritoneal space of mice. Moreover, mouse serum appeared conducive to *T. lewisi* growth. When injected into T. lewisi-infected gerbils, mouse serum strongly promoted growth of the parasites. In fact, mouse serum even promoted a threefold enhancement of parasitemia when injected into infected rats.

Mühlpfordt (1968) made a concerted effort to adapt *T. lewisi* to survival and growth in rat serum-treated mice. Serial passages over a period of 16 months failed to provide a mouse-adapted strain of parasite. Splenectomy of mice before inoculation of *T. lewisi* was without effect.

Our recent investigation was intended to provide insight concerning the mechanisms underlying murine resistance to *T. lewisi* (Albright and Albright, 1981*b*). An analysis of the kinetics of clearance of a standard inoculum of

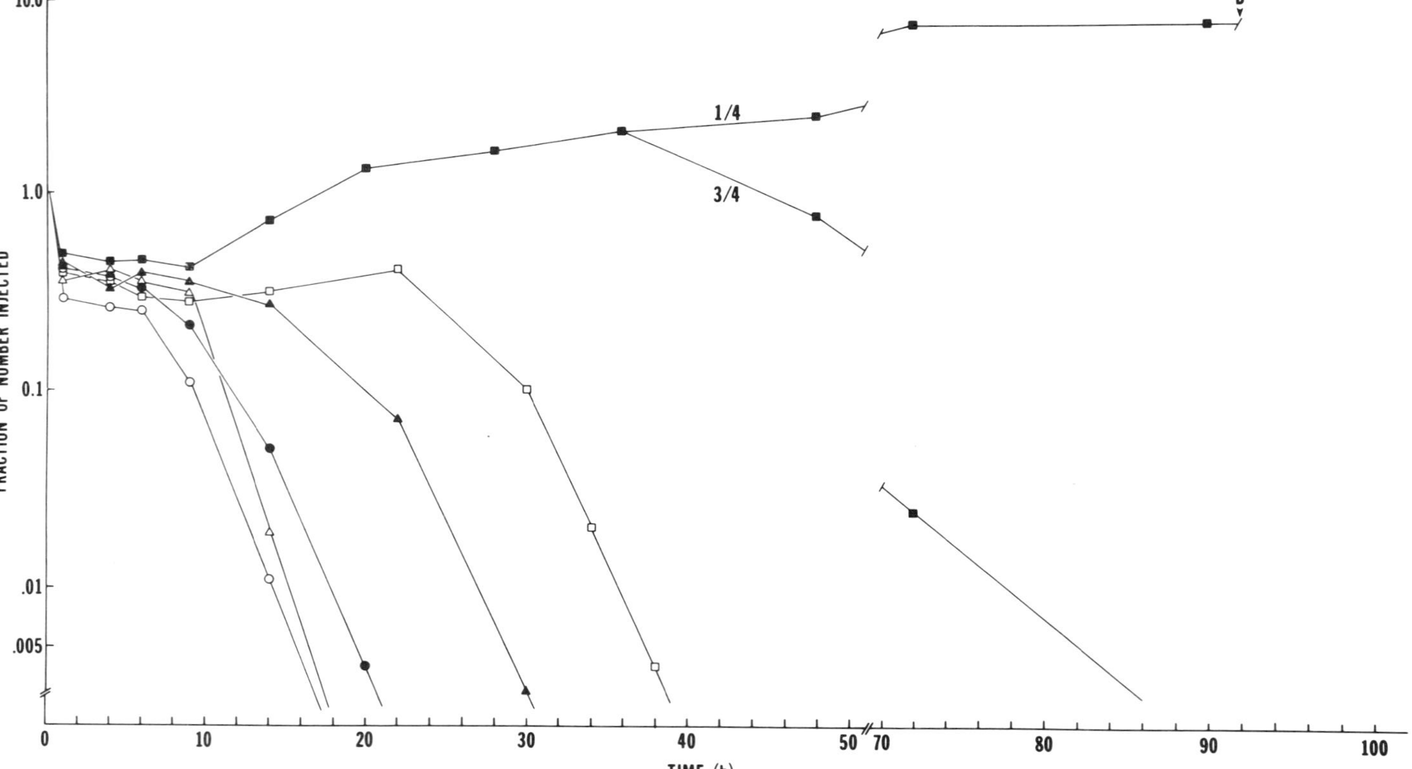

Figure 2. Survival of *T. lewisi* in the blood of mice. (○) No treatment; (△) one injection of *T. lewisi* sonicate; (▲) five injections of rabbit antiserum against mouse macrophages; (●) one injection of silica dust; (□) two injections of normal rat serum; (■) one injection of silica dust and two injections of normal rat serum. D, Death of mice. (From Albright and Albright, 1981*b*.)

T. lewisi by mice of various strains indicated only a slight difference; some strains (e.g., C57BL/6, PL, and DBA/1) eliminated parasites from the blood at a faster rate than did others (A/He, Balb/c C3H, CBA). In all strains a lag of 8–10 hr occurred postinoculation before rapid clearance was initiated. Studies on the ability of rat and mouse spleen cells to support growth of *T. lewisi in vitro* showed that mouse cells were even more satisfactory than rat cells in providing the required nutrients, stimulants, and so forth. The effect of treating mice with various substances on survival and growth of inoculated *T. lewisi* was investigated. The results of this study are depicted in Fig. 2. As expected, injection of rat serum prolonged survival, extending the time preceding the phase of rapid clearance to about 24 hr (8–10 hr in untreated mice). Treatment of mice with a sonicate of *T. lewisi* was without effect, as was a single injection of silica dust preceding parasite inoculation. Some prolongation of survival of the trypanosomes was afforded by multiple injections of rabbit antiserum specific for murine macrophages, but the effect was less than that provided by fewer injections of rat serum. A dramatic effect was provided by a combination of one injection of silica (before trypanosome inoculation) and two injections of rat serum (one before and one after parasite inoculation). This treatment resulted not only in prolonged survival, but in intravascular growth of the parasites as well. Some of the mice died from the infection, although most of them eventually cleared the parasitemia. Our attempts to demonstrate that the immunoglobulin fraction of rat serum was the active component were inconclusive.

These investigations concerned with the prolongation of survival of trypanosomes in NP hosts after treatment with serum of natural hosts have not yet provided firm conclusions. It does appear, however, that the explanation of the benefit provided by serum treatment lies in a blocking effect on host natural resistance mechanisms rather than some nutritional stimulus. We are inclined, for example, to interpret the results of our investigation as indicating that both macrophages and NK cells are at work in the resistance of mice to *T. lewisi*. If this is correct, we must ask why the effect of these two systems is so much less powerful when directed against *T. musculi* than against *T. lewisi*. This leads to the general question: Are the mechanisms of natural resistance used by NP hosts qualitatively the same as those employed by resistant, and to a lesser extent, susceptible, permissive hosts? If so, what is responsible for the quantitative differences in effectiveness of the systems? There is little insight at present but the questions are open to investigation.

IV. CONCLUSIONS

Natural resistance comprises a formidable array of toxic substances and aggressive cells, including the following:

Complement
Complement activators (excluding immune complexes)
Lipoproteins
Macrophages
NK and NC cells
Platelets (?)

Which of these substances or cells becomes involved in the defense against an invader, and with what intensity each is expressed, depends both on the composition of the parasite and on the match between genes of the parasite and the host. Except possibly in highly abnormal individuals, no prospective hosts are uniformly susceptible to all parasites. Nor are any prospective hosts uniformly resistant to all parasites. Some prospective hosts are prepared to recognize and react vigorously against a potential parasite using several of the available defense mechanisms; these are nonpermissive and avoid becoming hosts. Recognition involves complementary molecules. The search for recognition sites (receptors?) on host cells toward parasites should prove rewarding. Other prospective hosts either do not clearly recognize parasites or, if they do, are unable to respond forcefully. When this occurs and the acquired immune mechanisms also fail to respond vigorously, the host is both permissive and susceptible. It is the achievement of this last condition that parasites are constantly seeking during evolution.

In his review in 1976, Terry wrote, "In summary, an investigator faced with an example of innate resistance to an infectious agent must suspect many causes, all difficult to investigate, and faces the likely prospect of being unable to apply any discoveries he might make to other systems." Perhaps the climate has changed a bit since 1976 and conditions, tempered by new knowledge gained during the last few years, are now more favorable for the student of natural resistance. Even if they are not, the study of natural resistance is intellectually challenging and therefore rewarding. Terry (1976) also wrote, "The contribution made by innate resistance to freedom from infection is undoubtedly greater than that by acquired immune responses. But the last has been more extensively studied and our knowledge of its mechanisms is far greater." It is time to change this balance of knowledge to a state more favorable to natural resistance. We hope this chapter will further this cause.

ADDENDUM

Several papers have appeared recently in which the analysis of genetic control of natural resistance to bacterial infections has been continued. These studies are highly relevant to the subject of resistance to parasites. Strong evidence has been presented that the genes *Bcg, Ity* and *Lsh* are either the same gene, or that they are tightly linked on murine chromosome 1 (Skamene *et al.*, 1982*a*). In the case of resistance to *Mycobacterium bovis* (BCG), it has been demonstrated that resistant strains develop little or no acquired immunity when exposed to low, or even moderately high, numbers of bacteria; in contrast, acquired immunity, in the form of delayed hypersensitivity and other sequels, develops readily in BCG-sensitive strains (Pelletier *et al.*, 1982). Cross-protective immunity was studied in two strains that differed in sensitivity to BCG but displayed the same initial susceptibility to *Listeria monocytogenes*. The BCG-susceptible strain that developed acquired immunity to BCG displayed immunity to *Listeria* as well; in contrast, the BCG-resistant strain, which did not manifest acquired immunity to BCG upon exposure, showed no change in its susceptibility to *Listeria*. It seems apparent that naturally resistant mice are able to expel a BCG infection rapidly without developing typical cell-mediated immunity.

Recent reports of the continuing investigations of the resistance manifested through macrophages have provided evidence for two kinds of mechanisms—one intrinsic to the macrophage and a second extrinsic mechanism. The intrinsic mechanism is concerned with regulating the intracellular growth of organisms that multiply intracellulary and has been termed a *microbiostatic* mechanism (Skamene *et al.*, 1982*b*). The extrinsic mechanism is concerned with recruitment from bone marrow progenitors of monocytes and macrophages capable of microbiocidal activity. In the case of resistance to *L. monocytogenes*, controlled by an autosomal gene, *Lr*, mice that possess the dominant allele, Lr^{r}, are able to manifest a rapid monocyte/macrophage mobilization. Indeed, it appears that the Lr gene may regulate elaboration of an inducible monocytopoietin that is capable of markedly reducing the generation time of monocyte progenitors. (Skamene *et al.*, 1982*b*). Regarding cytodynamics of progenitor cells, it is interesting to speculate about the significance of an earlier report (Brodt *et al.*, 1981) whereby it was demonstrated that B-lymphocyte-deprived mice displayed substantially elevated resistance to *L. monocytogenes* and to the extracellular protozoan parasite, *Trypanosoma musculi.*

Another recent report (Wakelin and Donachie, 1983) deals with a thoughtful and comprehensive investigation of the role of MHC genes in the T-cell mediated inflammatory expulsion of intestinal stages of *Trichinella spiralis.* One reason for the interest in this report lies in the finding of a possible deficiency in the involvement of hemopoietic precursor cells in the inflammatory reaction against the adult worms. The presence of the H-2^{q} allele at either the *K* or *D* locus was sufficient to enhance the resistance to *T. spiralis* of mice otherwise endowed with a susceptible haplotype. In contrast, the H-2^{d} allele at the *D* locus was associated with substantially diminished resistance of mice carrying the *q* alleles (resistant) at all other loci. Two alternative explanations were offered for the effect of the *d* allele, one of which is especially relevant to the present discussion; *viz*. the possibility that H-2D^{d} moderates the response of myeloid precursor cells to lymphocyte-derived factors. The result could be deficient involvement of myelocytes in the intestinal inflammatory reaction.

Earlier in the chapter we referred to the fascinating studies of the relative resistance of the ndama breed of cattle, as compared to the zebu breed, to African trypanosomiasis. A thorough review of this subject, including a comprehensive report of their own investigations, was published recently by Murray *et al.* (1982). It is now established that "Trypanotolerance" (i.e., natural resistance) exists in cattle as an innate characteristic. Evidence has been presented by these investigators that natural resistance can be enhanced in cattle. They believe that it will be possible to produce resistant livestock through selective breeding and application of immunotherapeutic procedures.

ACKNOWLEDGMENTS

Our thanks go to Carter Diggs, John Finerty, Raymond Kuhn, Frank Richards, and James Strickler for helpful and informative discussions. We are grateful to Carter Diggs, John Finerty, Frank Hatcher, and Ray Kuhn for permitting us to see manuscripts not yet published. Our special gratitude goes to Phillip D'Alesandro and Raymond Kuhn, who have read the manuscript and offered many valuable comments. Our research has been supported by grants from the National Science Foundation (PCM 80-17183), the National Institutes of Health (AG 03267), and the Lee Foundation.

V. REFERENCES

Aaronovitch S., and Terry, R. J., 1972, The trypanolytic factor in normal human serum, *Trans. R. Soc. Trop. Med. Hyg.* **66**:344.

Ackert, J. E., Eisenbrandt, L. L., Wilmoth, J. H., Glading, B., and Pratt I., 1935, Comparative resistance of five breeds of chickens to the nematode *Ascarida lineata* (Schneider), *J. Agric. Res.* **50**:607.

Aikawa, M., Miller, L. H., Rabbege, J. R., and Epstein, N., 1981, Freeze fracture study on the erythrocyte membrane during malarial parasite invasion, *J. Cell. Biol.* **91**:55.

Albright, J. W., and Albright, J. F., 1980, Trypanosome-mediated suppression of murine humoral immunity independent of typical suppressor cells, *J. Immunol.* **124**:2481.

Albright, J. W., and Albright, J. F., 1981*a*, Differences in resistance to *Trypanosoma musculi* infection among strains of inbred mice, *Infect. Immun.* **33**:364.

Albright, J. W., and Albright J. F., 1981*b*, Basis of the specificity of rodent trypanosomes for their natural hosts, *Infect. Immun.* **33**:355.

Albright, J. W., and Albright J. F., 1981*c*, Inhibition of murine humoral immune responses by substances derived from trypanosomes, *J. Immunol.* **126**:300.

Albright J. W., and Albright J. F., 1982, Studies on the mechanisms involved in the decline of immunological resistance of aging mice to *Trypanosoma musculi*, *Mech. Ageing Dev.* **20**:315.

Albright, J. W., Huang, K. Y., and Albright, J. F., 1983, Natural killer activity in mice infected with *Trypanosoma musculi*, *Infect. Immun.* **40**:869.

Allen, E. M., Moore, V. L., and Stevens, J. O. 1977. Strain variation in BCG-induced chronic pulmonary inflammation in mice. I. Basic model and possible genetic control by non-H-2 genes, *J. Immunol.* **119**:343.

Amsbaugh, D. F., Hansen, C. T. Prescott, B., Stashak, P. W., Asofsky, R., and Baker, P. J., 1974, Genetic control of the antibody response to type III pneumococcal polysaccharide in mice. II. Relationship between IgM immunoglobulin levels and the ability to give an IgM antibody response, *J. Exp. Med.* **139**:1499.

Attallah, A. M., Lewis, F. A., Urritia-Shaw, A., Folks, T., and Yeatman, T. J., 1980, Natural killer cells (NK) and antibody-dependent-cell-mediated cytotoxicity (ADCC) components of *Schistosoma mansoni* infection, *Int. Arch. Allergy Appl. Immun.* **63**:351.

Bayne, C. J., Buckley, P. M., and DeWan, P. C., 1980, *Schistosoma mansoni*: Cytotoxicity of hemocytes from susceptible snail hosts for sporocysts in plasma from resistant *Biomphalaria glabrata*, *Exp. Parasitol.* **50**:409.

Behin, R., Mauel, J., and Sordat, B., 1979, *Leishmania tropica*: Pathogenicity and *in vitro* macrophase function in strains of inbred mice, *Exp. Parasitol.* **48**:81.

Biozzi, G., Stiffel, C., Mouton, D., and Bouthillier, Y., 1975, Selection of lines of mice with high and low antibody responses to complex immunogens, in: *Immunogenetics and Immunodeficiency* (B. Benacerraf, ed.), pp. 179–227, University Park Press, Baltimore.

Biozzi, G., Mouton, D., Sant'Anna, O. A., Passos, H. C., Gennari, M., Reis, M. H., Ferreira, V. C. A., Heumann, A. M., Bouthillier, Y., Ibanez, O. M., Stiffel, C., and Siqueria, M., 1979, Genetics of immunoresponsiveness to natural antigens in the mouse, *Curr. Top. Microbiol. Immunol.* **85**:31.

Blackwell, J., Freeman, J., and Bradley, D. J. 1980, Influence of H-2 complex on acquired resistance to *Leishmania donovani* infection in mice, *Nature* (*Lond.*) **283**:72.

Blum, K., and Cioli, D., 1978, Paradoxical behavior of Biozzi high and low responder mice upon infection with *Schistosoma mansoni*, *Eur. J. Immunol.* **8**:52.

Boiledieu, D., and Valembois, P., 1977, Natural cytotoxic activity of sipunculid leukocytes on allogeneic and xenogeneic erythrocytes, *Develop. Compar. Immunol.* **1**:207.

Bradley, D. J., 1972, Regulation of parasite populations: A general theory of the epidemiology and control of parasitic infections, *Trans. R. Soc. Trop. Med. Hyg.* **66**:697.

Bradley, D. J., 1977, Regulation of Leishmania populations within the host. II. Genetic control of acute susceptibility of mice to *Leishmania donovani* infection, *Clin. Exp. Immunol.* **30**:130.

Bradley, D. J., 1980, Genetic control of resistance to protozoal infections, in: *Genetic Control of Natural Resistance to Infection and Malignancy* (E. Skamene, P.A.L. Kongshavn, and M. Landy, eds.), pp. 9–28, Academic Press, New York.

Bradley, D. J., Taylor, B. A., Blackwell, J., Evans, E. P., and Freeman, J., 1979, Regulation of Leishmania populations within the host. III. Mapping of the locus controlling susceptibility to visceral leishmaniasis in the mouse, *Clin. Exp. Immunol.* **37**:7.

Brodt, P., Kongshavn, P., Vargas, F., and Gordon, J., 1981, Increased natural host resistance mechanisms in B lymphocyte-deprived mice, *J. Reticuloendothel. Soc.* **30**:283.

Brooks, B. O. and Reed, N. D., 1979, The effect of trypan blue on the early control of *Trypanosoma musculi* parasitemia in mice, *J. Reticuloendothel. Soc.* **25**:325.

Brown, A. R., Crandall, C. A., and Crandall, R. B.. 1977, The immune response and acquired resistance to Ascaris suum infection in mice with an X-linked B lymphocyte defect, *J. Parasitol.* **63**:950.

Brown, W. H., 1914, A note on the pathogenicity of *Trypanosoma lewisi*, *J. Exp. Med.* **19**:406.

Butterworth, A. E., David, J. R., Franks, D., Mahmoud, A. A. F., David, P. H., Sturrock, R. F., and Houbs, V., 1977, Antibody-dependent eosinophil-mediated damage to ^{51}Cr-labelled *Schistosoma mansoni.* Damage by purified eosinophils, *J. Exp. Med.* **145**:136.

Cauley, L. K., and Murphy, J. W., 1979, Response of congenitally athymic (nude) and phenotypically normal mice to a *Cryptococcus neoformans* infection, *Infect. Immun.* **23**:644.

Chang, K-P., and Chiao, J. W., 1981, Cellular immunity of mice to *Leishmania donovani in vitro*: Lymphokine-mediated killing of intracellular parasites in macrophages, *Proc. Natl. Acad. Sci. USA* **78**:7083.

Cheers, C., and McKenzie, I. F. C., 1978, Resistance and susceptibility of mice to bacterial infection: genetics of listeriosis, *Infect. Immun.* **19**:755.

Cheers C., McKenzie, I. F. C., Mandel T. E., and Chan, Y. Y., 1980, A single gene (*Lr*) controlling natural resistance to murine listeriosis, in: *Genetic Control of Natural Resistance to Infection and Malignancy* (E. Skamene, P. A. L. Kongshavn, and M. Landy, eds.), pp. 141–147, Academic Press, New York.

Cheng, T. C., 1968, The compatibility and incompatibility concept as related to trematodes and molluscs, *Pac. Sci.* **22**:141.

Civil, R. H., and Mahmoud, A. A. F., 1978, Genetic differences in BCG-induced resistance to *Schistosoma mansoni* are not controlled by genes within the major histocompatibility complex of the mouse, *J. Immunol.* **120**:1070.

Civil, R. H., Warren, K. S., and Mahmoud, A. A. F., 1978, Conditions for Bacillus-Calmette-Guerin-induced resistance to infection with *Schistosoma mansoni* in mice, *J. Infect. Dis.* **137**:550.

Clark, I. A., Allison, A. C., and Cox F. E. G., 1976, Protection of mice against *Babesia* spp. and *Plasmodium* spp. with BCG, *Nature* (*Lond.*) **259**:309.

Clark, I. A., Cox, F. E. G., and Allison, A. C., 1977*a*, Protection of mice against *Babesia* spp. and *Plasmodium* spp. with killed *Corynebacterium parvum*, *Parasitology* **74**:9.

Clark, I. A., Wills, E. J., Richmond, J. E., and Allison, A. C., 1977*b*, Suppression of babesiosis in BCG-infected mice and its correlation with tumor inhibition, *Infect. Immun.* **17**:430.

Crandall, C. A., and Crandall, R. B., 1971, *Ascaris suum*: Immunoglobulin responses in mice, *Exp. Parasitol.* **30**:426.

Crandall, R. B., 1975, Decreased resistance to *Trichinella spiralis* in aged mice, *J. Parasitol.* **61**:566.

Cudkowicz, G., and Hochman, P. S., 1979, Do natural killer cells engage in regulated reactions against self to ensure homeostasis? *Immunol. Rev.* **44**:13.

Cudkowicz, G., Landy, M., and Shearer, G. M. (eds.), 1978, *Natural Resistance Systems Against Foreign Cells, Tumors and Microbes*, Academic Press, New York.

Culbertson, J. T., 1941, *Immunity Against Animal Parasites*, Columbia University Press, New York.

Cunningham, D. S., Kuhn, R. E., and Rowland, E. C., 1978, Suppression of humoral responses during *Trypanosoma cruzi* infections in mice, *Infect. Immun.* **22**:155.

Curtiss, L. K., and Edgington, T. S., 1976, Regulatory serum lipoproteins: Regulation of lymphocyte stimulation by a species of low density lipoprotein, *J. Immunol.* **116**:1452.

Curtiss, L. K., DeHeer, D. H., and Edgington, T. S., 1977, *In vivo* suppression of the primary immune response by a species of low density serum lipoprotein, *J. Immunol.* **118**:648.

D'Alesandro, P. A., 1962, *In vitro* studies of ablastin, the reproduction-inhibiting antibody to *Trypanosoma lewisi*, *J. Protozool.* **9**:351.

D'Alesandro, P. A., 1979, Rodent trypanosomiases, in: *Pathogenicity of Trypanosomes* (G. Losos and A. Chouinard, eds.), pp. 63–70, International Development Research Centre, Ottawa.

Desowitz, R. S., and Watson, H. J. C., 1951, Studies on *Trypanosoma vivax*. I. Susceptibility of white rats to infection, *Ann. Trop. Med. Parasitol.* **45**:207.

Desowitz, R. S., and Watson, H. J. C., 1952, Studies on *Trypanosoma vivax*, III. Observation on the maintenance of a strain in white rats, *Ann. Trop. Med. Parasitol.* **46**:92.

Desowitz, R. S., and Watson, H. J. C., 1953, Studies on *Trypanosoma vivax*. IV. The maintenance of a strain in white rats without sheep serum supplement, *Ann. Trop. Med. Parasitol.* **47**:62.

DeTolla, L. J., Scott, P. A., and Farrell, J. P., 1980, Infection with *Leishmania tropica major*: Genetic control in inbred mouse strains, in: *Genetic Control of Natural Resistance to Infection and Malignancy* (E. Skamene, P.A.L. Kongshavn, and M. Landy, eds.), pp. 39–45, Academic Press, New York.

Djeu, J. Y., Heinbaugh, J. A., Holden, H. T., and Herberman, R. B., 1979, Augmentation of mouse natural killer cell activity by interferon and interferon inducers, *J. Immunol.* **122**:175.

Dluzewski, A. R., Rangachari, K., Wilson, R. J. M., and Gratzer, W. B., 1981, Entry of malaria parasites into resealed ghosts of human and simian erythrocytes, *Br. J. Haematol.* **49**:97.

Duca, C. J., 1939, Studies on age resistance against trypanosome infections. II. The resistance of rats of different age groups to *Trypanosoma lewisi*, and the blood response of rats infected with this parasite, *Am. J. Hyg.* **29**:25.

Ehrlich, P., 1907, *Berl. Klin. Wochnschr.* **44**:233 (cited in Taliaferro and Olsen, 1943).

Ellner, J. J., and Mahmoud, A. A. F., 1979, Killing of schistosomula of *Schistosoma mansoni* by normal human monocytes, *J. Immunol.* **123**:949.

Emmerling, P., Finger, H., and Hof, H., 1977 Cell-mediated resistance to infection with *Listeria monocytogenes* in nude mice, *Infect. Immun.* **15**:382.

Eugui, E. M., and Allison, A. C., 1980, Differences in susceptibility of various mouse strains to haemoprotozoan infections: Possible correlation with natural killer activity, *Parasite Immunol.* **2**:277.

Finerty, J. F., Rosenberg, Y. J., Kendrick, L., McKelvin, R. P., and Hansen, C. T., 1983, The use of CBA/N and nude mice to study the regulation of B-cell activation following infection with *Trypanosoma rhodesiense* (in press).

Fiser, R. H., Jr., Denniston, J. C., McGann, V. G., Kaplan, J., Adler, W. H., Kastello, M. D., and Beisel, W. R., 1973, Altered immune functions in hypercholesterolemic monkeys, *Infect. Immun.* **8**:105.

Frenkel, J. K., and S. A. Caldwell, 1975, Specific immunity and nonspecific resistance to infection: *Listeria*, protozoa, and viruses in mice and hamsters, *J. Infect. Dis.* **131**:201.

Gabrielson, A. E., Pickering, R. J., Linna, T. J., and Good, R. A., 1973, Haemolysis in chicken serum. II. Ontogenetic development, *Immunology* **25**:179.

Ganapati, P. N., 1948, Cultivation of *Trypanosoma cruzi* in the developing embryo, *Nature (Lond.)* **162**:963.

Gardner, I. D., and Remington, J. S., 1978*a*, Aging and the immune response. I. Antibody formation and chronic infection in *Toxoplasma gondii*-infected mice, *J. Immunol.* **120**:939.

Gardner, I. D., and Remington, J. S., 1978*b*, Aging and the immune response. II. Lymphocyte responsiveness and macrophage activation in *Toxoplasma gondii*-infected mice, *J. Immunol.* **120**:944.

Gasbarre, L. C., Finerty, F. F., and Louis, J. A., 1981, Non-specific immune responses in CBA/N mice infected with *Trypanosoma brucei*, *Parasite Immunol.* **4**:273.

Gidlund, M., Orn, A., and Wigzell, H., 1978, Enhanced NK cell activity in mice injected with interferon and interferon inducers, *Nature* (*Lond.*) **273**:759.

Gorczynski, R. M., and MacRae, D., 1982, Analysis of subpopulations of glass-adherent mouse skin cells controlling resistance/susceptibility to infection with *Leishmania tropica*, and correlation with the development of independent proliferative signals to Lgt-1+/Lyt-2+ T lymphocytes, *Cell. Immunol.* **67**:74

Greenblatt, C. L., and Lincicome, D. R., 1966, Identity of trypanosome growth factors in serum. II. Active globulin components, *Exp. Parasitol.* **19**:139.

Greenblatt, C. L., and Shelton, E., 1968, Mouse serum as an environment for the growth of *Trypanosoma lewisi*, *Exp. Parasitol.* **22**:187.

Greenblatt, C. L., Jori, L. A., and Cahnmann, H. J., 1969, Chromatographic separation of a rat serum growth factor required by *Trypansoma lewisi*, *Exp. Parasitol.* **24**:228.

Greenblatt, H. C., Rosenstreich, D. L., and Diggs, C. L., 1980, Genetic control of natural resistance to *Trypanosoma rhodesiense* in mice, in: *Genetic Control of Natural Resistance in Infection and Malignancy* (E. Skamene, P. A. L. Kongshavn, and M. Landy, eds.), pp. 89–95, Academic Press, New York.

Gros, P., Skamene, E., and Forget, A., 1981, Genetic control of natural resistance to *Mycobacterium bovis* (BCG) in mice, *J. Immunol.* **127**:2417.

Groves, M. G., Rosenstreich, D. L., and Osterman, J. V., 1980, Genetic control of natural resistance to *Rickettsia tsutsugamushi* infection in mice, in: *Genetic Control of Natural Resistance to Infection and Malignancy* (E. Skamene, P. A. L. Kongshavn, and M. Landy, eds.), pp. 165–171, Academic Press, New York.

Guckian, J., Christensen, W., and Fine, D., 1978, Trypan blue inhibits complement-mediated phagocytosis by human polymorphonuclear leucocytes, *J. Immunol.* **120**:1580.

Handman, E., Ceredig, R., and Mitchell, G. F., 1979, Murine cutaneous leishmaniasis: Disease patterns in intact and nude mice of various genotypes and examination of some differences between normal and infected macrophages, *Aust. J. Exp. Biol. Med. Sci.* **57**:9.

Haque, A., Lefebvre, M. N., Ogilvie, B. M., and Capron, A., 1978, *Dipetalonema viteae* in hamsters: Effect of antiserum of immunization with parasite extracts on production of microfilariae, *Parasitology* **76**:61.

Hatcher, F. M., and Kuhn, R. E., 1981, Spontaneous lytic activity against allogeneic tumor cells and depression of specific cytotoxic responses in mice infected with *Trypanosoma cruzi*, *J. Immunol.* **126**:2436.

Hatcher, F. M., Kuhn, R. E., Cerrone, M. C., and Burton, F. C., 1981, Increased natural killer cell activity in experimental American trypanosomiasis, *J. Immunol.* **127**:1126.

Hatcher, F. M., and Kuhn, R. E., 1982*a*, Modulation of natural killer cell activity during murine infection with *Trypanosoma cruzi*, in: *NK Cells and Other Natural Effector Cells* (R. B. Herberman, ed.), pp. 503–509, Academic Press, New York.

Hatcher, F. M., and Kuhn, R. E., 1982*b*, Natural killer (NK) cell activity against extracellular forms of *Trypanosoma cruzi*, in: *NK Cells and Other Natural Effector Cells* (R. B. Herberman, ed.), pp. 1091–1097, Academic Press, New York.

Hatcher, F. M., and Kuhn, R. E., 1982*c*, Destruction of *Trypanosoma cruzi* by natural killer cells, *Science* **218**:295.

Hauser, W. E., Jr., Sharma, S. D., and Remington, J. S., 1982, Natural killer cells induced by acute and chronic toxoplasma infection, *Cell. Immunol.* **69**:330.

Herberman, R. B., 1978, Natural cell-mediated cytotoxicity in nude mice, in: *The Nude Mouse in Experimental and Clinical Research* (J. Fogh and G. C. Giovanella, eds.), pp. 135–144, Academic Press, New York.

Herberman, R. B. (ed.), 1980, *Natural Cell-mediated Immunity Against Tumors*, Academic Press, New York.

Herberman, R. B., and Holden, H. T., 1978, Natural cell-mediated immunity, *Adv. Cancer Res.* **27**:305.

Herberman, R. B., Nunn, M. E., and Lavrin, D. H., 1975, Natural cytotoxicity reactivity of mouse lymphoid cells against syngeneic and allogeneic tumors. I. Distribution of reactivity and specificity, *Int. J. Cancer* **16**:216.

Herrick, C. A., and Cross, S. X., 1936, The development of natural and artificial resistance of young rats to the pathogenic effects of the parasite *Trypanosoma lewisi*, *J. Parasitol.* **22**:126.

Hibbs, J., 1975, Activated macrophages as cytotoxic effector cells. I. Inhibition of specific and nonspecific tumor resistance by trypan blue, *Transplantation* (*Baltimore*) **19**:77.

Hormaeche, C. E., 1979, Natural resistance to *Salmonella typhimurium* in different inbred mouse strains, *Immunology* **37**:311.

Hormaeche, C. E., Brock, J., and Pettifor, R., 1980, Natural resistance to mouse typhoid: possible role of macrophages, in: *Genetic Control of Natural Resistance to Infection and Malignancy* (E. Skamene, P.A.L. Kongshavn, and M. Landy, eds.), pp. 121–130, Academic Press, New York.

Howard, J. G., Hale, C., and Chan-Liew, W. L., 1980*a*, Immunological regulation of experimental cutaneous leishmaniasis: I. Immunogenetic aspects of susceptibility to *Leishmania tropica* in mice, *Parasite Immunol.* **2**:303.

Howard, J. G., Hale, C., and Chan-Liew, F. Y., 1980*b*, Immunological regulation of experimental cutaneous leishmaniasis. III. Nature and significance of specific suppression of cell-mediated immunity in mice highly susceptible to *Leishmania tropica*, *J. Exp. Med.* **152**:594.

Howard, J. G., Hale, C., and Liew, F. Y., 1981, Immunological regulation of experimental cutaneous leishmaniasis. IV. Prophylactic effect of sublethal inradiation as a result of abrogation of suppressor T cell generation in mice genetically susceptible to *Leishmania tropica*, *J. Exp. Med.* **153**:557.

Hsu, C-K., Hsu, S. H., Whitney, R. A., and Hausen, C. T., 1976, Immunopathology of schistosomiasis in athymic mice, *Nature* (*Lond.*) **262**:397.

Hsu, K-H. L., Hiramoto, R. N., and Ghanta, V. K., 1982, Immunosuppressive effect of mouse serum lipoproteins. II. *In vivo* studies, *J. Immunol.* **128**:2107.

Huang, K. Y., Schultz W. W., and Gordon, F. B., 1968, Interferon induced by *Plasmodium berghei*, *Science* **162**:123.

Hudson, K. M., 1971, Natural immunity of the cotton rat to *Trypanosoma vivax*, *Trans. R. Soc. Trop. Med. Hyg.* **65**:232.

Hudson, K. M., 1972, Further studies on the cotton rat and the action of its serum on *Trypanosoma vivax*, *Trans. R. Soc. Trop. Med. Hyg.* **66**:345.

Hudson, K. M., and Terry, R. J., 1970, Natural immunity of the cotton rat to *T. vivax*, *Trans. R. Soc. Trop. Med. Hyg.* **64**:170.

Ihara, I., Harada, Y., Ihara, S., and Kawakami, M., 1982, A new complement-dependent bactericidal factor found in nonimmune mouse sera: Specific binding to polysaccharide of Ra chemotype *Salmonella*, *J. Immunol.* **128**:1256.

Irvin, A. D., Young, E. R., Osborn, G. D., and Francis, L. M. A., 1981, A comparison of *Babesia* infections in intact, surgically splenectomized, and congenitally asplenic (Dh/+) mice, *Int. J. Parasitol.* **11**:251.

Jahiel, R. I., Vilcek, J., and Nussenzweig, R. S., 1970, Exogenous interferon protects mice against *Plasmodium berghei* malaria, *Nature* (*Lond.*) **227**:1350.

Jaouni, K. C., 1979, *Entamoeba histolytica*: Genetic control of susceptibility in chicken eggs, *Exp. Parasitol.* **47**:54.

Jayawardena, A., 1981, Immune responses in malaria, in: *Parasitic Diseases*, Vol. I: *The Immunology* (J. M. Mansfield, ed.), pp. 85–136, Marcel Dekker, New York.

Jayawardena, A. N., and Kemp, J. D., 1979, Immunity to *Plasmodium yoelii* and *Babesia microti*: Modulation by the CBA/N X-chromosome, *Bull. W.H.O.* 57(suppl. 1):255.

Johnson, G. R., Nicholas, W. L., Metcalf, D., McKenzie, I. F. C., and Mitchell, G. F., 1979, Peritoneal cell population of mice infected with *Mesocestoides conti* as a source of eosinophils, *Int. Arch. Allergy Appl. Immunol.* **59**:315.

Johnston, L. A. Y., 1967, Epidemiology of bovine babesiosis in northern Queensland, *Aust. Vet. J.* **43**:427.

Kaplan, M. H., and Volankis, J. E., 1974, Interaction of C-reactive protein complexes with the complement system. I. Consumption of human complement associated with the reaction of CRP with pneumococcal C-polysaccharide and with the choline phosphatides, lecithin, and sphingomyelin, *J. Immunol.* **112**:2135.

Kierszenbaum, F., and Howard, J. C., 1976, Mechanism of resistance against experimental *Trypanosoma cruzi* infection: The importance of antibodies and antibody forming capacity in the Biozzi high and low responder mice, *J. Immunol.* **116**:1208.

Kierszenbaum, F., and Sonnenfeld, G., 1982, Characterization of the antiviral activity produced during *Trypanosoma cruzi* infection and protective effects of exogenous interferon against experimental Chagas' disease, *J. Parasitol.* **68**:194.

Kierszenbaum, F., Ivanyi, J., and Budzko, D. B., 1976, Mechanisms of natural resistance to trypanosomal infection: Role of complement in avian resistance to *Trypanosoma cruzi* infection, *Immunology* **30**:1.

Kierszenbaum, F., Gottlieb, C. A., and Budzko, D. B., 1981, Antibody-independent, natural resistance of binds to *Trypanosoma cruzi* infection, *J. Parasitol.* **67**:656.

Kiessling, R., Klein, E., and Wigzell, H., 1975, Natural killer cells in the mouse. I. Cytotoxic cells with specificity for mouse Moloney leukemia cells. Specificity and distribution according to genotype, *Eur. J. Immunol.* **5**:112.

Kiessling, R., Hochman, P. S., Haller, O., Shearer, G. M., Wigzell, H., and Cudkowicz, G., 1977, Evidence for a similar or common mechanisim for natural killer cell activity and resistance to hemopoietic grafts, *Eur. J. Immunol.* **7**:655.

Kongshavn, P. A. L., Sadarangani, C., and Skamene, E., 1980, Genetically determined differences in antibacterial activity of macrophages are expressed in the environment in which the macrophage precursors mature, *Cell. Immunol.* **53**:341.

Krampitz, H. E., 1975, Ablastin: Antigen tolerance and lack of ablastin control of *Trypanosoma musculi* during host's pregnancy. *Exp. Parasitol.* **38**:317.

Landolfo, S., Martinotti, M. G., Martinetto, P., and Forni, G., 1980, Natural cell-mediated cytotoxicity against *Trichomonas vaginalis* in the mouse. I. Tissue, strain, age distribution, and some characteristics of the effector cells, *J. Immunol.* **124**:508.

Laveran, A., 1902, *C. R. Hebd. Seances Acad. Sci.* **134**:734 (cited in Rifkin, M. R., 1978).

Leventhal, R., and Soulsby, E. J. L., 1977, *Ascaris suum*: Cuticular binding of the third component of complement by early larval stages, *Exp. Parasitol.* **41**:423.

Lie, K. J., Heyneman, D., and Richards, C. S., 1977, *Schistosoma mansoni*: Temporary reduction of natural resistance in *Biomphalaria glabrata* induced by irradiated miracidia of *Echinostoma paraensei*, *Exp. Parasitol.* **43**:54.

Lincicome, D. R., 1955, Growth factor in normal rat serum for *Trypanosoma lewisi*, *J. Parasitol.* **41**(section 2):15.

Lincicome, D. R., 1958, Growth of *Trypanosoma lewisi* in the heterologous mouse host, *Exp. Parasitol.* **7**:1.

Lincicome, D. R., and Francis, E. M., 1961, Quantitative studies on heterologous sera inducing development of *Trypanosoma lewisi* in mice, *Exp. Parasitol.* **11**:68.

Lozzio, B. B., 1976, The Lasat mouse: A new model for transplantation of human tissues, *Biomedicine (Paris)* **24**:144.

Lust, J. A., Kumar, V., Burton, R. C., Bartlett, S. P., and Bennett, M., 1981, Heterogeneity of natural killer cells in the mouse, *J. Exp. Med.* **154**:306.

Mahmoud, A. A. F., Peters, P. A., Civil, R. H., and Remington, J. S., 1979, *In vitro* killing of schistosomula of *Schistosoma mansoni* by BCG and *C. parvum* activated macrophages, *J. Immunol.* **122**:1655.

McLaren, D. J., Bannister, L. H., Trigg, P. I., and Butcher, G. A., 1979, Freeze fracture studies on the interaction between the malaria parasite and the host erythrocyte in *Plasmodium knowlesi* infections, *Parasitology* **79**:125.

McLeod, R., Bensch, K. G., Smith, S. M., and Remington, J. S., 1980, Effects of human peripheral blood monocytes, monocyte-derived macrophages and spleen mononuclear phagocytes on *Toxoplasma gondii*, *Cell. Immunol.* **54**:330.

Miller, L. H., McAuliffe, F. M., and Mason S. J., 1977, Erythrocyte receptors for malaria merozoites, *Am. J. Trop. Med. Hyg.* **26**:204.

Mitchell, G. F., 1977, Studies on immune responses to parasite antigens in mice. V. Different susceptibilities of hypothymic and intact mice to *Babesia rodhaini*, *Int. Arch. Allergy Appl. Immunol.* **53**:385.

Mitchell, G. F., 1978, Metazoan and protozoan parasitic infections in nude mice, in: *Contemporary Topics in Immunobiology*, Vol. 8 (N. L. Warner and M. D. Cooper, eds.), pp. 55–67, Plenum Press, New York.

Mitchell, G. F., 1979*a*, Responses to infection with metazoan and protozoan parasites in mice, *Adv. Immunol.* **28**:451.

Mitchell, G. F. 1979*b*, Effector cells, molecules and mechanisms in host–parasite immunity to parasites, *Immunology* **38**:209.

Mitchell, G. F., Hogarth-Scott, R. S., Lewers, H. M., Edwards R. D., Cousins, G., and Moore, T., 1976, Studies on immune responses to parasite antigens in mice. I. *Ascaris suum* larvae numbers and antiphosphorylcholine responses in infected mice of various strains and hypothymic nu/nu mice, *Int. Arch. Allergy Appl. Immunol.* **56**:64.

Mitchell, G. F., Goding, J. W., and Rickard, M. D., 1977, Studies on immune responses to larval cestodes in mice. I. Increased susceptibility of certain mouse strains and hypothymic mice to *Taenia taenaeiformis* and analysis of passive transfer of resistance with serum, *Aust. J. Exp. Biol. Med. Sci.* **55**:165.

Moore, V. L. Sternick, J. L., Schrier, D. J., Taylor, B. A., and Allen, E. M. 1980, Genetic control of BCG-induced chronic granulomatous inflammation and anergy, in: *Genetic Control of Natural Resistance to Infection and Malignancy* (E. Skamene, P. A. L. Kongshavn, and M. Landy, eds.), pp. 191–200, Academic Press, New York.

Morrison, W. I., and Murray, M., 1972, *Trypanosoma congolense*: Inheritance of susceptibility to infection in inbred strains of mice, *Exp. Parasitol.* **48**:364.

Mosier, D. E., Zitron, I. M., Mond, J. J., Ahmed, S., Scher, I., and Paul S. E., 1977, Surface immunoglobulin D as a functional receptor for a subclass of B lymphocytes, *Immunol. Rev.* **37**:89.

Most, H., Nussenzweig, R. S., Vanderberg, J., Herman, R., and Yoeli, M., 1966, Susceptibility of genetically-standardized (JAX) mouse strains to sporozoite- and blood-induced *Plasmodium berghei* infections, *Mil. Med.* **131**:915.

Mühlpfordt, H., 1968, Untersuchungen über die experimentelle *Trypanosoma lewisi*-Infection der Maus, *Z. Tropenmed. Parasitol.* **19**:73.

Murphy, J. W., and McDaniel, D. O., 1982, *In vitro* effects of natural killer (NK) cells on *Cryptococcus neoformans*, in: *NK Cells and Other Natural Effector Cells* (R. B. Herberman, ed.), pp. 1105–1112, Academic Press, New York.

Murray, M., and Morrison, W. I., 1979*a*, Non-specific induction of increased resistance in mice to *Trypanosoma congolense* and *Trypanosoma brucei* by immunostimulants, *Parasitology* **79**:349.

Murray, M., and Morrison, W. I., 1979*b*, Parasitemia and host susceptibility to African trypanosomiasis, in: *Pathogenicity of Trypanosomes* (G. Losos and A. Chouinard, eds.), pp. 71–81, International Development Research Center, Ottawa.

Murray, M., Morrison, W. I., and Whitelaw, D. D., 1982, Host susceptibility to African trypansomiasis: Trypanotolerance, *Adv. Parasitol.* **21**:1.

Nappi, A. J., 1981, Cellular immune response of *Drosophila melanogaster* against *Asobara tabida*, *Parasitology* **83**:319.

Neilson, J. T. M., 1978, Primary infections of *Dipetalonema viteae* in an outbred and five inbred strains of golden hamsters, *J. Parasitol.* **64**:378.

Nery-Guimaraes, F., and Lage, H. A., 1972, A retratariedade das aves ao *Trypanosoma* (*Schizotrypanum*) *cruzi.* II. Refrateriedade das galinhas desde o nascimento: persistencia da refratariedade apos bursectomia: Infeccoes en ovos embrionados, *Mem. Inst. Oswaldo Cruz* **70**:97 (cited in Kierszenbaum, F., Ivanyi, J., and Budzko, 1976).

Nogueira, N., and Cohn, Z., 1978, *Trypanosoma cruzi*: *In vitro* induction of macrophage microbicidal activity, *J. Exp. Med.* **148**:288.

Nogueira, N., Chaplan, S., and Cohn, Z., 1980, *Trypanosoma cruzi*: Factors modifying ingestion and fate of blood form trypomastigotes, *J. Exp. Med.* **152**:447.

O'Brien, A. D., Rosenstreich, D. L., Metcalf, E. S., and Scher, I., 1980*a*, Differential sensitivity of inbred mice to *Salmonella typhimurium*: A model for genetic regulation of innate resistance to bacterial infection, in: *Genetic Control of Natural Resistance to Infection and Malignancy* (E. Skamene, P. A. L. Kongshavn, and M. Landy, eds.), pp. 101–114, Academic Press, New York.

O'Brien, A. D., Rosenstreich, D. L., Scher, I., Campbell, G. H., MacDermott, R. P., and Formal, S. B., 1980*b*, Genetic control of susceptibility to *Salmonella typhimurium* in mice: Role of the *Lps* gene, *J. Immunol.* **124**:20.

O'Brien, A. D., Rosenstreich, D. L., and Taylor, B. A., 1980*c*, Control of natural resistance to *Salmonella typhimurium* and *Leishmania donovani* in mice by closely linked but distinct genetic loci, *Nature* (*Lond.*) **287**:440.

Ogilvie, B. M., and MacKenzie, C. D., 1981, Immunology and immunopathology of infections caused by filarial nematodes, in: *Parasitic Diseases: The Immunology* (J. M. Mansfield, ed.), pp. 227–289, Marcel Dekker, New York.

Ojo, E., Haller, O., Kimura, A., and Wigzell, H., 1978, An analysis of conditions allowing *Corynebacterium parvum* to cause either augmentation of inhibition of natural killer cell activity against tumor cells im mice, *Int. J. Cancer* **21**:444.

Ojo-Amaize, E. A. Salimonu, L. S., Williams, A. I. O., Akinwolere, O. A. O., Shabo, R., Alm, G. V., and Wizgell, H., 1981, Positive correlation between degree of parasitemia, interferon titers, and natural killer cell activity in *Plasmodium falciparum*-infected children, *J. Immunol.* **127**:2296.

Osmond, A. P., Mortensen, R. F., Siegel, J., and Gewurz, H., 1975, Interaction of C-reactive protein with the complement system. III. Complement-dependent passive hemolysis initiated by CRP, *J. Exp. Med.* **142**:1065.

Pasvol, G., Weatherall, D. J., and Wilson, R. J. M., 1980, The increased susceptibility of young red cells to invasion by the malaria parasite *Plasmodium falciparum*, *Br. J. Haematol.* **45**:285.

Pearson, R. D., and Steigbigel, R. T., 1980, Mechanism of lethal effect of human serum upon *Leishmania donovani*, *J. Immunol.* **125**:2195.

Pelletier, M., Forget, A., Bourassa, D., Gros, P., and Skamene, E., 1982, Immunopathology of BCG infection of genetically resistant and susceptible mouse strains, *J. Immunol.* **129**:2179.

Perkins, M., 1981, Inhibitory effects of erythrocyte membrane proteins on the *in vitro* invasion of the human malarial parasite (*Plasmodium falciparum*) into its host cell, *J. Cell. Biol.* **90**:563.

Perlmutter, R. M., Nahm, M., Stein, K. E., Slack, J., Zitron, I., Paul, W. E., and Davie, J. M., 1979, Immunoglobulin subclass-specific in immunodeficiency in mice with an X-linked B-lymphocyte defect, *J. Exp. Med.* **149**:993.

Perrudet-Badoux, A., Binaghi, R. A., and Boussac-Aron, Y., 1978, *Trichinella spiralis* infection in mice: Mechanism of the resistance in animals genetically selected for high and low antibody production, *Immunology* **35**:519.

Phillips, S. M., Diconza, J. J., Gold, J. A., and Reid, W. A., 1977, Schistosomiasis in the congenitally athymic (nude) mouse. I. Thymic dependency of eosinophil granuloma formation and host morbidity, *J. Immunol.* **118**:594.

Plant, J., and Glynn, A. A., 1976, Genetics of resistance to infection with *Salmonella typhimurium* in mice, *J. Infect. Dis.* **133**:72.

Plant, J., and Glynn, A. A., 1979, Locating salmonella resistance gene on mouse chromosome 1, *Clin. Exp. Immunol.* **37**:1.

Plant, J., and Glynn, A. A., 1980, Control of resistance to *Salmonella typhimurium* in hybrid generations of inbred mice and Biozzi mice, in: *Genetic Control of Natural Resistance to Infection and Malignancy* (E. Skamene, P. A. L. Kongshavn, and M. Landy, eds.), pp. 133–140, Academic Press, New York.

Ranatunga, P., and Wanduragala, L., 1974, Reactions and haematology in imported Jersey cattle preimmunized in Ceylon, *Br. Vet. J.* **128**:9.

Richards, C. S., 1975, Genetic studies of pathologic conditions and susceptibility to infection in *Biomphalaria glabrata*, *Ann. N.Y. Acad. Sci.* **266**:394.

Rifkin, M. R., 1978, Identification of the trypanocidal factor in normal human serum: high density lipoprotein, *Proc. Natl. Acad. Sci. USA* **75**:3450.

Rivera-Ortiz, C-I, and Nussenzweig, R. S., 1976, *Trichinella spiralis*: Anaphylactic antibody formation and susceptibility in strains of inbred mice, *Exp. Parasitol.* **39**:7.

Rivnay, B., Bergman, S., Shinitsky, M., and Globerson, A., 1980, Correlations between viscosity, serum cholesterol, lymphocyte activation and aging in man, *Mech. Ageing Dev.* **12**:119.

Roberts-Thomson, I. C., and Mitchell, G. F., 1979, Protection of mice against *Giardia muris* infection, *Infect. Immun.* **24**:971.

Roder, J. C., 1979, The beige mutation in the mouse. I. A stem cell predetermined impairment in natural killer cell function, *J. Immunol.* **123**:2168.

Rogers, T. J., Balish, E., and Manning, D. D., 1976, The role of thymus-dependent cell mediated immunity in resistance to experimental disseminated candidiasis, *J. Reticuloendothel. Soc.* **20**:291.

Roudsky, D., 1910*a*, Sur la réceptivité de souris blanche à *Trypanosoma lewisi* Kent, *C.R. Soc. Biol. (Paris)* **68**:458 (cited in D'Alesandro, 1979).

Roudsky, D., 1910*b*, Sur le *Trypanosoma lewisi* Kent renforcé, *C. R. Soc. Biol. (Paris)* **69**:384 (cited in D'Alesandro, 1979).

Roudsky, D., 1911*a*, Action pathogène de *Trypanosoma lewisi* Kent renforcé sur la souris blanche, *C. R. Soc. Biol. (Paris)* **152**:56 (cited in D'Alesandro, 1979).

Roudsky, D., 1911*b*, Sur la possibilité de rendre le *Trypanosoma lewisi* virulent pour d'autres rongeurs que le rat, *C. R. Acad. Sci. (Paris)* **152**:56 (cited in D'Alesandro, 1979).

Rytel, M. W., and Marsden, P. D., 1970, Induction of an interferon-like inhibitor by *Trypanosoma cruzi* infection in mice, *J. Trop. Med. Hyg.* **19**:929.

Rytel, M. W., Rose, H. D., and Stewart, R. D., 1973, Absence of circulating interferon in patients with malaria and with American trypanosomiasis, *Proc. Soc. Exp. Biol. Med.* **144**:122.

Scher, I., Sharrow, S. O., and Paul, W. E., 1976, X-linked B-lymphocyte defect in CBA/N mice. III. Abnormal development of B-lymphocyte populations defined by their density of surface immunoglobulin, *J. Exp. Med.* **144**:507.

Scher, I., Steinberg, A. D., Berning, A. K., and Paul, W. E., 1975, X-linked B-lymphocyte immune defect in CBA/N mice. II. Studies of the mechanisms underlying the immune defect, *J. Exp. Med.* **142**:637.

Schultz, W. W., Huang, K. Y., and Gordon, F. B., 1968, Role of interferon in experimental mouse malaria, *Nature (Lond.)* **220**:709.

Shaw, G. L., and Dusanic, D. G., 1973, *Trypanosoma lewisi*: Termination of pregnancy in the infected rat, *Exp. Parasitol.* **33**:46.

Simpson, A. J. G., and Cioli, D., 1982, The key to the schistosome's success, *Nature (Lond.)* **296**:809.

Skamene, E., and Kongshavn, P. A. L., 1979, Phenotypic expression of genetically controlled host resistance to *Listeria monocytogenes*, *Infect. Immun.* **25**:345.

Skamene, E., Knogshavn, P. A. L., and Sachs, D., 1979, Resistance to *Listeria monocytogenes* in mice: Genetic control by genes that are not linked to the H-2 complex, *J. Infect. Dis.* **139**:228.

Skamene, E., Kongshavn, P. A. L., and Landy, M. (eds.), 1980, *Genetic Control of Natural Resistance to Infection and Malignancy*, Academic Press, New York.

Skamene, E., Forget, A., Gros, P., and Kongshavn, P. A. L., 1982*a*, Chromosome 1 locus: A major regulator of natural resistance to intracellular pathogens, in: *NK Cells and Other Natural Effector Cells* (R. B. Herberman, ed.), pp. 313–318, Academic Press, New York.

Skamene, E., Stevenson, M. M., and Kongshavn, P. A. L., 1982*b*, Natural cell-mediated immunity against bacteria, in: *NK Cells and Other Natural Effector Cells* (R. B. Herberman, ed.), pp. 1513–1520, Academic Press, New York.

Smith, H., Skehel, J. J., and Turner, M. H. (eds.), 1980, *The Molecular Basis of Microbial Pathogenicity*, Verlag Chemie, Weinheim.

Smith, M. A., and Clegg, J. A., 1976, Different levels of acquired immunity to *Schistosoma mansoni* in two strains of hamsters, *Parasitology* **73**:47.

Smith, P. D., Elson, C. O., Keister, D. B., and Nash, T. E., 1982, Human host response to *Giardia lamblia*. I. Spontaneous killing by mononuclear leukocytes *in vitro*, *J. Immunol.* **128**:1372.

Smithers, S. R., Terry, R. J., and Hockley, D. J., 1969, Host antigens in schistosomiasis, *Proc. R. Soc. Lond. B Biol. Sci.* **B171**:483.

Sonnenfeld, G., and Kierszenbaum, F., 1981, Increased serum levels of an interferon-like activity during the acute period of experimental infection with different strains of *Trypanosoma cruzi*, *Am. J. Trop. Med. Hyg.* **30**:1189.

Stevenson, M. W., Kongshavn, P. A. L., and Skamene, E., 1980, Macrophage inflammatory responses in *Listeria*-resistant and *Listeria*-sensitive mice, in: *Genetic Control of Natural Resistance to Infection and Malignancy* (E. Skamene, P. A. L. Kongshavn, and M. Landy, eds.), pp. 565–574, Academic Press, New York.

Stutman, O., Figarella, E. F., Paige, C. J., and Lattime, E. C., 1980*a*, Natural cytotoxic (NC) cells against solid tumors in mice: General characteristics and comparison to natural killer (NK) cells, in: *Natural Cell-Mediated Immunity Against Tumors* (R. B. Herberman, ed.), pp. 187–229, Academic Press, New York.

Stutman, O., Dien, P., Wisun, R. E., and Lattime, E. C., 1980*b*, Natural cytotoxic cells against solid tumors in mice: Blocking of cytotoxicity by D-mannose, *Proc. Natl. Acad. Sci. USA* **77**:2895.

Sullivan, J. T., and Richards, C. S., 1981, *Schistosoma mansoni*, NIH-SM-PR-2 strain, in susceptible and nonsusceptible stocks of *Biomphalaria glabrata*: Comparative histology, *J. Parasitol.* **67**:702.

Taliaferro, W. H., and Olsen, Y. P., 1943, The protective action of normal sheep serum against infections of *Trypanosoma duttoni* in mice, *J. Infect. Dis.* **72**:213.

Targett, G. A. T., 1973, Thymus dependency and chronic antigenic stimulation: immu-

nity to parasitic protozoans and helminths, in: *Contemporary Topics in Immunobiology*, Vol. 2 (A. J. S. Davies and R. L. Carter, eds.), pp. 217–238, Plenum Press, New York.
Taylor, B., 1980, Recombinant inbred strains of mice: Use in genetic analysis of disease resistance, in: *Genetic Control of Natural Resistance to Infection and Malignancy* (E. Skamene, P. A. L. Kongshavn, and M. Landy, eds.), pp. 1–7, Academic Press, New York.
Taylor, D. W., Bever, C. T., Rollwagen, F. M., Evans, C. B., and Asofsky, R., 1982, The rodent malaria parasite *Plasmodium yoelii* lacks both types 1 and 2 T-independent antigens, *J. Immunol.* **128**:1854.
Terry, R. J., 1957, Antibody against *Trypanosoma vivax* present in normal cotton rat serum, *Exp. Parasitol.* **6**:404.
Terry, R. J., 1976, Innate resistance to trypanosome infections, in: *Biology of the Kinetoplastida*, Vol. 1 (W. H. R. Lumsden and D. A. Evans, eds.), pp. 477–492, Academic Press, New York.
Thiroux, A., 1909, *C. R. Acad. Sci.* **149**:534 (cited in Taliaferro and Olsen, 1943).
Trischmann, T. M., and Bloom, B., 1982, Genetics of murine resistance to *Trypanosoma cruzi*, *Infect. Immun.* **35**:546.
Trischmann, T M., Tanowitz, H., Wittner, M., and Bloom, B., 1978, *Trypanosoma cruzi*: Role of the immune response in the natural resistance of inbred strains of mice, *Exp. Parasitol.* **45**:160.
Wakelin, D., 1975, Genetic control of immune responses to parasites: Immunity to *Trichuris muris* in inbred and random-bred strains of mice, *Parasitology* **71**:51.
Wakelin, D., 1978, Genetic control of susceptibility and resistance to parasitic infection, *Adv. Parasitol.* **16**:219.
Wakelin, D., 1980, Genetic control of immunologically mediated resistance to helminthic infections, in: *Genetic Control of Natural Resistance to Infection and Malignancy* (E. Skamene, P. A. L. Kongshavn, and M. Landy, eds.), pp. 55–66, Academic Press, New York.
Waki, S., and Suzuki, M., 1977, *Proceedings of the Second International Workshop on Nude Mice, 1976* (cited in Mitchell, 1979).
Wassom, D. L., David, C. S., and Gleich, G. J., 1979, Genes within the major histocompatibility complex influence susceptibility to *Trichinella spiralis* in the mouse, *Immunogenetics* **9**:491.
Weibust, R. S., 1972, Inheritance of plasma cholesterol levels in mice, *Genetics* **73**:303.
Weiss, L., Kedar, E., Weigensberg, M., and Slavin, S., 1982, Natural cell-mediated cytotoxicity in mice treated with total lymphoid irradiation (TLI), *Cell. Immunol.* **70**:188.
Weiss, N., 1978, Studies on *Dipetalonema viteae* microfilariae in hamsters in relation to worm burden and humoral immune response, *Acta Trop.* **35**:137.
Wiener, E., and Bandieri, A., 1974, Differences in antigen handling by peritoneal macrophages from the Biozzi high and low responder lines of mice, *Eur. J. Immunol.* **4**:457.
Wakelin, D., and Donachie, A. M., 1983, Genetic control of immunity to *Trichinella spiralis*: Influence of H-2 linked genes on immunity to the intestinal phase of infection, *Immunology* **48**:343.
Williams, D. M., Grumet, F. C., and Remington, J. S., 1978, Genetic control of murine resistance to *Toxoplasma gondii*, *Infect. Immun.* **19**:416.
Wolfe, S. A., Tracey, D. E., and Henney, C. S., 1976, Induction of "natural killer" cells by BCG, *Nature (Lond.)* **262**:584.
Wright, D. H., 1968, The effect of neonatal thymectomy on the survival of golden hamsters infected with *P. berghei*, *Br. J. Exp. Pathol.* **49**:379.
Yoshino, T. P., Cheng, T. C., and Renwrantz, L. R., 1977, Lectin and human blood group determinants of *Schistosoma mansoni*: Alteration following *in vitro* transformation of miracidium to mother sporocyst, *J. Parasitol.* **63**:818.

Yu, P, 1972, Some host–parasite genetic interaction models, *Theor. Pop. Biol.* **3**:347.

Zuckerman, Z., 1957, Blood loss and replacement in plasmodial infections. I. *Plasmodium berghei* in untreated rats of varying age and in adult rats with erythropoietic mechanisms manipulated before inoculation, *J. Infect. Dis.* **100**:172.

Zwart, D., and Brocklesby, D. W., 1979, Babesiosis: Non-specific resistance, immunological factors and pathogenesis, *Adv. Parasitol.* **17**:49.

Chapter 2

Intracellular Mechanisms of Killing

Nadia Nogueira

Department of Cellular Physiology and Immunology
The Rockefeller University
New York, New York 10021

I. INTRODUCTION

One of the principal physiologic functions of "professional" phagocytic cells (neutrophils, eosinophils, and mononuclear phagocytes) is to ingest and destroy microorganisms. Mononuclear phagocytes, however, may also provide a favorable environment for the survival and multiplication of a variety of intracellular pathogens, including many obligate intracellular parasites.

Several mechanisms have been reported to be involved in the survival or destruction of microorganisms within phagocytic cells, among which one can cite lysosomal hydrolases and the acid pH within the lysosomes, cationic proteins (Lehrer *et al.*, 1981), lysozyme, iron-binding proteins (Haurani *et al.*, 1973). This review focuses on oxidative mechanisms of mononuclear phagocytes, since these are the ones to which most evidence has been directed.

Phagocytosis of microorganisms by leukocytes is usually accompanied by a respiratory burst, resulting in the rapid utilization and reduction of molecular oxygen to water. In a series of biochemical events, unstable oxygen intermediates such as superoxide anion and hydrogen peroxide are formed. These compounds react with each other to form highly reactive moieties, such as hydroxyl radical and singlet oxygen, which have been implicated in the microbicidal activity displayed by the cells, as reviewed by Badwey and Karnovsky (1980) and Klebanoff (1980).

There is now considerable evidence that oxidative mechanisms are involved in the intracellular killing of a variety of intracellular parasitic protozoa, all of which use the macrophage as their host cell, i.e., *Trypanosoma cruzi, Toxoplasma gondii, Leishmania donovani, L. tropica,* and *L. enrietti.*

II. PARASITE INTERIORIZATION BY MONONUCLEAR PHAGOCYTES

A. Subcellular Localization

Interiorization of these parasites by macrophages takes place by a phagocytic process, probably mediated by receptorlike peptides on the macrophage surface (Fig. 1A). Protease treatment of the macrophage surface has been shown to abolish the uptake of *T. cruzi* by mouse peritoneal macrophages (Nogueira and Cohn, 1976). In the case of *Trypanosoma cruzi,* in which the parasite is able to replicate in other cell types, the avoidance of phagocytosis by the macrophage is an advantageous evasion mechanism, since the macrophage seems to be the only host cell potentially microbicidal for these organisms. Indeed, the blood form of this parasite possesses on its surface an antiphagocytic factor that dramatically reduces its uptake (Nogueira *et al.*, 1980). Reversal of this antiphagocytic activity can be achieved by trypsinization of the parasite surface or by opsonization with specific antibody.

Once in the parasitophorous vacuole, the organisms can prevent lysosomal fusion and live and multiply within the confines of the phagosomes, as is the case of *Toxoplasma gondii* (Jones and Hirsch, 1972). Alternatively, phagolysosomal fusion may occur, but the organisms may be resistant to lysosomal hydrolases and proceed to replicate normally, as is the case with *Leishmania* (Ebert *et al.*, 1976). In contrast, *T. cruzi*, permits phagolysosomal fusion but escapes from the phagocytic vacuole and replicates in the cytosol (Nogueira and Cohn, 1976).

B. Biochemical Events

As described in Sections III, IV, and V phagocytosis of the organisms could result in their destruction by reactive oxygen intermediates generated by the respiratory burst. However, the parasites seem to be able to survive in normal macrophages by either inhibiting the respiratory burst, or else by possessing antioxidant defenses that render them more resistant to the toxic products of the reduction of oxygen. Macrophage activation by lymphocyte factors resulting from the immune response to these microorganisms can result in either restoration of the respiratory burst or the ability to generate higher levels of reactive oxygen intermediates sufficient to kill the intracellular organisms, or both. Suppressor factors may also play a role in modulating macrophage function by interfering with the same two parameters.

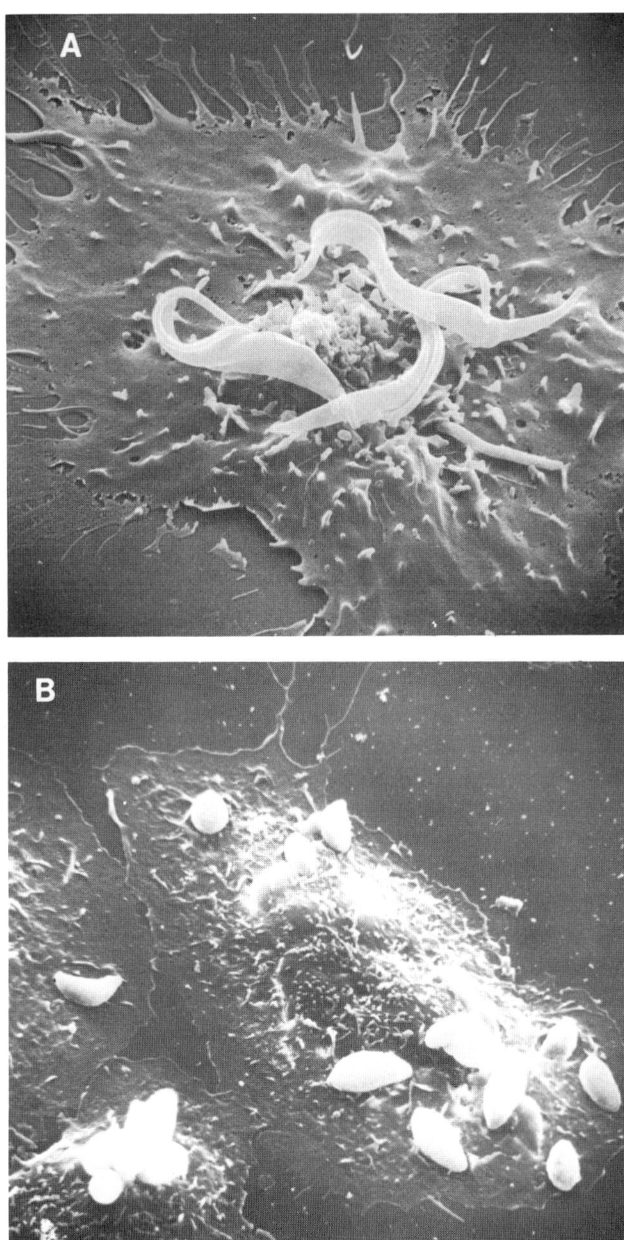

Figure 1. (A) *T. cruzi* trypomastigotes and (B) *Toxoplasma gondii* trophozoites in mouse macrophages activated *in vivo* by infection and boosting with *T. cruzi*. Note intense macrophage circumferential spreading. Scanning electron micrographs taken and kindly supplied by Dr. Gilla Kaplan.

III. *TRYPANOSOMA CRUZI:* MACROPHAGE ACTIVATION AND INTRACELLULAR KILLING

A. Correlation with H_2O_2 Generation in a Murine System

Infection and boosting of mice with live *bacillus Calmette-Guerin* (BCG) or *T. cruzi*, but not injection with thioglycollate or proteose peptone, has been demonstrated by Nogueira *et al.* (1977) to result in peritoneal macrophages capable of destroying trypomastigotes of *T. cruzi* (*in vivo* activation). The trypomastigote forms were capable of surviving and multiplying within normal resident peritoneal macrophages, In addition, Nogueira and Cohn (1978) have shown that normal resident peritoneal macrophages could be induced to display microbicidal activity against trypomastigotes of *Trypanosoma cruzi* by exposure to products from antigen-stimulated, sensitized spleen cell populations (lymphokines). The conditions required for optimal microbicidal activity were constant exposure of the macrophages to, and daily renewal of, the spleen cell factors (*in vitro* activation). Removal of the lymphokines or failure to add fresh supernatants resulted in reversal of the microbicidal effect.

In 1971 Nathan *et al.* had described that guinea pig peritoneal macrophages cultured for 72 hr with migration inhibitory factor (MIF)-rich supernatant fluids showed an enhanced conversion of $[1\text{-}^{14}C]$glucose to $^{14}CO_2$, when compared with untreated controls. Nathan and co-workers speculated that the increase in the hexose monophosphate shunt activity in these cells might be coupled to the activity of a H_2O_2 generating mechanism, which was potentially microbicidal. The increase in HMPS activity required 2-3 days of culture, was dependent on the dose of lymphocyte mediators, and could be reversed by the removal of the mediators.

Later, Nathan and Root (1977) reported that treatment of mice with BCG or *Corynebacterium parvum*, but not thioglycollate broth or proteose peptone, enhanced the ability of their peritoneal macrophages to release hydrogen peroxide after phagocytosis or treatment with PMA. Johnston *et al.* (1978) demonstrated that immunologically induced and inflammatory peritoneal macrophages released large amounts of superoxide anion when compared with control resident cells.

The remarkable correlation in the conditions required for expression of the two activated macrophage functions led Nathan and Nogueira to study the ability of the macrophages to release H_2O_2 as a biochemical correlate of their level of antimicrobial function. Indeed, Nathan *et al.* (1979) were able to demonstrate that the ability to release H_2O_2 and the ability to kill trypanosomes were closely correlated under a variety of experimental conditions. Under three different conditions, treatments that enhanced trypanocidal activity also increase H_2O_2-releasing capacity: (a) immunization and boosting of mice with *T. cruzi* or

BCG (Nogueira *et al.*, 1977) (Fig. 2) or (b) incubation of macrophages from nonimmune mice in lymphokines derived from antigen-induced immune spleen cells (Fig. 3) or (c) mitogen-induced normal spleen cells (Nogueira and Cohn, 1978). Both functions (H_2O_2 release and trypanocidal activity) were found to follow the same time course and dose-response curve, and both were reversible over a similar period of time (Nathan *et al.*, 1979). In contrast, five different protocols (single injections of heat-killed *T. cruzi* or BCG, injection of proteose peptone, or *in vitro* treatment of macrophages with control lymphokines or medium alone) failed to induce either trypanocidal activity or H_2O_2-releasing capacity.

An investigation of the sensitivity of epimastigotes and trypomastigotes of *T. cruzi* to H_2O_2 generated by glucose and glucose oxidase resulted in an LD_{50} of 6.5 nmol/min for epimastigotes, and of 9.4 nmol/min for trypomastigotes (Table I). Such concentrations of H_2O_2 could possibly be achieved within the confines of the vacuolar system. A new approach to this question has been recently provided by Tanaka *et al.* (1982) by using glucose oxidase coupled to zymosan in an attempt to quantitate the amount of H_2O_2 needed to kill one parasite. Their results allow them to estimate that this could be achieved by a flux of H_2O_2 of 2.7×10^{-7} nmol/min.

The lower resistance to H_2O_2 observed in epimastigotes of *T. cruzi* may also explain its higher sensitivity to killing by even normal peritoneal macrophages, which produce very low levels of reactive oxygen intermediates. In this respect, the recent publication by Tanaka *et al.* (1982) adds further support to this hypothesis. Epimastigotes were killed in a macrophage cell line, J774.16. but not in a mutant clone derived from it that lacked the respiratory burst and was markedly defective in the production of reactive oxygen intermediates.

B. Correlation with H_2O_2 Generation in Human Mononuclear Phagocytes

Similar studies have now been undertaken with human peripheral blood monocytes and with human monocyte-derived macrophages. Freshly harvested human blood monocytes release large amounts of reactive oxygen intermediates (Reiss and Roos, 1978; Nakagawara *et al.*, 1981). During differentiation *in vitro*, monocytes display a marked decrease in their capacity to secrete H_2O_2 or O_2^-, a decrease that does not seem related to changes in the specific activity of antioxidant enzymes (Nakagawara *et al.*, 1981). Supernatants from mitogen- or antigen-stimulated human blood mononuclear cells enhanced the ability of monocytes or monocyte-derived macrophages to release H_2O_2 or O_2^- in response to PMA (phorbol-myristate acetate) or zymosan (Nakagawara *et al.*, 1982) (Fig. 4). The magnitude of the lymphokine effect was dependent on whether the LK was added to fresh monocytes or to macrophages derived from them. Addition of LK on day 2 retarded the decline in H_2O_2-releasing capacity.

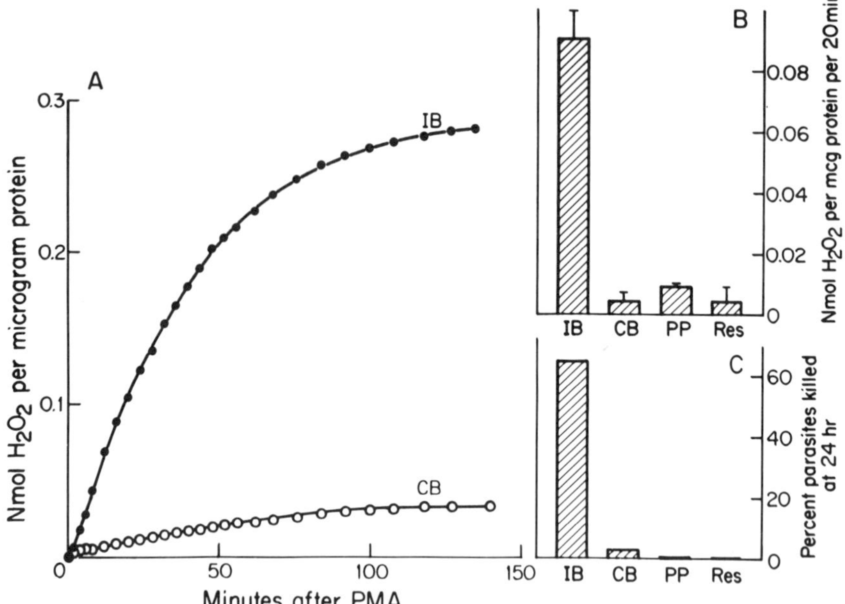

Figure 2. H_2O_2 release and trypanicidal activity of freshly explanted macrophages after various treatments of cell donors. Macrophages were adherent to glass coverslips for 2–4 hr before assays were started. (A) Time course of H_2O_2 release after addition of 100 ng/ml PMA (phorbol myristate acetate) to macrophages from mice immunized with viable *T. cruzi* and boosted with heat-killed *T. cruzi* (B) or from control mice given the boosting injection alone (CB). Results are expressed in terms of the adherent cell protein on matched cover-slips. (B) Initial rates of H_2O_2 release after release after addition of 100 ng/ml of PMA to macrophages from mice that were immunized and boosted (IB) or only boosted (CB) as in (A), or given proteose-peptone (PP), or untreated resident peritoneal macrophages (Res). Mean SEM for three experiments with each cell type, except two experiments with Res cells. (C) Percentage reduction of intracellular trypanosomes 24 hr after infection of freshly plated macrophages from mice treated, as in (B). Mean of two experiments. (Courtesy of *Journal of Experimental Medicine*.)

Addition of LK after the cells had lost their H_2O_2-generating ability (days 5–6) restored it to about 60% of the values seen with freshly explanted monocytes. Effects of LK were dose and time dependent, with maximal effects at 3 days' exposure. Release of H_2O_2 was suppressed by 95 ± 4% by exposure of the monocytes for 4 days to hydrocortisone at 1.9×10^{-7} *M*.

A similar modulation of human mononuclear phagocytes *in vitro* was also observed when trypanocidal activity was investigated (Nogueira *et al.*, 1982*b*) (Fig. 5). *T. cruzi*-induced lymphokines derived from patients with chronic Chagas disease, as well as BCG and concanavalin A (con A)-induced lymphokines, resulted in induction of microbicidal activity against trypomastigotes of

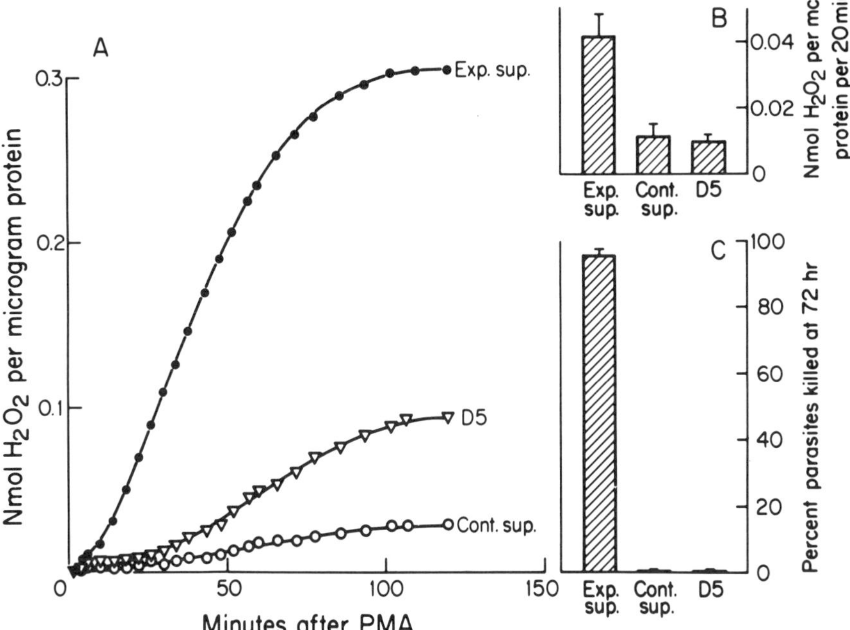

Figure 3. H_2O_2 release and trypanosome activity of macrophages after 3 days' culture in SCF (spleen cell factor) and control media. (A) Time course of H_2O_2 release after addition of 100 ng/ml PMA to CB macrophages incubated in 12.5% supernate of spleen cells from *T. cruzi*-immunized mice cultured with heat-killed *T. cruzi* [experimental supernate (exp. sup.)], in 12.5% supernate of normal spleen cells cultured with heat-killed *T. cruzi* [control supernate (cont. sup.)], or in D5 alone. (B) Initial rates of H_2O_2 release after addition of 100 ng/ml PMA to PP, CB, or Res macrophages incubated for 3–4 days in D5 alone or in 6–25% of experimental or control supernates prepared from spleen of normal mice or of mice immunized with either *T. cruzi* or BCG and challenged with the homologous antigen. Mean ± SEM for eight experiments. (C) Percentage reduction in intracellular trypanosomes 48 hr after infection of macrophages. Macrophages were cultured for 24 hr before infection. Thus the results pertain to 72 hr of culture, in similar media, as for (B). Mean ± SEM for eight experiments with PP cells. Initial numbers of parasites per 100 macrophages (mean ± SEM) were 53.1 ± 3.7 (exp. sup.), 53.3 ± 3.2 (control sup.), and 50.5 ± 3.2 (D5). Abbreviations as in Fig. 2. (Courtesy of the *Journal of Experimental Medicine*.)

T. cruzi. In contrast to the murine system, and as observed with the induction of H_2O_2-releasing capacity by human macrophages, optimal microbicidal activity in human macrophages could be achieved without the need for daily addition of the soluble factors. Induction of microbicidal activity was also dependent on the dose of lymphokine. Hydrocortisone 10^{-6} *M* added 24 hr before infection, and readded and maintained for 3 days after infection resulted in an considerable increase in the number of intracellular parasites. This result can be ex-

Table I. Susceptibility of Parasites to H_2O_2 and to Macrophage Microbicidal Activity[a]

Parasite	Susceptibility to H_2O_2 LD_{50} (nmol/min)	Anti-oxidant defenses: Catalase (U/mg)	Anti-oxidant defenses: GPO (nmol/min mg)	Anti-oxidant defenses: SOD (U/mg)	Intracellular killing: macrophage status
T. cruzi					
Epimastigote	6.5	None	None	0.54	Mouse-resident macrophages; Human macrophages
Trypomastigote	9.4	ND	ND	ND	*In vivo* activated mouse macrophages; LK activated human and mouse macrophages
L. tropica					
Promastigote	0.5	0.00011	0.2	6.4	LK activated J774G8; Resident mouse macrophages
L. donovani					
Promastigote	1.5	0.00015	0.5	9.1	Resident mouse macrophage; Human macrophages
Amastigote[b]	$8.1 + 10^{-4}$ *M*	0.00045	7.2	12.8	LK activated J774G8
T. gondii					
Trophozoites	20	0.013	11.7	6.1	(LK + endotoxin) activated mouse macrophage; *In vivo* activated mouse macrophage; Freshly explanted human monocytes

[a]Modified from Nathan and Nakagawara (1982). (See text for additional references.)
[b]Two independent determinations.

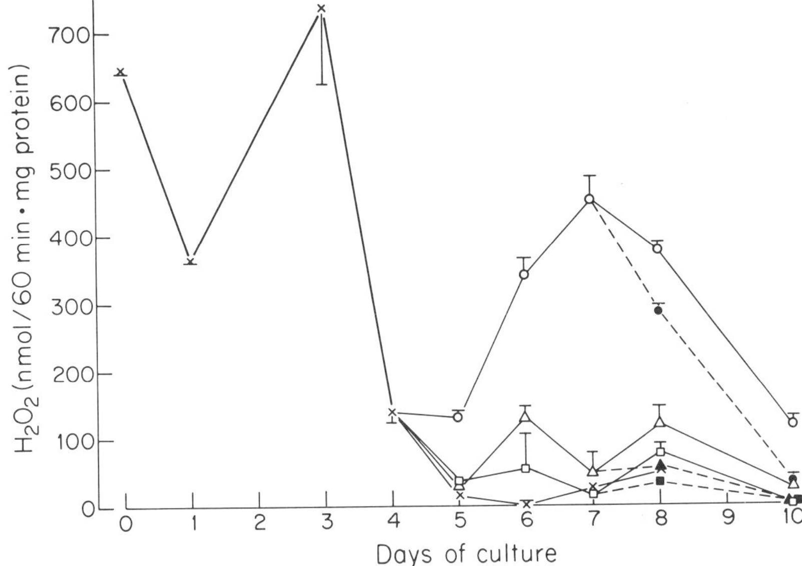

Figure 4. Kinetics of enhancement of H_2O_2 release by 10% BCG-induced lymphokines (○), by control lymphokine (△), or by control medium (□) added to human monocyte-derived macrophages on day 4 of culture. (x) Values for cells incubated in medium alone. On day 7, some macrophages were removed from lymphokines (●), (▲) and control media (■), rinsed, and placed in medium alone (■) for 1-3 additional days. Means ± SEM for triplicates in a representative experiment, using PMA as a stimulus. (Courtesy of *Journal of Clinical Investigation*.)

plained by the reduction in H_2O_2-generating capacity of hydrocortisone-treated cells. Hydrocortisone, however, had no effect on the induction of trypanocidal activity mediated by lympokines on human macrophages.

C. Triggering of H_2O_2 Release by Macrophages Phagocytizing *T. cruzi*

Yet another group of studies provided additional strength to the correlation between H_2O_2 release and microbicidal activity. Secretion of H_2O_2 by macrophages has been shown to be very slow until cells are exposed to membrane-perturbing agents. Among these, phorbol myristate acetate, calcium ionophores, immune complexes, and several phagocytic particles have been described (Nathan and Root, 1977; Johnston, 1978; Tomioka and Saibo, 1980).

Both epimastigotes and trypomastigotes of *T. cruzi* are capable of triggering the production of H_2O_2 during their ingestion by macrophages (Nogueira *et al.*,

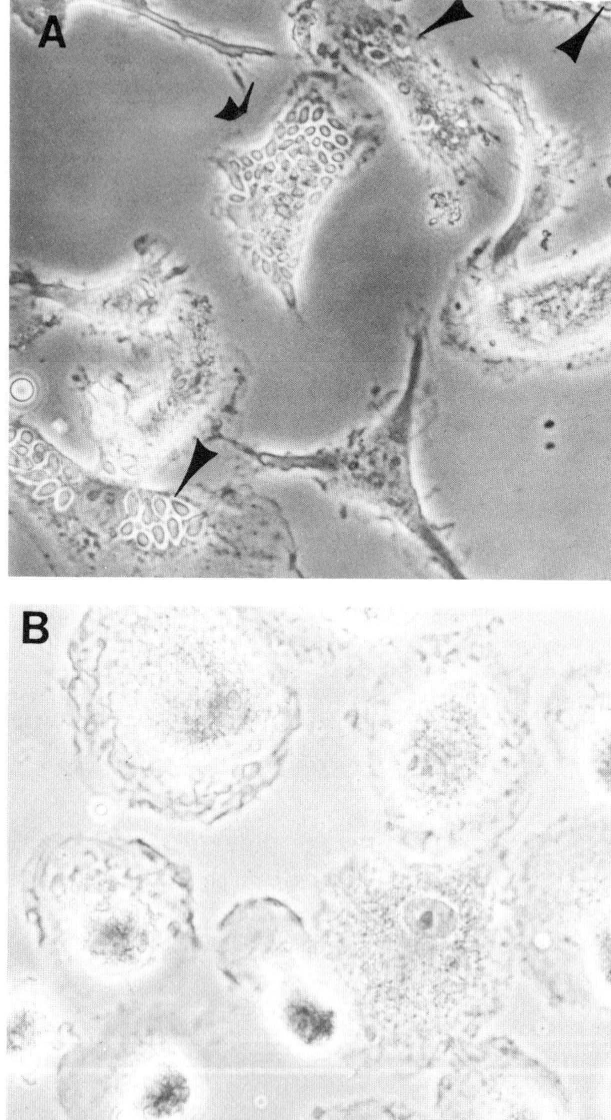

Figure 5. Human monocyte-derived macrophages exposed to 2.5% control lymphokine (A) or 2.5% *T. cruzi*-induced lymphokine from a patient with chronic Chagas disease (B) 72 hr after injection with *T. cruzi*. Note abundance of intracellular parasites (*arrows*) in samples exposed to control lymphokine, as compared with *T. cruzi*-lymphokine induced cells, where parasites are absent.

1982*a*). Epimastigotes, which are more sensitive to H_2O_2, are killed and digested within normal macrophages after ingestion, but trypomastigotes are not. These forms can, however, be rendered sensitive to killing by normal resident peritoneal macrophages or by human monocyte-derived macrophages by coating the microorganisms with eosinophil peroxidase (EPO), as shown by Nogueira *et al.* (1982*a*). This increases the cytotoxic efficiency of the H_2O_2 generated by the ingestion of the parasites. The intracellular killing of EPO-coated trypomastigotes could by inhibited by catalase and azide, an inhibitor of EPO, suggesting that toxicity is indeed mediated through the small amounts of H_2O_2 generated by the phagocytic event in normal macrophages and the peroxidatic activity of EPO. EPO-coated organisms could also be killed extracellularly when exposed to normal macrophages at high parasite-cell ratios or when a high phagocytic load of another particle was given simultaneously. This effect could also be inhibited by both azide and catalase, but not by superoxide dismutase, suggesting that enough H_2O_2 can be released by phagocytosis of a high number of organisms to generate toxic concentrations of H_2O_2 outside the confines of the vacuolar system. The exquisite susceptibility of epimastigotes of *T. cruzi* to H_2O_2 can be explained by their diminished antioxidant defenses. They are reported not to possess catalase or glutathione peroxidase (Boveris, 1980).

IV. *TOXOPLASMA GONDII*: MACROPHAGE ACTIVATION AND INTRACELLULAR KILLING

A. Correlation with the Generation of Reactive Oxygen Intermediates in a Murine System

The same approach used for *T. cruzi* was later applied to *Toxoplasma gondii* by Murray and Cohn (1980) (Fig. 1B). Macrophages were obtained from mice after injection of *Toxoplasma*, BCG, *C. parvum*, thioglycollate medium, proteose peptone, or heart infusion broth. In each case, the ability or failure of the experimental protocols to generate macrophages capable of inhibiting the replication of toxoplasma closely correlated with their ability to secrete H_2O_2. In addition, exposure to lymphokines under similar experimental conditions as those described for *T. cruzi* resulted in antimicrobial activity against *Toxoplasma gondii*, but only if the lymphokines were supplemented by endotoxin or other inflammatory agents, the activity of which may be speculated to be attributable to the presence of endotoxin as well. Antimicrobial activity also paralleled enhancement of the release of reactive oxygen intermediates (Murray *et al.*, 1979). This effect could be inhibited by glucose deprivation and partially reversed by the use of scavengers of oxygen intermediates, such as catalase, superoxide dismutase, mannitol, benzoate, histidine, and diazobicyclooctane, suggesting a role for the products of the interaction of O_2^- and H_2O_2, such as hydroxyl radical and perhaps singlet oxygen. The use of aminotriazole, a catalase inhibitor, rendered macrophages more microbicidal to *Toxoplasma gondii* (Murray *et al.*, 1980), as did immune macrophages from acatalasemic mice.

Toxoplasma gondii possess substantial amounts of catalase (4.8×10^{-2} Baudhuin units/mg protein) and it is quite resistant to H_2O_2 ($\leqslant 19$ nmol/min). Inhibition of killing by SOD (superoxide dismutase), catalase, and scavengers of O_2 and OH° implicates reduced forms of oxygen, particularly OH° and singlet oxygen, as mediators of killing (Murray and Cohn, 1979).

B. Failure to Trigger the Respiratory Burst

Wilson *et al.* (1980) recently demonstrated that trophozoites of *Toxoplasma gondii* are capable of triggering the respiratory burst and the production of reactive oxygen intermediates during their ingestion by *C. parvum* or *Toxoplasma*-activated mouse peritoneal macrophages and freshly explanted human monocytes. This ability was evaluated by measuring [1-^{14}C]glucose oxidation [a reflection of HMP (hexose monophosphate) shunt activity], nitroblue tetrazolium reduction (a reflexion of O_2^- production), or chemiluminescence (a reflection of radical formation). These cells are able to kill the parasites. By contrast, normal resident mouse peritoneal macrophages and monocyte-derived human macrophages had a much smaller respiratory burst and permitted the parasites to replicate intracellularly. However, ingestion of antibody-coated parasites induces triggering of the respiratory burst in these cells, allowing them to destroy the organisms. *Toxoplasma gondii* coated with EPO were killed much more efficiently by mononuclear phagocytes than were untreated ones (Locksley *et al.*, 1982).

These results suggest that microorganisms may evolve evasion mechanisms to prevent destruction within phagocytic cells by avoiding the triggering of the respiratory burst. The process of macrophage activation would involve their capacity to restore the ability to generate the respiratory burst, permitting parasite destruction.

V. *LEISHMANIA* SPP.: MACROPHAGE ACTIVATION AND INTRACELLULAR KILLING

A. *Leishmania enrietti*

Buchmuller and Mauel (1981) have presented similar evidence regarding the intracellular killing of *L. enrietti*. Macrophages activated by supernatants from Con A-stimulated lymphocytes displayed increases in microbicidal activity against *L. enrietti* amastigotes that closely correlated with an enhancement in oxidative metabolism. In contrast to what has been described for *Toxoplasma gondii*, however, aminotriazole decreased the leishmanicidal effect displayed by activated macrophages. No explanation is available for this discrepancy. Addition of exogenous horseradish peroxidase (HRP) in the medium markedly potentiated the killing of *L. enrietti* amastigotes.

B. *Leishmania tropica*

Promastigotes of *L. tropica* are very sensitive to H_2O_2 (LD_{50} = 0.5 nmol/min) and can survive within J774.G8 cells. These cells produce very little reactive oxygen intermediates. Activation of the cells by lymphokines resulted in increased production of reactive oxygen intermediates and in acquisition of the capacity to kill promastigotes. These parasite forms contain little, if any, catalase, and very low levels of glutathione peroxidase (Murray, 1981).

C. *Leishmania donovani*

Leishmania donovani amastigotes were reported to be destroyed by mouse peritoneal macrophages after several days' treatment with lymphokines (Haidaris and Bonventre, 1981). The killing was inhibited by 50% with the addition of catalase. Amastigotes have been described as failing to provide the membrane stimulus necessary to trigger the respiratory burst in nonactivated cells as well as in *in vivo* activated cells (Haidaris and Bonventre, 1982; Murray, 1982). Amastigotes were found to be killed only in activated cells capable of triggering the respiratory burst and high levels of oxidative intermediates, such as lymphokine-activated macrophages. *L. donovani* amastigotes were found in the last report to contain 3-fold more catalase and 14-fold more glutathione peroxidase and to be four times more resistant to enzymatically generated H_2O_2 than are promastigote forms. Promastigotes also elicited a much stronger respiratory burst than did amastigotes in nonactivated cells. These factors may contribute to the enhanced survival of amastigotes in macrophages as compared with promastigotes, as well as to their higher *in vivo* infectivity (Keithly, 1976).

Promastigote stages of *L. donovani* have also been shown to be killed by human polymorphonuclear leukocytes (Pearson and Steigbigel, 1981). In human monocyte-derived macrophages (7 days after explanting) promastigotes were found by Pearson *et al.* (1982) to induce an oxidative burst, as measured by both chemiluminescence or NBT (nitroblue tetrazolium) reduction. This oxidative response was comparable in intensity to that obtained by phagocytosis of opsonized zymosan by the cell population. These workers reported that it did not result in total killing of the ingested promastigotes as measured by microscopic observation of stained and fixed preparation within the first 12 hr post-infection. Amastigotes were also found to initiate an oxidative burst in monocyte-derived macrophages, but the magnitude of the response was less than that observed for promastigotes.

VI. CONCLUSIONS

Oxidative mechanisms have been implicated in the killing of many other pathogens by mononuclear phagocytes, including bacteria (*Mycobacterium microti:* Walker and Lowrie, 1981; *Listeria monocytogenes:* Cline, 1970; *Staphylococcus aureus:* Davis *et al.*, 1968; Sagone *et al.*, 1976; Ramsey and Klebanoff, 1982) and fungi (*Candida albicans* and *Candida parapsilosis:* Sasada and Johnston, 1980; *Candida pseudotropicalis:* Lehrer, 1975).

The inverse relationship between virulence of intracellular bacteria and catalase activity has long been noted (Huddleson and Stahl, 1942; (Rockenmacher, 1949; Knox *et al.*, 1956). We can now add to this list a sensitivity order to H_2O_2 or to intracellular killing by macrophages and levels of parasite antioxidant defenses (catalase, glutathione peroxidase, SOD). Variation in levels of these antioxidant defenses may also reflect different sensitivities to reactive oxygen intermediates of the various life stages of the organisms.

It is clear, however, that much more needs to be learned about antimicrobial mechanisms and their specific interactions with individual pathogens. Different effector molecules seem to be involved in antimicrobial activity against the various organisms, and synergism between oxidative and nonoxidative mechanisms is likely to occur. In addition, oxidative mechanisms may not play any role in antimicrobial activity against other pathogens, as seems to be the case with *Chlamydia psittaci* (Byrne and Faubion, 1982) and *Chlamydia trachomatis* (Yong *et al.*, 1982).

The rapid progress made during the last few years in the investigation of oxidative mechanisms of antimicrobial activity in mononuclear phagocytes brings hope that many of the questions still puzzling us today will soon be answered, in addition to introducing an array of new approaches to the therapy of parasitic diseases.

ACKNOWLEDGMENT

I thank Dr. Carl Nathan and Dr. Zanvil Cohn for reading the manuscript.

VII. REFERENCES

Badwey, J. A., and Karnovsky, M. L., 1980, Active oxygen species and the functions of phagocytic leukocytes, *Annu. Rev. Biochem.* **49**:695.

Boveris, A., Sies, H., Martino, E. E., DoCampo, R., Turrens, J. F., and Stoppani, A. D. M.,

1980, Deficient metabolic utilization of hydrogen peroxide in *Trypanosoma cruzi*, *Biochem. J.* **188**:643.

Buchmuller, Y., and Mauel, J., 1981, Studies on the mechanism of macrophage activation: Possible involvement of oxygen metabolites in killing of *Leishmania enriettii* by activated mouse macrophages, *J. Reticuloendothel. Soc.* **29**:181.

Byrne, G. I., and Faubion, C. L., 1982, Lymphokine-mediated microbistatic mechanisms restrict *Chlamydia psittaci* growth in macrophages, *J. Immunol.* **128**:469.

Cline, J. J., 1970, Drug potentiation of macrophage function, *Infect. Immun.* **2**:601.

Davis, W. C., Huber, H., Douglas, S. D., and Fudenberg, H. H., 1968, A defect in circulating mononuclear phagocytes in chronic granulomatous disease of childhood, *J. Immunol.* **101**:1093.

Ebert, F., Enriquez, G. L., and Muhlpfordt, H., 1976, Electron microscopic studies of the phagocytosis of *Leishmania donovani* by hamster peritoneal macrophages and its lysosomal activity *in vitro*, *Behr. Inst. Mitt.* **60**:65.

Haidaris, C. G., and Bonventre, P. F., 1981, Elimination of *Leishmania donovani* amastigotes by activated macrophages, *Infect. Immun.* **33**:918.

Haidaris, C. G., and Bonventre, P. F., 1982, A role for oxygen-dependent mechanisms in killing of *Leishmania donovani* tissue forms by activated macrophages, *J. Immunol.* **129**:850.

Haurani, F. I., Meyer, A., and O'Brien, R., 1973, Production of transferrin by the macrophage, *J. Recituloendothel. Soc.* **14**:309.

Huddleson, I. F., and Stahl, W. H., 1942, Catalase activity of the species of *Brucella* as a criterion of virulence, *Univ. Mich. Agric. Exp. Sta. Tech. Bull.* **182**:57.

Johnston, R. B., Jr., 1978, Oxygen metabolism and the microbicidal activity of macrophages, *Fed. Proc.* **37**:2759.

Johnston, R. B., Jr., Godzik, C. A., and Cohn, Z. A., 1978, Increased superoxide anion production by immunologically activated and chemically elicited macrophages, *J. Exp. Med.* **148**:115.

Jones, T., and Hirsch, J., 1972, The interaction between *Toxoplasma gondii* and mammalian cells. II. The absence of lysosomal fusion with phagocytic vacuoles containing living parasites, *J. Exp. Med.* **136**:1173.

Keithly, J. S., 1976, Infectivity of *Leishmania donovani* amastigotes and promastogotes and promastigotes for golden hamsters, *J. Protozool.* **23**:244.

Klebanoff, S. J., 1980, Oxygen intermediates and the microbicidal events, in: *Mononuclear Phagocytes: Functional Aspects*, (R. van Furth, ed.), p. 1105, Martinus Nijhoff, The Hague.

Knox, R., Meadow, P. M., and Worssan, R. H., 1956, The relationship between the catalase activity, hydrogen peroxidase sensitivity, and isoniazid resistance of mycobacteria, *Am. Rev. Tuberc. Pulm. Dis.* **73**:726.

Lehrer, R. I., 1975, The fungicidal mechanisms of human monocytes. I. Evidence for myeloperoxidase-linked and myeloperoxidase-independent candidacidal mechanisms, *J. Clin. Invest.* **55**:338.

Lehrer, R. I., Szklarek, D., Selsted, M. E., Fleischmann, J., 1981, Increased content of microbicidal cationic peptides in rabbit alveolar macrophages elicited by complete Freund adjuvant, *Infect. Immun.* **33**:775.

Locksley, R. M., Wilson, C. B., and Klebanoff, S. J., 1982, Role of endogenous and acquired peroxidase in the toxoplasmacidal activity of murine and human mononuclear phagocytes, *J. Clin. Invest.* (in press).

Murray, H. W., 1982, Cell-mediated immune response in experimental visceral leishmaniasis. II. Oxygen-dependent killing of intracellular *Leishmania donovani* amastigotes, *J. Immunol.* **129**:351.

Murray, H., 1981, Interaction of Leishmania with a macrophage cell line. Correlation between intracellular killing and the generation of oxygen intermediates, *J. Exp. Med.* **153**:1690.

Murray, H. W., and Cohn, Z. A., 1979, Macrophage oxygen-dependent antimicrobial activity, I. Susceptibility to *Toxoplasma gondii* to oxygen intermediates, *J. Exp. Med.* **150**: 938.

Murray, H. W., and Cohn, Z. A., 1980, Macrophage oxygen-dependent antimicrobial activity. III. Enhanced oxidative metabolism as an expression of macrophage activation, *J. Exp. Med.* **150**:950.

Murray, H. W., Juangbhanich, C. W., Nathan, C. F., and Cohn, Z. A., 1979, Macrophages oxygen-dependent antimicrobial activity. II. The role of oxygen intermediates, *J. Exp. Med.* **150**:950

Murray, H. W., Nathan, C. F., and Cohn, Z. A., 1980, Macrophage oxygen-dependent antimicrobial activity. IV. Role of endogenous scavengers of oxygen intermediates, *J. Exp. Med.* **152**:1610.

Nakagawara, A., Nathan, C. F., and Cohn, Z. A., 1981, Hydrogen peroxide metabolism in human monocytes during differentiation *in vitro*, *J. Clin. Invest.* **68**:1243.

Nakagawara, A., DeSantis, N. M., Nogueira, N., and Nathan, C. F., 1982, Lymphokines enhance the capacity of human monocytes to secrete reactive oxygen intermediates, *J. Clin. Invest.* **70**:1042-1048.

Nathan, C. F., and Nakagawara, A., 1982, Role of oxygen intermediates in macrophage killing of intracellular pathogens: A review, in: *Self-Defense Mechanisms. Role of Macrophages*, A Naito Foundation Symposium (D. Mizuno, Z. A., Cohn, K. Takeya, and N. Ishida, eds.), p. 279, University of Tokyo Press–Elsevier Biomedical Press, Tokyo.

Nathan, C. F., and Root, R. K., 1977, Hydrogen peroxide release from mouse peritoneal macrophages. Dependence on sequential activation and triggering, *J. Exp. Med.* **146**: 1648.

Nathan, C. F., Karnovsky, M. L., and David, J. R., 1971, Alterations of macrophage functions by mediators from lymphocytes, *J. Exp. Med.* **133**:1356.

Nathan, C. F., Nogueira, N., Juangbhanich, C., Ellis, J., and Cohn, Z. A., 1979, Activation of macrophages *in vivo* and *in vitro*. Correlation between hydrogen peroxide release and killing of *Trypanosoma cruzi*, *J. Exp. Med.* **149**:1056.

Nogueira, N., and Cohn, Z. A., 1976, *Trypanosoma cruzi:* Mechanisms of entry and intracellular fate in mammalian cells, *J. Exp. Med.* **143**:1402.

Nogueira, N., and Cohn, Z. A., 1978, *Trypanosoma cruzi*, *in vitro* induction of macrophage microbicidal activity, *J. Exp. Med.* **148**:288.

Nogueira, N., Gordon, S., and Cohn, Z., 1979, *Trypanosoma cruzi:* Modification of macrophage function during infection, *J. Exp. Med.* **146**:157.

Nogueira, N., Chaplan, S., and Cohn, Z. A., 1980, *Trypanosoma cruzi:* Factors modifying ingestion and fate of blood form trypomastigotes, *J. Exp. Med.* **152**:447.

Nogueira, N., Klebanoff, S., and Cohn, Z. A., 1982*a*, *Trypanosoma cruzi:* Sensitization to macrophage killing by eosinophil peroxidase, *J. Immunol.* **128**:1705.

Nogueira, N., Chaplan, S., Reesink, M., Tydings, J., and Cohn, Z. A., 1982*b*, *Trypanosoma cruzi*: Induction of microbicidal activity in human mononuclear phagocytes, *J. Immunol.* **128**:2142.

Pearson, R. D., and Steigbigel, R. T., 1981, Phagocytosis and killing of the protozoan *Leishmania donovani* by human polymorphonuclear leukocytes, *J. Immunol.* **127**:1438.

Pearson, R. D., Harcus, J. L., Symes, P. H., Romito, R., and Donowitz, G. R., 1982, Failure of the phagocytic oxidative response to protect human monocyte-derived macrophages from infection by *Leishmania donovani*, *J. Immunol.* **129**:128.

Ramsey, P.G., Martin, T., Chi, E., and Klebanoff, S. J., 1982, Arming of mononuclear phagocytes by eosinophil peroxidase bound to *Staphylococcus aureus*, *J. Immunol.* **128**:415.

Reiss, M., and Roos, D., 1978, Differences in oxygen metabolism of phagocytosing monocytes and neutrophils, *J. Clin. Invest.* **61**:480.

Rockenmacher, M., 1949, Relationship of catalase activity to virulence in *Pasteurella pestis*, *Proc. Soc. Exp. Biol. Med.* **71**:99.

Sagone, A. L., Jr., King, G. W., and Metz, E. N., 1976, A comparison of the metabolic response to phagocytosis in human granulocytes and monocytes, *J. Clin. Invest.* **57**: 1352.

Sasada, M., and Johnston, R. B., Jr., 1980, Macrophage microbicidal activity. Correlation between phagocytosis associated oxidative metabolism and the killing of Candida by macrophages, *J. Exp. Med.* **152**:85.

Tanaka, Y., Kiyotaki, C., Tanowitz, H., and Bloom, B., 1982, Reconstitution of a variant macrophage cell line defective in oxygen metabolism with a H_2O_2-generating system, *Proc. Natl. Acad. Sci. USA* **579**:2584.

Tomioka, H., and Saito, H., 1980, Hydrogen-peroxide releasing function of chemically elicited and immunologically activated macrophages: Differential response to wheat germ and Concanavalin A, *Infect. Immun.* **29**:469.

Walker, L., and Lowrie, D. B., 1981, Killing of *Mycobacterium microti* by immunologically activated macrophages, *Nature (Lond.)* **293**:69.

Wilson, C. B., Tsai, V., and Remington, J. S., 1980, Failure to trigger the oxidative metabolic burst by normal macrophages. Possible mechanism for survival of intracellular pathogens, *J. Exp. Med.* **151**:328.

Yong, E. C., Klebanoff, S. J., and Kuo, C.-C., 1982, Toxic effect of human polymorphonuclear leukocytes on *Chlamydia trachomatis*, *Infect. Immun.* **37**:422.

Chapter 3

Induction and Expression of Mucosal Immune Responses and Inflammation to Parasitic Infections

Dean Befus and John Bienenstock

Department of Pathology, Host Resistance Programme
McMaster University Health Sciences Centre
Hamilton, Ontario, Canada L8N 3Z5

I. INTRODUCTION

At least five million people die annually of gastrointestinal infection (Holmgren, 1981), and the morbidity level from such infections must be several-fold greater. In the veterinary field, an estimated 20% of pigs born in North America and Britain die before weaning of diarrheal disease caused by *Escherichia coli* and transmissible gastroenteritis (TGE) virus (Porter, 1979). The human and economic losses resulting from gastrointestinal infections are thus of staggering proportions; e.g., in Kenya, *Ascaris lumbricoides* infection, generally considered to be relatively nonpathogenic and unimportant, results in economic losses in excess of $5 million per year (Latham *et al.*, 1977). Similarly depressing figures could be provided for the mortality and morbidity statistics of respiratory infections in the international community (e.g., World Health Organization Technical Report, 1980). The magnitude of the problem of mucosal infection has stimulated research directed at its solution and, given the success of vaccination programs with various infectious diseases, the potential for immunoprophylaxis is being vigorously explored.

We have recently reviewed the current concepts of mucosal immunology (Bienenstock and Befus, 1980; McDermott *et al.*, 1982) and how these concepts are currently applied in the dissection of mechanisms of host resistance against parasitic infections at mucosal surfaces (Befus and Bienenstock, 1982*a*). For detailed discussion and references we refer the reader to these works. In this review we develop the conceptual framework on which efforts to dissect the mechanisms of host resistance at mucosal sites are currently based and highlight certain areas in which significant advances have been made recently, or that we believe promise to uncover significant information in the near future. We hope this approach will help researchers identify important areas in which to apply their expertise for continued progress toward successful immunoprophylactic or therapeutic intervention in mucosal infections.

Given our own interests and expertise, as well as the weight of current research, our coverage focuses on protozoan and helminthic infections of the gastrointestinal tract. Where appropriate other mucosae, including the respiratory and urogential tracts, and infectious agents other than protozoa and helminths are considered. Initially, we describe the present understanding of the uptake, processing, and presentation to responsive lymphocytes of parasite antigens in the mucosal lymphoid system. Next, the subsequent generation of various immune responses and mucosal inflammatory changes are described. Finally, we attempt to synthesize information from the many different experimental models of gastrointestinal infection and disease to provide insight into the effector mechanisms of host resistance and the mechanisms that minimize potentially pathogenic responses in the host and facilitate the symbiotic relationship of many host-parasite coexistences. In the three sections—antigen uptake, the generation of immune responses and inflammation, and effector mechanisms—we compartmentalize our discussions along anatomic lines, i.e., the mucosal epithelium, underlying lamina propria, and lymphoid tissues, to emphasize the roles of cells and responses in the different sites.

II. THE INTERFACE: HOST CELLS MEET PARASITE ANTIGENS

Although in recent years knowledge of parasite antigens has steadily increased and the development of monoclonal antibody technology has facilitated tremendous forward strides, e.g., with antigens of *Plasmodium* spp. (Beaudoin *et al.*, 1981), information about the antigenic makeup of protozoa and helminth parasites inhabiting mucosal surfaces is embarrassingly sparse (reviewed by Clegg and Smith, 1978; Pery and Luffau, 1979; Befus and Bienenstock, 1982*a*). Monoclonal antibody technology is just beginning to be applied to the study of antigens of some of these parasites; other workers have recently applied procedures of surface iodination or biosynthetic radiolabeling to the study of the antigenic makeup of the gastrointestinal nematodes *Trichinella spiralis* and *Nippostrongylus brasiliensis* (Parkhouse *et al.*, 1981; Maizels *et al.*, 1982). The power of these tools to identify antigens recognized by resistant and/or susceptible hosts and to establish the spectra of antigens expressed throughout the developmental stages of a particular parasite leads one to predict confidently that these procedures will be widely applied to study host–parasite interactions at mucosal sites. Undoubtedly, within a few short years considerable information will be available about parasite antigens, which will permit study of their early interaction with host cells.

A. Antigens and Epithelial Cells

Parasite antigens released in the lumen of the gastrointestinal tract do not encounter the epithelial cell barrier until they have penetrated the combined diffusion barrier of the unstirred water layer and the mucus coat (Smithson *et al.*, 1981). These physical barriers are obviously of less significance to antigens released by parasites that invade the intestinal tissues or inhabit the paramucosal lumen proximal to these barriers. The intestinal epithelial cell barrier is composed of large areas of columnar digestive/absorptive cells fused by tight junctions. Antigen uptake by these cells is generally considered minimal (Owen, 1977), although conclusive information about the quantitative significance of this uptake mechanism is lacking (Walker *et al.*, 1972). In contrast, however, specialized epithelial regions overlying mucosal lymphoid aggregates, such as intestinal Peyer's patches (PPs), appendix, or solitary lymphoid nodules, collectively called gut-associated lymphoid tissue (GALT) (Befus and Bienenstock, 1982*b*), seem especially designed to sample the antigenic makeup of the lumenal environment continuously. Similar specialized epithelial regions in the lung overly the bronchus-associated lymphoid tissue (BALT) (Bienenstock *et al.*, 1973).

This lymphoid follicle-associated epithelium (FAE) (Bockman and Cooper, 1973) is distinct from surrounding columnar epithelium in that it is heavily infiltrated with lymphocytes and is therefore called a lymphoepithelium. FAE contains a paucity of goblet cells and has a specialized cell type called an M (microfold/membrane) cell (Fig. 1) that is active in the pinocytosis of lumenal material including antigens (Owen, 1977). The ingested material can be passed rapidly to adjacent lymphocytes. The lack of goblet cells in the lymphoepithelium may be important in its sampling of the lumenal environment by allowing for gaps in the mucus barrier that aid access of antigen to M cells. There is a consensus that M cells originate from stem cells in the crypt epithelium (Smith and Peacock, 1980; Bhalla and Owen, 1982), but there is no knowledge of the factors controlling their differentiation or numbers. Similarly, whether M cells process antigen to particularly immunogenic forms, analogous to macrophage processing of antigen, has not been studied.

Paradoxically, although the lymphoepithelium appears to be highly specialized to function in facilitating immune responses, thereby providing host protection against infectious agents, it has also recently been identified as a site at which at least some organisms gain access to systemic sites, or a foothold for local colonization. For example, poliovirus, *Salmonella enteritidis*, and *Listeria*

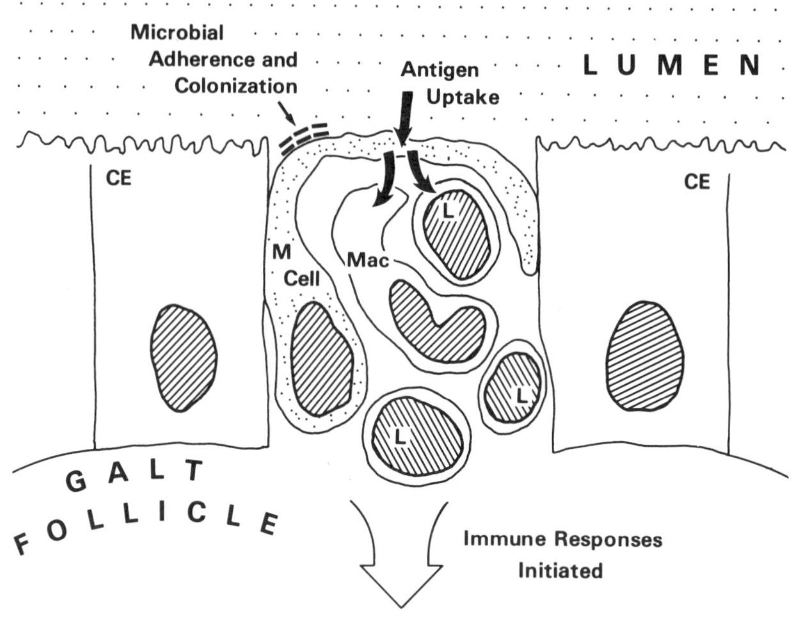

Figure 1. Diagrammatic representation of follicle-associated epithelium (FAE) in the intestine. A membrane (M) cell is surrounded by columnar epithelial (CE) cells, by lymphocytes (L), and by macrophages (MAC). M cells are important sites of antigen and microbial uptake and are substrates for microbial adherence and colonization. GALT, Gut-associated lymphoid tissue.

monocytogenes disseminate to systemic sites through PP (reviewed by Befus and Bienenstock, 1982*a*); *E. coli* strain RDEC-1 and reovirus type I colonize the mucosal lymphoepithelium (Cantey and Inman, 1981; Wolf *et al.*, 1981) and, with the latter at least, specifically the M cell. To what extent similar cell-parasite interactions exist with other organisms must be ascertained, and how these interactions relate to the generation of immunity or of a symbiotic relationship, must be investigated as well.

B. Induction Events in Mucosal Lymphoid Aggregates

The functional characteristics of mucosa-associated lymphoid tissue (MALT) including those in the gut [gut-associated lymphoid tissue (GALT)] and lung [bronchus-associated lymphoid tissue (BALT)] have been extensively reviewed recently (Ottaway *et al.*, 1979; Cebra *et al.*, 1980; Mestecky *et al.*, 1980; Befus and Bienenstock, 1982*a,b*; McDermott *et al.*, 1982). Briefly, in MALT such as PPs, antigens are processed and presented to local lymphoid cells that then respond appropriately through clonal expansion and differentiation to initiate protective host responses. Most cells produced in this manner do not remain in the PP, but migrate through afferent lymphatics to the draining mesenteric lymph nodes (MLNs), where further differentiation and expansion may occur. Cells leave the MLN, enter the thoracic duct and then the circulation, from which they localize in various sites as directed by their phenotypic and microenvironmental characteristics, such as antigens, hormones, chemotactic factors, blood flow, and specific, but unidentified, receptor–acceptor interactions (Bienenstock and Befus, 1980). Cell types known or thought to originate in PPs include macrophages, IgA and IgE precursors, helper and suppressor T cells, cytotoxic cells, and mucosal mast cells (MMC).

1. Macrophages: Accessory Cells

Despite abundant morphologic evidence for macrophages in GALT (LeFevre *et al.*, 1979), little knowledge of their functions has been forthcoming, with the exception of their apparent incompetence as accessory cells in *in vitro* humoral responses and in the induction of cytotoxicity to allogeneic cells (Kagnoff and Campbell, 1974). However, it was recently established using enzymatic methods to extract cells from GALT that PP macrophages are indeed capable of presenting antigen to lymphocytes (Richman *et al.*, 1981*a*; Frangakis *et al.*, 1982; Kiyono *et al.*, 1982; MacDonald and Carter, 1982). The heterogeneity of mucosal macrophages and their functional repertoire in mucosal lymphoid aggregates, including monokine production [interleukin-1, (IL-1)], cytotoxic capabilities, as well as antigen presentation, can be explored now that isolation procedures are available. Owen *et al.* (1981) clearly established that macrophages in PPs are also capable of phagocytosis and presumably destruction of parasites such as *Giardia muris.* The latter seems likely, since Smith *et al.* (1982) demonstrated that human peripheral blood monocytes spontaneously kill *G. lamblia in vitro.*

To date the migrations of macrophages to and from MALT have not been adequately examined, although BALT has been considered a major source of alveolar macrophages (McDermott *et al.*, 1982). Using mesenteric lymphadenectomy, MacPherson and Steer (1979) and Pugh and MacPherson (1982) have investigated the migration of nonlymphoid cells in intestinal lymphatic drainage of the MLN. These studies establish that such cells leave intestinal lymphatics on a regular basis but must be removed from the lymphatic drainage in the MLN and thus do not normally enter the efferent lymphatic drainage. These cells have potent accessory cell function in lymphocyte responses and appear to be related to dendritic cells rather than to mononuclear phagocytes, as evidenced by their expression of surface markers, phagocytic activity, and morphology (Pugh and MacPherson, 1982). Thus, the precise nature of the cells expressing accessory function in mucosal lymphoid aggregates and in the lamina propria remains to be clarified (Lipsky and Kettman, 1982). The model employing mesenteric lymphadenectomy could also be used to investigate the monokine production of cells in afferent and efferent lymph node lymphatics in normal animals or in animals with gastrointestinal infections to aid in dissecting the events associated with parasitic infections (Wasi *et al.*, 1982).

2. *Lymphocyte Subpopulations*

a. Plasma Cell Precursors. It is well established that PPs represent a rich source of IgA precursor cells, many of which appear to have been previously primed to lumenal antigens (Bienenstock and Befus, 1980; Cebra *et al.*, 1980). When appropriately stimulated, these IgA precursors proliferate and leave the PP to localize selectively in the lamina propria of many mucosal tissues by way of the lymphatics, MLN, thoracic duct, circulation, and spleen (Roux *et al.*, 1981; Tseng, 1981). In the mucosal lamina propria, many of the precursor cells complete their differentiation into IgA plasma cells producing specific antibodies (Husband, 1982; Pierce and Cray, 1982,). Precursor cells for other isotypes are also represented in GALT (Cebra *et al.*, 1980; Durkin *et al.*, 1981). The mucosa-seeking IgG-containing cells of MLN and bronchial lymph nodes (BLN) presumably originate in the local MALT (McDermott and Bienenstock, 1979). The pronounced IgE responses induced by many mucosal parasitic infections, especially with nematodes (Jarrett and Miller, 1982) and some viruses (Welliver *et al.*, 1981), undoubtedly have their origin in MALT because in GALT of germ-free animals there is a high frequency of IgE precursor cells (Durkin *et al.*, 1981). These precursors must be exquisitely regulated under normal circumstances by factors directly or indirectly derived from the microbial flora. However, parasitic infections bypass this negative regulation and stimulate the production of high levels of parasite-specific and heterologous IgE antibodies.

The reasons PP cells express the IgA isotype so prominently, and perhaps IgE as well, have been discussed extensively (e.g., Cebra *et al.*, 1980). For simplicity, two hypotheses have been proposed: an "instructive" one in which isotype switching with a predominance of IgA occurs locally, and a "selective" hypothesis in which IgA precursors generated elsewhere localize selectively in the PPs (Phillips-Quagliata *et al.*, 1982). Although conclusive evidence is lacking, current studies favor the "instructive" hypothesis, since they have identified a T cell from PPs that induces surface IgM^+ cells to switch to surface IgA^+ selectively (Strober and Kawanishi, 1982). If such observations are confirmed for IgA, it will be essential to search for similar switch T cells specific for other isotypes, especially IgE in the PP. Current evidence (Urban *et al.*, 1977) argues that such IgE switch T cells do not exist or are not always important because neonatal thymectomy of rats does not depress the generation of surface IgE-bearing cells after *N. brasiliensis* infection.

As evidenced by the paucity of comments made in this section about parasitic organisms, our knowledge about their interactions with cells in MALT is indeed incomplete. As parasite antigens become better characterized, and thus available for study, their ability to stimulate IgA, IgE, or other isotype precursors in PP either directly or indirectly through accessory cells or local regulatory T-cell circuits must be scrutinized.

b. T Cells. The characteristics of T cells in MALT has been most extensively investigated in PP. Recent studies suggest that about 35% of isolated PP cells express Thy 1 on their surface (Kiyono *et al.*, 1982). The following PP T cells have thus far been described: IgA-specific switch T cells, which induce switching to IgA isotype but do not facilitate maturation of IgA-secreting cells (Strober and Kawanishi, 1982); IgA-specific helper cells, which stimulate IgA antibody production (Elson *et al.*, 1979; Richman *et al.*, 1981*b*; Kiyono *et al.*, 1982); IgG and IgE suppressor cells (Richman *et al.*, 1981*b*; Ngan and Kind, 1978); T cells of the contrasuppressor circuit (Green *et al.*, 1982), which inhibit the activity of suppressor T cells locally; and Thy 1^+ cells, which will ultimately localize selectively in the mucosal epithelium (Guy-Grand *et al.*, 1978) and be capable of expressing natural killer activity (Guy-Grand and Vassalli, 1982; Tagliabue *et al.*, 1982).

After antigenic stimulation, IgA-specific helper T cells can be identified in PPs within 1 day and subsequently in the spleen (Strober *et al.*, 1981). IgG-specific suppressor cells are found by day 3 in the PPs, MLN, and spleen and are said to disseminate from the PPs to systemic sites (Mattingly and Waksman, 1978; Richman *et al.*, 1981*b*). These cellular migrations together with the presence of the contrasuppressor circuit of I-J$^+$ T cells in the PP (Green *et al.*, 1982) facilitate the generation and maintenance of local immune responsiveness and systemic depression of potentially immunopathogenic IgG responses. The presence of IgA helper T cells in spleen may be important as a host-protective mechanism as well, because IgA antibodies are important in the clearance of circulating antigens by the liver (Bienenstock and Befus, 1982).

To date, these recent concepts of T-cell functions in MALT have not been applied to dissect the responses induced by mucosal parasitic infections. The study of such regulatory cells and their network circuits will undoubtedly expose some of the mechanisms of host resistance, as well as those underlying the depression of local immune responses during mucosal parasitic infection (McElroy *et al.*, 1982).

3. *Mucosal Mast Cells*

In response to many parasitic infections and in some diseases of unknown etiology, such as Crohn's disease and ulcerative colitis, mucosal mast cells, (MMCs), hyperplasia occurs (Befus and Bienenstock, 1982*a*). There is an abundance of evidence that MMCs represent a functional unique mast cell population in rodents (Befus *et al.*, 1982*a*; Pearce *et al.*, 1982) and humans (Strobel *et al.*, 1981; Ruitenberg *et al.*, 1982; Befus *et al.*, 1982*b*).

Our studies on the development of intestinal mastocytosis have established that MLNs of parasitized rats contain both MMC precursors and presumably a T cell that produces mastopoietic factor(s) (Befus and Bienenstock, 1979; Denburg *et al.*, 1980). Nawa and Miller (1979) demonstrated that mastopoietic activity is also present in the surface Ig^- thoracic duct lymph cell population, but whether it contains either MMC precursors, inducer cells, or both was not established. Taken together, these results are analogous to aspects of the IgA cell life history and suggest that precursor and/or inducer cells for the unique MMC population reside in MALT such as PPs. However, there is no conclusive evidence to confirm this hypothesis.

4. Ontogeny of Mucosal Immunocompetence

Although it is obvious that the immune system of the newborn animal is immature and that maternally derived antibody is important in protecting the newborn against infection, there is a dearth of knowledge about the development of mucosal immunocompetence in the young. Putative M cells develop in the human fetus by 17 weeks' gestation (Moxey and Trier, 1978), and in the chicken specialized FAE develops by day 15 *in ovo* (Bockman and Cooper, 1973). However, the ontogeny of the functional capacity of FAE and M cells in antigen sampling and perhaps processing and presentation requires further study. Similarly, although surveys of the development of IgA^- and other isotype-containing cells in mucosae have been conducted in a variety of species, e.g., mouse (Crabbé *et al.*, 1970), pig (Brown and Bourne, 1976; Allan and Porter, 1977), and rat (Brown, 1979), knowledge of the ontogeny of antigen-binding lymphocytes in mucosal tissues of the fetus in newborn is virtually nonexistent (Dwyer and Mackay, 1972). Such knowledge is essential to the logical administration of vaccines during the fetal and neonatal periods, if, for example, the responsiveness to polysaccharide antigens develops more slowly than does responsiveness to protein antigens (Banks, 1982). Likewise, all the new knowledge of T cells in MALT must be subjected to ontogenetic analysis.

Infectious agents and the normal flora are important in the ontogeny of mucosal immunocompetence (Ottaway *et al.*, 1979), but this area lacks careful investigation. Bacterial lipopolysaccharide appears to be mitogenic for the development of the intestinal IgA cell population (Husband, 1980) and for the generation of spleen-seeking suppressor cells (Mattingly *et al.*, 1979; McGhee *et al.*, 1980). It will be both interesting and important to determine whether antigens that the newborn encounters in combination with maternal antibody generate long-lasting but different responses from antigens in the absence of maternal antibody. This is especially true given the evolving discussions on the role of breast-feeding in the health and productivity of the young.

C. Comment

Knowledge of the antigenic makeup of parasites that inhabit mucosae is not available; hence systematic analyses of the events in antigen uptake, processing, presentation, and the early phases of the induction of responses at mucosal sites have not been conducted. Nevertheless, the rapidly evolving concepts of mucosal regulatory circuits provide an exciting basis for such studies in the future.

III. IMMUNE RESPONSES AND MUCOSAL INFLAMMATION IN PARASITIC INFECTIONS

A. Epithelium

There are many different cell types within the intestinal epithelium overlying the normal villi, which can be divided arbitrarily into two groups: (1) those that arise endogenously, from stem cells in the crypt epithelium, and (2) those that originate in sites other than the intestinal epithelium. The former group includes columnar digestive/absorptive cells, goblet cells, enteroendocrine cells, and Paneth cells (Cheng and Leblond, 1974), whereas the latter includes cells of the immunologic system, such as intraepithelial lymphocytes (IELs), mast cells, and rarely eosinophils, neutrophils, and macrophages.

1. Cells Arising from Precursors in the Crypt

a. Columnar Digestive/Absorptive Cells. Villus atrophy and crypt hyperplasia are common features associated with intestinal parasitic infections (reviewed by Befus and Bienenstock, 1982*a*). Crypt mitotic rates have been shown to double during *Giardia* infections in the mouse, and cell production per villus increased comparably (MacDonald and Ferguson, 1978). Ferguson and Jarrett (1975) established that partial villus atrophy in *N. brasiliensis*-infected rats was thymus-dependent, this finding has been confirmed by others studying giardiasis (Roberts-Thomson and Mitchell, 1978) and trichinosis in mice (Manson-Smith *et al.*, 1979*a*). The mechanisms underlying this thymus dependency of epithelial pathology have not been determined, although it may be that T lymphocytes, directly or indirectly, control the cycling time of crypt stem cells as well as the factors that orchestrate their differentiation along different lines.

Castro (1981) has thoroughly reviewed the literature on alterations in brush-border enzymatic activities, fluid secretion, and absorption during intestinal parasitic infections. Whether any or all of these pathophysiologic changes are induced or modulated by the development of immune responses to the parasites in the intestine is presently unknown. We do not find it unreasonable to hypothesize that T-lymphocyte products, in addition to influencing cycling times of stem cells in the crypts, might also influence the expression of certain characteristics of the cells, such as disaccharidases or other enzyme systems. Information in these areas is sadly lacking and requires the development of systems to study the expression of epithelial cell characteristics *in vitro* under the influence of factors derived from various cell types, including antigen-stimulated immune lymphocyte populations.

b. Goblet Cells. Considerable attention has been focused recently on the hyperplasia of goblet cells in the intestinal epithelium during parasitic infection because of an analysis of the mechanisms involved in goblet cell hyperplasia previously established during gastrointestinal nematode infections (Wells, 1963). Miller and Nawa (1979*a*) showed that transfer of immune thoracic duct lymphocytes in *N. brasiliensis*-infected rats hastened the development of goblet cell hyperplasia in infected recipients. Subsequently, it was shown that putative T cells from the thoracic duct lymph were responsible for this transfer of goblet cell hyperplasia (Miller *et al.*, 1979) and that a factor in immune serum possessed similar activity (Miller and Nawa, 1979*b*). Taken together, these results are consistent with the hypothesis that products derived from immune T cells direct the differentiation of goblet cells from crypt stem-cell precursors.

c. Enteroendocrine Cells and Paneth Cells. There is disappointingly little information about changes in intestinal hormones during mucosal infections. The information available has been concisely reviewed by Castro (1981). To our knowledge, there is no information about changing levels of enteroendocrine cells in different parasitic infections of the intestine; such studies together with investigations of changes in the regulatory activity of intestinal hormones during such infections will go a long way to improve our understanding of pathophysiologic changes in the inflammatory events in the mucosae. The way in which neuroendocrine substances influence immunologic events and *vice versa* is largely unexplored, although there are some exciting suggestions that neuropeptides such as vasoactive intestinal peptide modulate lymphocyte function (O'Dorisio *et al.*, 1981) and stimulate the release of mast cell mediators (e.g., Fjellner and Hagermark, 1981).

There is a similar dearth of information about Paneth cell numbers and function during parasitic infection. Roberts-Thomson *et al.* (1976) have provided the only report of Paneth cell hyperplasia during mucosal parasitic infection; whether this is a thymus-dependent event, as has been established for changes in epithelial cell kinetics and the goblet cell hyperplasia, remains to be studied.

2. Intraepithelial Cells Derived from Exogenous Sites

a. Intraepithelial Lymphocytes. IELs normally represent 10–15% of the total epithelial cells in the human intestine (Ferguson, 1977), whereas in the pig they can represent as much as 30–40% of all cells (Chu *et al.*, 1982). The phenotypic characteristics and origin of IELs were reviewed in detail recently (Befus and Bienenstock, 1982*a*). In brief, depending on the techniques used, it is claimed the 45–95% bear T-lymphocyte markers, although only about 50% are thymus-dependent, as evidenced by studies in nude athymic or neonatally thymectomized mice. In rat, mouse, and humans some 70–90% of IELs have surface markers associated with the cytotoxic/suppressor phenotype, but few cells bear surface markers of the inducer phenotype (Janossy *et al.*, 1980; Lyscom and Brueton, 1982; Guy-Grand and Vassalli, 1982). Some 40-60% of IELs contain a few (1–20) metachromatic granules that appear similar to mast cell granules (Rudzik and Bienenstock, 1974); that is, in addition to their metachromasia, apparently they contain small quantities of histamine and incorporate ^{35}S into a proteoglycan core (Guy-Grand *et al.*, 1978). Recent studies have established that IELs possess NK activity (Tagliabue *et al.*, 1982), these results, together with questions of the relationships between mast cells and IELs, are directing current thoughts of the functional significance of IELs.

Giardiasis in the mouse induces significant elevations in numbers of IEL (MacDonald and Ferguson, 1978), but this is not generally true of other mucosal parasitic infections studied to date (Befus and Bienenstock, 1982*a*). However, this has been inadequately investigated; a recent study using TGE (Chu *et al.*, 1982) has suggested that the numbers of IELs may increase or decrease, depending on where in the epithelium and intestine one looks and at what time relative to infection. Furthermore, simple observations of cell numbers provide limited information about their functions. There are a number of reasons, in addition to their location, for considering this cell type important in host resistance at mucosal sites.

b. Intraepithelial Mast Cells (Globule Leukocytes). Many parasitic, especially helminthic, infections induce an infiltrate of mast cells or globule leukocytes into the epithelium. In the rat, the laboratory animal in which effect this has been most extensively studied, it occurs in association with the early developmental phases of a mucosal mast cell hyperplasia in the lamina propria and may represent up to 50% of the total mast cells per villus (Miller and Jarrett, 1971). In the mouse, intraepithelial mast cells arise early, as in the rat, but represent a greater proportion of the mast cell population, i.e., up to 90% of the total (Ruitenberg and Elgersma, 1976; Lee and Wakelin, 1982). In our comparisons of *N. brasiliensis* infection in both mice and rats, it appears that the intraepithelial mast cell population is relatively short-lived, whereas the population in the lamina propria is more long-lived. This finding corresponds well with the observations that the total mast cell population in the mouse, in which >90% is in the epithelium, is short-lived, whereas in the rat the total population has a much longer life span, and a smaller proportion is present in the epithelium. The postulated functions of these cells are discussed in Section IV.

There has been considerable controversy as to the origin and ontogeny of intraepithelial mast cells. Murray *et al.* (1968) provided ultrastructural evidence that cells previously called globule leukocytes in the intestine of rats, sheep, and cattle are actually mast cells in various stages of degranulation. However, using similar parasite models, Ruitenberg and co-workers proposed on the basis of kinetic analyses of cells in infected immunocompetent and athymic animals that in mice (Ruitenberg and Elgersma, 1979) and rats (Ruitenberg *et al.*, 1979) globule leukocytes represented a previously unrecognized cell type bearing no ontogenetic relationship to mast cells in the lamina propria (see also Mayrhofer, 1980). Given that this proposal has not received further support, together with an abundance of evidence that globule leukocytes and mast cells in the lamina propria share many characteristics (Murray *et al.*, 1968; Miller and Walshaw, 1972; Befus *et al.*, 1982*a*), we believe that the weight of existing evidence favors the earlier hypothesis that these two cells are homologous. However, this must be firmly established by determining that (1) mast cells in the lamina propria migrate into the epithelium, and (2) the factors that stimulate that movement.

B. Lamina Propria

1. Plasma Cell Responses

In the normal intestine IgA-containing cells represent > 80% of the plasma cells, with the remaining proportion made up of small numbers of IgG- and IgM-containing cells and virtually no IgE- or IgD-producing immunocytes (Brandtzaeg and Baklien, 1976). Both the cell proportions and their absolute numbers change after certain stimuli, although this has received less careful study after parasitic infection (reviewed by Befus and Bienenstock, 1982*a*) than in inflammatory bowel disease in humans (Brandtzaeg and Baklien, 1976). Such studies do provide knowledge of local events, but have limited power of resolution regarding functional significance because surveys of plasma cell numbers provide no information of antibody specificity, secretory rates, or availability.

Perhaps studies of IgE-containing cells in the mucosa of parasitized animals have been most enlightening. Given the tremendous elevations in serum IgE levels induced by some infections such as *N. brasiliensis* in the rat, the source of this IgE became a relevant question. Mayrhofer *et al.* (1976) established that large numbers of IgE-containing cells were present in the intestinal lamina propria as well as in the local draining lymph nodes. Careful study established that the IgE-containing cells in the lamina propria were not plasma cells but rather mucosal mast cells (MMCs) that had in some way incorporated IgE into their cytoplasm. The cells in the mesenteric lymph nodes appeared to be truly plasma cells. Various workers have confirmed that the MLNs as well as the bronchial lymph nodes (BLNs) (Allan and Mayrhofer, 1981; Befus *et al.*, 1982*c*) are the major sites of IgE antibody synthesis after *N. brasiliensis* infection. To our knowledge the gut mucosal lamina propria has not been proved to be a prominent site of IgE production in any parasitic infection.

Antibody levels against mucosal parasite antigens have been widely studied (Ogilvie and Love, 1974; Wakelin, 1978*a*; Befus and Bienenstock, 1982*a*) and in the serum appear to involve largely IgG and IgE isotypes. However, it is widely acknowledged that IgA antibody in secretions may be as relevant if not more so, and the few studies to date have confirmed the presence of IgA antibody in secretions and noted that it appears early in the parasitic infections (e.g., day 6 in *N. brasiliensis*-infected rats, Poulain *et al.*, 1976), whereas serum antibody such as IgE appears later (Ogilvie, 1967; Befus *et al.*, 1982*c*). Clearly, much greater evidence than the mere presence or level of antibody is needed before statements about its functional significance can be made. Unfortunately, in the past this was apparently less obvious, but efforts are now being made to analyze in greater detail the functional significance of individual antibodies such as anti-sporozoite antibodies in *Plasmodium berghei* (Yoshida *et al.*, 1980; Hollingdale *et al.*, 1982) and antibodies to nematode cuticular antigens (Maizels *et al.*, 1982). The logical development of such studies is to investigate not only the nature of the antigen and the protective role of the antibody, but also to determine how the antibody acts *in vivo* through, for example, complement fixation, or its inhibition by IgA (Griffiss, 1975) or by binding to Fc receptors such as those for IgA on neutrophils, lymphocytes, monocytes, and macrophages (Bienenstock and Befus, 1982).

Reports of antibody levels against parasites described earlier provide little evidence of the site of antibody synthesis in the host. We would predict that both local and systemic sites such as the spleen are involved, the latter through the dissemination of either antigens or sensitized cells from mucosal sites. In the absence of evidence to the contrary it can be assumed that the rules defined in other systems governing plasma cell isotype selection, proliferation, and maturation apply in mucosal parasitic infections as well.

2. T-Cell Responses

Knowledge of T cells in the intestinal lamina propria is surprisingly sparse. A number of studies have estimated that of the isolated lamina propria lymphocytes, T cells represent 35–85% (reviewed by Arnaud-Battandier, 1982). Whereas the IELs express largely the suppressor/cytotoxic surface phenotype, the lamina propria T cells bear the inducer markers (Janossy *et al.*, 1980; Lyscom and Brueton, 1982). Functional studies have, however, demonstrated that cytotoxic T cells do appear in the intestinal lamina propria (Arnaud-Battandier *et al.*, 1978; Davies and Parrott, 1981*a*; Arnaud-Battandier, 1982); we have demonstrated that the cytotoxic T-cell precursor and the cell(s) modulating its clonal expansion *in vitro* can be derived from the murine intestinal lamina propria (D. A. Clark, A. D. Befus, and J. Bienenstock, unpublished observations, 1983). Lamina proprial lymphocytes also produce migration inhibition factor (MIF) after oral administration of antigens (Frederick and Bohl, 1976; Huntley *et al.*, 1979), but whether this MIF was T-cell derived was not determined.

None of these aspects of lamina propria T cells has been studied using parasitized animals. Now that procedures are available whereby cells can be isolated from the intestine, it is essential that they be applied to dissect the T-cell arm of local responses to parasites. Presumably, gut T cells are responsive to parasite antigens, but their precise functional characteristics in the long list of T-cell-dependent antiparasite responses (Mitchell, 1980) remains to be studied. For example, do gut T cells make a variety of lymphokines such as IL-2 or eosinopoietic or mastopoietic factors in response to parasitic infection, and how do such responses differ, given different parasites or primary versus secondary infections? The observations of Rose *et al.* (1976) and Ottaway *et al.* (1980) that T-lymphoblast localization or retention, or both, in the parasitized intestine is enhanced in *T. spiralis*-infected mice, suggest that this infection may provide a fine model to investigate intestinal T-cell function. The correlation between this T-cell localization and the commencement of worm loss 2–3 days later (Manson-Smith *et al.*, 1979*b*) reinforces this suggestion.

3. Mucosal Mast Cell Hyperplasia

Helminthic infections such as *N. brasiliensis* or *T. spiralis* induce a marked, thymus-dependent intestinal MMC cell hyperplasia (Ruitenberg and Elgersma, 1976; Mayrhofer and Fisher, 1979). To our knowledge, protozoan infections do not induce intestinal mastocytosis. The intestinal MMC is functionally distinct from peritoneal mast cells (PMCs) in rats (Miller, 1980; Enerbäck, 1981; Befus *et al.*, 1982*a*; Pearce *et al.*, 1982). Unfortunately, apart from our own work (Denburg *et al.*, 1980), it is not clear whether any of the recent studies of mast cell development *in vitro* are relevant to MMC or only to the mast cell with functional characteristics of the PMC. If one assumes that both the MMC and PMC share a common precursor and are distinct because of separate inducer factors that stimulate the expression of different parts of the genome within the precursor, then one can conclude that the MMC precursor is a thymus-independent lymphoid-like cell that is driven to differentiate by a thymus-dependent inducer system (reviewed by Befus and Bienenstock, 1982*a*; Bienenstock *et al.*, 1982). There has been some controversy about the nature of the mastopoietic factor that stimulates mouse bone marrow mast cells; it is clearly distinct from IL-2 (Yung *et al.*, 1981) and may be identical to IL-3 (Ihle *et al.*, 1982). In our hands, cell-free supernatants that stimulate MMC growth *in vitro* have high levels of IL-3, but other supernatants with high IL-3 do not stimulate MMC growth. These results could be interpreted in many ways but may identify that multiple factors are involved in MMC differentiation and proliferation in the lamina propria of parasitized hosts. Undoubtedly, the modulation of MMC function in parasitic infection is equally, if not more, complex than the mechanisms controlling mastocytosis, as discussed in Section IV.A.2.c.

4. Other Responses

In the intestinal inflammatory responses generated during various parasitic infections a variety of cellular infiltrates and pathologic reactions can be seen. For example, *T. spiralis* infection induces a prominent neutrophil infiltrate by day 8 and subsequently a mixed mononuclear/polynuclear cell response (Ismail and Tanner, 1972; Larsh and Race, 1975). Intestinal eosinophilia is prominent in *T. spiralis* infection as well as with *N. brasiliensis* and *Strongyloides ratti* infection (Kelly and Ogilvie, 1972; Moqbel, 1980). Presumably, intestinal eosinophilia is under lymphocyte regulation, as is MMC hyperplasia (Befus and Bienenstock, 1979; Nawa and Miller, 1979) and peripheral eosinophilia (Basten and Beeson, 1970).

Granulomatous responses occur in the intestine during *Nematospiroides dubius* and *Schistosoma mansoni* infections (Jones and Rubin, 1974; Weinstock and Boros, 1981); presumably granulomas in the intestinal mucosa are induced by the T-cell-dependent mechanisms so well established by Warren and co-workers for liver and lung granulomas (Warren, 1976). In the liver and colon the granulomatous reactions around the schistosome eggs, which are large early in infection, are subsequently reduced in size by the actions of suppressor T cells (Colley, 1981; Weinstock and Boros, 1981) and perhaps local antibodies (Boros *et al.*, 1982; Olds *et al.*, 1982). However, granulomas in the ileal mucosa never become large as in the liver and colon and appear to be modulated by a mechanism(s) distinct from granulomas elsewhere (Weinstock and Boros, 1981). Clearly, the mechanism(s) underlying this dampening of potentially pathologic responses in the intestine require characterization, as they are undoubtedly relevant in other intestinal diseases and infections.

C. Intestinal Lumen

Elsewhere we have discussed extensively the presence and potential functions of cells found free in the intestinal lumen (Bienenstock and Befus, 1980; Befus and Bienenstock, 1982*a*; Heatley and Bienenstock, 1982). Suffice it to say that lymphocytes, macrophages, and neutrophils have been widely recognized in the luminal contents and, at least for neutrophils, their emigration into the lumen can be immunologically stimulated (Bellamy and Nielsen, 1974). Owen *et al.* (1979) showed some beautiful pictures of cells in the lumen adjacent, and some even attached, to trophozoites of *G. muris.* Such observations must be extended to other systems and reproduced before clear concepts of their significance can be established.

IV. MUCOSAL EFFECTORS OF HOST RESISTANCE

A. Parasite Elimination

The ultimate effectors of elimination of parasitic organisms from a mucosal site such as the intestine are still unknown. Obviously, however, these elusive effector substances must act on the parasite whether it is fully in the lumen or partially or wholly embedded in the intestinal wall. This section continues our discussion from the point of view of anatomic compartmentalization and discusses first candidates for the sources of these ultimate factor(s) in the epithelium and then in the lamina propria. We attempt to outline the complexity and integration of various components of the host response building to the climax of parasite elimination. In the development of these concepts, many host–parasite relationships have been studied, and our synthesis is structured from this diversity of systems (Miller *et al.*, 1982). Hence, no one system necessarily expresses the complete range of host mucosal responses discussed; in the different host–parasite relationships, the relative importance of the different components must be variable. Nevertheless, we hope that a discussion of the repertoire of potential host responses will stimulate further study on diverse fronts and ultimately lead to indentification of the important effectors of host resistance.

1. Epithelium

a. Columnar Epithelial Cells. There is little concrete knowledge that identifies the columnar epithelial cell as a source of effector substances for parasite elimination. However, these cells are affected by infection (Castro, 1981) and are sources of a variety of digestive enzymes, lysosomal enzymes (e.g., Danovitch *et al.*, 1972), and at least one complement component (Colten *et al.*, 1968). Perhaps the lactoperoxidase antimicrobial enzyme system of saliva (Tenovuo *et al.*, 1982) and milk has a source in the intestinal epithelium as well. Polymeric but not monomeric IgA is known to interact with both lysozyme and complement to express bacteriocidal activity (Hill and Porter, 1974); a recent report established a similar potentiation of lactoperoxidase-mediated antibacterial activity by IgA (Tenovuo *et al.*, 1982). Considerable uncertainty surrounds the question of a role for complement in the intestinal lumen (reviewed by Befus and Bienenstock, 1982*a*); in general it is concluded that complement levels are low there and lumenal contents are anticomplementary. However, observations of the localization of C3 in a granular pattern of immunofluores-

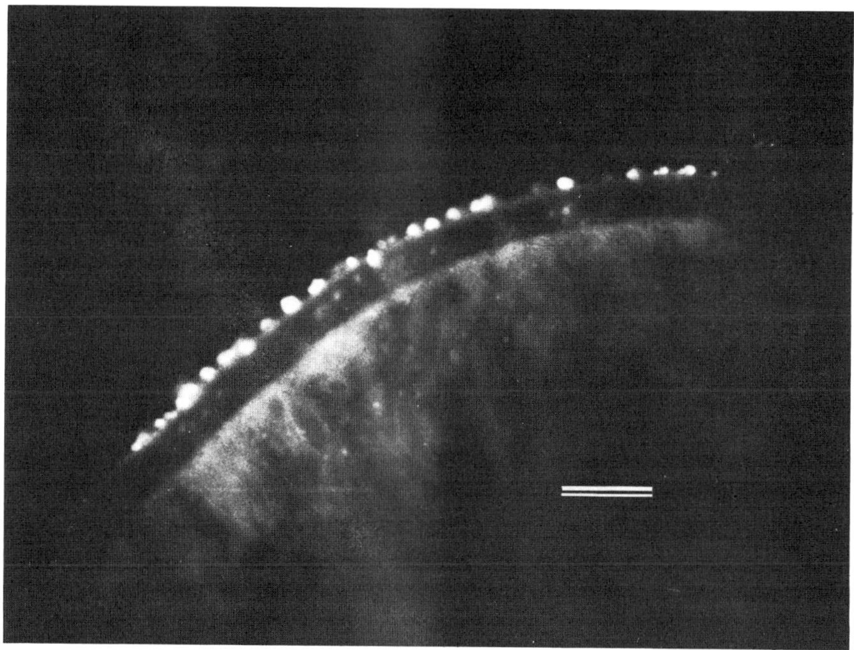

Figure 2. Granular distribution of complement component 3 (C3) on the surface of the adult tapeworm, *Hymenolepis diminuta*, in the lumen of the murine intestine. Bar, 10 μm. (From Befus, 1977, with permission of Academic Press.)

cence at the surface of the lumen-dwelling tapeworm, *Hymenolepis diminuta* (Fig. 2) by Befus (1977) and Threadgold and Befus (1977) have questioned this conclusion, which requires further investigation.

b. Goblet Cells. It has long been proposed that mucus plays a prominent role in protection against intestinal infection (Florey, 1933; Frick and Ackert, 1948); recently a strong correlation between mucus secretion by goblet cells and elimination of the nematode, *N. brasiliensis*, from previously sensitized animals was identified (Miller *et al.*, 1981). Similarly, during the last few years it has been established that immunologic stimuli, including antigen–antibody complexes and anaphylactic reactions perhaps mediated through histamine, induce mucus release from goblet cells (Walker *et al.*, 1977; Lake *et al.*, 1980). Although some of the parasites become entrapped in mucus (Miller *et al.*, 1981), the precise role of mucus in parasite elimination remains to be established. It could possibly be merely mechanical aspects of mucus that are important, but one suspects that such is not the complete mechanism. Associated with mucus or its secretion, or both, are likely to be other effector components. For example, IgA was recently shown to be associated with mucus and was found to enhance the adherence of bacteria to the mucus layer (Magnusson and Stjernstrom,

1982); IgA in the mucus layer may facilitate parasiticidal activities of substances, such as complement, lysozyme, or lactoperoxidase, also present in the mucus belt. In this way, mucus could act not only as a mechanical barrier, but also as a matrix in which other components to the host-effector response interact with the parasite.

Miller and Nawa (1979*a*), in their description of goblet cell hyperplasia in response to *N. brasiliensis* infection, showed that the mucin quality changed from an acidic to neutral mucopolysaccharide by day 14 of infection. The mechanisms underlying this qualitative change are unknown and may merely reflect some aspect of goblet cell maturity. However, such qualitative changes in mucins may be important in host resistance because, for example, mucus contains lectins (carbohydrate-binding glycoproteins, e.g., Beyer and Barondes, 1982) and many carbohydrate moieties that might bind to endogenous or exogenous lectins. *E. coli* adheres to human mucosal cells through a mannose-binding lectin (adhesin) on its surface (Ofek *et al.*, 1977), and *Vibrio cholerae* has a similar fucose-binding adhesin (Jones and Freter, 1976). Minor changes in the expression of mucin or epithelial cell carbohydrate moieties or lectins could influence bacterial adherence and colonization. Unfortunately, there is little information about lectins or carbohydrate moieties on helminth or protozoan surfaces, hence their possible relevance in host-parasite relationships is unknown. We would predict that such lectin–ligand interactions are relevant not only in initial colonization and site selection–(there are regional differences in lectin-binding activities of the intestine (Etzler and Branstrator, 1974; Freeman *et al.*, 1980) by parasites–but in host resistance as well. Recent studies by Castro and Harari (1982) suggest that *T. spiralis* infection induces changes in lectin-binding activities in the intestinal epithelium that correlate with sensitization to parasite antigens. Such innovative studies require confirmation and extension.

c. Intraepithelial Lymphocytes. IELs are abundant in the epithelium and at least in some infections increase significantly in number. In accordance with other cell types discussed in this chapter, an increase in number of IELs provides little information about their functional significance; the corollary holds that an absence of such an increase does not imply that they are functionally insignificant. IELs express natural killer (NK) function in both the guinea pig (Arnaud-Battandier *et al.*, 1978) and mouse (Tagliabue *et al.*, 1981, 1982), and it appears that the effector cell is the granulated IEL (Tagliabue *et al.*, 1982; Timonen *et al.*, 1982). In addition, IELs express antibody-dependent cell cytotoxicity (ADCC) activity in the guinea pig (Arnaud-Battandier *et al.*, 1978), and about 50% of human granulated lymphocytes express Fc receptors for IgG Timonen *et al.*, 1981). Recent investigations have demonstrated that murine IEL bear receptors for IgE, as do at least some clones of murine NK cells (K.L. Rosenthal, A.D. Befus, T. Ishizaka, and J. Bienenstock, in preparation, 1983). The latter observation of heterogeneity among NK clones is consistent with evidence for NK heterogeneity *in vivo* (Minato *et al.*, 1981). Our studies established that NK cells in the intestinal epithelium are largely of one subpopulation that expresses high levels of Thy-1 (Tagliabue *et al.*, 1982); it is interesting that IELs possess IgE receptors as well. Whether Fc receptors for other isotypes, especially IgA, are expressed by IEL must be determined and the functional significance explored.

High levels of IgE antibody in parasitic infections together with IELs possessing Fc receptors for IgE may provide an important parasiticidal activity through ADCC. The role of the granules of IEL and cytotoxic activity remains to be established unequivocally, but ultrastructural investigations (Carpen *et al.*, 1982) suggest that precisely directed granule exocytosis is important. This provides further links between NK cells and mast cells, in addition to the granule similarities including proteoglycan and histamine content, as well as the possession of IgE receptors. Mast cells also have cytotoxic potential (Farram and Nelson, 1980), especially combinations of mast cell granules and eosinophil peroxidase (Henderson *et al.*, 1981). It may be that granule contents from IELs are potentiated in the parasiticidal activities by local peroxidases or lysosomal enzyme systems that are focused and given specificity by interactions with IgA antibodies in a mucus matrix.

To date, the studies of NK activity by IEL have used almost exclusively classic targets such as the YAC lymphoma cell line (Tagliabue *et al.*, 1981, 1982); whether cytotoxic activity exists against other targets such as parasites must be established. IELs from the porcine intestine are capable of destroying TGE-infected cells (Cepica, 1982), and presumably this is a more generalized phenomenon. Both *G. muris* and *N. brasiliensis* infections enhance splenic NK activity (A. Tagliabue, A.D. Befus, and J. Bienenstock, unpublished observations, 1983), but the mechanisms of this enhancement, e.g., interferon (?) and the possible effects on gut NK function have not been studied. Smith *et al.* (1982) have shown that peripheral blood monocytes spontaneously kill *G. lamblia in vitro*; perhaps gut IELs have similar activities either spontaneously or in conjunction with IgA or IgE antibodies, because NK cells have some qualities of monocytes (Grossi *et al.*, 1982), and there is evidence that monocytes are bacteriocidal in collaboration with IgA antibody (Lowell *et al.*, 1980).

Paradoxically, some parasites (*Eimeria* spp.) appeared to undergo at least the early phases of their development inside granulated IELs (Millard and Lawn, 1982; Befus and Bienenstock, 1982*a*). It is also suggested that viruses such as *Pichinde* develop in NK cells (Gee *et al.*, 1981). The extent of this phenomenon must be established; also the question of whether other mucosal infectious agents develop in IELs merits study. It will be essential to determine whether previous exposure to an infectious agent influences its relationship with IELs on reexposure.

2. *Lamina Propria*

As the functions of IgA and IgE antibody have been reviewed in detail in recent publications, this section is brief and focuses on some of the more recent information that has spawned new concepts for investigation (Befus and Bienenstock, 1982*a*; Bienenstock and Befus, 1982; McDermott *et al.*, 1982).

a. IgA Antibody. In serum a high proportion of IgA exists as a monomer (Kutteh *et al.*, 1982), whereas in secretions IgA is polymeric (largely dimeric), with the additional polypeptides J chain, produced by plasma cells and secretory component (SC) produced by intestinal epithelial and some other cells (Bienenstock and Befus, 1982). Secretory IgA (sIgA) is actively transported into secretions and is resistant to degradation in the proteolytic environment there. IgA antibody inhibits antigen uptake, microbial toxins and bacterial adherence and thus colonization of the epithelium. It induces mucophilia when bound to bacteria (Magnusson and Stjernstrom, 1982) and facilitates the microbicidal activities of lactoperoxidase (Tenovuo *et al.*, 1982) and lysozyme and complement (Hill and Porter, 1974). IgA does not activate complement by the classic pathway but can initiate the alternative pathway, although this is apparently limited because C5 cleavage does not occur and the feedback amplification loop for C3 convertase is not activated (Pfaffenbach *et al.*, 1982).

A variety of cell types possess Fc receptors for IgA, including T lymphocytes, neutrophils, monocytes (reviewed by Bienenstock and Befus, 1982), and perhaps even Paneth cells (Erlandsen *et al.*, 1976). The functional consequences of activation of these receptors through IgA–antigen complexes has been incompletely researched, but it seems clear that both helper and suppressor lymphocytes may possess IgA receptors (Bienenstock and Befus, 1982). IgA inhibits neutrophil phagocytosis (van Epps *et al.*, 1978; Wilton, 1978; Magnusson *et al.*, 1979) and chemotactic activity (van Epps and Williams, 1976; Kemp *et al.*, 1980), but promotes IgG-mediated ADCC by these cells (Shen and Fanger, 1981). IgA itself, in conjunction with monocytes, mediates ADCC; whether or not IELs act similarly remains to be tested.

These aspects of IgA function have not been explored in parasitic infections of mucosae but could very well be important effector mechanisms of parasite elimination. In *N. brasiliensis* infection, alveolar macrophages become activated and can kill third-stage larvae *in vitro* (Egwang *et al.*, 1983);observations that this infection also induces expression of IgA receptors on alveolar macrophages (Gauldie *et al.*, 1983*a*) suggest that IgA-mediated ADCC is involved. Similar mechanisms may be expressed in the intestine against migrating newborn larvae of *T. spiralis* or trophozoites of *Giardia* spp. or sporozoites, merozoites, or gametocytes of *Eimeria* spp. It is clear that IgA antibody can induce parasite elimination (Leid and Williams, 1974; Lloyd and Soulsby, 1978; Jacqueline *et al.*, 1978), but the mechanisms of this effector function are unknown.

IgA can server a protective role in another sense as well (Fig. 3). Hepatocytes and biliary epithelium synthesize SC and transport IgA into bile just as intestinal epithelial cells transport IgA into the lumen (Bienenstock and Befus, 1982). This hepatic transport system supplies large amounts of sIgA antibody into the

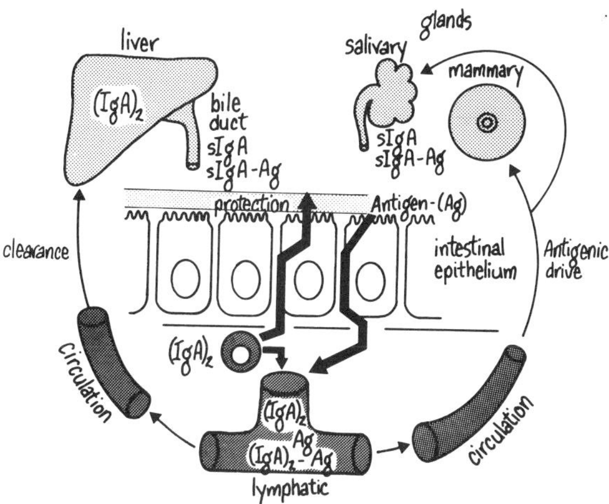

Figure 3. Possible functions of IgA antibodies in the mucosal immunologic network. Dimeric IgA produced in plasma cells in the intestinal lamina propria can be secreted into the intestinal lumen or passed into the circulation and subsequently transported into the bile or the secretions of other mucosal surfaces, such as the respiratory tract or salivary and mammary glands. IgA antibody transported from the circulation may be complexed with antigen and therefore function as either an antigen clearance mechanism or for the delivery of antigen to responsive lymphocytes in sites distant from its origin. (From Bienenstock and Befus, 1982, with permission of *Gastroenterology.*)

upper intestine of some species (Lemaître-Coelho *et al.*, 1978) and also serves as an efficient mechanism for the clearance of circulating antigens (Socken *et al.*, 1981; Russell *et al.*, 1981; Peppard *et al.*, 1981). Thus, an important role of IgA is not only to minimize antigen uptake at mucosae, but also to clear antigen that has managed to gain systemic access, thereby removing the source of potentially damaging immune complexes. Such a role might be particularly important in the maintenance of long-term, nonpathogenic host–parasite relationships. Because cells of other sites such as the respiratory and urogenital tracts and salivary and mammary glands make SC, transport of IgA–antigen complexes may occur in these sites as well (Bienenstock and Befus, 1982). This might be a method of distributing parasite and other antigens to distant mucosal sites for local immunocyte sensitization, which results in modification of local responses to antigen upon re-exposure (e.g., antigens in maternal milk influencing subsequent immune response of newborns).

b. IgE Antibody. In the past the association between high levels of IgE and parasitic infection was conceptually restricted almost exclusively to the mast cell and basophil-mediated immediate hypersensitivity reactions. However, Capron *et al.* (1975) showed that peritoneal macrophages express IgE receptors and that IgE–antigen complexes stimulate macrophage cytotoxicity against schistosomula (reviewed by Capron *et al.*, 1982). Eosinophils are now recognized as another cell type that expresses helminthic cytotoxic activity in conjunction with IgE antibodies (Capron *et al.*, 1982). In our recent observations, in which isolated IELs and some NK clones were found also to express IgE receptors (A. Petit, P. Ernst, K. Rosenthal, D. A. Clark, A. D. Befus, T. Ishizaka, and J. Bienenstock, in preparation, 1983; K. Rosenthal, T. Ishizaka, D. Befus, G. Dennert, H. Henngartner, and J. Bienenstock, in preparation, 1983) we have identified yet other mucosal cells, in addition to macrophages and eosinophils, for the study of IgE and ADCC against parasites at mucosae. It may be that IgE receptors on alveolar macrophages (Joseph *et al.*, 1980; Boltz-Nitulescu and Spiegelberg, 1981; Boltz-Nitulescu *et al.*, 1982; Gauldie *et al.*, 1983*a*) are fundamentally important in acquired resistance to helminths as the latter migrate through the lungs. The techniques now available for the isolation of cells from the intestinal lamina propria (Davies and Parrott, 1981*b*; Arnaud-Battandier, 1982; Befus *et al.*, 1982*a*; Tagliabue *et al.*, 1982) make such studies possible, and in the near future concrete information about the role of IgE and mucosally derived cells in parasitic infections should be available.

c. Mucosal Mast Cells. The role of MMC in host resistance to mucosal parasitic infection has been a controversial subject. At times almost illogical arguments were presented in attempts to make statements about function solely on the basis of estimates of cell numbers. Clearly, parasite elimination can occur in the congenital absence of recognizable mast cells (Uber *et al.*, 1980; Kojima *et al.*, 1980; Crowle and Reed, 1981), but how this relates to the events in normal animals is unknown. The potential functions of MMCs in parasitic infection are complex; for a more complete discussion, we refer the reader to two comprehensive reviews, one by Askenase (1980) and one by Befus and Bienenstock (1982*a*).

An important component of mast cell function involves changes in vascular permeability (Beaven, 1978) that facilitate the localization of serum and cellular components at the inflammatory site, i.e., leak lesion (Barth *et al.*, 1966). It is now widely recognized that mast cell products released locally modulate the activities of other cells. For example, histamine and eosinophil chemotactic factors attract and activate eosinophils to express potent cytocidal activities (Capron *et al.*, 1982). Histamine, through either H1-or H2-type histamine receptors, influences lymphocyte, neutrophil, macrophage, and goblet cell function as well (Befus and Bienenstock, 1982*a*). Mast cells stimulate delayed-type hypersensitivity reactions in murine skin through the release of serotonin (Askenase *et al.*, 1980) and can themselves mediate cytotoxicity (Farram and Nelson, 1980; Henderson *et al.*, 1981), perhaps through the proteases they contain (Woodbury and Miller, 1982).

How all these potential events are regulated and directed to a minimally pathogenic, but successful, outcome has not been dissected. This question is complicated by the knowledge that MMCs are functionally distinct from mast cells at many other sites (Enerbäck, 1981; Befus *et al.*, 1982*a*; Pearce *et al.*, 1982), yet most of the existing knowledge of mast cell function is based on those derived from nonmucosal tissues. Before clarity arises from the current information, greater emphasis must be placed on the study of MMCs either derived by isolation procedures (Befus *et al.*, 1982*a*) or by *in vitro* culture from precursors (Denburg *et al.*, 1980). For example, do neuropeptides such as neurotensin, substance P, secretin, vasoactive intestinal peptide, and somatostatin stimulate histamine secretion by MMCs, as has been reported for peritoneal mast cells (Johnson and Erdos, 1973; Fjellner and Hagermark, 1981; Kurose and Saeke, 1981; Theoharides *et al.*, 1981)? If this proves true, then such stimuli may be important in regulating host immune responses in the lamina propria in parasitic infection. Similarly, if the observation that different mediators from peritoneal mast cells can be released independently of one another (Theoharides *et al.*, 1982) applies to MMCs as well, this may be important in the orchestration of the many events taking place in the parasitized intestine.

d. Other Responses. Many other cell types are present in the lamina propria, including T lymphocytes, eosinophils, macrophages, and neutrophils; some of their functions have been discussed in Section III.B.4. Unfortunately, the activities of these cells in the intestine can only be postulated on the basis of studies on cells from other sites. The thymus dependency of many responses to parasitic infection is well known (Mitchell, 1980), but we lack information about cytoxic T cells in the intestine of parasitized animals or their potential specificity for parasite antigens. Similarly, whether intestinal T lymphocytes are activated to produce lymphokines such as lymphotoxin (Namba and Waksman. 1975*a*; Smith McMyne and Strejan, 1980) or inhibitor of DNA synthesis (IDS) (Namba and Waksman, 1975*b*) in response to infection is unknown.

Eosinophils are potent helminthocidal cells (Capron *et al.*, 1982); their peroxidase system provides at least some of this cytocidal activity (Klebanoff *et al.*, 1980). At mucosal sites eosinophil function remains to be defined in parasitic infections. Kazura and Grove (1978) showed that eosinophils killed newborn larvae of *T. spiralis in vitro* through an ADCC mechanism. Presumably, through this mechanism the large numbers of eosinophils in the infected intestine reduce the number of larvae that reach systemic sites. The observations of Dessein *et al.* (1981) that chronic treatment of rats with anti-IgE heavy-chain specific antibodies led to a significant reduction in tissue eosinophilia and a concurrent increase in muscle larvae are consistent with an important role for intestinal eosinophils in controlling the numbers of newborn larvae reaching systemic sites. To provide further evidence, intestinal eosinophils should be isolated from infected animals and their putative effects on newborn larvae tested directly.

Intestinal macrophage function in health or disease has been virtually neglected. Observations that alveolar macrophages have potent helminthocidal activities (Egwang *et al.*, 1983) must be applied to macrophages derived from intestinal tissues. Ultrastructural evidence suggests that intestinal macrophages kill protozoa (Owen *et al.*, 1981).

3. Summary

The spectrum of host responses to intestinal infections is diverse; Fig. 4 summarizes these responses and indicates the types of factors that may be ultimate effectors of parasite elimination. These factors include specific antibodies that may inhibit central metabolic pathways in the parasites, or together with various enzyme systems (peroxidases, lysozyme, complement) lead to direct parasite damage. IELs through either spontaneous cytocidal activities or ADCC mechanisms may damage parasites. Sensitization of lymphocytes after antigen uptake by the specialized FAE leads to the antibody and T-lymphocyte-dependent amplification mechanisms that generate the intestinal inflammatory responses. Many products derived from these inflammatory cells and other cells that they stimulate may act independently, but most likely in concert with other factors to damage the parasites. Mediators such as biogenic amines from mast cells and basophils may act directly on the parasites (Rothwell *et al.*, 1974), alter local permeability to facilitate serum and cell emigration or activate other cells to generate effectors of parasite elimination.

It is well established that in many infections parasite elimination involves

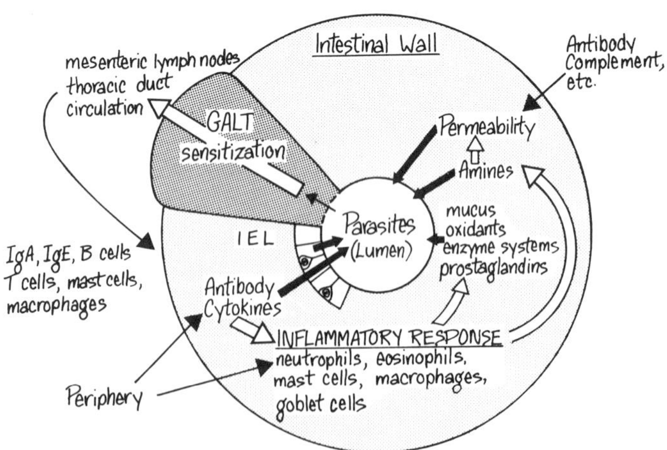

Figure 4. The complexity of mechanisms involved in resistance to intestinal infection. Antigens derived from intestinal parasites transit across the follicle-associated epithelium of the GALT and induce responses in lymphocytes there. These responses initiate, through multiple pathways, intestinal immune and inflammatory responses that generate various effector substances (thick, solid arrows) for the removal of the infectious agents. GALT, Gut-associated lymphoid tissue; IEL, intraepithelial lymphocyte. (From Befus and Bienenstock, 1982*a*, with permission of Karger.)

nonspecific events that will eliminate or damage other concurrent infections (Behnke *et al.*, 1977; Bruce and Wakelin, 1977; Dineen *et al.*, 1977; Kennedy, 1980) and that many parasites expelled from a host are not killed (e.g., Hopkins and Zajac, 1976; Wakelin and Wilson, 1979; Kennedy and Bruce, 1981), but are fully infective when transferred to naive hosts. The antigen-specific immune responses are essential to initiate and facilitate the generation of these nonspecific effectors and for the rapid recall or the generation of other effectors upon reexposure to the parasites (Mitchell *et al.*, 1982). During the next few years new information will be uncovered about the regulation by gastrointestinal hormones, neuroendocrine substances, and inflammatory mediators of the immunologically specific and the nonspecific events during infection and parasite elimination.

B. Host Pathology

In the continuing evolution of any host-parasite relationship, survival of both parties as a species is the primary objective. Thus, the parasite must develop strategies either (1) for rapid reproduction before the onslaught of host resistance that affects fecundity and survival, or (2) for persistent, relatively nonpathogenic infection of the host. The host, in turn, must eliminate a potentially pathogenic parasite or develop mechanisms to minimize pathologic events. Elsewhere, we have discussed in detail strategies adopted by parasites to facilitate the survival in immunized hosts (Befus and Bienenstock, 1982*a*). In this section we briefly consider some of the host immunolgic strategies to minimize pathologic events.

When an antigen is administered orally, a state of unresponsiveness to subsequent reexposure at systemic sites is often generated (Tomasi, 1980). This state is called "oral tolerance." We understand that oral tolerance can involve the generation of IgG- and IgE-specific suppressor cells in the PPs, which migrate to systemic sites (e.g., Richman *et al.*, 1981*b*; Section II.B.2.b). A recent report has suggested that other, unknown mechanisms are involved as well (Hanson and Miller, 1982). Concurrently, IgA helper cells are produced that mediate local IgA responses. To date no studies have been reported on the effects of intestinal infection on the generation or maintenance of oral tolerance. We believe that oral tolerance may be an important strategy adopted by the host to minimize systemic immune responses to mucosal parasites. This would reduce the possibility that mucosally derived antigens, especially in persistent infections, would initiate systemic immune complex-mediated disease. Moreover, the role of IgA antibody in clearing antigens from the circulation by hepatic transport into bile (Bienenstock and Befus, 1982) provides an additional safeguard.

Inflammatory responses are modulated in part by a variety of substances collectively called acute-phase reactants, which include α_1-proteinase inhibitor, serum amyloid A and P, complement components, C-reactive protein, α_2-

macroglobulin and many others (Fauve, 1980). The functions of these reactants are diverse and incompletely known (Befus and Bienenstock, 1982*a*), but many inhibit potentially tissue-damaging effects of inflammation. Acute-phase reactants are beginning to be investigated in parasitic infections (e.g., Pepys *et al.*, 1980; Gauldie *et al.*, 1983*b*); knowledge of their activities will be fundamentally important in dissecting the regulation of host responses to infection. Similarly, the study of intestinal neuropeptides and their effects on lymphocytes or other inflammatory cells (O'Dorisio *et al.*, 1981) promises to uncover many more connections in the network of regulation of mucosal immune and inflammatory responses.

V. CONCLUSIONS

The past 5 years have seen many developments in the fields of immunology, parasitology, and gastrointestinal neuroendocrine physiology. We have attempted to review some of the conceptual advances and to identify areas of interface among the fields that will uncover significant new knowledge of gastrointestinal responses to infection. The increasing awareness of the impact of parasitic diseases and the usefulness of model infections to study host responses promise to continue to drive the development of our knowledge. The area of genetics of resistance is rapidly evolving (Wakelin, 1978*b*; Skamene *et al.*, 1980; Mitchell *et al.*, 1982) and will provide new tools to dissect relevant responses in genetically distinct populations. This knowledge. together with the new tools and technologies, guarantee that the next decade will bring new understanding of the mechanisms of host resistance in intestinal and other mucosal sites.

ACKNOWLEDGMENTS

The Medical Research Council of Canada has provided research support for the authors. D.B. is a fellow with the Great Neglected Disease Program of the Rockefeller Foundation. We wish to thank Ms. N. Lyons and Ms. J. Robertson for their enthusiastic and meticulous assistance in the preparation of this manuscript.

VI. REFERENCES

Allan, W., and Mayrhofer, G., 1981, The distribution and traffic of specific homocytotropic antibody-synthesizing cells in *Nippostrongylus brasiliensis*-infested rats. Comparison of synthesis with the kinetics of antibody and IgE levels in serum and lymph, *Aust. J. Exp. Biol. Med. Sci.* **59**:723.

Allan, W. D., and Porter, P., 1977, The relative frequencies and distribution of immunglobulin-bearing cells in the intestinal mucosa of neonatal and weaned pigs and their significance in the development of secretory immunity, *Immunology* **32**:819.

Arnaud-Battandier, F., 1982, Immunologic characteristics of isolated gut mucosal lymphoid cells, in: *Recent Advances in Mucosal Immunity* (W. Strober, L. A. Hanson, and K. W. Sell, eds.), pp. 289-299, Raven Press, New York.

Arnaud-Battandier, F., Bundy, B. M., O'Neill, M., Bienenstock, J., and Nelson, D. L., 1978, Cytotoxic activities of gut mucosal lymphoid cells in guinea pigs, *J. Immunol.* **121**:1059.

Askenase, P. W., 1980, Immunopathology of parasitic diseases: Involvement of basophils and mast cells, *Sem. Immunopathol.* **2**:417.

Askenase, P. W., Bursztajn, S., Gershon, M. D., and Gershon, R. K., 1980, T-cell dependent mast cell degranulation and release of serotonin in murine delayed-type hypersensitivity, *J. Exp. Med.* **152**:1358.

Banks, K. L., 1982, Host defense in the newborn animal, *Am. J. Vet. Res.* **181**:1053.

Barth, E. E. E., Jarrett, W. F. H., and Urquhart, G. M., 1966, Studies on the mechanism of the self-cure reaction in rats infected with *Nippostrongylus brasiliensis*, *Immunology* **10**:459.

Basten, A., and Beeson, P. B., 1970, Mechanisms of eosinophilia. II. Role of the lymphocyte, *J. Exp. Med.* **131**:1288.

Beaudoin, R. L., *et al.*, 1981, Antigenic structure and related aspects of the biology of plasmodia: The present situation, *Bull. WHO* **59**:371.

Beaven, M. A., 1978, Histamine: Its role in physiological and pathological processes, in: *Monographs in Allergy*, Vol. 13, S. Karger, Basel, Munchen, Paris, London, New York, Sydney.

Befus, A. D., 1977, *Hymenolepis diminuta* and *H. microstoma*; mouse immunoglobulins binding to the tegumental surface, *Exp. Parasitol.* **41**:242.

Befus, A. D., and Bienenstock, J., 1979, Immunologically-mediated intestinal mastocytosis in *Nippostrongylus brasiliensis* infected rats, *Immunology*, **38**:95.

Befus, A. D., and Bienenstock, J., 1982*a*, Factors involved in symbiosis and host resistance at the mucosa–parasite interface, *Prog. Allergy* **31**:76.

Befus, A. D., and Bienenstock, J., 1982*b*, The mucosa-associated immune system of the rabbit, in: *Animal Models of Immunological Processes* (J. B. Hay, ed.), pp. 167–220, Academic Press, New York, London.

Befus, A. D., Pearce, F. L., Gauldie, J., Horsewood, P., and Bienenstock, J., 1982*a*, Mucosal mast cells. I. Isolation and functional characteristics of rat intestinal mast cells, *J. Immunol.* **128**:2475.

Befus, A. D., Pearce, F. L., Goodacre, R., and Bienenstock, J., 1982*b*, Unique functional characteristics of mucosal mast cells, *Adv. Exp. Med. Biol.* **149**:521.

Befus, A. D., Johnston, N., Berman, L., and Bienenstock, J., 1982*c*, Relationship between tissue sensitization and IgE antibody production in rats infected with the nematode, *Nippostrongylus brasiliensis*, *Int. Arch. Allergy Appl. Immunol.* **67**:213.

Behnke, J. M., Bland, P. W., and Wakelin, D., 1977, Effect of the expulsion phase of *Trichinella spiralis* on *Hymenolepis diminuta* infection in mice, *Parasitology* **75**:79.

Bellamy, J. E. C., and Nielsen, N. O., 1974, Immune-mediated emigration of neutrophils into the lumen of the small intestine, *Infect. Immun.* **9**:615.

Beyer, E. C., and Barondes, S. H., 1982, Secretion of endogenous lectin by chicken intestinal goblet cells, *J. Cell Biol.* **92**:28.

Bhalla, D. K., and Owen, R.L., 1982, Cell renewal and migration in lymphoid follicles of Peyer's patches and cecum–An autoradiographic study in mice, *Gastroenterology* **82**:232.

Bienenstock, J., and Befus, A. D., 1980, Mucosal immunology, *Immunology* **41**:249.

Bienenstock, J., and Befus, A. D., 1982, Some thoughts on the biologic role of immunoglobulin A, *Gastroenterology* **84**:178.

Bienenstock, J., Johnston, N., and Perey, D. Y. E., 1973, Bronchial lymphoid tissue. I. Morphological characteristics, *Lab. Invest.* **28**:686.

Bienenstock, J., Befus, A. D., Pearce, F., Denburg, J., and Goodacre, R., 1982, Mast cell heterogeneity: Derivation and function with emphasis on the intestine, *J. Allergy Clin. Immunol.* **70**:407.

Bockman, D. E., and Cooper, M. D., 1973, Pinocytosis by epithelium associated with lymphoid follicles in the bursa of Fabricius, appendix, and Peyer's patches. An electron microscopic study, *Am. J. Anat.* **136**:455.

Boltz-Nitulescu, B., and Spiegelberg, H. L., 1981, Receptors specific for IgE on rat alveolar and peritoneal macrophages, *Cell. Immunol.* **59**:106.

Boltz-Nitulescu, G., Plummer, J. M., and Spiegelberg, H. L., 1982, Fc receptors for IgE on mouse macrophages and macrophage-like cell lines, *J. Immunol.* **128**:2265.

Boros, D. L., Amsden, A. F., and Hood, A. T., 1982, Modulation of granulomatous hypersensitivity. IV. Immunoglobulin and antibody production by vigorous and immunomodulated liver granulomas of *Schistosoma mansoni*-infected mice, *J. Immunol.* **128**:1050.

Brandtzaeg, P., and Baklien, K., 1976, Immunoglobulin-producing cells in the intestine in health and disease, *Clin. Gastroenterol.* **5**:251.

Brown, P. J., 1979, The development of immunoglobulin-containing cell populations in intestine, spleen, and mesenteric lymph node of the young rat, *Vet. Immunol. Immunopathol.* **1**:15.

Brown, P. J., and Bourne, F. J., 1976, Development of immunoglobulin-containing cell populations in intestine, spleen, and mesenteric lymph node of the young pig as demonstrated by peroxidase-conjugated antiserums, *Am. J. Vet. Res.* **37**:1309.

Bruce, R. G., and Wakelin, D., 1977, Immunological interactions between *Trichinella spiralis* and *Trichuris muris* in the intestine of the mouse, *Parasitology* **74**:163.

Cantey, J. F., and Inman, L. R., 1981, Diarrhea due to *Escherichia coli* strain RDEC-1 in the rabbit: The Peyer's patch as the initial site of attachment and colonization, *J. Infect. Dis.* **143**:440.

Capron, A., Dessaint, J.-P., Capron, M., and Bazin, H., 1975, Specific IgE antibodies in immune adherence of normal macrophages to *Schistosoma mansoni* schistosomules, *Nature (Lond.)* **253**:474.

Capron, A., Dessaint, J.-P., Capron, M., Joseph, M., and Torpier, G., 1982, Effector mechanisms of immunity to schistosomes and their regulation, *Immunol. Rev.* **61**:41.

Carpen, O., Virtanen, I., and Saksela, E., 1982, Ultrastructure of human natural killer cells: Nature of the cytolytic contacts in relation to cellular secretion, *J. Immunol.* **128**:2691.

Castro, G., 1981, Physiology of the gastrointestinal tract in the parasitized host, in: *Physiology of the Gastrointestinal Tract* (L. R. Johnson, ed.), pp. 1381–1406, Raven Press, New York.

Castro, G. A., and Harari, Y., 1982, Intestinal epithelial membrane changes in rats immune to *Trichinella spiralis*, *Mol. Biochem. Parasitol.* **6**:191.

Cebra, J. J., Gearhart, P. J., Halsey, J. F., Hurwitz, J. L., and Shahin, R. D., 1980, Role of environmental antigens in the ontogeny of the secretory immune response, *J. Reticuloendothal. Soc.* **28**(suppl.):61.

Cepica, A., 1982, Antibody-dependent cell-mediated cytotoxicity and spontaneous cell-mediated cytotoxicity in porcine transmissible gastroenteritis, Ph.D. thesis, pp. 177, University of Guelph.

Cheng, H., and Leblond, C. P., 1974, Origin, differentiation and renewal of the four main epithelial cell types in the mouse small intestine. V. Unitarian theory of the origin of the four epithelial cell types, *Am. J. Anat.* **141**:537.

Chu, R. M., Glock, R. D., and Ross, R. F., 1982, Changes in gut-associated lymphoid tissues of the small intestine of eight week old pigs infected with transmissible gastroenteritis virus, *Am. J. Vet. Res.* **43**:67.

Clegg, J. A., and Smith, M. A., 1978, Prospects for the development of dead vaccines against helminths, *Adv. Parasitol.* **16**:165.

Colley, D. G., 1981, T lymphocytes that contribute to the immunoregulation of granuloma formation in chronic murine schistosomiasis, *J. Immunol.* **126**:1465.

Colten, H. R., Gordon, J. M., Rapp, H. J., and Borsos, T., 1968, Synthesis of the first component of guinea pig complement by columnar epithelial cells of the small intestine, *J. Immunol.* **100**:788.

Crabbé, P. A., Nash, D. R., Bazin, H., Eyssen, H., and Heremans, J. F., 1970, Immunohistochemical observations on lymphoid tissues from conventional and germ-free mice, *Lab. Invest.* **22**:448.

Crowle, P. K., and Reed, N. D., 1981, Rejection of the intestinal parasite *Nippostrongylus brasiliensis* by mast cell deficient W/W^{V} anemic mice, *Infect. Immun.* **33**:54.

Danovitch, S. H., Gallucci, A., and Shora, W., 1972, Colonic mucosal lysosomal enzyme activities in ulcerative colitis, *Am. J. Dig. Dis.* **17**:977.

Davies, M. D. J., and Parrott, D. M. V., 1981*a*, Cytotoxic T cell in small intestine epithelial, lamina propria and lung lymphocytes, *Immunology* **44**:367.

Davies, M. D. J., and Parrott, D. M. V., 1981*b*, Preparation and purification of lymphocytes from the epithelium and lamina propria of murine small intestine, *Gut* **22**:481.

Denburg, J. A., Befus, A. D., and Bienenstock, J., 1980, Growth and differentiation *in vitro* of mast cells from mesenteric lymph nodes of *Nippostrongylus brasiliensis* infected rats, *Immunology* **41**:195.

Dessein, A. J., Parker, W. L., James, S. L., and David, J. R., 1981, IgE antibody and resistance to infection. I. Selective suppression of the IgE antibody response in rats diminishes the resistance and the eosinophil response to *Trichinella spiralis* infection, *J. Exp. Med.* **153**:423.

Dineen, J. K., Gregg, P., Windon, R. G., Donald, A. D., and Kelly, J. D., 1977, The role of immunologically specific and non-specific components of resistance in cross-protection to intestinal nematodes, *Int. J. Parasitol.* **7**:211.

Durkin, H. G., Bazin, H., and Waksman, B., 1981, Origin and fate of IgE-bearing lymphocytes. I. Peyer's patches as differentiation site of cells simultaneously bearing IgA and IgE, *J. Exp. Med.* **154**:640.

Dwyer, J. M., and Mackay, I. R., 1972, The development of antigen-binding lymphocytes in foetal tissues, *Immunology* **23**:871.

Egwang, T. G., Gauldie, J. and Befus, A. D., 1983, Complement-dependent killing of *Nippostrongylus brasiliensis* infective larvae by rat alveolar macrophages, *Clin. Exp. Immunol.* (in press).

Enerbäck, L., 1981, The gut mucosal mast cell, *Monogr. Allergy* **17**:222.

Elson, C. O., Heck, J. A., and Strober, W., 1979, T-cell regulation of murine IgA synthesis, *J. Exp. Med.* **149**:632.

Erlandsen, S. L., Rodning, C. B., Montero, C., Parsons, J. A., Lewis, E. A., and Wilson, I. D., 1976, Immunocytochemical identification and localization of immunoglobulin A within Paneth cells of the rat small intestine, *J. Histochem. Cytochem.* **24**:1085.

Etzler, M. E., and Branstrator, M. L., 1974, Differential localization of cell surface and secretory components in rat intestinal epithelium by use of lectins, *J. Cell Biol.* **62**:329.

Farram, E., and Nelson, D. S., 1980, Mouse mast cells as anti-tumor effector cells, *Cell. Immunol.* **55**:294.

Fauve, R. M., 1980, Inflammation and natural immunity, in: *Progress in Immunology*, Vol. IV (M. Fougereau and J. Dausset, eds.), pp. 737–756, Academic Press, London.

Ferguson, A., 1977, Intraepithelial lymphocytes of the small intestine, *Gut* **18**:921.

Ferguson, A., and Jarrett, E. E. E., 1975, Hypersensitivity reactions in the small intestine. I. Thymus dependence of experimental "partial villous atrophy," *Gut* **16**:114.

Fjellner, B., and Hagermark, O., 1981, Studies on pruritogenic and histamine-releasing effects of some putative peptide neurotransmitters, *Acta Dermatovener (Stockholm)* **61**:245.

Florey, H. W., 1933, Observations on the functions of mucus and the early stages of bacterial invasion of the intestinal mucosa, *J. Pathol. Bacteriol.* **37**:283.

Frangakis, M. V., Koopman, W. J., Kiyono, H., Michalek, S. M., and McGhee, J. R., 1982, An enzymatic method for preparation of dissociated murine Peyer's patch cells enriched for macrophages, *J. Immunol. Meth.* **48**:33.

Frederick, G. T., and Bohl, E. H., 1976, Local and systemic cell-mediated immunity against transmissible gastroenteritis, an intestinal viral infection of swine, *J. Immunol.* **116**:1000.

Freeman, H. J., Lotan, R., and Kim, Y. S., 1980, Application of lectins for detection of goblet cell glycoconjugate differences in proximal and distal colon of the rat, *Lab. Invest.* **42**:405.

Frick, L. P., and Ackert, J. E., 1948, Further studies on duodenal mucus as a factor in age resistance of chickens to parasitism, *J. Parasitol.* **34**:192.

Gauldie, J., Richards, C., and Lamontagne, L. 1983*a*. Fc receptors for IgA and other immunoglobulins on resident and activated alveolar macrophages, *Mol. Immunol.* (in press).

Gauldie, J., Lamontagne, L., and Befus, A. D., 1983*b*, The acute phase response in parasite infection. I. Alpha-1-antitrypsin in *Nippostrongylus brasiliensis* infection of the mouse (submitted).

Gee, S. R., Chan, M. A., Clark, D. A., and Rawls, W. E., 1981, Role of natural killer cells in pichinde virus infection of Syrian hamsters, *Infect. Immun.* **31**:919.

Green, D. R., Gold, J., St. Martin, S., Gershon, R., and Gershon, R. K., 1982, Microenvironmental immunoregulation: Possible role of contrasuppressor cells in maintaining immune responses in gut-associated lymphoid tissues, *Proc. Natl. Acad. Sci. USA* **79**:889.

Griffiss, J. M., 1975, Bactericidal activity of meningococcal antisera. Blocking by IgA of lytic antibody in human convalescent sera, *J. Immunol.* **114**:1779.

Grossi, C. E., Cadoni, A., Zicca, A., Leprini, A., and Ferrarini, M., 1982, Large granular lymphocytes in human peripheral blood: Ultrastructural and cytochemical characterization of the granules, *Blood* **59**:277.

Guy-Grand, D., and Vassalli, P., 1982, Nature and function of gut granulated T lymphocytes, in: *Recent Advances in Mucosal Immunity* (W. Strober, L. A. Hanson, and K. W. Sell, eds), pp. 301–312, Raven Press, New York.

Guy-Grand, D., Griscelli, C., and Vassalli, P., 1978, The mouse gut T lymphocyte, a novel type of T cell. Nature, origin and traffic in mice in normal and graft-versus-host conditions, *J. Exp. Med.* **148**:1661.

Hanson, D. G., and Miller, S. D., 1982, Inhibition of specific immune responses by feeding protein antigens. V. Induction of the tolerant state in the absence of specific suppressor T cells, *J. Immunol.* **128**:2378.

Heatley, R. V., and Bienenstock, J., 1982, Luminal lymphoid cells in the rabbit intestine, *Gastroenterology* **82**:268.

Henderson, W. R., Chi, E. Y., Jong, E. C., and Klebanoff, S. J., 1981, Mast cell-mediated tumor-cell cytotoxicity. Role of the peroxidase system, *J. Exp. Med.* **153**:520.

Hill, I. R., and Porter, P., 1974, Studies of bacterial activity to *Escherichia coli* of porcine serum and colostral immunoglobulins and the role of lysozyme with secretory IgA, *Immunology* **26**:1239.

Hollingdale, M. R., Zavala, F., Nussenzweig, R. S., and Nussenzweig, V., 1982, Antibodies to the protective antigens of *Plasmodium berghei* sporozoites prevent entry into cultured cells, *J. Immunol.* **128**:1929.

Holmgren, J., 1981, Actions of cholera toxin and the prevention and treatment of cholera, *Nature (Lond.)* **292**:413.

Hopkins, C. A., and Zajac, A., 1976, Transplantation of *Hymenolepis diminuta* into naive, immune and irradiated mice, *Parasitology* **73**:73.

Huntley, J., Newby, T. J., and Bourne, F. J., 1979, The cell-mediated immune response of the pig to orally administered antigen, *Immunology* **37**:225.

Husband, A. J., 1980, Role of gram negative bacteria in ontogeny of gut immunity, *Aust. J. Exp. Biol. Med. Sci.* **58**:297.

Husband, A. J., 1982, Kinetics of extravasation and redistribution of IgA-specific antibody-containing cells in the intestine, *J. Immunol.* **128**:1355.

Ihle, J. N., Rebar, L., Keller, J., Lee, J. T., and Hapel, A., 1982, Interleukin 3: Possible roles in the regulation of lymphocyte differentiation and growth, *Immunol. Rev.* **63**:5.

Ismail, M. M., and Tanner, C. E., 1972, *Trichinella spiralis*: Peripheral blood, intestinal, and bone-marrow eosinophilia in rats and its relationship to the inoculating dose of larvae, antibody response and parasitism, *Exp. Parasitol.* **31**:262.

Jacquiline, E., Vernes, A., Bout, D., and Biguet, J., 1978, *Trichinella spiralis*: Facteurs immunitaires inhibiteurs de la production de larves. II. Première analyse *in vitro* des facteurs humoraux et secretoires actifs chez la souris, le rat, et le miniporc infestés ou immunisés, *Exp. Parasitol.* **45**:42.

Janossy, G., Tidman, N., Selby, W. S., Thomas, J. A., Granger, S., Kung, P. C., and Goldstein, G., 1980, Human T lymphocytes of inducer and suppressor type occupy different microenvironments, *Nature (Lond.)* **288**:81.

Jarrett, E. E. E., and Miller, H. R. P., 1982, The production and activities of IgE in helminth infection, *Prog. Allergy* **31**:178.

Johnson, A. R., and Erdos, E. G., 1973, Release of histamine from mast cells by vasoactive peptides, *Proc. Soc. Exp. Biol. Med.* **142**:1252.

Jones, C. E., and Rubin, R., 1974, *Nematospiroides dubius*: Mechanisms of host immunity. I. Parasite counts, histopathology and serum transfer involving orally or subcutaneously sensitized mice, *Exp. Parasitol.* **35**:434.

Jones, G. W., and Freter, R., 1976, Adhesive properties of *Vibrio cholerae*: Nature of the interaction with isolated rabbit brush border membranes and human erythrocytes, *Infect. Immun.* **14**:240.

Joseph, M., Tonnel, A. B., Capron, A., and Voisin, C., 1980, Enzyme release and superoxide anion production by human alveolar macrophages stimulated with immunoglobulin E, *Clin. Exp. Immunol.* **40**:416.

Kagnoff, M. F., and Campbell, S., 1974, Functional characteristics of Peyer's patch lymphoid cells. I. Induction of humoral antibody and cell-mediated allograft reactions, *J. Exp. Med.* **139**:398.

Kawanishi, H., and Strober, W., 1983, Regulatory T-cells in murine peyer's patches directing IgA-specific isotype switching, *Ann. N.Y. Acad. Sci.* **149**:243.

Kazura, J. W., and Grove, D. I., 1978, Stage-specific antibody-dependent eosinophil-mediated destruction of *Trichinella spiralis*, *Nature (Lond.)* **274**:588.

Kelly, J. D., and Ogilvie, B. M., 1972, Intestinal mast cell and eosinophil numbers during worm expulsion in nulliparous and lactating rats infected with *Nippostrongylus brasiliensis*, *Int. Arch. Allergy* **43**:497.

Kemp, A. S., Cripps, A. W., and Brown, S., 1980, Suppression of leucocyte chemokinesis and chemotaxis by human IgA, *Clin. Exp. Immunol.* **40**:388.

Kennedy, M. W., 1980, Immunologically mediated, non-specific interactions between the intestinal phases of *Trichinella spiralis* and *Nippostrongylus brasiliensis* in the mouse, *Parasitology* **80**:61.

Kennedy, M. W., and Bruce, R. G., 1981, Reversibility of the effects of the host immune response on the intestinal phase of *Trichinella spiralis* in the mouse, following transplanation to a new host, *Parasitology* **82**:39.

Kiyono, H., McGhee, J. R., Wannemuehler, M. J., Frangakis, M. V., Spalding, D. M., Michalek, S. M., and Koopman, W. J., 1982, *In vitro* immune responses to a T-cell-dependent antigen by cultures of disassociated murine Peyer's patch, *Proc. Natl. Acad. Sci. USA* **79**:596.

Klebanoff, S. J., Jong, E. C., and Henderson, W. R., Jr., 1980, The eosinophil peroxidase: Purification and biological properties, in: *The Eosinophil in Health and Disease* (A. A. F. Mahmoud and K. F. Austen, eds.), pp. 99–114, Grune & Stratton, New York.

Kojima, S., Kitamura, Y., and Takatsu, K., 1980, Prolonged infection of *Nippostrongylus brasiliensis* in genetically mast cell-depleted W/W^{v} mice, *Immunol. Lett.* **2**:159.

Kurose, M., and Saeki, K., 1981, Histamine release induced by neurotensin from rat peritoneal mast cells, *Eur. J. Pharmacol.* **76**:129.

Kutteh, W. H., Prince, S. J., and Mestecky, J., 1982, Tissue origins of human polymeric and monomeric IgA, *J. Immunol.* **128**:990.

Lake, A. M., Bloch, K. J., Sinclair, K. J., and Walker, W. A., 1980, Anaphylactic release of intestinal goblet cell mucus, *Immunology* **38**:173.

Larsh, J. E., Jr., and Race, G. J., 1975, Allergic inflammation as a hypothesis for the expulsion of worms from tissues: A review, *Exp. Parasitol.* **37**:251.

Latham, L., Latham, M., and Basta, S. S., 1977, The nutritional and economic implications of *Ascaris* infection in Kenya, *World Bank Staff Working Paper No. 271*, Sept. 1977, 86 pp., World Bank, Washington, D.C.

Lee, T. D. G., and Wakelin, D., 1982, The use of host strain variation to assess the significance of mucosal mast cells in the spontaneous cure response of mice to the nematode *Trichuris muris*, *Int. Arch. Allergy Appl. Immunol.* **67**:302.

LeFevre, M. E., Hammer, R., and Joel, D. D., 1979, Macrophages of the mammalian small intestine: A review, *J. Reticuloendothel. Soc.* **26**:553.

Leid, R. W., and Williams, J. F., 1974, The immunological response of the rat to infection with *Taenia taeniaeformis.* I. Immunoglobulin classes involved in passive transfer of resistance, *Immunology* **27**:195.

Lemaître-Coelho, I., Jackson, G. D. F., and Vaerman, J.-P., 1978, Relevance of biliary IgA antibodies in rat intestinal immunity, *Scand. J. Immunol.* **8**:459.

Lipsky, P. E., and Kettman, J. R., 1982, Accessory cells unrelated to mononuclear phagocytes and not of bone marrow origin, *Immunol. Today* **3**:36.

Lloyd, S., and Soulsby, E. J. L., 1978, The role of IgA immunoglobulins in the passive transfer of protection to *Taenia taeniaeformis* in the mouse, *Immunology* **34**:939.

Lowell, G. H., Smith, L. F., Griffiss, J. M., and Brandt, B. L., 1980, IgA-dependent, monocyte-mediated, antibacterial activity, *J. Exp. Med.* **152**:452.

Lyscom, N., and Brueton, M. J., 1982, Intraepithelial, lamina propria and Peyer's patch lymphocytes of the rat small intestine: Isolation and characterization in terms of immunoglobulin markers and receptors for monoclonal antibodies, *Immunology* **45**:775.

MacDonald, T. T., and Carter, P. B., 1982, Isolation and functional characteristics of adherent phagocytic cells from mouse Peyer's patches, *Immunology* **45**:769.

MacDonald, T. T., and Ferguson, A., 1978, Small intestinal epithelial cell kinetics and protozoal infection of mice, *Gastroenterology* **74**:496.

MacPherson, G. G., and Steer, H. W., 1979, Properties of mononuclear phagocytes derived from the small intestinal wall of rats, in: *Function and Structure of the Immune System* (W. Muller-Ruchholtz and K. Muller-Hermelink, eds.), pp. 433–438, Plenum, New York.

Magnusson, K.-E., and Stjernstrom, I., 1982, Mucosal barrier mechanisms. Interplay between secretory IgA (sIgA), IgG and mucins on the surface properties and association of salmonellae with intestine and granulocytes, *Immunology* **45**:239.

Magnusson, K.-E., Stendahl, O., Stjernstrom, I., and Edebo, L., 1979, Reduction of phagocytosis, surface hydrophobicity and charge of *Salmonella typhimurium* 395 MR10 by reaction with secretory IgA (sIgA), *Immunology* **36**:439.

Maizels, R. M., Philipp, M., and Ogilvie, B. M., 1982, Molecules on the surface of parasitic nematodes as probes of the immune response in infection, *Immunol. Rev.* **61**:109.

Manson-Smith, D. F., Bruce, R. G., and Parrott, D. M. V., 1979*a*, Villous atrophy and expulsion of intestinal *Trichinella spiralis* are mediated by T cells, *Cell. Immunol.* **47**:285.
Manson-Smith, D. F., Bruce, R. G., Rose, M. L., and Parrott, D. M. V., 1979*b*, Migration of lymphoblasts to the small intestine. III. Strain differences and relationship to distribution and duration of *Trichinella spiralis* infection, *Clin. Exp. Immunol.* **38**:475.
Mattingly, J. A., and Waksman, B. H., 1978, Immunologic suppression after oral administration of antigen. I. Specific suppressor cells formed in rat Peyer's patches after oral administration of sheep erythrocytes and their systemic migration, *J. Immunol.* **121**:1878.
Mattingly, J. A., Eardley, D. D., Kemp, J. D., and Gershon, R. K., 1979, Induction of suppressor cells in rat spleen: Influence of microbial stimulation, *J. Immunol.* **122**:787.
Mayrhofer, G., 1980, Thymus-dependent and thymus-independent subpopulations of intestinal intraepithelial lymphocytes: A granular subpopulation of probable bone marrow origin and relationship to mucosal mast cells, *Blood* **55**:532.
Mayrhofer, G., and Fisher, R., 1979, Mast cells in severely T-cell depleted rats and the response to infestation with *Nippostrongylus brasiliensis*, *Immunology* **37**:145.
Mayrhofer, G., Bazin, H., and Gowans, J. L., 1976, Nature of cells binding anti-IgE in rats immunized with *Nippostrongylus brasiliensis*: IgE synthesis in regional nodes and concentration in mucosal mast cells, *Eur. J. Immunol.* **6**:537.
McDermott, M. R., and Bienenstock, J., 1979, Evidence for a common mucosal immunological system. I. Migration of B immunoblasts into intestinal, respiratory and genital tissues, *J. Immunol.* **122**:1892.
McDermott, M. R., Befus, A. D., and Bienenstock, J., 1982, The structural basis for immunity in the respiratory tract, *Int. Rev. Exp. Pathol.* **23**:47.
McElroy, P. J., Szewczuk, M. R., and Befus, A. D., 1982, Regulation of heterologous IgM, IgG and IgA antibody responses in mucosal-associated lymphoid tissues of *Nippostrongylus brasiliensis*-infected mice, *J. Immunol.* **130**:435.
McGhee, J. R., Kiyono, H., Michalek, S. M., Babb, J. L., Rosenstreich, D. L., and Mergenhagen, S. E., 1980, Lipopolysaccharide (LPS) regulation of the immune response: T lymphocytes from normal mice suppress mitogenic and immunogenic responses to LPS, *J. Immunol.* **124**:1603.
Mestecky, J., McGhee, J. R., Crago, S. S., Jackson, S., Kilian, M., Kiyono, H., Babb, J. L., and Michalek, S. M., 1980, Molecular-cellular interactions in the secretory IgA response, *J. Reticuloendothel. Soc.* **28**(Suppl.):45.
Millard, B. J., and Lawn, A. M., 1982, Parasite–host relationships during the development of *Eimeria dispersa* Tyzzer 1929, in the turkey (*Meleagris gallopavo gallopavo*) with a description of intestinal intraepithelial leukocytes, *Parasitology* **84**:13.
Miller, H. R. P., 1980, The structure, origin and function of mucosal mast cells. A brief review, *Biol. Cell.* **39**:229.
Miller, H. R. P., and Jarrett, W. F. H., 1971, Immune reactions in mucous membranes. I. Intestinal mast cell response during helminth expulsion in the rat, *Immunology* **20**:277.
Miller, H. R. P., and Nawa, Y., 1979*a*, *Nippostrongylus brasiliensis*: Intestinal goblet-cell response in adoptively immunized rats, *Exp. Parasitol.* **47**:81.
Miller, H. R. P., and Nawa, Y., 1979*b*, Immune regulation of intestinal goblet cell differentiation. Specific induction of nonspecific protection against helminths?, *Nouv. Rev. Fr. Hematol.* **21**:31.
Miller, H. R. P., and Walshaw, R., 1972, Immune reactions in mucous membranes. IV. Histochemistry of intestinal mast cells during helminth expulsion in the rat, *Am. J. Pathol.* **69**:195.
Miller, H. R. P., Nawa, Y., and Parish, C. R., 1979, Intestinal goblet cell differentiation in *Nippostrongylus*-infected rats after transfer of fractionated thoracic duct lymphocytes, *Int. Arch. Allergy Appl. Immunol.* **59**:281.
Miller, H. R. P., Huntley, J. F., and Wallace, G., 1981, Immune exclusion and mucus-trapping

during the rapid expulsion of *Nippostrongylus brasiliensis* from primed rats, *Immunology* **44**:419.

Minato, N., Reid, L., and Bloom, B. R., 1981, On the heterogeneity of murine natural killer cells, *J. Exp. Med.* **154**:750.

Mitchell, G. F., 1980, T cell dependent effects in parasitic infection and disease, in: *Progress in Immunology*, Vol. IV (M. Fougereau and J. Dausset, eds.), pp. 794–808, Academic Press, London.

Mitchell, G. F., Anders, R. F., Brown, G. V., Handman, E., Roberts-Thomson, I. C., Chapman, C. B., Forsyth, K. P., Kahl, L. P., and Cruise, K. M., 1982, Analysis of infection characteristics and antiparasite immune responses in resistant compared to susceptible hosts, *Immunol. Rev.* **61**:137.

Moqbel, R., 1980, Histopathological changes following primary, secondary and repeated infections of rats with *Strongyloides ratti*, with special reference to tissue eosinophils, *Parasite Immunol.* **2**:11.

Moxey, R. C., and Trier, J. S., 1978, Specialized cell types in the human fetal small intestine, *Anat. Rec.* **191**:269.

Murray, M., Miller, H. R. P., and Jarrett, W. F. H., 1968, The globule leukocyte and its derivation from the subepithelial mast cell, *Lab. Invest.* **19**:222.

Namba, Y., and Waksman, B. H., 1975*a*, Regulatory substances produced by lymphocytes. II. Lymphotoxin in the rat, *J. Immunol.* **115**:1018.

Namba, Y., and Waksman, B. H., 1975*b*, Regulatory substances produced by lymphocytes. I. Inhibitor of DNA synthesis in the rat, *Inflammation* **1**:5.

Nawa, Y., and Miller, H. R. P., 1979, Adoptive transfer of the intestinal mast cell response in rats infected with *Nippostrongylus brasiliensis*, *Cell. Immunol.* **42**:225.

Ngan, J., and Kind, L. S., 1978, Suppressor T cells for IgE and IgG in Peyer's patches of mice made tolerant by the oral administration of ovalbumin, *J. Immunol.* **120**:861.

O'Dorisio, M. S., Hermina, N. S., O'Dorisio, T. M., and Balcerzak, S. P., 1981, Vasoactive intestinal polypeptide modulation of lymphocyte adenylate cyclase, *J. Immunol.* **127**:2551.

Ofek, I., Mirelman, D., Sharon, N., 1977, Adherence of *Escherichia coli* to human mucosal cells mediated by mannose receptors, *Nature (Lond.)* **265**:623.

Ogilvie, B. M., 1967, Reagin-like antibodies in rats infected with the nematode parasite *Nippostrongylus brasiliensis*, *Immunology* **12**:113.

Ogilvie, B. M., and Love, R. J., 1974, Co-operation between antibodies and cells in immunity to a nematode parasite, *Transplant. Rev.* **19**:147.

Olds, G. R., Olveda, R., Tracy, J. W., and Mahmoud, A. A. F., 1982, Adoptive transfer of modulation of granuloma formation and hepatosplenic disease in murine *Schistosomiasis japonica* by serum from chronically infected animals, *J. Immunol.* **128**:1391.

Ottaway, C. A., Rose, M. L., and Parrott, D. M. F., 1979, the gut as an immunological system, in: *International Review of Physiology. Gastrointestinal Physiology. III*, Vol. 19, (R. K. Crane, ed.), pp. 323–356, University Park Press, Baltimore.

Ottaway, C. A., Manson-Smith, D. F., Bruce, R. G., and Parrott, D. M. V., 1980, Regional blood flow and the localization of lymphoblasts in the small intestine of the mouse. II. The effect of a primary enteric infection with *Trichinella spiralis*, *Immunology* **41**:963.

Owen, R. L., 1977, Sequential uptake of horseradish peroxidase by lymphoid follicle epithelium of Peyer's patches in the normal unobstructed mouse intestine: an ultrastructural study, *Gastroenterology* **72**:440.

Owen, R. L., Nemanic, P. C., and Stevens, D. P., 1979, Ultrastructural observations on giardiasis in a murine model. I. Intestinal distribution attachment and relationship to the immune system of *Giardia muris*, *Gastroenterology* **76**:757.

Owen, R. L., Allen, C. L., and Stevens, D. P., 1981, Phagocytosis of murine *Giardia* by macrophages in Peyer's patch epithelium, *Infect. Immun.* **33**:591.

Parkhouse, R. M. E., Philipp, M., and Ogilvie, B. M., 1981, Characterization of surface antigens of *Trichinella spiralis* infective larvae, *Parasite Immunol.* **3**:339.

Pearce, F. L., Befus, A. D., Gauldie, J., and Bienenstock, J., 1982, Mucosal mast cells. II. Effects of anti-allergic compounds on histamine secretion by isolated intestinal mast cell, *J. Immunol.* **128**:2481.

Peppard, J., Orlans, E., Payne, A. W. R., and Andrew, E., 1981, The elimination of circulating complexes containing polymeric IgA by excretion in the bile, *Immunology* **42**:83.

Pepys, M. B., Baltz, M. L., Mussalam, R., and Doenhoff, M. J., 1980, Serum protein concentrations during *Schistosoma mansoni* infection in intact and T-cell deprived mice. I. The acute phase proteins, C3 and serum amyloid P-component (SAP), *Immunology* **39**:249.

Pery, P., and Luffau, G., 1979, Antigens of helminths, in: *The Antigens*, Vol. V (M. Sela, ed.), pp. 83–172, Academic Press, New York.

Pfaffenbach, G., Lamm, M. E., and Gigli, I., 1982, Activation of the guinea pig alternative complement pathway be mouse IgA immune complexes, *J. Exp. Med.* **155**:231.

Phillips-Quagliata, J. M., Roux, M. E., Arny, M., Kelly-Hatfield, P., McWilliams, M., and Lamm, M. E., 1983, Migration and regulation of B cells in the mucosal immune system, *Ann. N.Y. Acad. Sci.* **409**:194.

Pierce, N. F., and Cray, W. C., Jr., 1982, Determinants of the localization, magnitude, and duration of a specific mucosal IgA plasma cell response in enterically immunized rats, *J. Immunol.* **128**:1311.

Porter, P., 1979, Immunobiology of the alimentary tract in relation to oral immuno-prophylaxis against enteric infections in the piglet and calf, in: *Immunology of the Gastro-intestinal Tract* (P. Asquith, ed.), pp. 268–287, Churchill Livingstone, New York.

Poulain, J., Luffau, G., and Pery, P., 1976, *Nippostrongylus brasiliensis* in the rat: Immune response in serum and intestinal secretions, *Ann. Immunol. Inst. Pasteur* **127C**:215.

Pugh, C. W., and MacPherson, G. G., 1982, Non-lymphoid cells from rat intestinal lymph, *Adv. Exp. Med. Biol.* **149**:781.

Richman, L. K., Graeff, A. S., and Strober, W., 1981*a*, Antigen presentation by macrophage-enriched cells from the mouse Peyer's patch, *Cell. Immunol.* **62**:110.

Richman, L. K., Graeff, A. S., Yarchoan, R., and Strober, W., 1981*b*, Simultaneous induction of antigen-specific IgA helper T cells and IgG suppressor T cells in the murine Peyer's patch after protein feeding, *J. Immunol.* **126**:2079.

Roberts-Thomson, I. C., and Mitchell, G. F., 1978, Giardiasis in mice. I. Prolonged infections in certain mouse strains and hypothymic (nude) mice, *Gastroenterology* **75**:42.

Roberts-Thomson, I. C., Grove, D. I., Stevens, D. P., and Warren, K. S., 1976, Suppression of giardiasis during the intestinal phase of trichinosis in the mouse, *Gut* **17**:953.

Rose, M. L., Parrott, D. M. V., and Bruce, R. G., 1976, Migration of lymphoblasts to the small intestine. I. Effect of *Trichinella spiralis* infection on the migration of mesenteric lymphoblasts and mesenteric T lymphoblasts in syngeneic mice, *Immunology* **31**:723.

Rothwell, T. L. W., Prichard, R. K., and Love, R. J., 1974, Studies on the role of histamine and 5-hydroxytryptamine in immunity against the nematode *Trichostrongylus colubriformis.* I. *in vivo* and *in vitro* effects of the amines, *Int. Arch. Allergy Appl. Immunol.* **46**:1.

Roux, M. E., McWilliams, M., Phillips-Quagliata, J. M., and Lamm, M. E., 1981, Differentiation pathway of Peyer's patch precursors of IgA plasma cells in the secretory immune system, *Cell. Immunol.* **61**:141.

Rudzik, O., and Bienenstock, J., 1974, Isolation and characteristics of gut mucosal lymphocytes, *Lab. Invest.* **30**:260.

Ruitenberg, E. J., and Elgersma, A., 1976, Absence of intestinal mast cell response in congenitally athymic mice during *Trichinella spiralis* infection, *Nature* (*Lond.*) **264**:258.
Ruitenberg, E. J., and Elgersma, A., 1979, Response of intestinal globule leucocytes in the mouse during a *Trichinella spiralis* infection and its independence of intestinal mast cells, *Br. J. Exp. Pathol.* **60**:246.
Ruitenberg, E. J., Elgersma, A., and Kruizinga, W., 1979, Intestinal mast cells and globule leucocytes: Role of the thymus on their presence and proliferation during a *Trichinella spiralis* infection in the rat, *Int. Arch. Allergy Appl. Immunol.* **60**:302.
Ruitenberg, E. J., Gustowska, L., Elgersma, A., and Ruitenberg, H. M., 1982, Effect of fixation on the light microscopical visualization of mast cells in the mucosa and connective tissue of the human duodenum, *Int. Arch. Allergy Appl. Immunol.* **67**:233.
Russell, M. W., Brown, T. A., and Mestecky, J., 1981, Role of serum IgA: Hepatobiliary transport of circulating antigen, *J. Exp. Med.* **153**:968.
Shen, L., and Fanger, M. W., 1981, Secretory IgA antibodies synergize with IgG in promoting ADCC by human polymorphonuclear cells, monocytes, and lymphocytes, *Cell. Immunol.* **59**:85.
Skamene, E., Kongshavn, P. A. L., and Landy, M. (eds.), 1980, *Genetic Control of Natural Resistance to Infection and Malignancy*, Academic Press, New York.
Smith, M. W., and Peacock, M. A., 1980, "M" cell distribution in follicle-associated epithelium of mouse Peyer's patch, *Am. J. Anat.* **159**:167.
Smith, P. D., Elson, C. O., Keister, D. B., and Nash, T. E., 1982, Human host response to *Giardia lamblia.* I. Spontaneous killing by mononuclear leukocytes *in vitro*, *J. Immunol.* **128**:1372.
Smith McMyne, P., and Strejan, G. H., 1980, Relationships between cell-mediated immunity and the IgE antibody response. I. Lymphotoxin production to DNP-*Ascaris* conjugates, *Cell. Immunol.* **54**:140.
Smithson, K. W., Millar, D. V., Jacobs, L. R., and Gray, G. M., 1981, Intestinal diffusion barrier: Unstirred water layer or membrane surface mucous coat? *Science* **214**:1241.
Socken, D. J., Simms, E. S., Nagy, B., Fisher, M. M., and Underdown, J., 1981, Transport of IgA antibody–antigen complexes by the rat liver, *Mol. Immunol.* **18**:345.
Strobel, S., Miller, H. R. P., and Ferguson, A., 1981, Human intestinal mucosal mast cells: Evaluation of fixation and staining techniques, *J. Clin. Pathol.* **34**:851.
Strober, W., Richman, L. K., and Elson, C. O., 1981, The regulation of gastrointestinal immune responses, *Immunol. Today* **2**:156.
Tagliabue, A., Luini, W., Soldateschi, D., and Boraschi, D., 1981, Natural killer activity of gut mucosal lymphoid cells in mice, *Eur. J. Immunol.* **11**:912.
Tagliabue, A., Befus, A. D., Clark, D. A., and Bienenstock, J., 1982, Characteristics of natural killer cells in the murine intestinal epithelium and lamina propria, *J. Exp. Med.* **155**:1785.
Tenovuo, J., Moldoveanu, Z., Mestecky, J., Pruitt, K. M., and Rahemtulla, B.-M., 1982, Interaction of specific and innate factors of immunity: IgA enhances the antimicrobial effect of the lactoperoxidase system against *Streptococcus mutans*, *J. Immunol.* **128**:726.
Theoharides, T. C., Betchaku, T., and Douglas, W. M., 1981, Somatostatin-induced histamine secretion in mast cells. Characterization of the effect, *Eur. J. Pharmacol.* **69**:127.
Theoharides, T. C., Bondy, P. K., Tsakalos, N. D., and Askenase, P. W., 1982, Differential release of serotonin and histamine from mast cells, *Nature* (*Lond.*) **297**:229.
Threadgold, L. T., and Befus, A. D., 1977, *Hymenolepis diminuta:* Ulstrastructural localization of immunoglobulin-binding sites on the tegument, *Exp. Parasitol.* **43**:169.

Timonen, T., Ortaldo, J. R., and Herberman, R. B., 1981, Characteristics of human large granular lymphocytes and relationship to natural killer and K cells, *J. Exp. Med.* **153:** 569.

Timonen, T., Ortaldo, J. R., and Herberman, R. B., 1982, Analysis by a single cell cytotoxicity assay of natural killer (NK) cell frequencies among human large granular lymphocytes and of the effects of interferon on their activity, *J. Immunol.* **128:**2514.

Tomasi, T. B., Jr., 1980, Oral tolerance, *Transplantation (Baltimore)* **29:**353.

Tseng, J., 1981, Transfer of lymphocytes of Peyer's patches between immunoglobulin allotype congenic mice: Repopulation of the IgA plasma cells in the gut lamina propria, *J. Immunol.* **127:**2039.

Uber, C. L., Roth, R. L., and Levy, D. A., 1980, Expulsion of *Nippostrongylus brasiliensis* by mice deficient in mast cells, *Nature (Lond.)* **287:**226.

Urban, J. F., Jr., Ishizaka, T., and Ishizaka, K., 1977, IgE formation in the rat after infection with *Nippostrongylus brasiliensis.* II. Proliferation of IgE-bearing cells in neonatally thymectomized animals, *J. Immunol.* **118:**1982.

Van Epps, D. E., and Williams, R. C., 1976, Suppression of leukocyte chemotaxis by human IgA myeloma components, *J. Exp. Med.* **44:**1227.

Van Epps, D. E., Reed, K., and Williams, R. C., 1978, Suppression of human PMN bactericidal activity by human IgA paraproteins, *Cell. Immunol.* **36:**363.

Wakelin, D., 1978*a*, Immunity to intestinal parasites, *Nature (Lond.)* **273:**617.

Wakelin, D., 1978*b*, Genetic control of susceptibility and resistance to parasitic infection, *Adv. Parasitol.* **16:**219.

Wakelin, D., and Wilson, M. M., 1979, *Trichinella spiralis*: Immunity and inflammation in the expulsion of transplanted adult worms from mice, *Exp. Parasitol.* **48:**305.

Walker, W. A., Cornell, R., Davenport, L. M., and Isselbacher, K. J., 1972, Macromolecular absorption. Mechanism of horseradish peroxidase uptake and transport in adult and neonatal rat intestine, *J. Cell Biol.* **54:**195.

Walker, W. A., Wu, M., and Bloch, K. J., 1977, Stimulation by immune complexes of mucus release from goblet cells of the rat small intestine, *Science* **197:**370.

Warren, K. S., 1976, Immunopathology due to cell-mediated (type IV) reactions, in: *Immunology of Parasitic Infections* (S. Cohen and E. Sadun, eds.), pp. 448–467, Blackwell Scientific, Oxford.

Wasi, S., Vadas, P., Chin, G. W., Movat, H. Z., and Hay, J. B., 1982, The production of plasminogen activator by afferent but not efferent lymph cells emigrating from chronic granulomatous lesions in sheep, *J. Immunol.* **128:**1076.

Weinstock, J. V., and Boros, D. L., 1981, Heterogeneity of the granulomatous response in the liver, colon, ileum and ileal Peyer's patches to schistosome eggs in murine *Schistosomiasis mansoni*, *J. Immunol.* **127:**1906.

Welliver, R. C., Wong, D. T., Sun, M., Middleton, E., Jr., Vaughan, R. S., and Ogra, P. L., 1981, The development of respiratory syncytial virus-specific IgE and the release of histamine in nasopharyngeal secretions after infection, *N. Engl. J. Med.* **305:**841.

Wells, P. D., 1963, Mucin-secreting cells in rats infected with *Nippostrongylus brasiliensis*, *Exp. Parasitol.* **14:**15.

Wilton, J. M. A., 1978, Suppression by IgA of IgG-mediated phagocytosis by human polymorphonuclear leucocytes, *Clin. Exp. Immunol.* **34:**423.

Wolf, J. L., Rubin, D. H., Finberg, R., Kaufmann, R. S., Sharpe, A. H., Trier, J. S., and Fields, B. N., 1981, Intestinal M cells: A pathway for entry of reovirus into the host, *Science* **212:**471.

Woodbury, R. G., and Miller, H. R. P., 1982, Quantitative analysis of mucosal mast cell protease in the intestine of *Nippostrongylus*-infected rats, *Immunology* **46:**487.

World Health Organization Scientific Group, 1980, *Viral Respiratory Diseases*, Technical Report Series 642, WHO, Geneva.

Yoshida, N., Nussenzweig, R. S., Potocnjak, P., Nussenzweig, V., and Aikawa, M., 1980, Hybridoma produces protective antibodies directed against the sporozoite stage of malaria parasite, *Science* **207**:71.

Yung, Y.-P., Eger, R., Tertian, G., and Moore, M. A. S., 1981, Long-term *in vitro* culture of murine mast cells. II. Purification of a mast cell growth factor and its dissociation from TCGF, *J. Immunol.* **127**:794.

Chapter 4

Antigenic Characterization of Plasmodia

Luc H. Perrin, L. Rodriguez da Silva, and R. Dayal

WHO Research and Training Centre and *Department of Medicine*
Geneva Blood Centre *Geneva University Hospital*
1211 Geneva 4, Switzerland

I. INTRODUCTION

One hundred years ago, Laveran identified plasmodia as the causative agent of malaria; the concept of vector transmission was demonstrated at the beginning of the century. *Plasmodium* is a unicellular protozoan parasite transmitted to humans and other vertebrates by the bite of female *Anopheles* mosquitoes. An estimated one-third of the world's population lives in malaria endemic areas. About 200 million people are infected annually, and more than one million children die each year of malaria infection (World Health Organization, 1976). Although malaria infection was successfully eradicated in large areas through extensive eradication programs through the use of insecticides and chemoprophylaxis, the disease is showing a resurgence in areas in which it was formerly under reasonable control, e.g., in large areas of South America, Central America, India, and Sri Lanka (Gopalan, 1976). The current resurgence of malaria has been explained partly by operational failures and partly by biologic factors related to the development of resistance of the mosquito vector to insecticides (Pal, 1973) and of the parasite to the available chemotherapeutic agents, such as chloroquine and sulfonamides (McGregor, 1978).

This crisis has stimulated the establishment of research programs coordinated and funded by WHO and by national agencies for the development of additonal means of malaria control, particularly of vaccines. Malaria vaccines using intact or denatured parasites have been successful in experimental animal models (Kreier, 1980). However, these vaccines are unsuitable for humans for various reasons, including the difficulty of large-scale production of the parasite and the presence of host proteins in the vaccine preparations likely to induce adverse immunologic effects. An alternative approach would be to identify and produce the individual malarial antigens capable of inducing a protective immune response in the vaccinated host. Recent advances in chemistry and molecular biology indicate that it may be possible to synthetize such antigens chemically or to produce them by recombinant DNA technology. The development of immunodiagnostic tests for mass screening after standard eradication programs and vaccination is directly linked to the development of vaccine.

This review of antigenic characterization of plasmodia focuses on the identification and characterization of antigens that may be involved in host protection and/or that may be used as target antigens for the development of immunodiagnostic tests designed to detect ongoing or recent malaria infection. Particular emphasis is given to studies on human malaria parasites.

II. PLASMODIAL LIFE CYCLE

More than 100 species of plasmodia are known, all having a very restricted host specificity. For example, only three rare monkey species are highly susceptible to *P. falciparum*, the most lethal of human malaria species. Other plasmodial species infecting humans include *P. vivax*, *P. malariae*, and *P. ovale*. Because of the scarcity and the high cost of the rare monkeys susceptible to *P. falciparum*, numerous studies have been conducted on nonhuman plasmodial species, and particularly on *P. berghei* and *P. yoleii* in mice, *P. lophurae* and *P. gallinaceae* in birds, and *P. knowlesi* and *P. cynomolgi* in monkeys. The life cycle of these various plasmodia, although differing in the time course of development and in the host cell types invaded by sporozoites, is otherwise quite similar in the different species (Fig. 1). All species require both a vertebrate and an invertebrate host. Commonly an infected mosquito bites and injects sporozoites in the vertebrate host's blood. These sporozoites bind to and invade liver parenchymal cells in mammals and endothelial cells in birds, in which they divide asexually to form daughter progeny called merozoites. This extraerythrocytic development takes ~10 days for *P. falciparum* and is followed by rupture of hepatocytes, which release thousands of merozoites. These merozoites enter the circulation and are infectious only for erythrocytes (specific ligand-receptor interaction). After penetration in the erythrocytes, the merozoites first have a ring-shaped appearance and then enlarge to form trophozoites. Asexual division gives rise to 8 to 32 merozoites with a periodicity of 24-72 hr, depending on the plasmodial species. The erythrocytes containing merozoites, called schizonts, lyse at the end of the maturation and release merozoites that invade new erythrocytes. This erythrocytic asexual cycle is responsible for the clinical manifestations of malaria, and the rupture of schizonts is associated with periodic fever access.

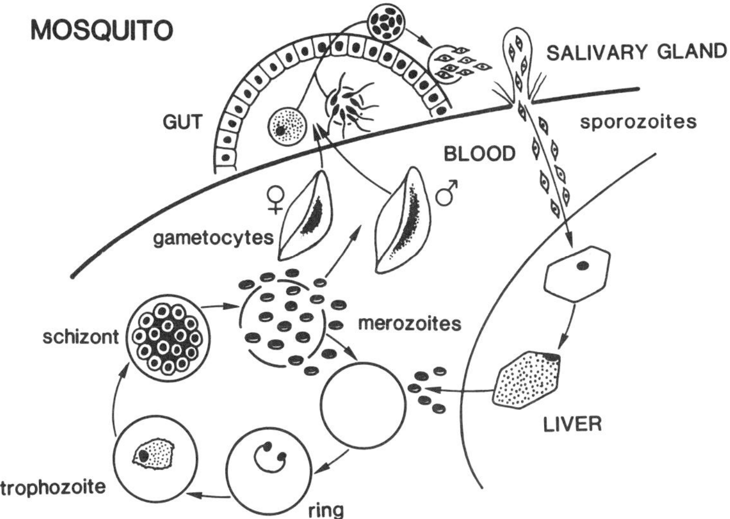

Figure 1. Life cycle of *Plasmodium* from mosquito to vertebrate host.

Most of the merozoites infect new erythrocytes and multiply again asexually, but others differentiate over a period of 7-12 days into sexual stages called male and female gametocytes. The gametocytes are surrounded by erythrocyte membrane and undergo further development in the mosquito that has ingested these forms during feeding. Fertilization occurs in the mosquito's stomach; the resulting zygote develops as an oocyst in the gut wall. After maturation of the oocyst, sporozoites are released and migrate to the mosquito's salivary glands.

There are several limitations to studies on plasmodia, including the low amount of parasite material available, the close association of the parasite with host components, and the complexity of the parasites's life cycle. The complex life cycle of the malaria parasite involves continuous morphologic enyzmatic and antigenic changes that are probably linked to the parasite's environmental adaptation to the host. Immunologic cross-reactivity has been shown by serologic methods between various plasmodia species (Ingram *et al.*, 1961) and between different developmental stages within the same species (Sodeman and Jeffery, 1964; Golenser *et al.*, 1977). However, serologic data and vaccination experiments have demonstrated that a number of antigenic determinants–notably those responsible for the induction of protective immunity–are species and stage specific (Nussenzweig *et al.*, 1969; Zuckerman, 1977). It appears, therefore, that the antigenic composition of plasmodia should be studied independently for each species and for each of the four developmental sequences of the plasmodial life cycle (sporogony, exoerythrocytic and erythrocytic schizongony, gametocytogony).

III. SPOROZOITES

Most studies concerning sporozoites have been conducted in mice using *P. berghei* for practical reasons. Immunity to sporozoites is strictly stage specific and is induced only by vaccination with mature sporozoites isolated from the mosquito's salivary glands; immature sporozoites and oocyst sporozoites are ineffective (Nussenzweig *et al.*, 1969). Sporozoites are difficult to purify from contaminating mosquito material, and it has not been possible to make a direct analysis of their components. Some of their components have been identified and characterized using surface or internal labeling techniques followed by immunoprecipitation with antisporozoite sera. Using these techniques and autoradiography of SDS-PAGE, the major surface immunogenic component of *P. berghei* sporozoites was identified as a polypeptide of 41 kd (Gwadz *et al.*, 1978). An immunologically identical polypeptide of 44 kd (Pb 44) was precipitated by a monoclonal antibody prepared against *P. berghei* sporozoites (Yoshida *et al.*, 1980). Pulse-chase experimants have detected two intracellular precursors of Pb 44 having a M_r of 54 and 52 kd (Yoshida *et al.*, 1981). Immunofluorescence and immunoelectron microscopic studies revealed this polypeptide on the entire surface of mature sporozoites and in lower densities, in a patchy distribution, on some oocyst preparations (Aikawa *et al.*, 1981) Pb 44 is stage specific and is not expressed in the erythrocytic stage of *P. berghei.* Pb 44 can be separated from other surface-labeled material using two-dimensional electrophoresis; it has been obtained in purified form from sporozoite extracts by combining isoelectric focusing and affinity chromatography using monoclonal antibodies as specific immunoabsorbant. Monoclonal antibodies against Pb 44 neutralize the infectivity of sporozoites *in vitro* and, after passive transfer, protect mice from a lethal challenge infection with viable sporozoites of the same species (Potocnjak *et al.*, 1980). Monoclonal antibodies reacting with the major surface protein of *P. knowlesi, P. vivax*, and *P. falciparum* sporozoites have also been prepared. Preincubation *in vitro* of sporozoites with specific monoclonal antibodies considerably decreases their infectivity *in vivo* in rhesus monkeys and chimpanzees, respectively. Preliminary experiments indicate that polypeptides of 67 and 58 kd and of 51 and 45 kd may be the major surface antigens, respectively, of *P. falciparum* and *P. vivax* sporozoites (Nardin *et al.*, 1982).

IV. EXOERYTHROCYTIC STAGES

The lack of an *in vitro* culture method for the exoerythrocytic (EE) stages has seriously restricted the analysis of this stage of the parasite's life cycle. Avian EE stages have been cultivated *in vitro* since 1966, but the complete *in vitro* EE cycle of *P. berghei* was established in culture in 1981 only, and the *in vitro* culture of the EE cycle of the plasmodial species infecting humans is as yet unsuccessful (Hollingdale *et al.*, 1981). All studies concerning antigenic characterization of EE stages have used antibodies as tools. In the *P. berghei* system, it has been shown *in vitro*, using sporozoites and WI 38 cells as acceptor cells for the EE cycle, that the Pb 44 antigen is shed from the surface of sporozoites after they attach to WI 38 cells (Hollingdale *et al.*, 1982). The WI 38 engulf the sporozoites into a parasitophorous vacuole, and further development occurs leading to the release of EE merozoites. When monoclonal antibodies against Pb 44 are incubated directly with the sporozoites and WI 38 cells, sporozoites do not attach and shedding does not occur. These experiments suggest that a specific receptor for sporozcite Pb 44 is present on WI 38 cells and probably on hepatic cells. Indirect immunofluorescence studies using liver section of EE stages of *P. berghei* as antigens; a variety of characterized antimalarial antisera have demonstrated the presence of common antigens between EE and erythrocytic stages. Interestingly, the sporozoite antigens predominate in early EE stages, and erythrocytic antigens predominate in late EE stages. Using specific monoclonal antibodies against Pb 44, it was shown that either Pb 44 or its precursors, or both, are expressed throughout the asexual EE cycle, indicating that EE and erythrocytic merozoites have antigenic differences.

V. ASEXUAL ERYTHROCYTIC STAGES

Three main parasite sources have been used for antigenic characterization of the asexual erythrocytic stages: blood from infected vertebrates parasitized erythrocytes from *in vitro* cultures, and plasma or sera of infected vertebrate hosts. For *P. falciparum*, the development of an *in vitro* culture technique has provided a continuous source of parasites for antigenic characterization of the erythrocytic stages (Traeger and Jensen, 1976). In this section, we review the literature concerning (1) the antigenic composition of the various asexual erythrocytic stages, (2) immune recognition of plasmodial antigens by polyclonal and monoclonal antibodies, and (3) soluble or circulating *P. falciparum* antigens in sera of infected individuals and antigens released in the culture medium of *in vitro* cultures.

A. Antigenic Characterization of the Various Asexual Erythrocytic Stages

Malarial antigens of the various asexual blood stages have been analyzed from several plasmodial species, including *P. falciparum, P. knowlesi, P. cyno-*

molgi, P. berghei, and *P. chabaudi.* Successive erythrocytic developmental stages have been prepared using a combination of techniques, such as differential lysis by mannitol, free-flow electrophoresis, density-gradient separation, and by limiting the time of invasion of erythrocytes by merozoites. Direct antigenic analysis of these stages has been hampered by difficulties in obtaining adequate amounts of parasites and by the presence of host-contaminant material, i.e., close association of the parasite with erythrocytes in these preparations. Some of these difficulties have been solved by metabolic labeling parasites *in vitro* with radiolabeled amino acids; since mammalian erythrocytes are depleted of their gene pool, all the *de novo* synthesized radiolabeled polypeptides are parasite gene products (Kilejian, 1979; Perrin *et al.*, 1979).

Figure 2 presents autoradiographs of metabolically labeled erythrocytic stages analyzed by SDS-PAGE. Synchronized cultures of *P. falciparum* were sequentially labeled with [^{35}S]methionine at ring, trophozoite, and schizont developmental stages. The lysed cells from each stage resolved in more than 40 radiolabeled polypeptides by SDS-PAGE with apparent relative molecular weights ranging from 200 to 30 kd (Perrin *et al.*, 1981*b*). Most of the polypeptide bands were common to all stages. Three bands of 96, 76, and 50 kd are more intensely labeled in the ring and trophozoite preparations. Such antigens may be the ideal targets for immunodiagnostic tests aiming at the detection of malaria antigens in parasitized blood, since only ring forms are present in high concentrations in the peripheral blood of individuals suffering from *P. falciparum* infection.

Early and late schizont preparations (late schizonts or segmenters representing schizonts containing well-defined merozoites) contained all polypeptides synthesized by rings and trophozoites. In addition, seven polypeptide bands of apparent $M_r > 200$, 160, 140, 105, 82, 55 (diffuse-band), and 41 kd were specifically present in both schizont preparations; however, these bands were more intensely labeled in late schizonts, suggesting that synthesis of these polypeptides begins during schizogony and increases with maturation of the schizonts to segmented forms. The latter preparation also contained one weakly labeled band of 250 kd. In some preparations, a 96-kd band common to all stages was identified as a doublet in the schizont preparations; one of these components may be schizont specific.

Similar results have been reported by other investigators with minor discrepancies concerning the molecular weight of the schizont-specific components (Kilejian, 1979; Brown *et al.*, 1981; Perkins, 1982). Schizont or merozoite components of apparent molecular weight of > 200, 140, 82, 55 and 41 also incorporate [^{3}H]glucosamine (Kilejian, 1980).

When *P. falciparum* parasites are cultivated *in vitro* for long periods, they lose some protrusions called "knobs," which are characteristic of infected

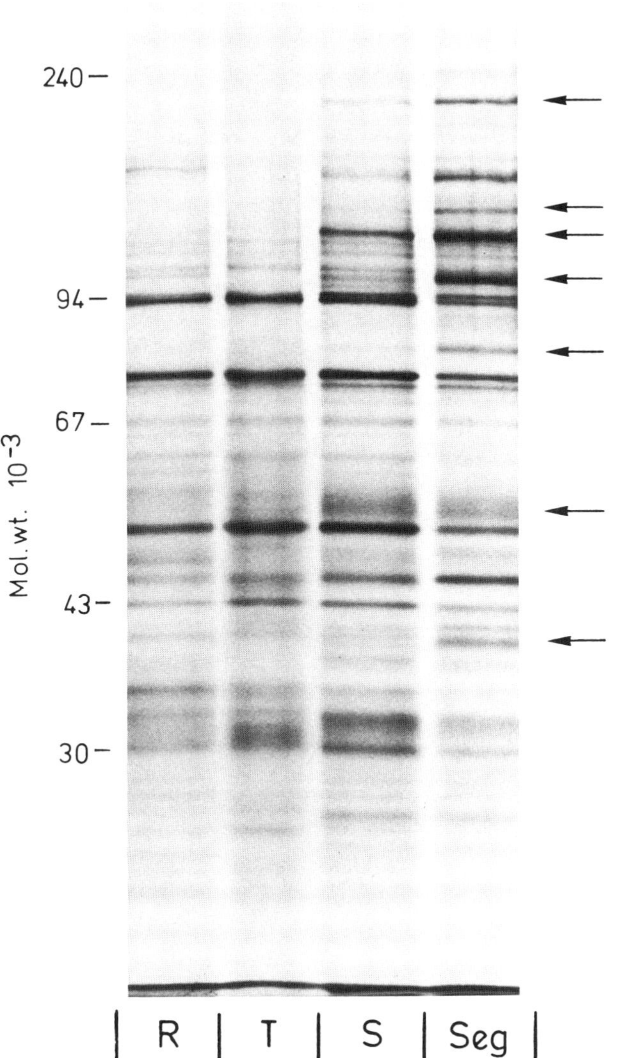

Figure 2. Autoradiographs of SDS-PAGE of various erythrocytic stages of *P. falciparum*: R, ring forms; T, trophozoites; S, schizonts; Seg., segmenters (mature schizonts). Synchronized cultures were sequentially labeled *in vitro* with [^{35}S] methionine (50 μCi/ml) for 6 hr at various intervals during schizogony. The cells were washed and treated with NP-40 (0.5% in TNE), and equal numbers of TCA precipitable counts per minutes were analyzed on 10% SDS-PAGE. *Arrows*, polypeptide bands synthesized at the schizont stage.

erythrocytes containing trophozoites and schizonts. Using strains both with and without knobs, a polypeptide of 80 kd was shown to be associated with the knobs (Kilejian, 1979) and expressed on the surface of parasitized red blood cells (RBC) containing trophozoites and schizonts. This polypeptide contains a high proportion of histidine, and may be analogous to a histidine-rich protein of *P. lophurae*. Ducklings immunized with this polypeptide are protected from a challenge infection with *P. lophurae* (Kilejian, 1978).

Surface labeling of RBC-containg *P. falciparum* schizonts, revealed 3 polypeptides of 200, 82 and 41 kd in addition to normal RBC components (Perrin *et al.*, 1982). Components expressed at the surface of *P. falciparum* merozoites have also been identified and their apparent M_r is 200 (3 polypeptides), 130 (possibly identical to the 140 kd schizont-merozoite specific polypeptide) and 96 kd (Perkins, 1982). Other investigators found that the 82 and 41 kd polypeptides are the predominant polypeptides expressed at the surface of merozoites (H. Heindrich, personal communication). Recent studies have shown that merozoites attach to glycophorin A and B on the erythrocyte surface and that the 200 and 140 kd merozoite polypeptides bind to glycophorins (Jungery *et al.*, 1983).

B. Identification of Plasmodial Antigens Relevant to Protection Using Polyclonal and Monoclonal Antibodies

Antibodies play a major role in immune protection against the asexual blood stages of plasmodia (Cohen *et al.*, 1961). Conventional antisera from experimental animals or humans semi-immune to malaria and antimalarial monoclonal antibodies, have been used to identify and, in some instances, purify plasmodial antigens that are relevant to protection.

Simian plasmodial antigen(s) relevant to protection have been identified by crossed immunoelectrophoretic analysis (Deans *et al.*, 1979) using *P. knowlesi* and *P. cynomolgi bastianellii* schizonts and hyperimmune anti-*P. knowlesi* serum. The extent of cross-reactivity between the two plasmodial species was determined by addition of antigens and by crossed-line immunoelectrophoresis. All but one of the *P. knowlesi* antigens identified by the hyperimmune serum were shared by *P. cynomolgi bastianellii*. The *P. knowlesi* specific antigen was a stage-dependent lipid antigen present in mature schizonts and merozoites only. Since, in this system, acquired immunity is species-specific, this unique antigen may be the antigen inducing protective antibodies. Another characteristic of "protective" antigens is their localization a the surface of schizonts, since these antigens may be the target of antibodies. Sera from rhesus monkeys immune to *P. knowlesi* selectively precipitated two antigens of 65 and 90 kd from a preparation of surface labeled *P. knowlesi* schizonts; moreover, the presence of antibodies to the 65-kd antigen correlated with clinical immunity in naturally infected monkeys (Schmidt-Ulrich *et al.*, 1981).

Immunoprecipitation of *P. falciparum* isolates metabolically labeled *in vitro*, using human antibodies from individuals with varying degrees of exposure and immunity to *P. falciparum*, has been used to identify the antigens likely to induce a protective immunity. Sera from adult inhabitants of endemic areas, having high resistance to malaria, preferentially precipitate six of the schizont-specific antigens of > 200, 140, 105, 82, 55, and 41 kd, both at a higher frequency and at a higher serum dilution than sera from adults recovering from a first *P. falciparum* infection (low-grade resistance). In contrast, antigens common to the various asexual developmental stages are precipitated at a similar frequency by both groups of sera (Perrin *et al.*, 1981*b*, 1982).

A representative analysis of immunoprecipitates by SDS-PAGE and autoradiography is shown in Fig. 3. In other studies using the same methodology,

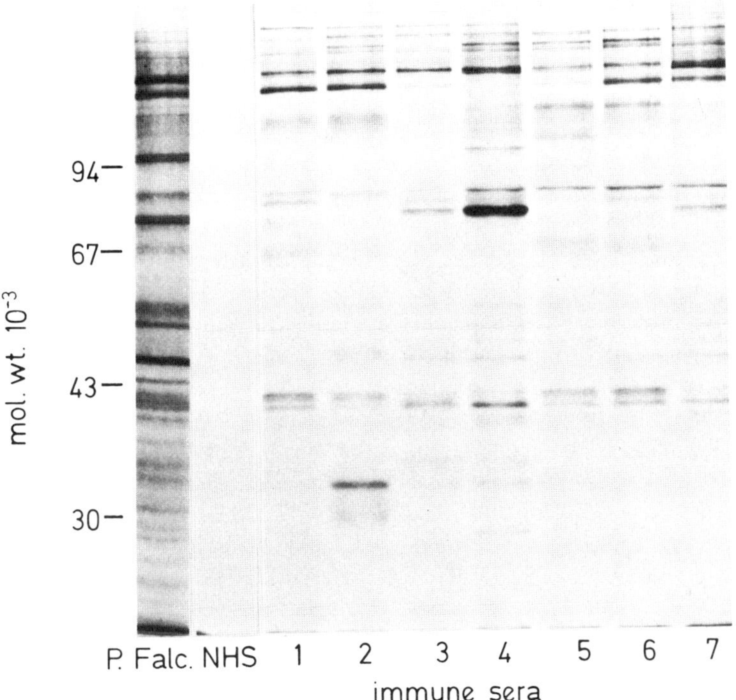

Figure 3. Autoradiographs of SDS-PAGE of ^{35}S-methionine-labeled *P. falciparum* polypeptides precipitated by seven adult sera from the *Gambia* and by normal human serum (NHS) from a normal European blood donor as control. *P. falc.*, starting labeled material provided by an asynchronous *P. falciparum* culture labeled with [^{35}S]methionine. *Arrows*, Polypeptides of 200, 160, 140, 105, 82, and 41 kd.

variable recognition patterns were seen during the course of the same infection and also for different *P. falciparum* isolates. It was also shown that antibodies to some schizont polypeptides may be linked with the capacity of human sera to inhibit the growth of *P. falciparum in vitro* (Brown *et al.*, 1981). The extent and importance of antigenic diversity in *P. falciparum* is difficult to appreciate in relation to the host/parasite relationship. Extensive serological studies and parasite enzyme typing have clearly established the antigenic diversity among various *P. falciparum* isolates (McGregor *et al.*, 1966, 1968). More recently, it has been shown that antigenic switch may occur for polypeptides expressed at the surface of schizonts during the course of blood-induced *P. falciparum* infection in monkeys, and it was postulated that this antigenic switch may explain the long time-span needed for the acquisition of immunity towards *P. falciparum* infection. Up to now, the antigenic determinants involved in the antigenic switch have not been characterized (Hommel, 1983).

Monoclonal antibodies prepared against asexual blood stages and exhibiting protective activity have been used to directly characterize and purify defined antigens likely to be involved in host protection. Passive transfer of monoclonal antibodies directed against merozoites of a murine malaria species, *P. yoleii*, protect mice from a lethal challenge infection (Freeman *et al.*, 1980). Mice immunized with affinity-purified antigens recognized by these monoclonal antibodies (M_r 230 and 235 kd) are protected against a lethal challenge infection (Holder and Freeman, 1981). Some monoclonal antibodies raised against *P. knowlesi* merozoites could agglutinate merozoites and prevent invasion of erythrocytes *in vitro* (Epstein *et al.*, 1981). These monoclonal antibodies recognized a merozoite component of 250 kd.

In human malaria, monoclonal antibodies directed against *P. falciparum* schizonts and merozoites inhibit the *in vitro* growth of *P. falciparum* in culture (Perrin *et al.*, 1981*a*). These inhibitory monoclonal antibodies react with a 41-kd, or an 82- and a 41-kd or a 140-kd antigen (Fig. 4) (Perrin and Dayal 1982). The 41- and 82-kd polypeptides appear as doublets on SDS PAGE and may be glycoproteins. The three antigens recognized by the inhibitory monoclonal antibodies are merozoite- and schizont-specific and have been identified in seven *P. falciparum* isolates from different geographic areas. Moreover, these polypeptides are among the six polypeptides preferentially precipitated by the sera of adults living in endemic areas and displaying a high degree of resistance to malaria (Perrin and Dayal, 1982). Taken together, these two observations strongly suggest that these antigens play a role in the induction of protective immunity *in vivo*. These *P. falciparum* antigens are currently being purified for immunization studies. Monoclonal antibodies have also been raised against various isolates of *P. falciparum* in order to identify antigens of the erythrocytic stages of *P. falciparum* associated with various developmental stages or isolate specific polypeptides (McBridge *et al.*, 1982; Hall *et al.*, 1983). The bulk of these studies are aimed at identifying the localization of *P. falciparum* antigens (sur-

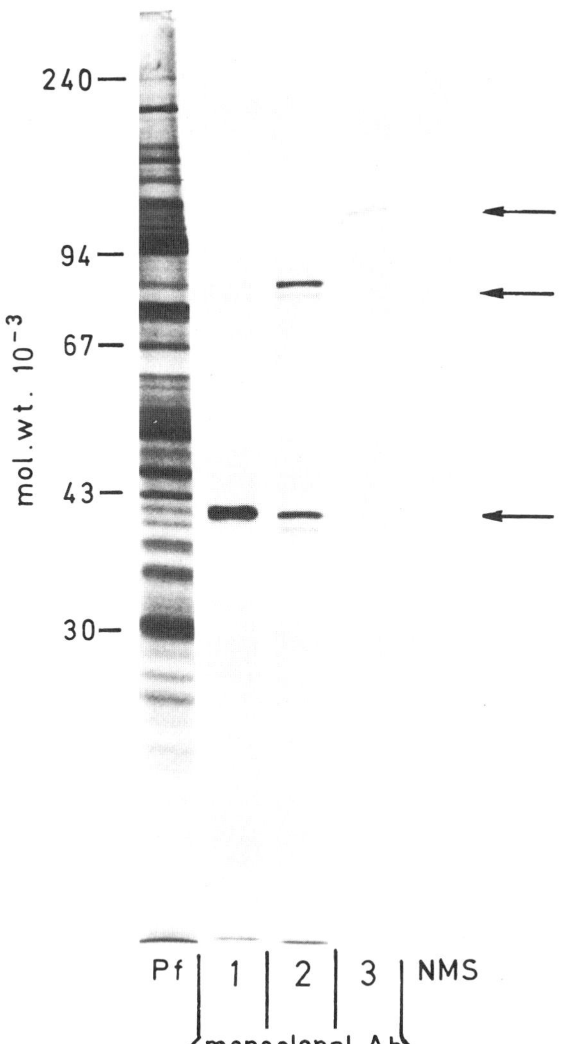

Figure 4. Autoradiographs of SDS–PAGE of *P. falciparum* polypeptides precipitated by monoclonal antibodies produced by three hybridome cell lines (1, 2, 3) and normal mouse serum (NMS). These monoclonal antibodies inhibit the growth of *P. falciparum* cultures. *Arrows*, polypeptides of 140, 82, and 41 kd.

face, internal antigens) and purifying and characterizing these antigens at the molecular level. Recombinant DNA technology and chemical synthesis of polypeptides may in the near future provide an alternative approach to the production of malaria antigens.

C. Circulating *P. falciparum* Antigens and Polypeptides Released in Culture Medium

Circulating *P. falciparum* antigens have been identified by double diffusion in sera of infected individuals (McGregor *et al.*, 1966, 1968). The antimalarial antisera for these studies were provided by immune sera of adults living in endemic areas and relatively resistant to reinfection. These sera yielded several precipitin bands against sera from individuals suffering from malaria. The antigens recognized by the various immune sera were classified according to their thermostability; antigens that were destroyed when the test sera were incubated at 56°C for 30 min were termed labile (L) antigens, those destroyed only by heating sera at 100°C for 5 min were termed resistant (R) antigens, and those that resist heating at 100°C for 5 min were termed stable (S) antigens (Wilson *et al.*, 1969; McGregor and Wilson, 1971). Using a panel of adult immune sera it was shown that strains, isolates and clones of *P. falciparum* could be characterized according to the various S antigens (Wilson, 1980). Other evidence of the heterogeneity of the *P. falciparum* strains have been obtained by morphology, infectivity, pathogenicity, drug sensitivity, isoenzyme typing, two-dimensional gel electrophoresis and recently by using monoclonal antibodies (Sanderson *et al.*, 1981).

Plasmodium falciparum polypeptides released during *in vitro* culture have been studied by analyzing the culture medium of metabolically labeled synchronous plasmodial cultures (McColm and Trigg, 1980). Most of the parasite-specific polypeptides are released at the end of schizogony and at the time of invasion of erythrocytes by merozoites (Wilson, 1974). When merozoites invade erythrocytes, they lose their surface coat, which should be at least one of the parasitic components present in culture medium. For *P. falciparum*, polypeptides of 200, 180, 82, and 45 kd are the major components released (Rodriguez da Silva *et al.*, 1983), polypeptides of 140 and 41 kd are also found, but in lesser amounts. Malaria culture medium has been used as a source of parasitic material for the purification of malaria antigens by affinity chromatography (Jepsen and Andersen, 1980). Soluble parasitic antigens released in culture may be a source of antigens for immunization, as shown by successful vaccination achieved in cattle using antigens purified from culture supernatant of *Babesia bovis* (Smith *et al.*, 1981) and in chickens, rodents, and primates using soluble plasmodial antigens (Todoric *et al.*, 1967; Jerusalem, 1969).

VI. GAMETES

Experimental studies showing that antibodies produced in a vertebrate host after vaccination with gametes can block fertilization in the gut of mosquitoes (Carter and Chen, 1976; Gwadz, 1976), and therefore stop transmission of the infection (transmission blocking immunity), have stimulated research aiming at the characterization of gamete antigens susceptible to raise a protective immune response. The antigamete antibody response is stage specific and does not affect gametocyte development in the vertebrate host in which gametocytes are protected from antibody attack by the erythrocyte membrane. Three polypeptides, of 230, 210, and 55 kd have been identified as exposed surface components of *P. gallinaceum* male gametes; two of them, those of 230 and 55 kd, have also been identified on the female gamete surface. Several monoclonal antibodies reacting with antigens represented at the surface of gametes have been characterized. Most of these antibodies were of the IgG isotype and reacted with the 230-kd antigen, but none had a suppressive effect on the infectivity of *P. gallinaceum* mosquitoes. One of these antibodies, of the IgM isotype, suppressed infectivity by about 99%, but the antigen recognized by this particular monoclonal antibody has not been identified (Rener *et al.*, 1980). Studies on gametes of other plasmodial species are in progress, alghough they are limited owing to technical reasons related to culture or purification of gametocytes, or both.

VII. CONCLUSIONS AND PROSPECTS

During the past decade, several recently developed technologies, including the *in vitro* culture of various developmental stages, metabolic labeling, and the hybridoma technique, have been applied to studies on malaria and have contributed greatly to our knowledge on plasmodial antigens.

These studies have focused on the identification of malarial antigens with functional characteristics and particularly of those antigens involved in the generation of a host-protective immune response. Immunologic and biochemical studies have emphasized the complexity of plasmodial antigens. There is serologic evidence of immunologic cross-reactivity between various plasmodial species and strains, as well as between different developmental stages of the same isolate. However, there is conclusive evidence from vaccination and *in vitro* growth inhibition studies of the species and stage-specific nature of plasmodial antigens involved in protection. Antigens exhibiting protective characteristics have been identified in several plasmodial species including *P. falciparum*. Characteristically, these antigens are expressed or synthesized late in each of the developmental sequence, e.g., in mature sporozoites, at the end of schizogony, and after maturation of the gametocytes.

Passive injection of monoclonal antibodies directed against a defined surface antigen of *P. berghei* sporozoites has been shown to protect mice from a lethal challenge infection. More recently, a single *P. yoleii* merozoite surface polypeptide affinity purified using host-protective monoclonal antibodies was shown to protect vaccinated mice against a challenge blood stage infection. Similar studies are in progress on other plasmodia, including human malaria species. The main limitation of this experimental approach is still the low-production capacity of pure plasmodial antigens. There is, however, increasing evidence that recombinant DNA technology and chemical synthesis of immunogenic polypeptides will provide a breakthrough for large-scale production of antigens suitable for human malaria vaccine.

ACKNOWLEDGMENTS

The original work reported in this chapter is supported by grant 3.862.081 of the Swiss National Foundation and by WHO, the UNDP/World Bank/WHO Special Program for Research and Training in Tropical Diseases. We thank Ms. G. Fiaux for her secretarial assistance.

VIII. REFERENCES

Aikawa, M., Yoshida, N., Nussenzweig, R. S., and Nussenzweig, V., 1981, The protective antigen of malarial sporozoites (*Plasmodium berghei*) is a differentiation antigen, *J. Immunol.* **126**:2494.

Brown, G. V., Anders, R. F., Stace, J. D., Alpers, M. P., and Mitchell, G. F., 1981, Immunoprecipitation of biosynthetically-labelled proteins from different Papua New Guinea *Plasmodium falciparum* isolates by sera from individuals in the endemic area, *Parasite Immunol.* **3**:283.

Carter, R., and Chen, D. H., 1976, Malaria transmission blocked by immunisation with gametes of the malaria parasite, *Nature* (*Lond.*) **263**:57.

Cohen, S., McGregor, I. A., and Carrington, S. C., 1961, Gamma globulin and acquired immunity to human malaria, *Nature* (*Lond.*) **192**:733.

Deans, J. A., 1979, Species-specific *Plasmodium knowlesi* antigens identified by crossed-line immunoelectrophoresis, in: *Immonodiagnostic Techniques in Malaria, Transactions of the Third Meeting of the SWG on the Immunology of Malaria, Panama,* p. 1, WHO/TDR, Geneva.

Epstein, H., Miller, L. H., Kaushel, D. C., Udeinya, I. J., Rener, J., Howard, R. J., Asofsky, R., Aikawa, M., and Hess, R. L., 1981, Monoclonal antibodies against a specific surface determinant on malarial (*Plasmodium knowlesi*) merozoites block erythrocyte invasion, *J. Immunol.* **127**:212.

Freeman, R. R., Trejdosiewicz, A. J., and Cross, G. A. M., 1980, Protective monoclonal antibodies recognizing stage-specific merozoite antigens of rodent malaria parasite, *Nature* (*Lond.*) **284**:366.

Golenser, J., Heeren, J., Verhave, J. P., Kaay, H. J. V. D., and Meuwissen, J. H. E., 1977, Crossreactivity with sporozoites, exoerythrocytic forms and blood schizonts of *Plasmodium berghei* in indirect fluorescent antibody tests with sera of rats immunised with sporozoites or infected blood, *Clin. Exp. Immunol.* **29**:43.

Gopalan, C., 1976, Annual report of director-general, Indian Council on Medical Research, Leipsig Press, New Delhi.

Gwadz, R. W., 1976, Malaria; successful immunisation against the sexual stages of *Plasmodium gallinaceum*, *Science* **193**:1150.

Gwadz, R. W., Cochrane, A. H., Nussenzweig, V., and Nussenzweig, R. S., 1978, Preliminary studies on vaccination of rhesus monkeys with irradiated sporozoites of *Plasmodium knowlesi* and characterisation of surface antigens of these parasites, *Bull. W.H.O.* **57** (Suppl. 1):165.

Hall, R., McBride, J., Morgan, G., Tait, A., Zolg, J. W., Walliker, D. and Scarfe, J., 1983, Antigens of the erythrocytic stages of the human malaria parasite *Plasmodium falciparum* detected by monoclonal antibodies, *Mol. Biochem. Parasitol.* **7**:247.

Holder, A. A., and Freeman, R. R., 1981, Immunisation against blood stage rodent malaria using purified parasite antigens, *Nature (Lond.)* **294**:361.

Hollingdale, M. R., Leef, J. L., McCullough, M., and Beaudoin, R. L., 1981, In vitro cultivation of the exoerythrocytic stage of *Plasmodium berghei* from sporozoites, *Science* **213**:1021.

Hollingdale, M. R., Zavalo, F., Nussenzweig, R. S., and Nussenzweig, V., 1982, Antibodies to the protective antigen of *Plasmodium berghei* sporozoites prevent entry into cultured cells, *J. Immunol.* **128**:1929.

Hommel, M., David, P. H., and Oligino, L. D., 1983, Surface alterations of erythrocytes in *Plasmodium falciparum* malaria, *J. Exp. Med.*, **157**:1137.

Ingram, R. L., Otken, L. B., and Jumper, J. R., 1961, Staining of malaria parasites by FA technique, *Proc. Soc. Exp. Biol. Med.* **106**:52.

Jepsen, S., and Andersen, B. J., 1980, Immunoabsorbent isolation of antigens from the culture medium of in vitro cultivated *Plasmodium falciparum*, *Acta Pathol. Microbiol. Scand.* Sect. C, **88**:263.

Jerusalem, C., 1969, Active immunization against malaria (*Plasmodium berghei*), 1. Definition of antimalarial immunity, *Z. Tropenmed. Parasitol.* **19**:171.

Jungery, M., Boyle, D., Patel, T., Pasvol, G. and Weatherall, D. J., Lectin-like polypeptides of *P. falciparum* bind to red cell sialoproteins, *Nature* **301**:704.

Kilejian, A., 1978, Histidine-rich protein as a model malaria vaccine, *Science* **201**:922.

Kilejian, A., 1979, Characterisation of a protein correlated with the production of knob-like protrusions on membranes of erythrocytes infected with *Plasmodium falciparum*, *Proc. Natl. Acad. Sci. USA* **76**:4650.

Kilejian, A., 1980, Stage specific proteins and glycoproteins of *Plasmodium falciparum*: Identification of antigens unique to schizonts and merozoites, *Proc. Natl. Acad. Sci. USA* **77**:3695.

Kreier, J. P., 1977, The isolation and fractionation of malaria infected cells, *Bull. WHO* **55**:317.

Kreier, J. P., 1980, chapters 4, 5, 6, 7, in: *Malaria*, Vol. 3: *Immunology and Immunisation* (J. P. Krier, ed.), Academic Press, London.

McColm, A. A., and Trigg, P. I., 1980, Release of radio-isotope labelled antigens from *Plasmodium knowlesi* during merozoite re-invasion in vitro, *Parasitology* **81**:199.

McGregor, I. A., 1978, Tropical aspects of the epidemiology of malaria, *Isr. J. Med. Sci.* **14**:523.

McGregor, I. A., and Wilson, R. J. M., 1971, Precipitating antibodies and immunoglobulins in *P. falciparum* infection in the Gambia, *Trans. R. Soc. Trop. Med. Hyg.* **65**:136.

McGregor, I. A., Hall, P. J., Williams, K, Hardy, C. L. S., and Turner, M. W., 1966, Demonstration of circulating antibodies to *Plasmodium falciparum* by gel-diffusion techniques, *Nature (Lond.)* **210**:1384.

McGregor, I. A., Turner, M. W., Williams, K., and Hall, P., 1968, Soluble antigens in the blood of African patients with severe *Plasmodium falciparum* malaria, *Lancet* **1**:881.

Nardin, E. H., Nussenzweig, V., Nussenzweig, R. S., Collins, W. E., Harinasuta, K. T.,

Tapchaisri, P., and Chomcharn, Yo, 1982, Circumsporozoite proteins of human malaria parasites *Plasmodium falciparum* and *Plasmodium vivax*, *J. Exp. Med.* **156**:20.

Nussenzweig, R. S., Vanderberg, J., Most, H., and Orton, C., 1969, Specificity of protective immunity induced by x-irradiated *P. berghei* sporozoites, *Nature* (*Lond.*) **222**:488.

Pal, R., 1973, The present state of insecticide resistance in anopheline mosquitoes, *WHO Mal.* 73:815.

Perkins, M., 1983, Surface proteins of schizont-infected erythrocytes and merozoites of *Plasmodium falciparum*, *Mol. Biochem. Parasitol.* **5**:55.

Perrin, L. H., and Dayal, R., 1982, Immunity to asexual erythrocytic stage of Plasmodium falciparum: Role of defined antigens in the humoral response, *Immunol. Rev.* **61**:245.

Perrin, L H., Magazin, M., Rieder, H., and Lambert, P. H., 1979, Antigens of *P. berghei* and *P. falciparum*: Characterisation using immune sera and monoclonal antibodies, in: *Immunodiagnostic techniques in Malaria, Transaction of the Third Meeting of the SWG on the Immunology of Malaria, Panama*, p. 27, WHO/TDR, Geneva.

Perrin, L. H., Ramirez, E., Lambert, P. H., and Miescher, P. A., 1981*a*, Inhibition of *P. falciparum* growth in human erythrocytes by monoclonal antibodies, *Nature* (*Lond.*) **289**:301.

Perrin, L. H., Dayal, R., and Rieder, H., 1981*b*, Characterization of antigens from erythrocytic stages of *Plasmodium falciparum* reacting with human immune sera, *Trans. R. Soc. Trop. Med. Hyg.* 75:1, 163.

Potocnjak, P., Yoshida, N., Nussenzweig, R. S., and Nussenzweig, V., 1980, Monovalent fragments (Fab) of monoclonal antibodies to a sporozoite surface antigen (Pb44) protect mice against malarial infection, *J. Exp. Med.* **151**:1504.

Rener, J., Carter, R., Rosenberg, Y., and Miller, L. H., 1980, Anti-gamete monoclonal antibodies synergistically block transmission of malaria by preventing fertilization in the mosquito, *Proc. Natl. Acad. Sci, USA* 77:6797.

Rodriguez da Silva, L., Loche, M., Dayal, R. and Perrin, L. H., 1983, *Plasmodium falciparum* polypeptides released during *in vitro* cultivation, *Bull. WHO* **61**:105.

Sanderson, A., Walliker, D. and Molez, J. F., 1981, Enzyme typing of *Plasmodium falciparum* from African and some other Old World countries, *Trans. R. Soc. Trop. Med. Hyg.* **75**:263.

Schmidt-Ulrich, R., Miller, L. H., Wallach, D. F., Lightholder, J., Powers, K. G., and Gwadz, R. W., 1981, Rhesus monkeys protected against *Plasmodium knowlesi* malaria produce antibodies against a 65,000-Mr*P. knowlesi* glycoprotein at the surface of infected erythrocytes, *Infect. Immun.* **34**:519.

Smith, R. D., James, M. A., Ristic, M., Aikawa, M., and Vega y Murguia, C. A., 1981, Bovine babesiosis: Prtection of Cattle with Culture-Derived Soluble *Babesia bovis* Antigen, *Science* **212**:335.

Sodeman, W. A., and Jeffery, G. M., 1964, Immunofluorescent staining of sporozoites of Plasmodium gallinaceum, *J. Parasitol.* **50**:477.

Todorovic, R., Ferris, D., and Ristic, M., 1967, Immunogenic properties of serum antigens from chickens acutely infected with *Plasmodium gallinaceum, Ann. Trop. Med. Parasitol.* **61**:117.

Trager, W., and Jensen, J. B., 1976, Human malaria parasites in continuous culture, *Science* **193**:673.

Wilson, R. J. M., 1974, The production of antigens by *Plasmodium falciparum* in vitro, *Int. J. Parasitol.* **4**:537.

Wilson, R. J., 1980, Serotyping P. falciparum malaria with S-antigens, *Nature* (*Lond.*) **284**: 451.

Wilson, R. J. M., McGregor, I. A., Hall, P., Williams, K., and Bartholomew, R., 1969, Antigens associated with *Plasmodium falciparum* infections in man, *Lancet* 2:201.

World Health Organization, 1976, Special Program for Research and Training in Tropical diseases No. 76.6, Malaria, WHO, Geneva.

Yoshida, N., Nussenzweig, R. S., Potocnjak, P., Nussenzweig, V., and Aikawa, M., 1980, Hybridoma produces protective antibodies directed against the sporozoite stage of malaria parasite, *Science* **207**:71.

Yoshida, N., Potocnjak, P., Nussenzweig, V., and Nussenzweig, R. S., 1981, Biosynthesis of Pb44, the protective anitgen of sporozoites of *Plasmodium berghei*, *J. Exp. Med.* **154**:1225.

Zuckerman, A., 1977, A review: Current studies of the immunology of blood and tissue protozoan. II. Plasmodium, *Exp. Parasitol.* **42**:374.

Chapter 5

Roles of Surface Antigens on Malaria-Infected Red Blood Cells in Evasion of Immunity

Russell J. Howard and John W. Barnwell

Malaria Section, Laboratory of Parasitic Diseases
National Institute of Allergies and Infectious Diseases
National Institutes of Health
Bethesda, Maryland 20205

I. INTRODUCTION

The life cycle of malaria parasites is extremely complex, involving multiple stages in the invertebrate vector and vertebrate host. However, the clinical manifestations of this disease in humans, and its mortality, are solely attributable to the asexual blood stages of the parasite. Furthermore, the blood stages are the forms of the parasite most susceptible to naturally acquired immunity. This chapter focuses on the surface antigens of infected red blood cells, their potential roles in inducing parasiticidal immune responses upon infection or immunization, their structure when known, and most importantly, recently obtained evidence that some of these surface antigens are involved in different mechanisms for evasion of host-protective immunity. Sections I–IV review the published literature and our own recent results implicating new surface antigens on infected cells as targets of antibody-dependent parasiticidal immune responses. Section V deals with the established capacity of malaria parasites to evade parasiticidal immunity and reviews the range of ways, both theoretical and indicated by experiment, in which this might be achieved. Finally, Sections VI and VII describe in detail two mechanisms for parasite evasion of immunity, with emphasis on recent experimental results from this laboratory. It is beyond the scope of this chapter to review mechanisms of malarial immunity not involving new antigens on infected erythrocytes. Excellent descriptions of alternative mechanisms are given elsewhere

(K. N. Brown, 1976; Cohen, 1979; Kreier and Green, 1980; McGregor, 1981; Allison, 1981; Jayawardena, 1981).

II. IMMUNITY TO ASEXUAL MALARIA PARASITES

A. Natural Infections in Humans

Epidemiologic studies in areas hyperendemic for malaria have shown that young infants remain relatively refractory to malaria for some months after birth, suggestive of some passive transfer of protection from the mother (Clark, 1939). After this period, young children bear the greatest density of parasites and highest clinical incidence over all age groups. Acute infections at this stage are largely responsible for the mortality of *Plasmodium falciparum* infections. Within a few years, young children show evidence for an ability to restrict the clinical effects of infection. They are able to harbor relatively high levels of blood parasitemia without the life-threatening symptoms of fever, anemia, and circulatory lesions (Sinton, 1939). An ability to restrict the density of blood parasites takes much longer to develop. Generally, by young adulthood the number of blood-stage parasites is controlled to low levels (McGregor, 1960); the symptoms of the disease are less dramatic, but no less insidious, as often they include persistent anemia and nephrotic complications (reviewed by I. N. Brown, 1969). Adults living in endemic areas not only have less frequent episodes of parasitemia than do young children living in the same areas (McGregor and Williams, 1978), they can control and rapidly clear their parasitemia upon reinfection (McGregor, 1960; McGregor and Williams, 1978). These observations indicate that protective immunity can be developed against the malaria parasite during natural infections. It is notable, however, that an extremely long period of repeated infection is required before host immunity is capable of clearing the bloodstream of all but a very low number of parasites (Table I, statement 1).

B. Experimentally Induced Infections in Human Subjects and Animals

Experimentally induced malarial infections permit more precise definition of the infective material and control of the timing of infection and reinfection. Furthermore, by initiating infection with asexual blood stage parasites via passage of infected blood, it is possible to avoid the complicating factor of uncontrolled relapses of new waves of asexual parasites from the liver that ensue after mosquito-borne inoculations with *vivax* and *malariae*-type malarias. A variety of malaria species in laboratory animals and birds, as well as infections of human volunteers with the human malarias (Redmond, 1941; Jeffery, 1966; Powell, *et al.*, 1972)

Table I. Experimental Evidence and Assumptions Involved in Support of a Role for New Antigens on Infected Erythrocytes in Antimalarial Immunity

Statements based on initial experimental evidence

1. Acquired protective immunity can be developed by repeated infection or immunization.
2. This immunity is generally malaria strain or species specific.
3. Antibody alone can confer protective immunity.[a]

Assumptions

1. Specific antigens exist that are the targets of host protective antibody.
2. Antibody binding to these antigens activates parasiticidal immune clearance.

Question

1. Where are the target antigens located?

Statements based on further experimental evidence

4. Malarial antigens are exposed on the surface of schizont-infected erythrocytes and on extracellular merozoites.
5. These antigens bind antibody from immune sera *in vitro*.
6. Antibody-dependent parasiticidal mechanisms operative against infected cells (e.g., opsonization) and merozoites (e.g., inhibition of reinvasion) have been demonstrated *in vitro*.

Assumption

3. Similar parasiticidal mechanisms are part of the protective immune response *in vivo*.

Question

2. What is the relative importance of immune responses to antigens on schizont-infected erythrocytes versus those on merozoites in different states of immunity?

[a]Protective immunity as defined in Section II.4.

have been used to study the development of acquired immunity to malaria. Often multiple infections are required to develop immunity, with drug-assisted clearance of parasites necessary in early inoculations. The effectiveness of such immunity depends to a large extent on the degree of previous infection. Resistance is generally most marked when subsequent infections involve the same parasite species, there being less protection or even no protection against challenge infections with other species (Voller and Rossan, 1969*b*).

These studies have led to the realization that a given species of *Plasmodium* may include a number of antigenically distinct strains of the parasite. For example, acquired immunity is most effective to challenge with the homologous strain, but there is often some degree of cross-protection when the challenge inoculum is a heterologous strain of the same species (Taliaferro, 1949; Jeffery, 1966; I. N. Brown, 1969; McGregor, 1981) (Table I, statement 2). These studies confirmed the epidemiologic findings that many repeated infections are required to achieve effective parasiticidal immunity.

C. Active Immunization

It is possible to induce some form of immunity to malarial infections by immunization using infected erythrocytes, purified merozoites, or subcellular fractions of infected cells, adjuvants being required for optimal effectiveness (Table I, statement 1). The extensive literature on this subject is reviewed elsewhere (Cohen and Mitchell, 1978; Desowitz and Miller, 1982). These immunization studies further support the postulated existence of particular target antigens of asexual parasites that can elicit potent parasiticidal immune responses upon immunization.

D. Nature of Immune Responses

1. Protective Immunity

Immunity to malaria parasites is characterized by a wide variety of host responses. Most have little or no effect on the survival of asexual parasites, as they are generated against parasite antigens that, for most of the parasite's life, are sheltered from the immune system, as they are within the infected erythrocyte. In addition, many host responses constitute a poorly understood polyclonal B-cell activation that leads to the production of numerous antibodies of unrelated specificities (Jayawardena, 1981). Nevertheless, some immune responses appear to exert a host-protective effect by interrupting in some way the proliferation of asexual parasites. When either the potency, or variety, or both, of these particular responses is sufficient, parasite destruction balances or exceeds multiplication and, provided the level of parasitemia and pathologic effects are not too great, the host survives.

Within the general terminology for malarial immunity, the terms protective immunity or parasiticidal immunity have a meaning that may differ from that used in other infections. These terms are used here to describe a state of partial or complete resistance in a previously susceptible host (I. N. Brown, 1969). Partial resistance is evident when the host is still infected but able to control, to varying extents, either the level of parasitemia, or the pathologic effects of infection, or both. A partially resistant host might therefore still suffer clinical symptoms of the disease and continue to harbor blood parasites. Complete resistance describes the capacity of the host to remove asexual blood parasites totally, i.e., sterile immunity. Both partial and complete protective immunity frequently show specificity for a particular strain or malaria species, reinforcing the concept that specific antigens exist that can be the targets of protective immunity (see Cohen and Mitchell, 1978, and Table I, statement 2).

Acquired protective immunity to malaria parasites may be evident as a lengthened period before which parasites appear in detectable numbers in the bloodstream, as a shortening of the period of the primary acute infection, a reduced parasitemia of the acute infection, a shortening of the period of relatively asymptomatic low-level parasitemia that may occur after resolution of the primary attack, or as a complete clearance of parasites from the bloodstream. I. N. Brown (1969) has pointed out that these manifestations of antiparasitic immunity may act quite independently via different mechanisms. Parasiticidal immune responses that completely prevent or tightly control the establishment of infection in a naive or immunized host may be different from those responsible for limitation of parasite growth during chronic infection.

The involvement of both cell-mediated and humoral immunity in malaria has been well established and extensively reviewed (Kreier and Green, 1980; Jayawardena, 1981; Allison, 1981). With respect to the focus of this chapter i.e., neoantigens on infected erythrocytes, only the latter has been documented. It is therefore appropriate to provide a brief overview of our current understanding of general antibody-mediated events only.

During malarial infections there is a marked increase in serum immunoglobulin levels. Specific antimalarial antibody can be demonstrated in these sera (reviewed by Sadun, 1966); however, there is a poor correlation between immune status of the host and the levels of antimalarial antibody. Much of the antimalarial antibody is therefore nonprotective (Targett and Voller, 1965). As mentioned previously, a large component of the total immunoglobulin in malarial sera is believed to result from nonspecific activation of B cells, since it appears not to react with malarial antigens (Abele *et al.*, 1965; Tella and Maegraith, 1965; Houba *et al.*, 1974; Greenwood *et al.*, 1971*a*; see Cohen and Butcher, 1971). There is evidence to support this view from the identification of a malarial antigen that causes B-cell activation *in vivo* (Greenwood, 1974; Freeman and Parish, 1981).

Despite this explosive humoral response of largely irrelevant antibody specificities, there is convincing evidence that antibodies are involved in some forms of acquired resistance to malaria.

Passive transfer of serum from immune donors has been shown to confer partial or complete immunity to infection in human, simian, and rodent malarias (reviewed by I. N. Brown, 1969; Phillips, 1970; Cohen and Butcher, 1971; also Phillips and Jones, 1972; Jayawardena *et al.*, 1978; Kreier and Green, 1980; Freeman and Parish, 1981) (Table I, statement 3). Protection conferred by passive serum transfer is species specific. Administration of immune adult immunoglobulin G (IgG) to 12 Gambian infants with severe malaria cured them of the disease and markedly reduced parasitemias (Cohen *et al.*, 1961). Passive transfer of serum immunoglobulins from adults living in endemic areas caused a dramatic fall in asexual forms of *P. falciparum* and *P. malariae* in children (McGregor *et al.*, 1963; Cohen *et al.*, 1969). IgG from immune adults living in West Africa was shown to control or reduce *P. falciparum* infections in children living in either West or East Africa (McGregor *et al.*, 1963). However, passive transfer experiments with serum from immune adults in Africa and Asia showed that reciprocal immunity is not transferred to these infections, providing further evidence for multiple strains within a particular malaria species (see Cohen and Butcher, 1971). In addition, human IgG purified from the sera of patients from endemic areas in Nigeria was shown to provide partial protection to *Aotus* monkeys infected with *P. falciparum* (Diggs *et al.*, 1972).

In reviewing the evidence for transfer of antimalarial immunity with serum or immunoglobulin, Jayawardena (1981) concluded that serum antibody can have a protective role, but that immunity was unlikely to depend exclusively on humoral mechanisms. It was pointed out that although hyperimmune sera, i.e., multiple immunizations, are effective, immune sera from animals recovered from a single infection are generally ineffective. Furthermore, the protective effects of serum transfer are not absolute except against low-challenge inocula, and protection is generally short-lived. It has also been shown that sera with protective activity are less effective when transferred into T-cell-deprived or splenectomized recipients, suggesting that protective effects in normal animals depend on parallel activation of cell-mediated responses (see Jayawardena, 1981).

The importance of B-cell responses for protective immunity is also evident from observations that chickens rendered agammoglobulinemic by chemical bursectomy (Rank and Weidanz, 1976) and mice rendered B-cell deficient by anti-μ treatment (Weinbaum *et al.*, 1976) succumb to normally nonlethal malarial infections.

This evidence leads us to infer that recognition of specific antigens by antibody can lead to the activation of antibody-dependent immune effector mechanisms that in turn lead to the destruction of asexual parasites (Table I, assumptions 1 and 2).

2. Target Antigens

The evidence for antibody-mediated parasiticidal immunity leads us to ask: What are the potential locations in the asexual cycle of parasite multiplication at which this antibody may act? Where are the target antigens? (Table I, question 1). Since antimalarial antibody levels in immune sera do not necessarily correlate with the capacity of these sera to confer immunity on passive transfer, many workers have sought to establish the nature of protective antimalarial antibody and its sites of action by performing *in vitro* assays for parasiticidal effects. These assays have led to the identification of at least two phases of the asexual parasites growth cycle that appear to be susceptible to antibody-dependent parasiticidal mechanisms *in vitro* and that serve as the current focus of attempts to demonstrate similar parasiticidal effects *in vivo.*

Figure 1 summarizes our current perception of the locations of target antigens. One phase is the intraerythrocytic parasite undergoing nuclear division, i. e., the schizont. At this stage of the parasite's growth, there is evidence for new antigens on the surface of the infected erythrocyte and increasing evidence for parasiticidal immune responses activated by antibody binding to these antigens. The second phase is the extracellular invasive form of the parasite, the merozoite, which is liberated from mature schizont-infected cells in order to infect new erythrocytes. Antigens on the surface of the merozoite have been shown to be recognized by antibody in immune sera, the consequences of which are decreased invasiveness and lowered rates of asexual proliferation. The target antigens of both phases of the asexual cycle that are considered subject to parasiticidal immunity are found on the surface of cells, making them accessible to antibody (Table I, statement 4).

The nature of new antigens on the surface of infected erythrocytes is discussed at length in Section IV. Evidence for the second type of target antigen, i.e., merozoite-borne determinants, was first provided by the studies of Cohen and colleagues. Their initial studies with *P. knowlesi* demonstrated inhibition of *in vitro* proliferation by sera from immune monkeys (Cohen *et al.*, 1969; Cohen and Butcher, 1970). Inhibition was found to be species-specific, complement independent, and dose dependent for the amount of Ig and could be produced by $F(ab')_2$ antibody fragments from immune sera.

One of the most illuminating findings of these studies was that antibody appeared, to have no effect on parasite growth and intraerythrocytic multiplication, but exerted its inhibitory effect only after schizogony (Fig. 1, mechanisms G, H, A, B). Surface antigens on the only extracellular form of the asexual parasite, the merozoite, were thereby implicated as targets for parasiticidal immunity (Table I, statements 5 and 6). Antibody bound to the merozoite surface was suggested to prevent attachment and invasion of erythrocytes and thus prevent continuing infection.

Several studies later confirmed the existence of antibody in immune human (Mitchell *et al.*, 1976; Wilson and Phillips, 1976; J. Brown and Smalley, 1981; G. V. V. Brown *et al.*, 1982) or monkey (Butcher and Cohen, 1972; Reese and Motyl, 1979; Chulay *et al.*, 1981) sera, which similarly inhibited multiplication of *P. falciparum in vitro*. Several different monoclonal antibodies that inhibit *P. falciparum* growth *in vitro* have also been identified (Perrin *et al.*, 1980, 1981). Sera from monkeys immune to *P. knowlesi* were shown to decrease *in vitro* proliferation by causing agglutination of free merozoites, thereby decreasing their invasiveness (Table I, statements 5 and 6). Inhibition was maximal with sera from animals immune to the homologous *P. knowlesi* strain (Miller *et al.*, 1975) (Fig. 1, mechanism H.).

Another mechanism of antibody inhibition of invasion was found with immune monkey serum and *P. falciparum* (Green *et al.*, 1981). In this case, the dispersal of merozoites from mature schizonts was inhibited by serum, preventing their interaction with new erythrocytes (Fig. 1, mechanism G.). Evidence for phagocytosis of agglutinated merozoites (Chow and Kreier, 1972; Green and Kreier, 1978; Brookes and Kreier, 1978) was consistent with the proposal that antibody could prevent merozoite reinvasion by agglutination, before or after

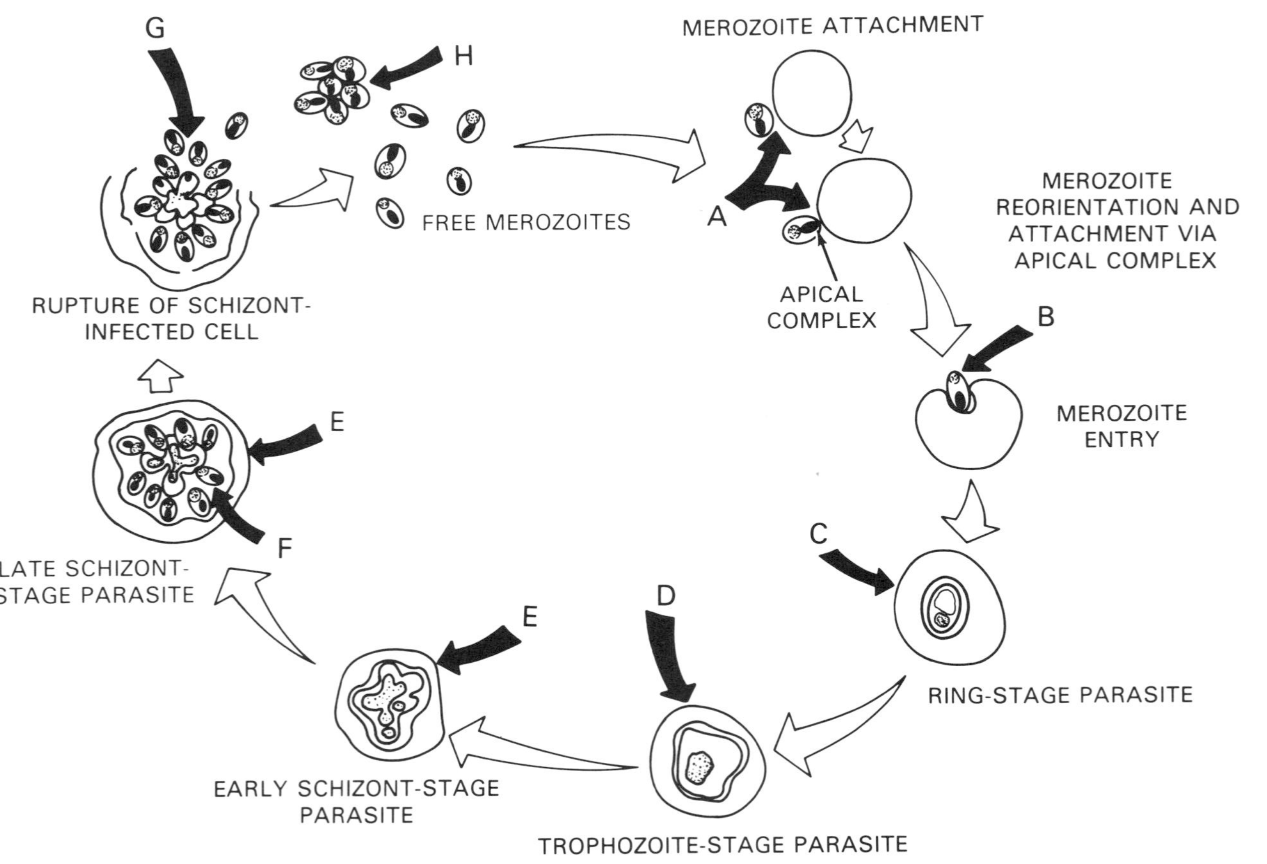
G
H
MEROZOITE ATTACHMENT
RUPTURE OF SCHIZONT-INFECTED CELL
FREE MEROZOITES
A
MEROZOITE REORIENTATION AND ATTACHMENT VIA APICAL COMPLEX
APICAL COMPLEX
B
MEROZOITE ENTRY
E
F
LATE SCHIZONT-STAGE PARASITE
C
D
E
RING-STAGE PARASITE
EARLY SCHIZONT-STAGE PARASITE
TROPHOZOITE-STAGE PARASITE

Figure 1. Hypothetical scheme showing points in the asexual cycle of parasite multiplication at which parasiticidal antibody might act. (A) Blockade of erythrocyte invasion by antibody against merozoite receptors for attachment, or against receptors for junction formation via the merozoites apical complex (Dvorak *et al.*, 1975; Miller *et al.*, 1979). (B) Antibody bound to the merozoite surface might enter with the invading parasite and disrupt intraerythrocytic growth (e.g., by disruption of membrane function). (C) Ring-stage parasites may express new antigens on the surface of the infected cell that are essential for parasite function. Antibody might either sensitize these cells for destruction, or affect intraerythrocytic growth by binding to antigens with metabolic functions. (D) Trophozoite-infected cells are known to express new antigens on their surface that appear to increase in density on schizont-infected cells, as shown at (E) (reviewed in Section IV). Antibody might sensitize these cells for destruction or act against the intraerythrocytic parasite as described for (C). (F) Erythrocytes infected by mature intraerythrocytic parasites, either schizonts or segmenters containing separated daughter parasites, appear to have increased membrane permeability (reviewed by Sherman, 1979). It is possible that antibody could enter the infected cell just before rupture. (G) Antibody could bind to the surface of merozoites before they are dispersed from the residual body, thereby preventing their release as individual parasites (Green *et al.*, 1981). (H) Extracellular free merozoites might be agglutinated by antibody, thereby reducing the extent of erythrocyte reinvasion (Miller *et al.*, 1975). The potential immune effector mechanisms activated by antibody binding on the surface of infected cells (C–E) are discussed in detail in Section III.B.1. The nature of the new antigens on the surface of infected cells (C–E) is discussed in Section IV. Evidence for an additional way in which antibody binding to the surface of infected erythrocytes of the *falciparum*-type malarias might result in parasite destruction is presented in Section VII.D.

dispersal from the schizont and that macrophages in liver and spleen would eliminate the antibody-coated parasites from the circulation.

Using the *in vitro* tests for antibody-dependent inhibition of merozoite reinvasion, attempts have been made to correlate the inhibitory capacity of different sera with the immune status of the serum donor. Whereas serum inhibitory activity and host-protective immunity showed good correlation in some rhesus monkeys infected with *P. knowlesi* (Butcher and Cohen, 1972), other animals with functional immunity failed to show inhibitory antibody in this test (Miller *et al.*, 1977). Since considerable variations in immune responses to infection are expected and documented for *P. knowlesi* (Voller and Rossan, 1969*b*), these results do not invalidate the potential role of antibody blockade of merozoite reinvasion as a protective mechanism *in vivo*. Clearly, other mechanisms of protective immunity must exist.

We expect that in some states of immunity the merozoite will be the major target of protective immune responses in the asexual cycle of parasite multiplication. This concept was confirmed by the demonstration that a monoclonal antibody that binds to a surface-membrane determinant on *P. yoelii yoelii* merozoites dramatically reduced the blood level of asexual parasites when injected *in vivo* (Freeman *et al.*, 1980). Furthermore, mice immunized with a merozoite-specific protein of M_r 235,000 purified from *P. yoelii yoelii* using this monoclonal antibody, were also protected against infection with this parasite (Holder and Freeman, 1981). Identification of the surface antigens on merozoites is therefore a critical step in understanding their role as targets for parasiticidal immunity (see Johnson *et al.*, 1980; Epstein *et al.*, 1981; Holder and Freeman, 1981). However, in other states of immunity, in which antibody does not appear to be capable of blocking merozoite reinvasion, other antigens must constitute the targets of parasiticidal immune responses (Table I, question 2). The remainder of this chapter describes new antigens on infected erythrocytes, some that might be potential targets for parasiticidal immunity to asexual malaria parasites, and others that appear to be linked to the capacity of malaria parasites to evade host immunity.

III. ANTIGENS ON MALARIA-INFECTED ERYTHROCYTES—TARGETS OF PARASITICIDAL IMMUNITY?

The asexual malaria parasite remains inside the host erythrocyte for many hours (~24–72 hr, depending on the *Plasmodium* species) as it grows in size, differentiates, and divides to produce 6 to 24 daughter merozoites, depending on the species. Once the infected cell ruptures, the released merozoites reinvade new erythrocytes within 30 sec (Dvorak *et al.*, 1975). The extracellular merozoite is therefore accessible to the humoral arm of host-protective immunity only for a very brief period, prompting the speculation that perhaps other sites, such as the surface of the infected erythrocyte, might elicit parasiticidal immunity as well. In this section we summarize the published evidence for the appearance of new antigens on the surface of schizont-infected cells. It will be seen that these antigens are accessible to antibody *in vitro*. The prolonged period over which these new antigens on infected cells would be exposed to the appropriate humoral antibodies *in vivo* has led to the proposal that these antigens should be potential targets for parasiticidal immunity.

A. New Antigens Expressed on Malaria-Infected Erythrocytes

1. Evidence

There are at least six lines of published evidence showing that new antigens must be expressed on the surface of malaria-infected erythrocytes (Table I, statements 4 and 5). Most methodologic approaches have employed antibody binding to the surface of these cells as an indication of the presence of new antigens. More recently, immunochemical methods for identifying new protein antigens have also been used.

1. *Agglutination*: Schizont-infected erythrocytes of *P. knowlesi* are specifically agglutinated by antibody (Eaton, 1938). This agglutination reaction, schizont-infected cell agglutination (SICA), defines antigens on these cells, called the SICA antigens (K. N. Brown and I. N. Brown, 1965; K. N. Brown, 1977). Their biochemical nature is described in Section IV. C. 4.
2. *Immunoelectron microscopy*: Antibody from *P. berghei*-infected mice (Poels *et al.*, 1977), *P. vivax*-infected monkeys (Aikawa *et al.*, 1975), and *P. falciparum*-infected monkeys (Langreth and Reese, 1979) has been shown by immunoelectron microscopy to bind to the surface of erythrocytes infected by these parasites. After adsorption with uninfected erythrocytes, rat antisera to *P. falciparum* was also found to contain antibody that bound specifically to infected cells (Kilejian *et al.*, 1977).
3. *Labeled secondary antibody*: Fluorescent or radiolabeled secondary antibody has been used to detect binding of antibody from infected animals to the surface of infected cells (Lustig *et al.*, 1977; Kim *et al.*, 1980; Hunter *et al.*, 1980).
4. *In vitro opsonization and phagocytosis*: Opsonization dependent on antimalarial antibody and phagocytosis of infected erythrocytes has been demonstrated for rodent (Chow and Kreier, 1972; Lustig-Shear *et al.*, 1979; Hunter *et al.*, 1979; Tosta and Wedderburn, 1980), avian (Zuckerman, 1945), simian (K. N. Brown *et al.*, 1970*b*; K. N. Brown and Hills, 1971), and human (Celada *et al.*, 1982) malarias (Table I, statement 6).
5. *Inhibition or reversal of attachment of P. falciparum-infected erythrocytes to endothelial cells in vitro*: Antibody blocks this specific interaction of infected erythrocytes and endothelial cells (see Section VII. D

for a discussion of this new assay) and confirms the existence of new antigens on *P. falciparum*-infected cells (I. J. Udeinya and L. H. Miller, unpublished data, 1983).

6. *Binding of antibody-coated infected cells to protein A-coated beads*: When blood from *P. falciparum*-infected squirrel monkeys was preincubated in immune serum and then perfused through a column of protein A-coated beads, infected erythrocytes were specificially retained by the column (David *et al.*, 1982). Since only infected erythrocytes containing mature trophozoites and more mature parasite stages were retained, it was concluded that these cells express new antigens on their surface.
7. *Biochemical studies*: Schizont-infected erythrocytes have been radiolabeled by methods claimed to be specific for cell-surface components and radiolabeled antigens have been identified in detergent extracts by crossed immunoelectrophoresis or isoelectric focusing/immunoelectrophoresis (Schmidt-Ullrich and Wallach, 1978; Deans *et al.*, 1978; Deans and Cohen, 1979; Schmidt-Ullrich *et al.*, 1979*a*,*b*). Finally, recent studies from our laboratory identifying the biochemical nature of the SICA-variant antigens have provided definitive evidence for malarial protein antigens on the surface of *P. knowlesi*-infected cells (R. J. Howard *et al.*, 1983, see Section IV. C. 4).

2. *Some New Antigens Shared by Infected and Uninfected Erythrocytes*

Most of the above-described studies showed the new antigens on infected cells to be absent from the surface of uninfected cells in the same blood sample, e.g., the SICA antigens of *P. knowlesi*, the antigens detected by immunoelectron microscopy or by biochemical radiolabeling studies, and the *P. falciparum* antigens related to reversal of endothelial attachment. However, some antigens detected by binding of labeled secondary antibody or by *in vitro* antibody-dependent phagocytosis appear to be shared by both infected and uninfected erythrocytes from infected blood, yet are absent from the surface of normal erythrocytes (Lustig *et al.*, 1977; Hunter *et al.*, 1980). Since the sera used in these studies were polyspecific, it is not yet possible to determine whether some antibody specificities in these cases were specific for surface antigens restricted to infected cells. In any case, a sharing of new antigens by infected and uninfected erythrocytes from infected animals is more likely to be linked to the excessive destruction of uninfected cells and to the consequent anemia that develops during many malarial infections (reviewed in Roth and Herman, 1979) than to parasiticidal immune responses.

3. *Potential Orgins of New Antigens on Infected Erythrocytes*

The above-described experiments clearly show that antimalarial antibody is capable of recognizing new antigens on the surface of malaria-infected erythrocytes *in vitro*. What potential mechanisms exist for the creation of new antigens on this membrane?

New antigens on the surface of infected erythrocytes could conceivably arise by insertion of parasite-derived proteins, glycoproteins, or glycolipids into the membrane. These components could be derived from within the infected cell, reflecting parasite synthesis and export to the surface of the infected cell. It should be noted that the parasite is separated from the host cell membrane by the parasitophorous vacuole membrane and by the host cell cytoplasm between these two membranes. Alternatively, parasite components released into plasma from rupturing cells at schizogony could attach to infected cells. New antigens could also arise by modification of host cell-membrane components; e.g., host proteins could be partially cleaved by proteases; glycoproteins could be subject to glycosidase cleavage; enzymes in plasma or on the membrane of the infected cell could create new antigenic determinants on host cell components by phosphorylation, glycosylation, acylation, or methylation; and changes in the membrane of infected erythrocytes could also expose formerly cryptic autoantigens.

Section III. B provides definitive evidence for the creation of new antigens on the surface of *P. knowlesi*-infected cells, by one of these mechanisms, i.e., insertion into the membrane of parasite proteins derived from the intracellular parasite. This result raises the possibility that other new antigens identified on infected cells, such as knob antigens and caveola antigens, are also synthesized by the intracellular parasite.

B. Parasiticidal Mechanisms—Activated *in Vivo* by Antibody Binding to Infected Cells

1. *Introductory Remarks*

Classic studies on antibody-mediated clearance of bacteria (Biozzi *et al.*, 1975), viruses (Mims, 1964), and altered erythrocytes (Jandl *et al.*, 1957; Jandl and Kaplan, 1960) from the circulation led to the identification of several immune mechanisms for clearance of antibody-coated particles. These mechanisms include binding of antibody-coated particles to macrophages via macrophage receptors for immunoglobulin (i.e., opsonization), followed by phagocytosis; binding of complement components to antibody attached to the foreign particle, followed by phagocytosis (in this case, macrophage receptors for complement components mediate macrophage binding); binding of complement components; activation of the complement cascade; and complement-mediated lysis of the antibody-coated particle. By analogy with these studies, one would expect that antibody-coated malaria-infected erythrocytes would be rapidly cleared from the circulation.

Before evaluating relevant data from the malaria literature, we should bear in mind an important caveat that reflects our ignorance of antimalarial immunity *in vivo*. Although antibody binding to the surface of malaria-infected erythrocytes can be readily demonstrated *in vitro*, these studies have generally used antimalarial serum collected either from an immunized animal or from an infected animal several days after resolution of acute infection, together with infected erythrocytes taken from another animal during acute infection. It should therefore, be pointed out that direct evidence for antibody binding to infected cells *in vivo* during resolution of malarial infection is not available. As monoclonal antibodies become available to defined surface antigens on infected cells, and once some of these antigens are purified, it may be possible to measure accurately both specific antibody levels in serum during all phases of infection as well as levels of specific antibodies bound to the surface of infected cells taken directly from the host. Such studies may require transfer of infected cells from one animal to another, since the clearance of antibody-coated cells *in vivo* might be very rapid.

2. *In Vivo Observations*

Histologic examinations of spleen and liver during both primary malaria infection and clearance of parasites from immune animals (Cannon and Taliaferro, 1931; Taliaferro and Cannon, 1936; Taliaferro and Mulligan, 1937) showed a marked enhancement of erythrocyte phagocytosis in malaria. Furthermore, the extent of phagocytosis of infected cells correlated with the degree of immunity. It is not clear whether this phagocytosis reflects opsonization, recognition of complement bound to parasitized cells or some other unidentified mechanism that might not even require antibody binding.

3. *In Vitro Experiments*

a. Opsonization. One series of *in vitro* studies found immune serum to have little opsonizing effect on phagocytosis of *P. berghei*-infected rat erythrocytes by rat macrophages (Chow and Kreier, 1972; Hamburger and Kreier, 1976; Brookes and Kreier, 1978). In contrast, other workers using mouse erythrocytes infected with *P. yoelii* (Tosta and Wedderburn, 1980) or *P. berghei* (Hunter *et al.*, 1979; Lustig-Shear *et al.*, 1979) demonstrated significant opsonizing effects with immune serum. With *P. knowlesi*-infected rhesus monkey erythrocytes the opsonizing activity of immune sera was found to correlate with immunity (K. N. Brown *et al.*, 1970*b*; K. N. Brown and Hills, 1971). Later, the same investigators showed that protection could develop in the absence of serum opsonins (K. N. Brown and Hills, 1974), suggesting that other factors may be involved in different states of protective immunity. This view was corroborated by studies with another malarial infection when Diggs and Osler (1975) showed that serum that protects rats against *P. berghei* infection failed on passive transfer to sensitize parasitized erythrocytes *in vitro* for *in vivo* destruction. From the kinetics and organ distribution of clearance of ^{51}Cr-labeled *P. berghei*-infected cells in immune and nonimmune animals, Quinn and Wyler (1979) concluded that the increased clearance rate of these cells over homologous uninfected erythrocytes was caused by rheologic alterations of parasitized cells rather than by antibody-dependent clearance mechanisms.

A more recent study using human monocytes from normal blood donors and *P. falciparum*-infected human erythrocytes from *in vitro* culture has reinforced the potential importance of opsonization and phagocytosis as a mechanism of parasite elimination (Celada *et al.*, 1982). The phagocytic activity of monocytes for infected cells was greatly enhanced by addition of immune sera from persons living in areas with endemic malaria. However, serum from subjects recovering from first infection or pooled normal human sera failed to enhance phagocytosis in this *in vitro* assay. The IgG fraction of immune sera was responsible for enhancement of phagocytosis. Adsorption of sera with infected erythrocytes, but not normal cells, removed the opsonic activity. These workers also demonstrated that infected cells containing schizonts and trophozoites were preferentially phagocytosed over infected cells containing ring forms thereby implicating new antigens on mature parasitized cells as the targets of opsonic antibody (Celada *et al.*, 1982). It was concluded that while opsonization appeared unlikely to be a major immune-protective mechanism in primary infections, it may prove to be a major protective mechanism in immune persons subject to repeated infection (Table I, statement 6).

b. Complement-Mediated Clearance of Infected Cells. There is no direct evidence for specific clearance of infected cells mediated by binding of antibody and complement components. Such evidence might not be expected from studies on infected cells collected from the circulation, since their clearance might be very rapid were they to be sensitized in this manner. In contrast, there is a growing body of evidence for nonspecific sensitization of uninfected erythrocytes, and presumably of infected-erythrocytes as well, by binding of parasite–antigen complexes that can then fix complement. Rupture of schizont-infected cells and resultant release of schizont antigens into plasma has been associated temporally with consumption of complement components (Glew *et al.*, 1975; Atkinson *et al.*, 1975). These schizont antigens are believed to form immune complexes that can bind to erythrocytes, and remain in the circulation or bind to other tissues (reviewed by Nussenzweig *et al.*, 1978). Binding of these antigen–antibody complexes to erythrocytes activates complement consumption and results in binding of complement to the cell surface. Binding of these complexes to other tissues also activates complement and causes glomerulonephritis, especially with *P. malariae* and *P. falciparum.* These postulates are supported by evidence for sensitization of uninfected erythrocytes from *P. falciparum* infections in humans with complement alone, with IgG, or with IgG plus complement (McGregor, 1978; Facer *et al.*, 1979; Facer, 1980*a,b*). Most importantly, immunoglobulin eluted from uninfected erythrocytes in the circulation of malaria patients has been shown to be specific for schizont antigens (Facer, 1980*a*).

Finally, complement binding to the surface of erythrocytes has been associated with pronounced phagocytosis (Woodruff *et al.*, 1973). This chain of events is generally cited as one of the causes of excessive anemia and destruction of uninfected erythrocytes in malaria (reviewed by Zuckerman, 1977; Facer, 1980*a*).

Since the binding of antigen–antibody complexes to uninfected erythrocytes appears to be a passive attachment (Facer, 1980*a*), is it possible that these complexes also bind to infected cells and activate complement binding, followed by phagocytosis of these cells? This nonspecific mechanism of clearance of infected erythrocytes will occur as a function of such factors as the degree of synchrony of the parasites, since the mechanism is activated by rupture of schizont-infected cells, the lifetime of circulating immune complexes, and the relative affinity of these complexes for uninfected cells, as opposed to erythrocytes infected by immature or mature parasites.

c. Complement-Mediated Lysis. Complement-mediated lysis of antibody-coated infected erythrocytes could not be demonstrated *in vitro* using *P. berghei* (Diggs and Osler, 1969) or *P. knowlesi* (K. N. Brown *et al.*, 1970*b*) -infected erythrocytes.

4. Summary

It should be apparent that our understanding of *in vivo* parasiticidal mechanisms for destruction of asexual malaria parasites in acquired immunity is minimal. While it is clear from histologic examination that phagocytosis of infected, and often uninfected, cells is enhanced in immune hosts, we only have *in vitro* evidence, which itself is incomplete, on the possible mechanisms of immune clearance (Table I, assumption 3). Infected cells clearly express new antigens on their surface that can be recognized *in vitro* by antibody from infected or immunized animals. There is little evidence for antibody binding to these cells either during recovery from infection or on challenge infections when the host is able to exercise rapid control over the extent of parasite multiplication. The rest of this chapter is concerned with the nature of the new antigens on infected cells, as it is our belief that identification of these antigens will be required before we can explore the consequences of recognition of these antigens *in vivo*.

IV. NATURE OF NEW ANTIGENS ON INFECTED CELLS

A. New Antigens Associated with Morphologic Alterations of the Membrane

1. Knobs

Plasmodium falciparum-infected erythrocytes develop alterations in their plasma membranes visible by electron microscopy as inverted cuplike plaques of electron-dense material just beneath the membrane (Langreth *et al.*, 1979) (Fig. 2). These alterations, called knobs or excrescences, have been identified on *P. falciparum*-infected erythrocytes of human and monkey origin, both *in vivo* and *in vitro* (Trager *et al.*, 1966; Luse and Miller, 1971; Aikawa *et al.*, 1972; Miller, 1972; Langreth *et al.*, 1978). More recently, knobs have been visualized with living parasite material by video-enhanced differential interference contrast microscopy (Trager *et al.*, 1982). *In vitro*, knobs are seen to appear first on the surface of trophozoite-infected cells and to increase in number as the parasite matures to schizont and segmenter stages (Langreth *et al.*, 1978). Immunoelectron microscopy has been used to show that different antigens are located on the erythrocyte surface at the knob compared with the surrounding areas (Kilejian *et al.*, 1977). Immune *Aotus* monkeys produce antibodies to the knobs as visualized by immunoelectron microscopy (Langreth and Reese, 1978, 1979).

Long-term *in vitro* culture of *P. falciparum* in human erythrocytes has been shown to lead to loss of knobs form the surface of trophozoite and schizont-infected cells (Langreth *et al.*, 1979). The parasites that express knobs have been described as knobby or K^+ to distinguish them from parasites, such as those from long-term culture, which are knobless or K .

Knobs are also found on the surface of erythrocytes of other *falciparum*-type malarias, such as *P. fragile* and *P. coatneyi. P. malariae*- and *P. brasilianum*-

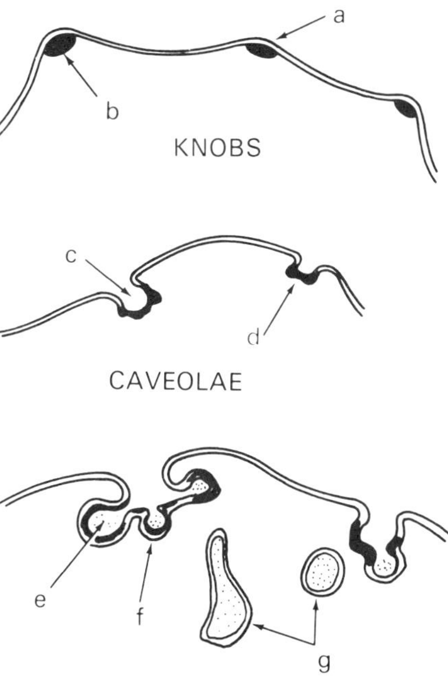

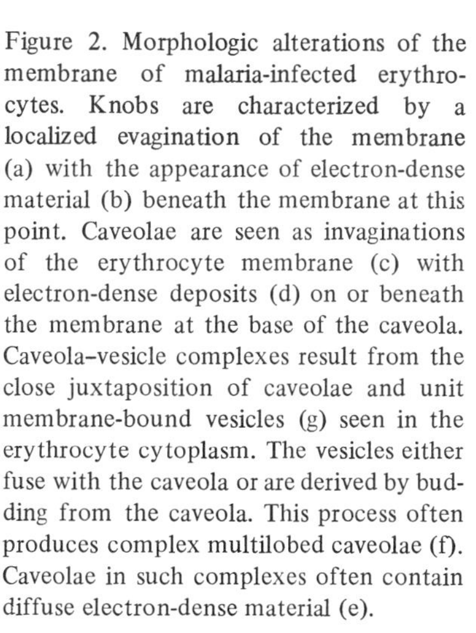
Figure 2. Morphologic alterations of the membrane of malaria-infected erythrocytes. Knobs are characterized by a localized evagination of the membrane (a) with the appearance of electron-dense material (b) beneath the membrane at this point. Caveolae are seen as invaginations of the erythrocyte membrane (c) with electron-dense deposits (d) on or beneath the membrane at the base of the caveola. Caveola–vesicle complexes result from the close juxtaposition of caveolae and unit membrane-bound vesicles (g) seen in the erythrocyte cytoplasm. The vesicles either fuse with the caveola or are derived by budding from the caveola. This process often produces complex multilobed caveolae (f). Caveolae in such complexes often contain diffuse electron-dense material (e).

infected erythrocytes also express knobs on the surface of the infected cell. Gametocyte-infected erythrocytes of the latter two *malariae*-type infections also have knobs on the cell surface. Other malaria parasites, such as *P. knowlesi* or *P. vivax*, do not exhibit knobs on the surface of infected erythrocytes (reviewed by Aikawa *et al.*, 1975).

Biochemical studies on the composition of knobs in these different parasites are under way. A protein of ~ 80,000 molecular weight has been identified by Coomassie Blue staining and by incorporation of radiolabeled proline or histidine, as a component of *P. falciparum* schizont-infected cells of K^+ but not K^- parasites. This protein is also absent from ring-stage K^+ parasites that do not express knobs (Kilejian, 1979, 1980, 1981). These results suggest that this 80,000-molecular-weight protein is associated in some way with the knob structure. It is not yet known whether this protein is on the erythrocyte surface at the knob, is part of the electron-dense material below the erythrocyte membrane at the knob, or is located elsewhere in the infected cell.

Taken together, it is clear that knobs represent a special type of morphologic alteration in the surface membrane of erythrocytes infected by a variety of species of malaria parasites. This morphologic alteration is known to be associated with the appearance of new antigens at the surface of the cell and the presence of at least one new protein on or in these cells. Discussion of the known and supposed functions of knobs *in vivo* is deferred until Section VII, since their function appears to be closely related to a special mechanism for evasion of parasiticidal immune responses.

2. *Caveolae and Caveola-Vesicle Complexes*

Giemsa-stained blood films of *vivax*- or *ovale*-type malaria parasites examined by light microscopy exhibit a pattern of stippling called Schüffner's dots. An ultrastructural counterpart of the pattern of stained dots is seen by electron microscopy. The plasmalemma of erythrocytes infected with these parasites exhibits many small invaginations or caveolae of ~ 90-nm diameter. The caveola base is often flattened and coated with fuzzy electron-dense material. Small vesicles of ~ 50-nm diameter are often seen surrounding the base of the caveola, forming a caveloa-vesicle complex (Sterling *et al.*, 1975; Aikawa *et al.*, 1975, 1977). Vesicles are also seen scattered free in the cytoplasm of erythrocytes infected by these parasites (Fig. 2). The presence of ferritin particles in the caveola-vesicle complexes of *P. vivax* or *P. cynomolgi*-infected erythrocytes suspended in solution containing cationized ferritin suggests that these complexes originate as endocytotic rather than exocytotic vesicles (Aikawa *et al.*, 1975). Most importantly, horseradish peroxidase-labeled antibody from monkeys chronically infected with *P. vivax*, but not labeled antibody from uninfected monkeys, bound to the vesicle membrane when incubated with intact cells. These morphologic alterations in the membrane of infected erythrocytes therefore appear to result in the expression of new antigens at the cell surface, as is true with knobs. The nature of these antigens is unknown. It should be noted that other species of malaria parasites, such as *P. knowlesi*, *P. fragile*, *P. coatneyi*, and *P. inui*, also exhibit caveloae on the surface of infected erythrocytes (reviewed in Aikawa *et al.*, 1975; see also Wunderlich *et al.*, 1982), but in these infections caveola–vesicle complexes are not observed, suggestive of functional differences in the caveolae of different species.

B. New Antigens Associated with Antigenic Variation: SICA Antigens

1. Antigenic Variation

Rhesus monkey erythrocytes containing schizont forms of *P. knowlesi* can be specifically agglutinated by antibody from infected animals (Eaton, 1938). The agglutination reaction was later called the SICA test and the antigen(s) recognized by this antibody SICA antigens. With the aid of this assay, it was discovered that these antigens appeared to vary during chronic infection (K. N. Brown and I. N. Brown, 1965). Chronic infection can be established in rhesus monkeys by suppressing initial infections with antimalarial drugs and then reinfecting when the animal's immunity can prevent the parasitemia from rising to life-threatening levels. Successive waves of parasites appear in the bloodstream over prolonged periods of relatively asymptomatic chronic infection. These parasite populations were cryopreserved and then inoculated into naive animals, in which they caused lethal infections unless suppressed by chemotherapy, in an effort to obtain serum from animals infected with only one population. The SICA phenotype of these different parasite populations was found to vary such that parasitized cells of a particular population were agglutinated only by sera from animals infected with the same population. Sera taken from animals that had been chronically infected over months or years were found to agglutinate many parasite populations of different SICA-antigen phenotypes (Brown and Brown, 1965; I. N. Brown *et al.*, 1968; Barnwell *et al.*, 1982*a*). Figure 3 summarizes observations on changes in SICA antigens in chronic parasitemia.

In contrast to the antigens of knobs and caveolae, which are part of distinct structures on the cell surface, the appearance of SICA antigens on the surface of rhesus monkey erythrocytes has not yet been associated with any particular morphologic change in the outer membrane of these cells.

2. Cloned Parasites of Different Variant Antigen Phenotypes

We have been able to clone *P. knowlesi* by micromanipulation of schizont-infected erythrocytes and thereby determine whether the genome of this parasite encodes for more than one variant antigen phenotype. A line of *P. knowlesi* Pk1(A) was cloned to produce Pk1(A+). Animals were infected with Pk1(A+),

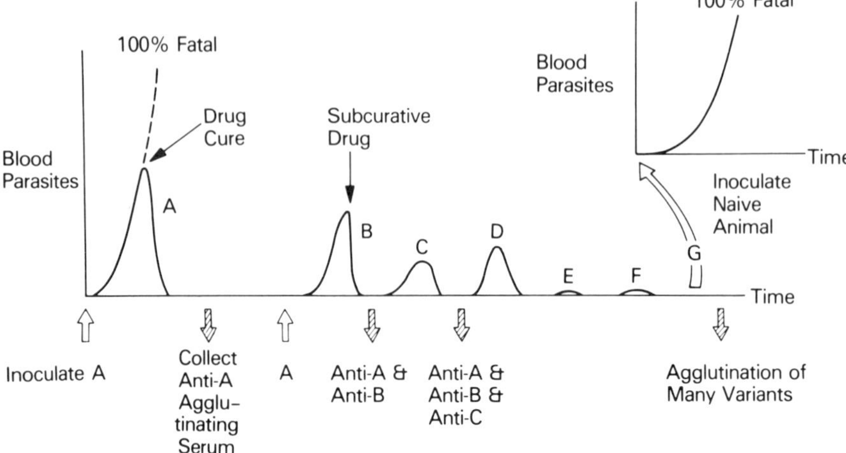

Figure 3. Antigenic variation of *P. knowlesi.* Schematic diagram summarizing the variation of SICA antigens on schizont-infected erythrocytes during chronic infection of rhesus monkeys and the variation in antibody response to SICA antigens (Brown and Brown, 1965; Brown *et al.*, 1968; Butcher and Cohen, 1972). Inoculation of naive animals with *P. knowlesi* is uniformly fatal in the absence of drug treatment. If the animal is drug-cured, serum collected after infection will agglutinate schizont-infected erythrocytes with the SICA phenotype of the inoculum (A). Reinoculation of an animal with the same SICA phenotype results in the appearance of parasites which again would kill the host if not drug cured. However, these parasites are not agglutinated by anti-A serum, being of another SICA phenotype (B). Subcurative drug treatment often leads to chronic, relatively low-grade relapsing parasitemia, shown by successive peaks of blood parasites. Each peak can be shown to consist of new SICA phenotypes of *P. knowlesi.* Serum collected after many such parasite relapses usually agglutinates many variants of the parasite strain. Parasites subinoculated by passage of blood from chronically infected animals produce lethal infections in naive hosts (e.g., peak G), indicating that the later antigenic variants are not avirulent.

drug cured, and then reinfected with Pk1(A+) after the appearance of serum antibody, which agglutinated these parasitized cells. The parasites recovered after reinfection were of different SICA phenotype from Pk1(A+) in that they were not agglutinated by the homologous antibody. This new line of *P. knowlesi*, Pk1(B+), was then recloned to produce the clone Pk1(B+)1+, an antigenic variant of Pk1(A+). This experiment showed that the genotype of *P. knowlesi* encodes for at least two different phenotypes (Barnwell *et al.*, 1982*b*; and Table V).

3. Nonagglutinable SICA[−] P. knowlesi

Some of the most illuminating properties of the SICA antigen (discussed in Section VI. B) have come from the identification of a phenotype of *P. knowlesi* that fails to express variant antigen on the surface of schizont-infected cells. Parasites passaged in splenectomized monkeys cannot be agglutinated by the SICA test. This phenotype is designated SICA[-] to distinguish it from parasites of the agglutinable (SICA[+]) phenotype, which express variant antigen on the infected cell (Barnwell *et al.*, 1982*a*). Comparison of the *in vivo* properties of parasites of the SICA[-] and SICA[+] phenotypes furnishes information on the immunologic role of the variant antigens (Section VI). Table II summarizes how SICA[-] parasites can be produced, the evidence for their lack of variant antigen expression, and some of their properties. We have also observed that under certain circumstances SICA[-] parasites reexpress variant antigens on the surface of the infected cell as summarized in Table III. Although the *in vivo* conditions required for this conversion of phenotype are poorly understood, it is clear that SICA[-] parasites can reexpress SICA-variant antigen, indicating that the genes for variant antigen are not lost by SICA[-] parasites.

Our observations of a dependence of *P. knowlesi* parasites on the presence of the host spleen for interconversion from one SICA[+] phenotype to another

Table II. The SICA[-] Phenotype[a]

Origin: Passage of SICA[+] parasites (cloned or noncloned) in splenectomized rhesus monkeys leads to gradual loss of variant antigen expression until eventually none can be detected on the surface of the infected cells.

Properties

1. Schizont-infected cells of this phenotype are not agglutinated by the following sera:
   a. Cross-reactive sera capable of agglutinating many SICA[+] variants of the same *P. knowlesi* strain
   b. Sera from splenectomized or intact animals infected multiple times with SICA[-] parasites
   c. Incubation with sera from (a) or (b) followed by secondary anti-rhesus monkey immunoglobulin
   d. Sera from rabbits immunized many times with SICA[-] parasites
2. Schizont-infected cells of this phenotype do not fluoresce when incubated with cross-reactive anti-SICA[+] sera or sera from animals infected or immunized with SICA[-] parasites, followed by fluorescein-conjugated anti-rhesus monkey immunoglobulin.
3. Biochemical studies show that SICA[-] parasites lack surface molecules on the schizont-infected cell in the molecular weight range shown to include the SICA antigens of SICA[+] parasites.
4. Comparison of the malarial antigens synthesized by SICA[+] and SICA[-] parasites (protein labeling with radioactive amino acids, detergent extraction, immunoprecipitation and 1D and 2D gel analysis) shows that aside from expression of variant antigen, these phenotypes are very similar, with very few differences in antigenic composition.
5. The initial rates of multiplication of SICA[-] and SICA[+] parasites are indistinguishable in splenectomized or intact animals. Expression of the SICA variant antigen *per se* is therefore not essential for parasite growth.

[a]Barnwell *et al.* (1982*a*,*b*); R. J. Howard and J. W. Barnwell (unpublished results, 1983).

Table III. Conversion of SICA[−] Parasites to SICA[+][a]

1. SICA[−] parasites passaged multiple times in splenectomized animals have never been observed to revert to SICA[+].
2. SICA[−] parasites (cloned and noncloned) passaged into naive nonsplenectomized animals generally remain SICA[−]. Reversion to the SICA[+] phenotype is occasionally seen after such passage.
3. SICA[−] parasites (cloned and noncloned) passaged into sterile nonsplenectomized animals previously infected with SICA[+] *P. knowlesi* have become SICA[+] (three experiments).
4. The genes for expression of SICA-variant antigen are therefore not lost from SICA[+] parasites when they become SICA[−] in splenectomized animals.

[a]Barnwell *et al.* (1982*a*,*b*).

(Barnwell *et al.*, 1982*b*) and of a dependence of these parasites on the host spleen for expression of variant antigen per se (Barnwell *et al.*, 1982*a*) open up fascinating questions at the molecular and biologic levels of host factors that appear to control the phenotype of malaria parasites. These questions are discussed at length elsewhere (Barnwell *et al.*, 1982*b*).

4. *Identification of SICA-Variant Antigens*

We have identified the SICA-variant antigens of *P. knowlesi* using the cloned parasites Pk1(A+) and Pk1(B+)1+ and a panel of rhesus monkey and rabbit antisera of defined agglutination specificity (Howard *et al.*, 1983). Our criteria for establishing the nature of the variant antigens were the following:

1. The only sera capable of immunoprecipitating variant antigen from cells of a particular SICA phenotype should be sera capable of agglutinating those cells.
2. Variant antigen should not be immunoprecipitated from uninfected erythrocytes that are not agglutinated.
3. The variant antigen of a particular clone should not be immunoprecipitated from another clone that exhibits completely different agglutination specificity with the panel of sera.

Additional properties expected of the SICA-variant antigen were the following:

1. Variant antigen should not be immunoprecipitated from SICA[−] nonagglutinable parasites.
2. There would be biochemical differences in the properties of variant antigens from clones of different SICA phenotype, other than the required differences in antigenicity, that would confirm the identification of these different antigens as distinct entities.

It is evident from these criteria that the agglutination specificity of the panel of sera that were used is the crux of this approach. The specificities of the sera used are given in Table IV and can be summarized as follows:

1. Sera 2, 3, and 4 are rhesus monkey sera specific for agglutination of Pk1(A+). Serum 18 is a rabbit antiserum specific for agglutination of Pk1(A+) also. These sera should only immunoprecipitate Pk1(A+) variant antigen.
2. Sera 5, 6, and 7 are rhesus monkeys sera specific for agglutination of Pk1(B+)1+. Serum 19 is a rabbit antiserum specific for agglutination of this clone. These sera should only immunoprecipitate Pk1(B+)1+ variant antigen.
3. Sera 9 and 10 are cross-reactive sera that agglutinate parasites of many different SICA phenotypes from the Malaysian H strain of *P. knowlesi.* Both clone Pk1(A+) and clone Pk1(B+)1+ are from a line of Malaysian H-strain parasites and are agglutinated with these sera. These sera should immunoprecipitate both Pk1(A+) and Pk1(B+)1+ variant antigens.

   The sera in these first three groups (i.e., sera 2–4 and 18, 5–7 and 19, and 9 and 10) are the most critical to these experiments, since each group contains agglutinating antibody, but should show patterns of immunoprecipitation that correlate with their distinct agglutination specificities.
4. Serum 8 is from a rhesus monkey infected and immunized with a non-cloned line of Malaysian H-strain parasites. It strongly agglutinates Pk1 (B+)1+ and displays variable, incomplete agglutination of Pk1(A+).
5. Sera 11, 12, and 15 are from animals infected or immunized with parasites/infected cells of SICA[–] *P. knowlesi.* These sera do not agglutinate either clone and should not immunoprecipitate variant antigens from either clone.
6. Sera 13 and 14 are from animals chronically infected with other strains of *P. knowlesi.* Only serum 13 exhibits weak agglutination, and only with Pk1(A+).
7. Serum 16 is a monoclonal antibody raised against merozoites of SICA[–] *P. knowlesi*, which immunoprecipitates a protein of M_r 230,000 from biosynthetically labeled schizont-infected cells (Epstein *et al.*, 1981). It agglutinates merozoites but does not agglutinate schizont-infected erythrocytes and should therefore not immunoprecipitate the variant antigens.

Two other pieces of information were used for these experiments. We showed that schizont-infected erythrocytes labeled by the lactoperoxidase–H_2O_2–Na[^{125}I] method were agglutinable by appropriate rhesus monkey antisera with little or no change in agglutination titer. This method of cell-surface labeling

Table IV. Agglutination Specificity in the SICA Test of Rhesus Monkey and Rabbit Antisera and a Monoclonal Antibody to Cloned *P. knowlesi* Parasites

Serum *N*	Source of antibody[a]	Animal treatment before serum collection[b]	Reciprocal agglutination titer of schizont-infected erythrocytes (H strain)	
			Clone Pk1(A+)	Clone Pk1(B+)1+
1	Normal rhesus monkeys	None	<10	<10
2	Rhesus monkey A25A	Pk1(A+) infection and immunization	40,960	<10
3	Rhesus monkey 398D	Pk1(A+) infection and immunization	20,480	<10
4	Rhesus monkey 184B	Pk1(A+) immunization	10,240	<10
5	Rhesus monkey 189B	Pk1(B+)1+ infection and immunization	<10	40,960
6	Rhesus monkey 529R	Pk1(B+)1+ infection and immunization	<10	40,960
7	Rhesus monkey 195E	Pk1(B+)1+ infection and immunization	<10	81,920
8	Rhesus monkey 608H	Line 1 (noncloned Malaysian H strain) infection and immunization	<10–160[c]	20,480
9	Rhesus monkey 331	Malaysian H strain (noncloned) chronic infection	5,120	20,480
10	Rhesus monkey 332	Malaysian H strain (noncloned) chronic infection	2,560	5,120
11	Rhesus monkey 780F	Clone S[c] (SICA[–]) infection	<10	<10
12	Rhesus monkey 705D	Clone S[c] (SICA[–]) infection	<10	<10
13	Rhesus monkey 125	Philippine strain (noncloned) chronic infection	160[c]	<10
14	Rhesus monkey 621	Hackery strain (noncloned) chronic infection	<10	<10
15	Rhesus monkey Q148B	Malaysian H strain SICA[–] merozoite immunization	<10	<10
16	Monoclonal antibody 13C11[d]		<10	<10
17	Normal rabbits	None	<10	<10
18	Rabbit 3	Pk1(A+) immunization	10,240	<10
19	Rabbit 7	Pk1(B+)1+ immunization	<10	>20,480

[a]Antibody identified by animal number.
[b]Immunizations for sera 2–8 and 18, 19 with schizont-infected erythrocytes and incomplete Freund's adjuvant and for serum 15 with merozoites and complete Freund's adjuvant.
[c]Agglutination of Pk1(A)+ schizont-infected erythrocytes with these sera was either undetectable or very weak with an incomplete mat of agglutinated cells.
[d]This monoclonal antibody immunoprecipitates a malarial protein of M_r 230,000 from extracts of schizont-infected cells. It binds to the surface of extracellular *P. knowlesi* merozoites, resulting in agglutination and reduced invasion (Epstein *et al.*, 1981).

could therefore be used in an attempt to label the SICA-variant antigens without destroying their reactivity with antibody. Second, all the agglutinating activity in rhesus monkey sera could be removed by passage of sera through protein A-Sepharose. Material eluted from the protein A-Sepharose with acid buffer completely retained the agglutination titer of the initial serum, suggesting that IgG antibodies are agglutinating. Most importantly, it should be possible, on the basis of these results, to purify complexes of SICA-variant antigen and agglutinating antibody by adsorption to protein A-Sepharose. This method was used throughout these experiments.

Two protocols proved suitable for identification of SICA-variant antigen (Howard *et al.*, 1983).

1. In one protocol we radioiodinated the surface proteins of schizont-infected erythrocytes of each clone, extracted the antigens with detergent, and performed immunoprecipitation analysis using the panel of sera (Table IV). Figure 4 shows the results.
2. The second protocol allowed us to test whether the variant antigens were proteins synthesized by the malarial parasite. Malarial proteins of each clone were radiolabeled by biosynthetic incorporation of [^{35}S]methionine during *in vitro* growth from the ring- to schizont-stage. Sera were briefly incubated with the intact infected cells, the cells washed to remove unbound antibody and then antigen-antibody complexes solublized with detergent for immunoprecipitation. The results of this approach are shown in Fig. 5.

When the patterns of radioiodinated antigens immunoprecipitated from Pk1(A+) are examined (Fig. 4A), it is apparent that antigens of M_r 210,000 and 190,000, together with minor antigens of M_r 170,000 and 182,000, were only immunoprecipitated by sera that agglutinate this clone. The major doublet of M_r 210,000 and 190,000 was not immunoprecipitated from Pk1(B+)1+ cells (Fig. 4B).

With radioiodinated infected cells from clone Pk1(B+)1+, however, a doublet of ^{125}I-antigens of M_r 205,000 and 200,000 was immunoprecipitated from this clone only by agglutinating sera (Fig. 4B). The variant-specific sera that agglutinate Pk1(A+) cells and that precipitate the M_r 210,000–190,000 doublet from Pk1(A+) cells did not immunoprecipitate the M_r 205,000–200,000 doublet from Pk1(B+)1+.

Several other ^{125}I-antigens are evident in Fig. 4A, B, i.e., M_r 230.000, 165,000, 97,000, 45,000; however they were immunoprecipitated from these clones by nonagglutinating as well as agglutinating sera. These antigens do not appear to be related to variant-specific agglutination.

The doublets of antigens precipitated from the two clones in a variant-specific manner, i.e., M_r 210,000 and 190,000 for Pk1(A+) and M_r 205,000 and 200,000 for Pk1(B+)1+, were not immunoprecipitated from radioiodinated

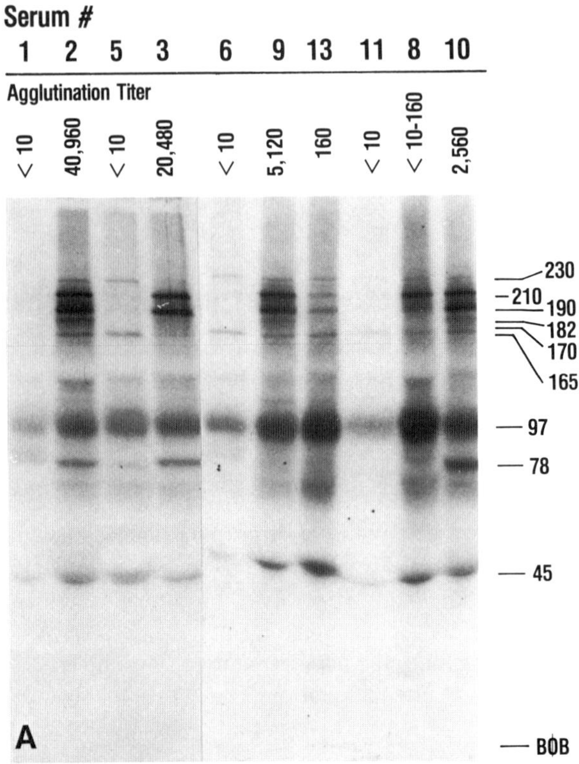

Figure 4. ^{125}I-lactoperoxidase-labeled variant antigens of cloned *P. knowlesi*, identified by immunoprecipitation of radioiodinated schizont-infected erythrocytes using sera of defined agglutination specificity. The properties of each serum, identified by number above each gel lane, are given in Table IV. The reciprocal agglutination titer of each serum with the cells from which the ^{125}I-antigens were extracted is also shown above each lane. (A) Immunoprecipitation of ^{125}I-antigens from clone Pk1(A+).

uninfected erythrocytes. Furthermore, electrophoresis of immunoprecipitated antigens from each clone on the same gel established that the antigens from each clone had different molecular weights, indicative of biochemical differences other than differences in antigenicity.

These ^{125}I-antigens therefore fulfill the criteria discussed for the properties of SICA-variant antigens. We were also able to show that these antigens are synthesized by the malaria parasite (Howard *et al.*, 1983).

^{35}S-methionine-labeled antigens were identified by binding antibody to intact cells before detergent solubilization (Fig. 5). With Pk1(A+) a complex group of malarial proteins were only immunoprecipitated by sera that agglutinate infected erythrocytes of this clone (Fig. 5A). The two major ^{35}S-labeled

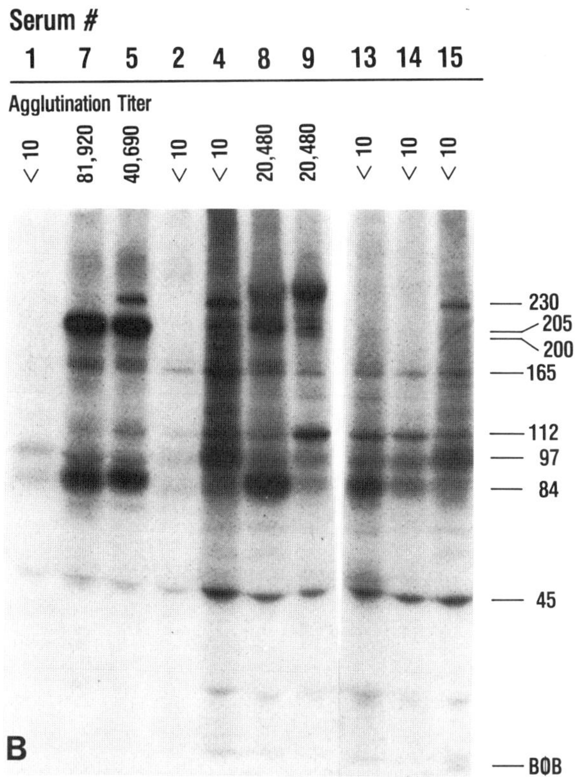

Figure 4. (*Continued*) (B) Immunoprecipitation of ^{125}I-antigens from clone Pk1(B+)1+. Antigens were separated by SDS-PAGE and identified by autoradiography. The M_r of major ^{125}I-antigens is given in kilodaltons. BØB, Bromphenol blue. (From Howard *et al.*, 1983.)

antigens had M_r of 210,000 and 190,000, identical to the M_r of radioiodinated antigens also specifically immunoprecipitated from these cells. Other minor ^{35}S-labeled antigens were also specifically immunoprecipitated from Pk1(A+): M_r 216,000, 213,000, 207,000, 197,000, and 183,000 (Fig. 5A).

With ^{35}S-methionine-labeled infected cells of clone Pk1(B+)1+, a major parasite antigen of M_r 205,000 was only immunoprecipitated by sera that agglutinate Pk1(B+)1+. The 5–15% acrylamide gradient gel used in this experiment does not always resolve the M_r 205,000–200,000 doublet of Pk1(B+)1+. There was also a minor ^{35}S-labeled antigen of M_r 225,000 specifically immunoprecipitated from Pk1(B+)1+ only by agglutinating sera (Fig. 5B).

These data prove that the variant antigens of *P. knowlesi* are synthesized by the malaria parasite and are not modified host components.

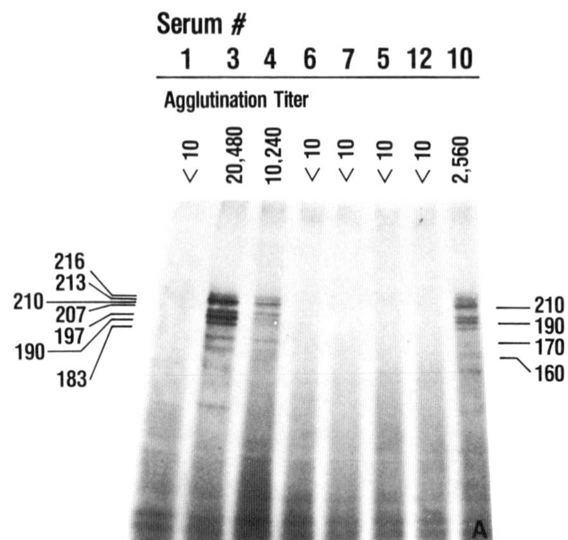
Serum #
1 3 4 6 7 5 12 10
Agglutination Titer
< 10
20,480
10,240
< 10
< 10
< 10
< 10
2,560
216
213
210
207
197
190
183
210
190
170
160
A

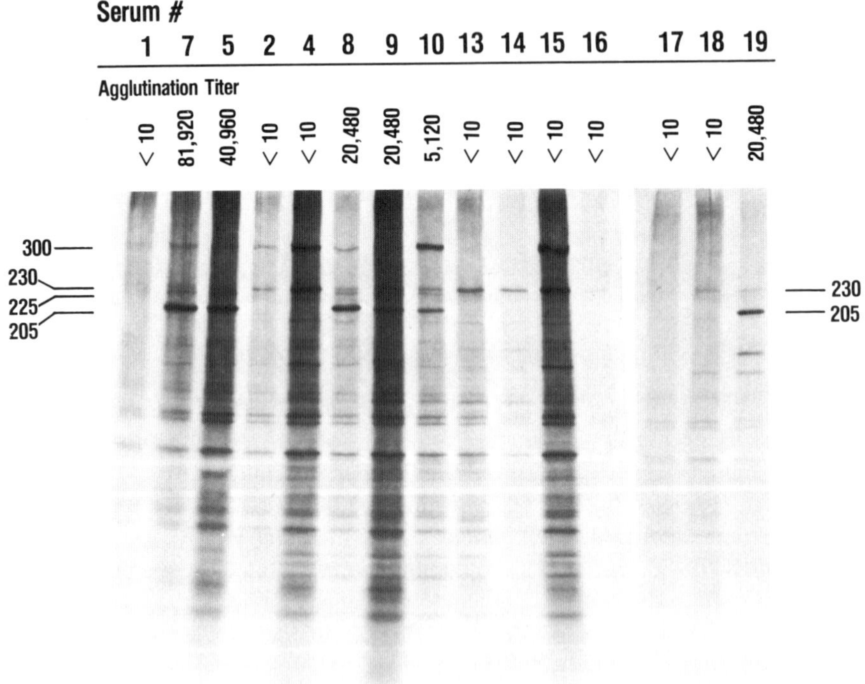
Serum #
1 7 5 2 4 8 9 10 13 14 15 16 17 18 19
Agglutination Titer
< 10
81,920
40,960
< 10
< 10
20,480
20,480
5,120
< 10
< 10
< 10
< 10
< 10
< 10
20,480
300
230
225
205
230
205
B

As is obvious in Fig. 5B, many other parasite antigens reacted with antibody in this particular experiment, but none was precipitated only by agglutinating sera. We suspect that the variability we observe in the detection of these additional antigens reflects the occasional presence of broken schizont-infected cells that permit antibody to react with intracellular parasites. Even so, most malarial antigens are inaccessible even when broken cells are present, since many more malarial antigens are immunoprecipitated when detergent solubilization precedes addition of antibody. The malarial antigen of M_r 300,000 in Fig 5B is of interest, as it appears to be immunoprecipitated only by sera against the same strain, i.e., Malaysian H strain of *P. knowlesi*. Sera 13 and 14 are from animals chronically infected with Phillipine or Hackery strains of this parasite and fail to precipitate this M_r 300,000 bond. We are investigating whether parasites of these other strains possess a similar high M_r antigen that is precipitated only by antisera to the homologous strain.

Our current understanding of the SICA-variant antigens of *P. knowlesi* is summarized in Tables V, VI, and VII. Several important questions remain to be answered concerning the molecular mechanisms involved in antigenic variation of *P. knowlesi*:

1. What is the origin of multiple antigens specifically immunoprecipitated from cloned parasites by agglutinating sera? It is possible that cloned parasites undergo antigenic variation during expansion of the clone in naive animals before antigenic analysis, i.e., in the presumed absence of variant-specific immunity. Each specifically immunoprecipitated band would then represent the variant antigen of individual clones of different SICA phenotype. Alternatively, each clone may have a high M_r variant antigen that is incompletely converted to lower M_r forms by proteolysis. Such cleavage could be a normal consequence of antigen expression on the erythrocyte membrane, or it could be attributable to artifactual proteolysis occurring during antigen solubilization and immunoprecipitation.
2. Somewhat related to the first question is the question of the degree of structural relatedness of the different bands specifically immunoprecipitated from each clone. Initial analysis of tryptic peptides or CNBr-

Figure 5. ^{35}S-labeled variant antigens of cloned *P. knowlesi* identified by immunoprecipitation of schizont-infected erythrocytes after 22 hr growth in medium containing [^{35}S]methionine. The sera used are described in Table IV and the reciprocal agglutination titer for each serum is given above each gel lane of immunoprecipitated antigens. Intact radiolabeled schizont-infected cells were incubated with various sera, and washed to remove unbound antibody, and antigen–antibody complexes were solubilized with Triton X-100. Protein A-Sepharose was used to purify these complexes and the ^{35}S-antigens analyzed by SDS-PAGE and fluorography. (A) Immunoprecipitation of ^{35}S-antigens from clone Pk1(A+). (B) Immunoprecipitation of ^{35}S-antigens from clone Pk1(B+)1+. The M_r of major ^{35}S-antigens is given in kilodaltons. (From Howard *et al.*, 1983.)

Table V. Antigenic Variations in *P. knowlesi*: Genotype and Phenotype

Genotype	
1. The genome of cloned *P. knowlesi* contains information for expression of several different variant antigen phenotypes.	Barnwell *et al.* (1982*b*) J. W. Barnwell (unpublished results, 1983)
2. Cloned parasites of SICA[+] phenotype can be converted to the SICA[−] phenotype, and in some cases this can be reversed to reexpression of SICA[+] phenotype. These changes appear to represent control of the expression of genes coding for variant antigens.	Barnwell *et al.* (1982*b*)

Definition of phenotype

1. Variant-specific antibody-mediated agglutination of schizont-infected erythrocytes:

Parasites +	Antisera from	Reference
Noncloned lines	Immunized/infected rhesus monkeys	R. N. Brown and I. N. Brown (1965); I. N. Brown *et al.* (1968)
Noncloned lines	Immunized rabbits	Hommel and David (1981)
Clones	Immunized/infected rhesus monkeys	Barnwell *et al.* (1982*a*,*b*)
Clones	Immunized rabbits or mice	R. J. Howard and J. W. Barnwell (unpublished observations, 1983)

2. Variant-specific fluorescence of schizont-infected erythrocytes:

Parasites +	Primary antisera from	Reference
Noncloned lines	Immunized rabbits and secondary fluorescent-labeled antiserum	Hommel and David (1981)
Clones	Immunized monkeys and secondary fluorescent-labeled antiserum	Barnwell *et al.* (1982*b*)

cleaved peptides would be expected to demonstrate whether these antigens share peptide sequences or whether they appear to be structurally unrelated. It will also be of great interest to compare the structure of antigens immunoprecipitated from clones of different SICA phenotype. The fact that polyspecific mouse, rabbit, and monkey sera raised against cloned parasites of different SICA phenotype specifically precipitate the variant antigens from each clone suggests, as a first approximation, that these proteins will be found to be structurally quite different from each other.

3. Is the generation of diverse SICA-antigen phenotypes—presumably the result of expression of diverse variant antigen genes—attributable to the controlled gene expression of a single variant antigen gene from a large repertoire of variant antigen genes within a given clone? Or are diverse

Table VI. Location of SICA-Variant Antigens

Observation	Reference
Presence	
1. *P. knowlesi*-infected erythrocytes containing late trophozoites and schizonts, the amount of antigen increasing as the parasite matures	Eaton (1938); K. N. Brown and I. N. Brown (1965)
a. ~ 1.25×10^4 molecules of variant antigen/cell on schizont-infected erythrocytes	Vincent and Wilson (1980)
b. Uneven distribution of SICA antigen over the cell surface as determined by immunoelectron microscopy	Hommel and David (1981)
2. Washed membranes prepared from infected cells as determined by either serologic analysis or immunoprecipitation of the specific SICA antigens	Vincent and Wilson (1980); S. B. Aley, J. W. Barnwell, and R. J. Howard (unpublished observations, 1983)
Absence	
1. Uninfected erythrocytes or immature parasitized cells as determined by agglutination	Eaton (1938); K. N. Brown and I. N. Brown (1965)
2. The surface of erythrocytes infected with schizonts of SICA[−] clones or lines of *P. knowlesi*, as determined by agglutination, fluorescent analysis, and immunoprecipitation analysis	Barnwell *et al.* (1982*a*); R. J. Howard and J. W. Barnwell (unpublished observations, 1983)
3. The surface of merozoites on immunocytochemical analysis	Hommel and David (1981)

phenotypes the result of controlled mutation or gene rearrangement occurring during parasite growth? A combination of both processes for generation of diverse SICA phenotypes is clearly possible. We are currently attempting to explore this question using molecular biologic techniques.

4. The route of transport of variant antigen synthesized by the intracellular parasite to the surface of the infected erythrocyte is unknown. It is generally believed that membrane-bound clefts and vesicles seen in the cytoplasm of infected cells are involved in the transport of antigens from the parasitophorous vacuole membrane surrounding the parasite to the cell surface (reviewed in Sherman, 1979). Models involving membrane-bound and soluble forms of parasite antigen have been proposed (Howard, 1982). Methods for isolating and identifying the intraerythrocytic membranes that may bear SICA antigen (vesicle, parasitophorous vacuole, and parasite membranes) must be developed. The preparation of monoclonal antibodies to SICA-variant antigens might also permit exploration of the route of transport by immunoelectron microscopy.
5. Are the variant antigens glycoproteins? So far we have been unable to

Table VII. Structural Analysis of SICA-Variant Antigens

Associated with proteins	
Treatment of schizont-infected erythrocytes with various proteases markedly decreased their agglutinability.	Hommel and David (1981)
The ^{125}I-variant antigens of one clone have been shown directly to be cleaved by trypsin treatment of intact cells.	Howard and Barnwell (unpublished)
Not associated with sialic acid	
Neuraminidase treatment did not affect agglutinability.	Hommel and David (1981)
Biochemical identification	
The variant antigens are proteins specifically immunoprecipitated only by agglutinating antisera.	

Clone	M_r of ^{125}I-variant antigens	
	Major	Minor
Pk1(A+)	210,000 and 190,000	182,000 and 170,000
Pk1(B+)1+	205,000 and 200,000	

^{125}I-antigens in the M_r range 180,000–230,000 have also been specifically immunoprecipitated from noncloned lines of *P. knowlesi* by agglutinating sera. Several major bands are seen in these cases.	Howard, Barnwell, and Kao (1983)
Origins of the SICA antigens	
The variant antigens are synthesized by the malaria parsite.	Howard, Barnwell, and Kao (1983)

Clone	M_r of ^{35}S-variant antigens labeled by [^{35}S]methionine incorporation	
	Major	Minor
Pk1(A+)	210,000 and 190,000	216,000, 213,000, 207,000, 197,000, and 183,000
Pk1(B+)1+	205,000 and 200,000	225,000

The differences in apparent M_r on SDS-PAGE of SICA variant antigens from different clones imply structural differences other than differing antigenicity.

demonstrate incorporation of [^{3}H]glucosamine into these antigens, although this monosaccharide is readily incorporated into several malarial glycoproteins (R. J. Howard and J. W. Barnwell, unpublished observations, 1983).

6. Why does the intraerythrocytic malaria parasite express highly immunogenic variant antigens on the surface of the infected erythrocyte, thereby

Table VIII. New Components on *P. knowlesi* Schizont-Infected Cells Other Than SICA-Variant Antigens

Membrane component	Comments	References
^{125}I-Lactoperoxidase-labeled proteins	Loss of some bands and the appearance of new bands with schizont-infected cells	Wallach and Conley (1977)
	No apparent changes with schizont-infected cells versus normal erythrocytes	Shakespeare *et al.* (1979)
	No changes in major proteins and the appearance of minor new ^{125}I-proteins,[a] some of which may be internal	Howard *et al.* (1982*a*)
^{125}I-Lactoperoxidase-labeled antigens	Four antigens identified by crossed immunoelectrophoresis, two of which are specific for mature parasitized erythrocytes	Deans *et al.* (1978); Deans and Cohen (1979)
	Two antigens (M_r 90,000, pI 5.2 and M_r 65,000, pI 4.5) identified by crossed immunoelectrophoresis using sera from immune monkeys. There is a correlation between antibody to the M_r 65,000 band and protective immunity. Other new protein antigens that were not iodinated were also identified.	Schmidt-Ullrich and Wallach (1978); Schmidt-Ullrich *et al.* (1979*a*, *b*, 1981)
	Several new antigens identified by immunoprecipitation,[b] some of which may be internal	Howard *et al.* (1982*a*)
Pyridoxal phosphate/ NaB^3H_4-labeled antigens	New antigens of M_r 125,000, 50,000 45,000, and 30,000 may be surface proteins	Howard *et al.* (1982*b*)

[a] M_r 230,000, 180,000, 165,000, 155,000, 135,000, 107,000, 72,000, and 65,000.
[b] M_r 230,000, 200,000, 180,000, 165,000, 155,000, 135,000, 130,000, 107,000, 72,000, 65,000, and 47,000.

signaling its presence to the host's immune system? It is possible that some part of the variant-antigen molecule must be inserted into the membrane of the infected erythrocyte in order to fulfill some aspect of parasite metabolism and homeostasis. Transport proteins required for import of essential metabolites might be inserted into this membrane. However, if this is the case, the essential part of the variant molecule is clearly not the variant-antigenic regions, not only because they vary and would therefore be unlikely to exert some constant metabolic function, but because SICA [−] parasites that lack variant antigen appear to multiply early in infection of splenectomized animals at a rate identical to that in SICA[+] parasites (Table II). It is possible that one of the radioiodinated antigens we have identified as potential surface antigens on schizont-infected cells containing SICA[-] parasites (R. J. Howard and J. W. Barnwell, unpublished observations, 1983) represents the constant portion of the variant-antigen molecule found on SICA[+] cells. There are no ^{125}I-antigens in the molecular-weight region for variant antigens (180,000–220,000) on SICA[-] schizont-infected cells. An alternative function of the SICA-variant antigens is suggested from the different *in vivo* properties of SICA[+] and SICA[-] parasites, as discussed further in Section VI.

C. Other New Antigens on Infected Cells

Evidence that the SICA-variant antigens of *P. knowlesi* and knob antigens of *P. falciparum* are involved in evasion of immunity by malaria parasites is discussed in Sections VI and VII. Numerous other biochemical and immunologic studies have identified additional antigens that may also be on the surface of malaria-infected erythrocytes. For these antigens there is as yet no evidence that they have a role in immune evasion. The very existence of these antigens exposed on the surface of infected cells suggests that malaria parasites must have evolved mechanisms designed to avoid the induction of immunity to these antigens. For the purpose of this review, in which we are concentrating on antigens believed to be important for a direct role in evasion of immunity, it is convenient to summarize the biochemical studies on *P. knowlesi* antigens other than the variant antigen in Table VIII. There have also been some studies on new proteins or antigens in the membrane of erythrocytes infected with rodent, avian, and other primate malarias. For example, with the rodent malaria *P. chabaudi*, a parasite antigen of M_r 250,000 was identified by antibody reaction with intact schizont-infected cells after [^{35}S]methionine uptake (Newbold *et al.*, 1982). Several parasite proteins labeled by uptake of [^{35}S]methionine, [^{3}H]glucosamine, or [^{3}H]mannose were cleaved by protease treatments of intact human erythrocytes infected with *P. falciparum* schizonts (Perkins, 1982). These parasite proteins would most probably constitute new erythrocyte surface antigens.

V. PARASITIC EVASION OF IMMUNITY–GENERAL CONCEPTS

Although it is clear that protective immunity to malaria parasites can often be acquired naturally by repeated infection or by appropriate immunization, the capacity of malaria parasites to survive and reproduce for prolonged periods in natural infections is well established (McGregor, 1972). Chronic malarial infections that evolve after the host overcomes initial acute attacks clearly demonstrate the host's capacity to restrict overwhelming parasite proliferation; however, they also demonstrate the parasite's capacity to evade host-defense mechanisms. How do some parasites manage to escape host defenses to survive chronic infection?

Chronic infections resulting from blood inoculation of asexual parasites do not involve the liver stages of the malaria parasite. The reappearance of parasites in the bloodstream after such an inoculation is termed a recrudescence, to distinguish this phenomenon from a relapse, which involves release of a new population of asexual parasites from liver stages. Recrudescences reflect the prolonged survival of a small number of asexual parasites that later expand in numbers to produce a detectable blood parasitemia.

Several hypothetical mechanisms could account for chronic asexual parasitemia. First, successive recrudescences may be genotypically and phenotypically identical. This would imply that the host-defense mechanisms that cleared most parasites of the preceding wave of parasitemia are relatively short-lived in effectiveness. According to this view, the parasites that survived the parasiticidal mechanisms induced by the first expanding parasite population could eventually initiate another wave of parasitemia. This would provide an antigenic stimulus to the host's immune system and reactivate processes of parasite destruction. This hypothesis is compatible with the suggestion that protective antimalarial immunity requires continuous antigenic stimulation (Shortt *et al.*, 1983).

Second, successive recrudescences may be genotypically identical, but phenotypically different. If malaria parasites had the genetic ability to express different variants of antigens that are the targets of parasiticidal immunity, it would be possible for newly formed variant phenotypes to escape immune responses directed against preceding variant phenotypes in a chronic infection. Variation of the phenotype of a malarial antigen is demonstrated by the SICA-variant antigens of *P. knowlesi*. Evidence for the evasion of variant-specific parasiticidal immune responses by alteration of SICA variant antigens is summarized in Section VI. The cycle of expansion of newly formed variants of the parasite and removal of these populations through variant-specific parasiticidal responses would continue as long as the parasite phenotype could produce a repertoire of new antigenic variants. The host's capacity to control and eventually eliminate such chronic parasitemia would depend on other variant-transcending immune responses directed specifically, or nonspecifically, against all variants of the parasite. The properties of this model, which appear to be fulfilled by *P. knowlesi* malaria in rhesus monkeys (K. N. Brown and I. N. Brown, 1965; K. N. Brown, 1971, 1976, 1977), pose another question: How would new variants arise? Would a cloned parasite of a particular genotype and phenotype produce progeny of each different variant phenotype with a certain frequency per division independent of the selection pressure of the host's immune system? Or does the development of variant-specific immune responses direct in some way the formation of new variants? Our own studies with cloned *P. knowlesi* (Section

VI. A) and earlier work by Brown and colleagues (K. N. Brown *et al.*, 1970*a*; K. N. Brown, 1973) suggest that the latter mechanism is appropriate in this case.

A third possible mechanism for the appearance of new parasite populations in chronic infections arises if successive populations prove to be genotypically as well as phenotypically different. As with the mechanism just presented, phenotypic variation of critical antigens would enable the parasite to evade host immunity. Two possibilities within this mechanism depend on the properties of the initial inoculum. If the original inoculum were clonal, then genotypic variation, and consequent phenotypic variation could arise through a process of somatic mutation of genes coding for target antigens of host-protective immunity. The generation of diverse variants of the SICA antigen during chronic *P. knowlesi* infection could conceivably reflect somatic mutation of the variant antigen gene(s) in a similar manner to that believed to operate for generation of new genotypes of Ig genes during the lifetime of an organism. In contrast, if the initial inoculum were noncloned, a sequence of parasite populations of different phenotypes could arise in an order dependent on such factors as their proportions in the inoculum and their growth rates.

Antigenic variation of parasites taken from a chronic malarial infection induced by a cloned parasite could clearly arise by a combination of the last two mechanisms–the genotype of the inoculum may have the capacity to express different phenotypes, and it may also undergo somatic mutation.

Equally intriguing to the phenomenon of chronicity in malaria is the capacity of malaria parasites to reestablish infection upon reinfection of a host that has eliminated or controlled earlier infections. The capacity of *P. knowlesi* to establish an often lethal infection upon reinoculation of an animal previously infected with the same inoculum is linked to the appearance of new variants of the SICA antigen on infected erythrocytes (K. N. Brown, and I. N. Brown, 1965). Is this also the case in *P. falciparum* infections in humans, which require that the host survive multiple infections over several years in order to achieve protection from lethal infection?

Considerable evidence has accumulated for diversity of the phenotype of malaria parasites of particular species of malaria parasites. Although in many cases this evidence has not yet been linked to the parasite's capacity to evade immunity, other studies show good evidence for a relationship between an alteration in parasite phenotype and prolonged survival of the parasite *in vivo*. For example, splenectomized primates recovered from an initial infection are more susceptible to challenge with heterologous strains of *P. falciparum* than with the homologous strain (Cadigan and Chaicumpa, 1969; Voller and Richards, 1970). It is also apparent from these studies that *P. falciparum* isolates collected within a very small geographic area were in fact heterogeneous when tested in infection and challenge experiments (Cadigan and Chaicumpa, 1969). Similar experiments showed that immunity to *P. falciparum* infection in humans is also strain-specific (Jeffery, 1966; Sadun *et al.*, 1966). Passive transfer of immunity to *P. falciparum* with serum also displays strain specificity (McGregor *et al.*, 1963); the capacity of sera to inhibit *in vitro* growth of this parasite is strain specific as well (Wilson and Phillips, 1976). Immunity to *P. knowlesi* clearly demonstrates a variant-specific component presumed to act against the SICA antigen on schizont-infected erythrocytes (K. N. Brown and I. N. Brown, 1965; Voller and Rossan, 1969*b*; K. N. Brown, 1971; Butcher and Cohen, 1972). Challenge experiments with different relapse populations of *P. cynomolgi* collected from sporozoite-induced infections strongly suggest antigenic variation in this simian malaria (Voller and Rossan, 1969*a*).

There is also evidence for phenotypic variation related to immune evasion from studies with rodent parasites. Relapse variants of *P. berghei* in mice, collected after drug treatment of the initial population or collected from animals partially immune because of previous infection, showed different virulence and immunogenicity to the parent population (Cox, 1957, 1959, 1962; Wellde and Sadun, 1967). Differences in virulence of different isolates of *P. berghei* have also been demonstrated more recently (Branton, 1978; Wery *et al.*, 1979). In *P. yoelii yoelii*, virulence is clearly established as a variable genetic character (Knowles and Walliker, 1980).

Several variable phenotypic characters of malaria parasites have been identified more specifically by various biochemical and/or immunochemical methods:

1. Malarial isozymes displaying heterogeneity in electrophoretic mobility have been identified in natural populations of rodent malarias (Carter, 1970) and of *P. falciparum* in man (Carter and Voller, 1975). These isozymes are useful markers for parasite typing conceivably related to differences in innate parasite virulence but probably unrelated to evasion of immunity.
2. Heat-stable antigens (S antigens) have been identified in the sera of infected and convalescent patients recovering from *P. falciparum*. They display antigenic heterogeneity on double-diffusion precipitin analysis versus malarial sera and define individual parasite populations (Wilson, 1981).
3. A panel of monoclonal antibodies has been used with several strains of *P. falciparum* to define antigens on the surface of intraerythrocytic schizonts and merozoites. Different strains of this parasite exhibit considerable antigenic diversity as defined by reaction with these antibodies (McBride *et al.*, 1982).
4. Two-dimensional gel analysis of malarial proteins of different *P. falciparum* strains has identified several proteins that display heterogeneity in molecular weight and/or pI characteristic of each particular isolate (Tait, 1981).
5. Evidence has very recently been obtained for antigenic variation of surface-membrane components on *P. falciparum*-infected erythrocytes.

Schizont-infected cells can be specifically agglutinated *in vitro* by antibodies in sera from infected monkeys or humans (J. W. Barnwell, unpublished data, 1983). Testing a variety of *P. falciparium* isolates with a panel of heterologous and homologous antisera suggests that the surface antigens responsible for agglutination apparently vary with different isolates. Furthermore, studies using a fluorescent secondary antibody to detect binding of antibody from infected squirrel monkeys to the surface of *P. falciparum*-infected cells, have also demonstrated antigenic heterogeneity in surface antigens (Hommel *et al.*, 1983). With different relapse parasites obtained from animals repeatedly infected with the same isolate, it was found that red cell surface fluorescence resulted only with homologous monkey serum.

The above antigens may each play a role in evasion of immunity to *P. falciparum* malaria.

6. The biochemical identification of the SICA-variant antigens of *P. knowlesi* was described in Section IV. B.4. Variation in the antigenic structure of these proteins is almost certainly linked to evasion of variant-specific parasiticidal immune responses to this parasite.

Thus far we have discussed in very general terms the possibility that the capacity of malaria parasites to evade host immunity involves either the survival of a small number of parasites coupled to rapid waning of antimalarial immunity or phenotypic variation of critical target antigens. Evasion of host immunity may act also at the induction of immunity. Since it is clear from immunization studies that strain-and species-specific antigenic targets must exist that can be recognized by strongly parasiticidal strain-or species-specific immune responses (K. N. Brown *et al.*, 1970*a*), it follows that during the normal course of most natural malaria infections, the parasite does not elicit these responses. What prevents induction of parasiticidal immune responses upon initial infection or reinfection of the host? This question is important regardless of the capacity of malaria parasites to undergo antigenic variation and thereby escape variant-specific immune responses. It should be noted that whether the balance of anti-malarial immune responses and parasite growth in each infection leads to expansion or contraction of the parasite population will depend not only on the qualitative nature of the immune responses, but also on the magnitude of these responses. The malaria parasite could conceivably escape host defenses by either limiting in some way the magnitude of parasiticidal immune responses or by preventing the induction of these immune responses.

Aside from the possibility that antigenic variation of critical antigens permits evasion of immunity, several other evasion mechanisms implicated in different malarial infections have been summarized by Cohen (1976) and Kreier and Green (1980):

1. Release of soluble blocking antigens that might block immune effector mechanisms directed against schizont-infected erythrocytes or merozoites.
2. Interference of the parasite with the host's immune responses: There is extensive evidence for immune depression in malaria; it has been suggested that K-cell action might be blocked, and there is evidence for suppression of various immune responses by malaria infection. Direct evidence for an important role of such parasite-induced modulations of immunity in evasion of immunity is lacking.
3. Growth of the malaria parasite in a protected location within the erythrocyte: This mechanism of evasion of immunity would apply for immature intraerythrocytic parasites that may not express new antigens on the surface of the infected cell. We have pointed out that such is not the case for mature intraerythrocytic parasites that express new malarial antigens on the surface of the host cell in *P. knowlesi* infection (Section IV. B.4), and most probably in other malarias as well (see Section III.A).

These three mechanisms for evasion of parasiticidal immune responses do not necessarily involve antigenic variation of the asexual parasite. Evidence that these immune evasion mechanisms play a role in malaria is discussed elsewhere by Cohen (1976), Playfair (1978), and Kreier and Green (1980).

Why do new surface antigens on malaria-infected erythrocytes, which are known to be recognized by antibody from infected animals, fail to elicit powerful immune responses *in vivo* that would clear infected cells from the circulation? Or, if these antigens do elicit parasiticidal immune responses, how does the parasite manage to evade these responses? Antigens on the surface of infected erythrocytes have been studied for their role in evasion of immunity by several workers in this laboratory. The next two sections summarize these new results.

VI. SICA ANTIGEN–IMMUNE EVASION

We have already described the appearance of different variants of *P. knowlesi* in chronically infected animals (Fig. 3), definition of the variant-antigen phenotype by the antibody-mediated agglutination reaction (Table V), the biochemical nature of the SICA-variant antigens (Table VII), and the identification of the nonagglutinable SICA[-] phenotype (Table II). In this section we concentrate first on the previously published information regarding the SICA antigen and immune responses in *P. knowlesi* and then follow with new data comparing the *in vivo* properties of SICA[-] and SICA[+] parasites. The latter results suggest an additional role of the variant antigen in evasion of parasiticidal immunity other than by antigenic variation per se.

A. Immune Evasion by Antigenic Variation

It has been suggested that antigenic variation in uncloned *P. knowlesi* malaria is induced by and dependent on the presence of antibody of the appropriate variant specificity (K. N. Brown *et al.*, 1970*a*; K. N. Brown, 1973). We reexamined the role of antibody to the SICA antigen in antigenic variation using cloned *P. knowlesi* (Barnwell *et al.*, 1983).

Three nonsplenectomized monkeys were reinfected several weeks after drug cure of their initial infection with the same cloned parasites (thawed from cryopreservation) with which they had first been infected. Previous studies with noncloned parasites (I. N. Brown *et al.*, 1968; K. N. Brown *et al.*, 1970*a*) showed that in such cases of homologous variant challenge, reinfection with the same variant type results in antigenic variation. In all three monkeys used in our study, the parasites from the second inoculation changed their variant type. Sera from each animal before reinoculation agglutinated the parasitized cells of the challenge inoculum, but not the parasitized cells of the resulting infection. They were agglutinated, however, by cross-reactive sera that agglutinate many SICA[+] variants of the same *P. knowlesi* strain. Cloned parasites are clearly able to change their variant type as readily as are noncloned parasites.

Two monkeys were first infected with either of two different SICA[+] clones. Several weeks after their primary infections had been drug cured, they were reinfected with either of two different cloned variant types. Sera obtained from these monkeys just before reinfection were found to agglutinate erythrocytes infected with the clone of the first inoculum, but not to agglutinate erythrocytes infected with either of the cloned parasites used for reinfection. In these two cases of heterologous variant challenge, the parasite populations that emerged on reinfection were of the same variant phenotype as the challenge (i.e., second) inoculum. As in the case of uncloned parasites (K. N. Brown *et al.*, 1970*a*), we found that the variant type remains unchanged if the monkey has no agglutinating antibodies specific to the variant type of the infecting parasite population (Barnwell *et al.*, 1982*b*).

Furthermore, the time course of parasitemia in monkeys challenged with a homologous variant was identical to that of the monkeys challenged with heterologous variants. In the first case, the variant type of the challenge inoculum changed, and in the second case it did not change. This result suggests that in the presence of variant-specific agglutinating antibody, the antibody induces a switch in phenotype of the variant antigen expressed on the surface of infected cells. Antibody-dependent destruction of those parasitized cells expressing the same variant antigen as the initial inoculum (the majority), coupled with selection of any parasitized cells that had spontaneously switched to another variant antigen phenotype, would be expected to lead to a delay in the appearance of parasites. These ideas were originally put forward by Brown and co-workers using noncloned parasites after they had shown that the breakthrough parasite population from homologous variant challenge appeared with ⩽ 0.5-day delay in growth rate (K. N. Brown, 1973).

We do not known the role of variant-specific parasiticidal immune responses in immunity to *P. knowlesi.* It is clear that the SICA-variant antigen changes in the presence of homologous agglutinating antibody. We do not yet know, however, what would have happened had the antigen not changed. Since the variant-specific antibody measured *in vitro* binds to the surface of infected cells, it is not unreasonable to propose that *in vivo* such antibody would sensitize infected cells for immune clearance by any one of a variety of mechanisms (elaborated in Section III. B.1). One could argue that the very fact that successive parasite populations express different variant antigens constitutes strong circumstantial evidence for the existence of variant-specific parasiticidal responses. Why else would the parasite have evolved the capacity to express multiple variant-antigen phenotypes? Given this parasite's capacity to vary the SICA antigen, the agglutinating variant-specific antibody response is not parasiticidal, but rather inductive for expression of another variant-antigen phenotype.

Brown and colleagues measured variant-specific opsonizing antibody responses and attempted to correlate the presence of these antibody responses with protective immunity. Schizont-infected cells were incubated with mouse peritoneal macrophages in the presence of serum from infected monkeys and the extent of phagocytosis was measured (K. N. Brown *et al.*, 1970*b*). By measuring opsonizing and agglutinating antibodies during the course of infection/reinfection it was evident that variant-specific opsonizing antibodies were not found without the presence of variant-specific agglutinating antibodies. However, agglutinating antibodies appeared *before* opsonizing antibodies. The opsonizing activity in these monkey sera was equally variant-specific to that of antibodies mediating agglutination. It was suggested that the opsonizing antibodies might be of a different immunoglobulin class, subclass, or affinity/avidity from agglutinating antibodies. There was a correlation between the appearance of variant-specific opsonizing antibody and protective immunity in animals with chronic infection after repeated cycles of infection and drug cure (K. N. Brown *et al.*, 1970*b*; K. N. Brown, 1971). However, the correlation between opsonizing response and host protection was not absolute. Sera from chronically infected animals displayed opsonizing activity that was not variant-specific, yet parasites were still present (K. N. Brown and Hills, 1974). Furthermore, animals immunized with a particular SICA variant in complete Freund's adjuvant (CFA) develop high titers of agglutinating antibody only to the specific variant (K. N. Brown *et al.*, 1970*a,b*; Phillips *et al.*, 1970). On challenge, even with heterologous SICA variants, these animals show an increased prepatent period (time to detect parasites in the circulation) and a sterilizing immune response. But they do not develop nonvariant-specific opsonizing activity. These experiments showed that sterilizing immunity to *P. knowlesi* can operate without either variant-specific antibodies homologous to the challenge inoculum or opsonizing antibodies. It still remains possible that a direct correlation might exist between variant-specific agglutinating antibody (which induces antigenic variation) and the levels of antibody that can sensitize erythrocytes *in vitro* (where, in the absence of the spleen, antigenic variation does not occur) (Barnwell *et al.*, 1982*b*) for opsonization and phagocytosis by cells from infected monkeys. Antigenic variation could then be seen as a mechanism for evasion of such phagocytosis (known to be a major feature of protective immunity in rhesus monkeys) (Taliaferro, 1949), that evolved in such a way as to be activated by binding of variant-specific antibody.

B. Comparison of SICA[+] and SICA[-] Parasites: Evidence for Immunosuppression

1. Rationale for Invoking Immunosuppression

Earlier we mentioned the fact that some rhesus monkeys immunized with noncloned variants of *P. knowlesi* in FCA develop sterilizing immunity to infection with other antigenic variants of the same strain. (K. N. Brown *et al.*, 1979*a,b*; Phillips *et al.*, 1970). Some of these animals even possess sterilizing immunity when challenged with different *P. knowlesi* strains (K. N. Brown *et al.*, 1970*a*); however, these animals died when challenged with such simian malaria species as *P. cynomologi bastianelii* or *P. inui.* These two examples show that host-protective immunity that transcends variant-specific immune responses can be developed in rhesus monkeys to *P. knowlesi.* The immunity developed after long-term chronic infection with noncloned parasites may similarly restrict parasite growth by an immune response that transcends different antigenic variants. It remains possible that in this situation an accumulation of parasiticidal variant-specific immune responses, plus cross-reactivity of some of these responses with any new variant-antigen phenotypes that may arise, serves to control parasite growth. Whatever the case in the third example, it is clear that host-protective immunity that is strain and/or species specific in nature can be developed against *P. knowlesi* malaria (K. N. Brown and I. N. Brown, 1965; Voller and Rossan, 1969*b*; K. N. Brown *et al.*, 1970*a*). We therefore infer, as in our introductory review of the literature with human malaria (Section II), that strain and species-specific antigenic determinants of *P. knowlesi* must constitute the targets of such immunity. The actual effector mode (cellular versus humoral) that destroys parasites in these states of immunity is unimportant for this discussion. Why does this potent parasiticidal immunity not develop after repeated infection and drug cure with SICA[+] *P. knowlesi*? If the antigens are present on or in these cells that constitute the targets of variant-transcending immunity, what prevents the development of such responses during initial *P. knowlesi* infection? Immunosuppression of these and other responses can be suggested to account for these facts. Provided variant-transcending parasiticidal immune responses were suppressed, the parasite could establish and repeatedly reestablish chronic infection. The suppression necessary for parasite survival could be specific or nonspecific in nature, as long as the potent parasiticidal responses were suppressed. The parasite's capacity to vary the SICA-variant antigen also allows it to evade variant-specific immune responses. We can postulate on this basis that *P. knowlesi* parasites must have evolved powerful mechanisms for evasion of such variant-transcending parasiticidal immune responses; otherwise they would not survive.

Numerous studies with other malarial infections, including infections in humans, support the concept that suppression of immune responsiveness contributes to the lethality and chronicity of this disease (see Jayawardena, 1981; Kreier and Green, 1980). Immunosuppression is generally directly related to the degree of parasitemia (Greenwood *et al.*, 1971*b*). Loss of responsiveness to T- and B-cell mitogens has been documented, antibody responses to unrelated thymus-independent and -dependent antigens are suppressed, delayed-type-hypersensitivity is suppressed, and in severe infections cell-mediated immune responses are reduced; however, as yet there is no evidence for suppression of parasite-specific immune responses (see Jayawardena, 1981). Although several mechanisms for immunosuppression during malaria infection have been proposed (Terry, 1978), studies using purified parasite antigens and measuring specific antimalarial immune responses are needed to demonstrate convincingly a role of specific immunosuppression in immune evasion by plasmodia. Our *in vivo* studies comparing SICA[+] and SICA[-] parasites, led us to suggest that the SICA antigen is involved in immunosuppression of variant-transcending immune responses to *P. knowlesi.*

2. *Virulence of SICA[+] and SICA[-] Parasites*

P. knowlesi is highly virulent in nonsplenectomized rhesus monkeys and usually kills this host (Garnham, 1966). We also found this to be true of SICA[+] parasites, but not always with SICA[-] (Barnwell *et al.*, 1982*b*). In four intact monkeys infected with SICA[+] parasites, the parasitemia rose 8- to 10-fold daily and would have killed the animals had antimalarial drugs not been administered (Fig. 6A). At the time of treatment, the peak parasitemias were 16%, 19%, 21%, and 25%.

At low parasitemias (<1%), infection of intact rhesus monkeys with SICA[-] parasites also gave 8- to 10-fold daily increases in parasitemia (Fig. 6B–D). However, in four animals the parasitemias began to decline after reaching a peak of 1–6%. Two monkeys were given chloroquine after the initial decline in parasitemia (Fig. 6B), but in two other monkeys the parasitemias were followed for several days, during which the parasitemia fell even further, before institution of antimalarial chemotherapy (Fig. 6C). In four other monkeys infected with SICA[-]

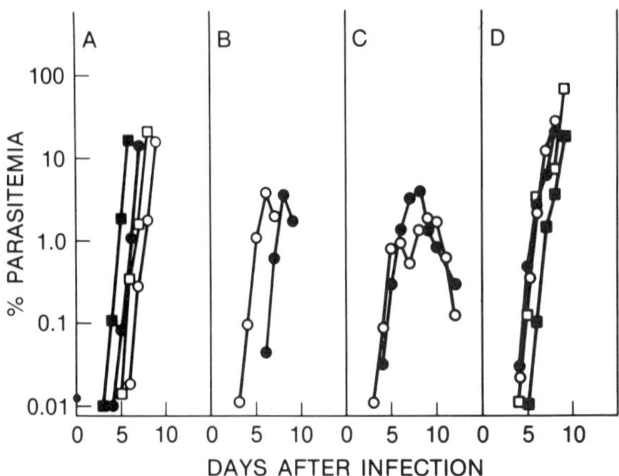

Figure 6. Comparison of the virulence of SICA[+] and SICA[-] *P. knowlesi* infections in intact (i.e., with a spleen) rhesus monkeys. The peripheral blood parasitemia, expressed on a logarithmic scale as the percentage of red blood cells containing parasites, was measured at the same time on successive days after inoculation. (A) Parasitemias of SICA[+] parasites in four animals. (B,C) Parasitemias of SICA[-] parasites in four animals where parasites remained SICA[-]. (D) Parasitemias of SICA[-] parasites in four animals in which parasites converted to SICA[+]. The approximately 10-fold daily multiplication of SICA[+] parasites (A,D) should be noted. Although SICA[-] parasites show a similar multiplication rate during initial stages, the parasitemia declines after reaching a maximum of 1-5% (B,C).

parasites, the parasitemias continued to rise, reaching 21%, 25%, 27%, and 60% (Fig. 6D) before initiation of drug therapy. At the time of peak parasitemia in all these animals, SICA tests were performed on the parasitized erythrocytes. In those cases of SICA[-] inoculation in which infection was controlled (Fig. 6B, C) the parasites remained SICA[-], whereas in monkeys that did not control their infections, the parasites were found to have converted from SICA[-] to the SICA[+] phenotype. The SICA[-] phenotype derived by passage in splenectomized animals therefore appears to be less virulent than the SICA[+] parasite in nonsplenectomized monkeys. In splenectomized rhesus monkeys, neither SICA [+] nor SICA[-] infections are controlled (Barnwell *et al.*, 1982*b*).

Other hemoprotozoa have been shown to be less virulent after passage through splenectomized hosts, e.g., *Plasmodium cynomolgi* in rhesus monkeys (L. H. Schmidt, unpublished data) and *Babesia bovis* in cattle (Callow, 1977). However, the most intriguing aspect of our results with *P. knowlesi* is the correlation between virulence and SICA phenotype expression. If the difference in virulence of SICA[+] and SICA[-] parasites is related to the loss of SICA antigen expression on the cell surface that we have demonstrated biochemically (R. J. Howard and J. W. Barnwell, unpublished results, 1983), and not some other unidentified change accompanying conversion from SICA[+] to [-], these results suggest that parasites that express the variant antigen (i.e., SICA[+]) have increased survival advantage in the presence of the spleen.

It must be emphasized that the differences observed in virulence of SICA[+] and SICA[-] parasites (Fig. 6) were seen in primary infections. In primary infections when parasitemias rise rapidly to produce a fatal infection, the variant-antigen phenotype of SICA[+] parasites does not change. Consequently, the virulence of SICA[+] parasites under these conditions is not related to antigenic variation per se. We suggest that the variant antigen expressed on the surface of infected cells functions to depress or suppress spleen-dependent immune responses to a target(s) of parasiticidal immunity (variant or strain transcending?) or in some way prevents access of the immune response to the parasite.

The nature of the early spleen-dependent immune response elicited by SICA [-] parasites—but apparently not by SICA[+] parasites—is unknown. It need not be antibody dependent. In primary infections of the rodent malaria *P. chabaudi*, for example, the parasite was controlled equally well in normal mice and in mice injected from birth with goat antimouse μ-chain, which made no antibody to this parasite (Grun and Weidanz, 1981).

3. *Immunogenicity of SICA[+] and SICA[-] Parasites*

We have preliminary results with a limited number of monkeys that indicate a dramatic difference in the nature of immune responses elicited by infection with SICA[-] and SICA[+] parasites (J. W. Barnwell, unpublished results, 1983). The results shown in Fig. 7B reconfirm the earlier observation (K. N. Brown and I. N. Brown, 1965; Cohen and Butcher, 1970) that reinoculation with *P. knowlesi* is usually fatal unless antimalarial drugs are used. The SICA antigen phenotype of the parasite population arising after reinoculation will be altered in the case of homologeous challenge, and unaltered when an heterologous variant is used for challenge. In either case, the emerging parasite population grows rapidly with 8- to 10-fold daily increases in parasitemia. In contrast, when SICA[+] parasites were used for challenge after an initial infection with SICA[-] parasites (Fig. 7A), the parasitemia rose to only 1-5% before beginning to decline. The parasitemia declined to about 0.1% over the next few days without application of antimalarial chemotherapy. We then drug treated the animals to ensure their survival. Although this experiment with SICA[-] infection followed by SICA[+] challenge has been performed to date with only two animals, the results are so dramatically different from those recorded for SICA[+] infection and challenge, in which such a parasitemic decline is never seen on second challenge, that we are optimistic that experiments currently being performed with a larger number of animals will provide statistical support for this finding.

We interpret the results shown in Fig. 7 as follows: SICA[+] parasites do not induce a parasiticidal immune response on infection and drug cure sufficient to

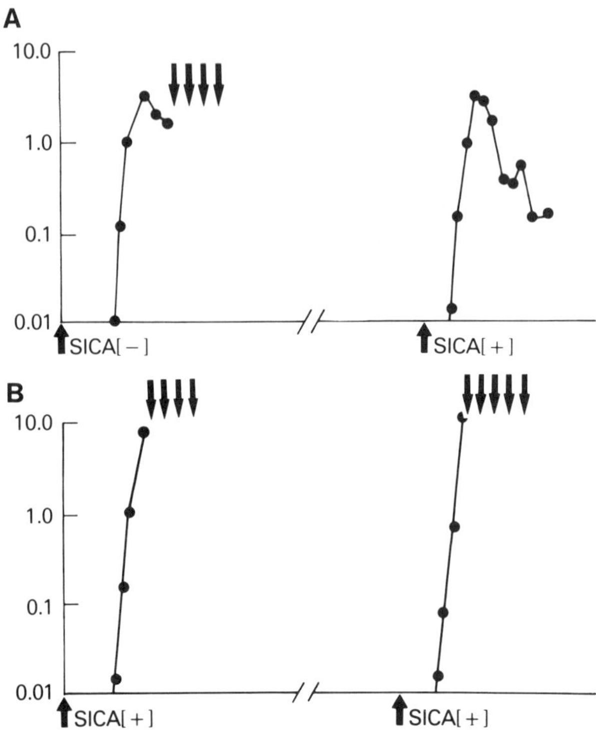

Figure 7. Infection of nonsplenectomized rhesus monkeys with (A) SICA[−] or (B) SICA[+] *P. knowlesi* and challenge with SICA[+] parasites. This schematic representation shows the results for single animals. Very similar results were obtained to those of (A) with one other animal, and with six other animals for results in (B). The vertical axis shows blood parasitemia on a logarithmic scale, while the horizontal axis shows data points collected on successive days. Animals were infected as shown by the upward vertical arrows (↑) and administered with antimalarial drugs as shown by the downward vertical arrows (↓). After drug cure of the initial infection, a 3- to 4-week period elapsed before challenge infection. The challenge inoculum of SICA[+] parasites was approximately 10 times the minimal lethal dose for naive recipients. The animals infected first with SICA[−] parasites showed a maximal parasitemia of 5%, after which the parasitemia rapidly declined to ~0.1%. In contrast, animals infected first with SICA[+] parasites showed daily 10-fold increases in parasitemia necessitating drug cure (to save the animal) at ≧10% parasitemia. It must be emphasized that failure to drug cure infections of ≧10% parasitemia is uniformly lethal within 24 hr, since the next cycle of parasite reinvasion produces a parasitemia of ~100%.

affect significantly the rise of parasitemia on subsequent SICA[+] challenge infection. However, a SICA[−] infection and drug cure induces spleen-dependent immune responses that are then very effective against subsequent SICA[+] challenge. The latter result suggests that SICA[+] and SICA[−] parasites share an antigen that is accessible on both parasites to an immune response that is developed only after SICA[-] infection. Since this immune response is not developed after SICA[+] infection, induction of this response appears to be prevented by the presence of the SICA variant antigen on SICA[+] parasites.

Until we have completely characterized all components of SICA[+] and SICA[-] parasites, it remains possible that some other as yet unidentified component of SICA[+] parasites is also lost upon conversion to the SICA[-] phenotype. Such a component could prevent induction of parasiticidal immune respones by SICA[+] parasites rather than the variant antigen. However, on comparative one-dimensional SDS-PAGE of total malarial proteins labeled with various radioactive amino acids, or of total immunoprecipitated malarial antigens, we saw no difference in SICA[-] and SICA[+] parasites (R. J. Howard and J. W. Barnwell, unpublished results, 1983). If other proteins are lost in this conversion, they must represent a very small proportion of the total malarial proteins, as does the variant antigen.

4. General Comments on SICA Antigen and Immune Evasion

It is well established that the capacity of *P. knowlesi* to alter the SICA variant antigen on schizont-infected erythrocytes is linked to the capacity of this parasite to evade variant-specific immunity (K. N. Brown and I. N. Brown, 1965). We have confirmed this with cloned SICA[+] *P. knowlesi* parasites and conclude from challenge experiments with homologous cloned variants that antigenic variation occurs via an antibody-induced change in parasite phenotype. Elsewhere we show that antigenic variation is dependent not only on homologous variant agglutinating antibody, but on some spleen-dependent factor(s), as well, since it does not occur in splenectomized hosts (Barnwell *et al.*, 1982*b*). The capacity of *P. knowlesi* to evade variant-specific parasiticidal immune responses does not account for the capacity of this parasite either to infect rhesus monkeys repeatedly on reinoculation, or establish long-term chronic infections during which potent strain- and/or species-specific immune responses are clearly either prevented or evaded in some manner.

By *in vivo* comparison of the virulence and immunogenicity of SICA[+] parasites, which express variant antigen on the surface of infected cells, and SICA[-] parasites, which lack identifiable variant antigen on the red cell surface, we have implicated the variant antigen in immunosuppression of spleen-dependent parasiticidal immune responses. SICA[-] parasites are less virulent in nonsplenectomized monkeys than are SICA[+] parasites, indicative of the induction of parasiticidal immune responses by SICA[-] parasites. Furthermore, the immunity developed after SICA[-] infection and drug cure is effective against virulent SICA[+] parasites. Since the variant antigen is by definition the phenotypic character that defines the difference between SICA[+] and SICA[-] parasites, and no other biochemical differences have been identified to date, we suggest that in the presence of the variant antigen, the target molecules of variant-transcending parasiticidal immunity fail to elicit these immune responses. The variant antigen on infected erythrocytes therefore appears to have a dual role: evasion of variant-specific immune responses and suppression of variant-transcending parasiticidal immune responses. It will be of interest to determine the structural and immunochemical basis for these roles, both of which enable the parasite to evade host immunity.

These studies are also of importance in the search for suitable antigenic targets on which to base an antimalarial vaccine. Of greatest interest in this regard is the identification of the antigenic targets of parasiticidal immunity, whether they be strain- or species-specific. We are attempting to identify the antigen of SICA[-] parasites that elicits parasiticidal immunity effective against both SICA [+] and SICA[-] phenotypes after SICA[-] infection.

Finally, there may be a parallel between the functions of the SICA variant antigens suggested here and those of the variant-surface glycoproteins (VSGs) of the African trypanosomes. There is evidence for a dual role of VSGs in immune evasion by this other protozoan parasite: Not only does the parasite's capacity to express different VSGs enable it to evade variant-specific immunity, but VSGs appear to exert an immunosuppressive function. Membrane fractions bearing the VSGs cause immunosuppression (Clayton *et al.*, 1979), and the degree of immunosuppression is related to the differences in parasite virulence and pathogenicity of clones of different VGA phenotype (Sacks *et al.*, 1980). It has therefore been suggested that the VSGs may be immunosuppressive and that differences in their structure account for differences in the degree of immunosuppression (Sacks *et al.*, 1982). With purified membranes from erythrocytes infected with SICA[+] or SICA[-] *P. knowlesi*, it may prove possible to demonstrate directly an analogous immunosuppressive role for the malarial variant antigens.

VII. KNOBS AND IMMUNE EVASION BY SEQUESTRATION

A. Sequestration *in Vivo*

Several primate malarias, such as *P. falciparum*, *P. fragile*, and *P. coatneyi*, are characterized by the virtually complete absence of mature asexual parasites (mature trophozoites, schizonts, and segmenters) from the peripheral circulation. Ring-stage and very young trophozoite-stage parasites are the predominant or exclusive stages seen in the blood. In other malarias, such as *P. knowlesi* in monkeys and *P. berghei* in the rat, mature stages are seen in the blood, although it is again notable that their numbers are fewer than in immature stages. This phenomenon is termed either deep vascular schizogony, describing the fact that parasite nuclear division (schizogony) is restricted to internal organs and tissues, or sequestration. The latter term emphasizes the separation or isolation of mature parasitized cells from the circulation. The phenomenon was first described with *P. falciparum* (Bignami and Bastianelli, 1890).

In malarial infections characterized by relatively synchronous intraerythrocytic parasite development, e.g., *P. fragile*, *P. coatneyi*, or *P. knowlesi* in subhuman primates or *P. falciparum* in humans and monkeys, sequestration is evident as the cyclical appearance of ring and early trophozoite stage parasites in the peripheral blood. During the intervening periods, lower numbers of more mature parasites are seen with *P. knowlesi* (Miller *et al.*, 1971), while with the *falciparum*-type malarias the blood is almost devoid of mature parasites (Miller *et al.*, 1971). It is of interest to note that in low parasitemias of *P. knowlesi* ($<5\%$) the decrease in peripheral blood parasitemia as parasites mature is quite marked (80-90% sequestration), but at higher parasitemias ($>5\%$) the decline is much less (a maximum of 50% of the cells sequester versus $\geq 95\%$ of *P. falciparum* or *P. coatneyi*), suggestive of saturation of a limited number of sequestration sites at higher parasitemias (Miller *et al.*, 1971) (Table IX).

The organ and tissue distribution of mature parasitized erythrocytes is characteristic of each malaria species and host and also depends to some extent on the total blood parasitemia (Table IX).

In the primate infections of *P. falciparum* in *Aotus trivirgatus* (Miller, 1969; Voller *et al.*, 1969), *P. fragile* (Fremount and Miller, 1975), or *P. coatneyi* (Desowitz *et al.*, 1969; Miller *et al.*, 1971*a*) in *Macaca mulatta* the heart and adipose tissue are the major sites of deep vascular schizogony. The patterns of secondary organ involvement are different, however, with the spleen, skeletal muscle, and small intestine having the next highest levels with *falciparum* malaria (Miller, 1969), and liver, lung, and spleen being the next most important with *coatneyi* malaria. Furthermore, in *falciparum* malaria the mature trophozoites and schizonts were seen on the venous side of cardiac capillaries, whereas in *coatneyi* malaria these parasitized cells were confined to capillaries and venules. The tissue distribution of parasites in fatal *falciparum* malaria of humans has been examined at autopsy. Parasitized red cells were seen lining the capillaries of the heart and intestine and the margins of veins (Ash and Spitz, 1945; Clark and Tomlinson, 1949; Spitz, 1946) in a manner similar to that observed in the monkey infection. Extensive occlusion of cerebral capillaries by parasitized cells was noted in a fatal case of *P. falciparum* malaria in humans (Rigdon, 1942); however, overwhelming natural *P. falciparum* infections in humans exhibit a spectrum of brain lesions. In some dying patients there is an almost total lack of parasites in the brain, while in others infected cells are packed around the capillary walls, causing microthrombi and gray matter hemorrhages (Spitz, 1946). These hemorrhages have been suggested to account for the convulsive symptomatology seen in some terminal patients with *P. falciparum* infections (Spitz, 1946). Monkey infections with *P. falciparum* exhibit very few changes in brain and much lower sequestration of infected cells at this site (Miller, 1969; Jervis *et al.*, 1972; Gutierrez *et al.*, 1976), but this is most probably attributable to the lower parasitemias of the

Table IX. Comparison of Deep Vascular Schizogony in Different Malarias[a]

Parasite	Deep vascular schizogony (DVS)	Primary organs of DVS at low parasitemia	Presence of knobs
Primate			
Vivax type			
P. vivax, *P. simium*, *P. cynomolgi*	—	—	No
Ovale type			
P. simiovale, *P. fieldi*	—	—	Occasionally seen
Falciparum type			
P. falciparum, *P. fragile*, *P. coatneyi*	++++	Heart, adipose tissue	Yes[b]
Malariae type			
P. malariae, *P. brasilianum*	—	—	Yes[b]
P. inui	—	—	No
Other			
P. knowlesi	+ to ++	Liver, small intestine	No
Rodent			
P. berghei	++ to +++[c]	Bone marrow, liver, lung, spleen	No

[a]Adapted from tables of Aikawa *et al.* (1975) and Miller *et al.* (1971*a*) together with other references described in the text.
[b]*Falciparum*-type parasites have knobs only on erythrocytes infected with asexual parasites. The *Malariae*-type parasites, *P. brasilianum* and *P. malariae*, have knobs on erythrocytes infected with asexual parasites (both immature and mature) and also on gametocytes.
[c]*P. berghei* infection is also characterized by sequestration of erythrocytes infected with immature parasites in kidney and ventricular myocardial vasculature (Desowitz and Barnwell, 1976).

monkey infection and the brevity of these infections, rather than to a difference in properties of the parasite or host endothelial cells.

An additional well-documented site of sequestration of *P. falciparum* in human infections is the placenta of primigravida (Jilly, 1969; McGregor, 1978; Bray and Sinden, 1979). Mature trophozoite- and schizont-infected erythrocytes are seen lining the capillary walls and clumped within the capillary lumen.

In *P. knowlesi* infections of *M. mulatta*, the organ distribution is quite different from that seen with *P. falciparum*, *P. fragile*, and *P. coatneyi* (Miller *et al.*, 1971*a*). In low parasitemia, schizont-infected cells were trapped mainly in the periportal hepatic sinusoids and submucosal venules of the small intestine. As the parasitemia increased, they were sequestered in other organs, many cerebral capillaries and venules being filled with parasitized cells.

Sequestration of erythrocytes infected with rodent malaria parasites has also been described. *Plasmodium berghei* schizont-infected cells sequester in bone marrow, the major site, and in liver, lung, and spleen (Alger, 1963; Jacobs and Warren, 1967; Miller and Fremount, 1969; Desowitz and Barnwell, 1976). This organ distribution is therefore quite distinct from that seen with the primate malarias. An additional interesting observation with *P. berghei* infection of the white rat was the sequestration of young forms (i.e., newly invaded merozoites) in the kidney and ventricular myocardial vasculature, equal to or greater than the degree of sequestration of schizont-infected cells in bone marrow, liver, and other sites (Desowitz and Barnwell, 1976).

Sequestration in primate malarias of the *falciparum* type can be seen to share some properties with sequestration in the primate malaria *P. knowlesi*, while in other respects it is quite different. In both cases mature parasitized cells are sequestered via attachment to endothelial cells lining the blood vessels of various organs and tissues. Three properties that show a difference are (1) the degree of disappearance of parasitized red cells from the peripheral blood, (2) their distribution in various organs and tissues, and (3) the presence of knobs on the surface membrane of erythrocytes infected with *falciparum*-type malarias (*P. falciparum*, *P. fragile*, *P. coatneyi*). Knobs are absent from erythrocytes infected with the primate malarias *P. knowlesi*, *P. vivax*, and *P. cynomolgi.* They are also not found on erythrocytes infected with rodent parasites, such as *P. berghei.* The occurrence of knobs with different malarias and evidence that they are associated with the appearance of new antigens on the cell surface are discussed in Section IV.A.

Trager and colleagues (Trager *et al.*, 1966; Rudzinska and Trager, 1968) first suggested that knobs might be the sites of attachment of infected erythrocytes, since asexual *P. falciparum* or *P. coatneyi*-infected erythrocytes were found to express knobs and to be sequestered, whereas gametocytes of these species had none and were observed in the peripheral circulation. Ultrastructural studies with material taken from infected monkeys supported this concept (Luse and

Miller, 1971; Aikawa *et al.*, 1972; Miller, 1972; Fremount and Miller, 1975; Gutierrez *et al.*, 1976), since the knobs were the closest points of contact of parasitized erythrocytes to endothelial cells. Electron microscopic study shows the regions of erythrocyte membrane between the knobs to be separated by a narrow gap from the endothelial cell membrane, whereas the knobs appeared to be in tight contact. Since those malaria species that exhibit the organ distribution of sequestration and extremely high efficiency of sequestration characteristic of *P. falciparum* all possess knob protrusions, and these protrusions are seen to be the points of closest contact to endothelial cells, it has been concluded that the knobs mediate endothelial cell binding.

How can the sequestration of *P. knowlesi* be accounted for? This parasite does not express knobs on the infected erythrocyte. The fact that its pattern of sequestration in different organs and tissues is unlike that of *P. falciparum* suggests that the mechanism might be different. Rheologic studies with *P. knowlesi*-infected cells showed that they are less deformable than uninfected erythrocytes (Miller *et al.*, 1971*b*) and suggested that this may account for their obstruction in capillaries. This does not account for their attachment to endothelia of the periportal hepatic sinusoids and submucosal venules of the small intestine (Miller *et al.*, 1971), although it may account for the occlusion of cerebral capillaries in high parasitemia when other sequestration sites are filled and mature infected erythrocytes appear to spill over into the peripheral circulation. An alternative hypothesis for sequestration of malaria parasites within cerebral capillaries, originally proposed with respect to *P. falciparum*, is that cerebral edema resulting from increased capillary permeability narrows the capillary lumen, which leads to entrapment of infected cells (Maegraith, 1969). As yet there is no satisfactory explanation for sequestration of *P. knowlesi* in low parasitemia other than the proposition that some change in the surface properties of the infected erythrocyte causes binding to endothelial cells of certain tissues.

In vivo studies and electron microscopic examination have illuminated another important property of the knob protrusions on infected erythrocytes. The quartan malarias *P. malariae* and *P. brasilianum* express knobs on the surface of erythrocytes infected with young and mature asexual parasites and on erythrocytes infected with gametocytes. These knobs are morphologically indistinguishable from those of *falciparum* malaria; however, *P. malariae* and *P. brasilianum* do not undergo deep vascular schizogony (Smith and Theakston, 1970; Sterling *et al.*, 1972). The presence of knobs on the surface of these infected erythrocytes is presumably linked to some functional change in membrane properties other than endothelial attachment. These observations suggest that knobs may have multiple functions important for parasite survival—if not, their expression would appear to be entirely detrimental for the parasite, since they result in the expression of new antigens on the infected cell and can probably act as targets for parasiticidal immune responses. Indeed, in *P. brasilianum* infections of splenectomized squirrel monkeys (*Saimiri sciureus*) (Sterling *et al.*, 1972) it has been proposed that the entrapment of parasitized cells by phagocytic cells in the liver in the immune cure of infection is related to the recognition of knob components. Other functions of knobs have not been identified—they might include altered membrane transport to permit uptake of essential metabolites and release of waste metabolites or release of antigens into the plasma for modulation of the immune response. If it is eventually proved that the knobs on erythrocytes infected with the quartan malarias perform some other functional role besides sequestration, it will be of interest to determine whether knobs of the *falciparum*-type malarias have the same function(s) as well as endothelial attachment. Attempts to account for the roles of knobs with *P. malariae* and *P. brasilianum* on the basis of the biologic data alone are further confounded by the fact that another quartan malaria, *P. inui*, does not have knobs (see Aikawa *et al.*, 1975). A molecular analysis of knob components and their functions is required to explain these observations.

Electron microscopic studies of two *ovale*-type primate malarias, *P. simiovale* and *P. fieldi*, in splenectomized *M. mulatta* demonstrated knobs on the membrane of infected cells (Aikawa *et al.*, 1977). As in the case of *falciparum*- and *malariae*-type malarias, these morphologic alterations of the membrane consisted of a projection of the membrane with an electron-dense deposit beneath. Erythrocytes infected with uninucleate trophozoites had knobs randomly distributed over the membrane. Schizont-infected erythrocytes had the knobs concentrated at the tip of angular projections of the erythrocyte membrane. *P. ovale*-type malarias are similar to the *malariae* parasites in that they also do not sequester, schizont-infected erythrocytes being found in the peripheral blood. The knobs on *P. simiovale*- and *P. ovale*-infected cells therefore do not function in endothelial attachment. As with the knobs on *malariae* parasites, their role(s) are not yet known.

These relationships among parasite type, the presence of knobs, sequestration of schizont-infected erythrocytes, and the sites of sequestration are summarized in Table IX.

B. Binding to Endothelial Cells and Melanoma Cells *in Vitro*

An *in vitro* binding assay for attachment of *P. falciparum*-infected erythrocytes to human endothelial cells was recently developed (Udeinya *et al.*, 1981). Endothelial cells were obtained from human umbilical vein and cultured as a layer of cells in plastic culture dishes. Erythrocytes infected with trophozoite- and schizont-stage parasites were added to the cultures and incubated, and unattached cells were removed by aspiration and washing. After glutaraldehyde fixing and Giemsa staining, parasitized erythrocytes could be seen attached to the endothelial cells. Between 22 and 44% of the endothelial cells were found to bind erythrocytes, the number of erythrocytes per cell ranging from one to >100. Binding was specific for trophozoite- and schizont-infected cells: Only 1.4–3.8% of the initial erythrocyte suspension consisted of parasitized erythrocytes, but >93% of the bound erythrocytes were parasitized. Electron microscopic studies confirmed that the *P. falciparum*-infected erythrocytes were attached to endothelial cells *in vitro* via knobs. Specificity controls for attachment included the demonstration that parasitized erythrocytes did not attach to human dermal fibroblasts, together with the failure of uninfected erythrocytes to attach (Udeinya *et al.*, 1981). Since this assay established an *in vitro* correlate of adhesion of parasitized erythrocytes to endothelium *in vivo*, it now becomes feasible to explore the molecular mechanism of adhesion.

The usefulness of the *in vitro* binding assay has been expanded by the establishment of an amelanotic melanoma cell line as an alternative substrate for attachment of infected erythrocytes (Schmidt *et al.*, 1982). The difficulties of obtaining fresh umbilical cord and limitations on the scale of the binding assay inherent in this source of endothelial cells are obviated by the use of rapidly growing amelanotic melanoma cells. Several criteria were used to establish these melanoma cells convincingly as a substrate with a very similar, if not identical, binding mechanism for infected erythrocytes as endothelial cells. Binding to the amelanotic melanoma was specific for *P. falciparum*-infected erythrocytes: >98% of bound cells contained trophozoites or schizonts, whereas <4% of the erythrocytes in the starting suspension contained mature parasites. Nearly all (96–100%) of the melanoma cells were found to bind infected erythrocytes. Human erythrocytes infected with a Brazilian strain (It) of *P. falciparum* and *Aotus trivirgatus* erythrocytes infected with another strain (Malaysian Camp/CHQ), both of which express knobs on the erythrocyte surface, bound to melanoma cells as well as to endothelial cells. Only trophozoite- and schizont-infected cells were able to bind. Two nonsequestering malaria parasites were also used to test whether the specificity of the amelanotic melanoma paralleled that of human endothelial cells: A K^- variant of the St. Lucia strain of *P. falciparum* that does not sequester *in vivo* (schizont-infected cells remain in the peripheral blood) did not attach to the melanoma or endothelial cells; *A. trivirgatus* erythrocytes infected with *P. vivax* also failed to bind. The parasite stage specificity of attachment and correlation of attachment with the presence of knobs are therefore identical for endothelial cells and the amelanotic melanoma. Transmission electron microscopy confirmed that infected erythrocytes attached to amelanotic melanoma cells via knobs (Schmidt *et al.*, 1982).

C. Sequestration and Parasite Survival

Sequestration in the *falciparum*-type malarias involves the attachment of erythrocytes infected with mature asexual parasites to endothelial cells via the knobs on the erythrocyte membrane. The specificity of this interaction, together with the fact that new antigens are expressed at the knobs (Section IV.A), suggests that the knobs contain a molecule(s) that mediate binding to a component on the endothelial cell membrane. In the case of other malaria species, such as *P. knowlesi* and *P. berghei*, which also sequester, but by a different mechanism depending on the tissue specificity and degree of sequestration (Section VII.A), attachment is also to endothelial cells. In these latter cases, there are no morphologic correlates with sequestration, such as knobs on the erythrocyte membrane, although caveolae have been identified in the membrane of *P. knowlesi*-infected erythrocytes (Section IV.B). The presence of new antigens has been demonstrated on the membrane of erythrocytes infected with *P. berghei* or *P. knowlesi* (Section IV.B–D), and it is possible that one of these new antigens also binds to a component on endothelial cells. We would predict that the binding molecules of the *falciparum*-type malarias and those of other sequestering parasites would be different in some way to account for the different tissue distributions and efficiencies of erythrocyte sequestration. Different host species may have different distributions within the blood vessels of their tissues of various classes of endothelial cells, which have different affinities for infected erythrocytes. Such differences could account in part for the different patterns of sequestration in the infections described earlier. However, there must also be specific differences in different species of malaria parasite in the molecular basis for endothelial cell attachment: *P. knowlesi* and *P. coatneyi* infections of the same host species (*M. mulatta*) result in markedly different types of sequestration.

These observations, together with the fact that several malaria species do not appear to sequester, lead us to ask: What is the significance of sequestration for parasite survival? Why do only some parasites appear to cause alterations in the membrane of the infected erythrocyte, which presumably create new membrane antigens, such that mature parasitized cells bind? One potential molecular explanation for this functional difference in different malaria parasites is that all malaria species express similar new antigenic components in the erythrocyte membrane, but that only in some species, such as *P. falciparum*, do other aspects of membrane topography and structure favor a molecular configuration/distribution that results in a functional capacity to bind to endothelial cells of particular tissues. However, this argument and those developed in the following discussion still do not explain why some parasites have evolved the capacity to alter membrane structure such that endothelial binding ensues, whereas others have not.

Since reduced levels of oxygen have been shown to favor development of *P. falciparum in vitro* (Scheibel *et al.*, 1979), it has been suggested that the hypoxic venular environment in which mature parasitized cells are sequestered permits optimal parasite growth. Such conditions might be essential for late trophozoite and schizont stages of some malaria species, since parasite differentiation and division are maximal at these stages.

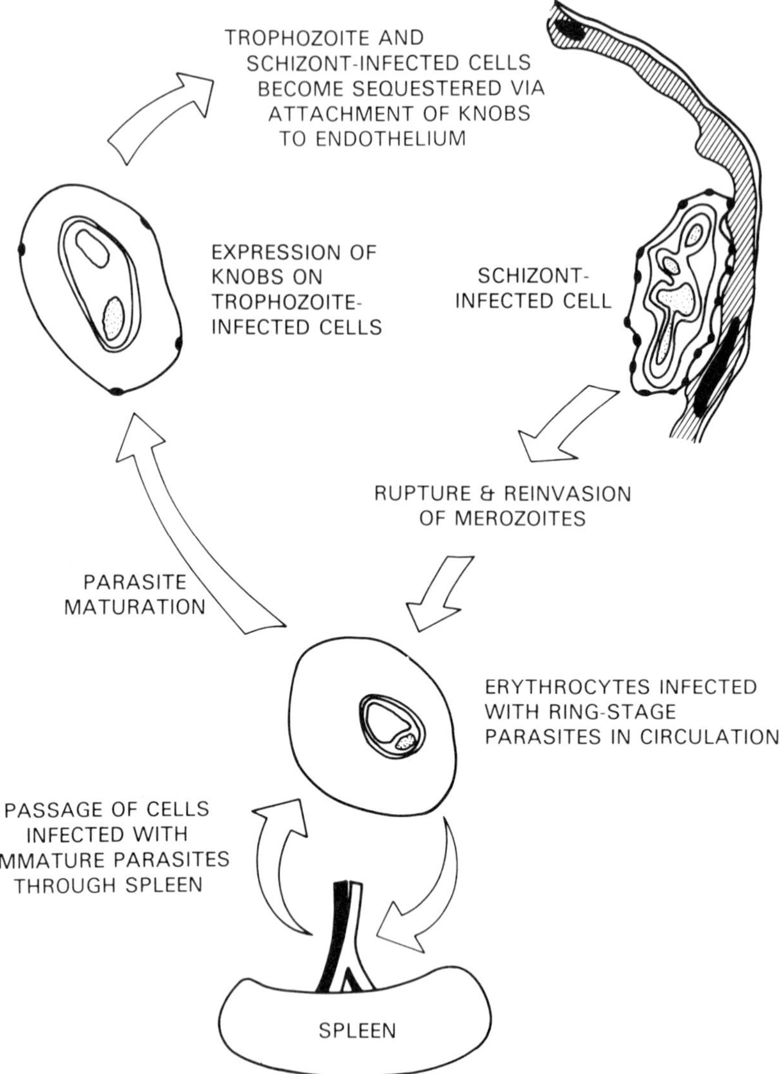

Figure 8. Evasion of localized spleen-dependent parasiticidal immune responses by sequestration of mature *falciparum*-type parasites. Erythrocytes infected with immature (ring-stage) parasites are found in the peripheral circulation. These cells pass through the spleen. Since ring-stage parasites appear to express either very low levels of new surface antigens on the surface of the infected cell, or none at all, and since these cells are rheologically little changed from normal erythrocytes, most survive splenic passage. As the parasite enlarges to the trophozoite stage knobs are expressed on the surface of the infected erythrocyte. The density of knobs increases as the parasite matures further and undergoes schizogony (nuclear budding). The knobs mediate attachment of trophozoite- and schizont-infected erythrocytes to endothelium. Erythrocytes infected with these parasite stages are not normally seen in

Several workers have proposed that attachment of *P. falciparum*-parasitized erythrocytes to endothelial cells removed from the spleen enables mature parasites to avoid passage through this organ (Udeinya *et al.*, 1981; Barnwell *et al.*, 1982*b*). The spleen is a major organ of host defenses to malaria infection (reviewed by Kreier and Green, 1980) with the capacity for a variety of localized specific and nonspecific mechanisms of parasite destruction. Specific mechanisms could include opsonization and phagocytosis of antibody-coated cells by splenic macrophages and destruction by antigen-specific cytotoxic T-cells. Nonspecific mechanisms could include release of lymphokines that might destroy intracellular parasites and the removal or pitting of parasites from infected erythrocytes as they pass through the small fenestrations of the basement membrane between spleen cords and sinuses (Schnitzer *et al.*, 1972). Erythrocytes infected with ring- and early trophozoite-stage parasites bear lower levels of new surface antigens than do late trophozoite- and schizont-infected erythrocytes (Section IV.A, C, D) and may therefore be able to pass through the spleen without suffering antigen-specific immune clearance. Cells infected with immature parasites would also be more deformable, since the parasite is a much smaller inclusion at this stage. This property would result in a lower rate of loss of ring stages through pitting, as compared with schizont stages. These several arguments related to localized spleen-dependent mechanisms of parasite destruction lead us to suggest that the capacity of mature asexual malaria parasites to sequester might have evolved in order to avoid these mechanisms at precisely the most susceptible parasite stages. This concept is summarized schematically in Fig. 8 for *P. falciparum* malaria. Since knobs mediate sequestration with this malaria, this scheme suggests that evasion of spleen-dependent immune responses is caused by an endothelial cell-binding component in the infected erythrocyte's membrane at the knobs. There is no direct molecular evidence that such a binding component is of parasite origin, although the facts that knobs express new parasite-dependent antigens (Section IV.A) and that insertion of a parasite protein, the SICA antigen (Section IV.C) into the outer membrane of infected cells has been demonstrated with another malaria parasite make such a possibility very likely. Additional evidence, also indirect, for a role of new parasite-dependent antigens on infected

---

the peripheral circulation because of this attachment (i.e., they are sequestered). By sequestration in this manner, these mature parasitized cells, which express new antigens on the cell surface and which are considerably altered in deformability compared to uninfected cells, effectively evade passage through the spleen. Any parasiticidal immune responses (both specificially and nonspecificially activated) that are localized in spleen are thereby avoided by these mature parasites. Note, however, that humoral antibody that might mediate blockade of attachment via knobs to endothelium (Fig. 9B) or reversal of attachment (Fig. 9C) could exert a potent parasiticidal immune response by releasing mature infected cells into the peripheral circulation, leading to antibody-dependent or antibody-independent immune destruction of these cells in spleen.

erythrocytes in attachment to endothelial cells comes from attempts to block attachment with antibody, described in Section VII.D.

It must be stressed that although this concept might prove valid for *P. falciparum* infection, it is obviously inadequate when one considers infections with the primate malarias *P. brasilianum* and *P. malariae*, both of which express knobs on infected erythrocytes that do not sequester. It is most probable that these knobs expose new antigens on the cell surface, as with *P. falciparum*-infected erythrocytes (Section IV.A). Given the notorious persistence of these quartan malarias, one must explain why nonspecific clearance mechanisms in spleen and specific clearance mechanisms activated by antibody to presumed knob antigens fail to remove these parasitized cells. Perhaps these malarias have evolved completely different mechanisms for suppression and/or evasion of host-immune responses to the *falciparum*-type malarias that do not require sequestration of mature parasitized erythrocytes. In addition, changes in splenic microcirculation might have profound effects on the capacity of parasitized erythrocytes to pass unharmed through the spleen. It would be of interest to examine whether during *falciparum*-malaria the splenic microinoculation favors destruction of parasitized erythrocytes, while that of the quartan malarias might favor their survival.

D. Role of Antibody to Knob Components in Immunity

There is no direct evidence that specific immune responses directed against the new antigens on knobs (Kilejian *et al.*, 1977; Langreth and Reese, 1979) are a component of host immunity. During immunologic crisis and cure of *P. falciparum* and *P. coatneyi* infections in monkeys, mature parasitized cells that are normally sequestered are detected in the circulation. It has been suggested that binding of antibody to knob components at this time is associated with the appearance of mature parasitized erythrocytes in the periphery (WHO Technical Report, 1975). Although such a mechanism of *in vivo* parasite destruction remains conjectural, there is some recent supportive evidence for such a possibility from *in vitro* studies with the endothelial cell/amelanotic melanoma binding assay that clearly demonstrates that anti-malarial antibody can block attachment of mature parasitized erythrocytes. Preincubation of *P. falciparum*-parasitized erythrocytes with serum from an immune *Aotus* monkey has been shown to abolish attachment (Udeinya *et al.*, 1981).

In other experiments (I. J. Udeinya and L. H. Miller, unpublished data, 1983), serum was collected at various times from *Aotus* monkeys immunized by repeated infections with the Malaysian Camp HQ strain of *P. falciparum.* Sera collected from these animals before infection had no effect on binding. Sera obtained after protective immunity had been established inhibited binding completely when preincubated with the infected cells. This is illustrated schematically in Fig. 9B. These workers also showed that such sera could be added *after* parasitized cells had been allowed to attach to endothelial cells (or the amelanotic mel-

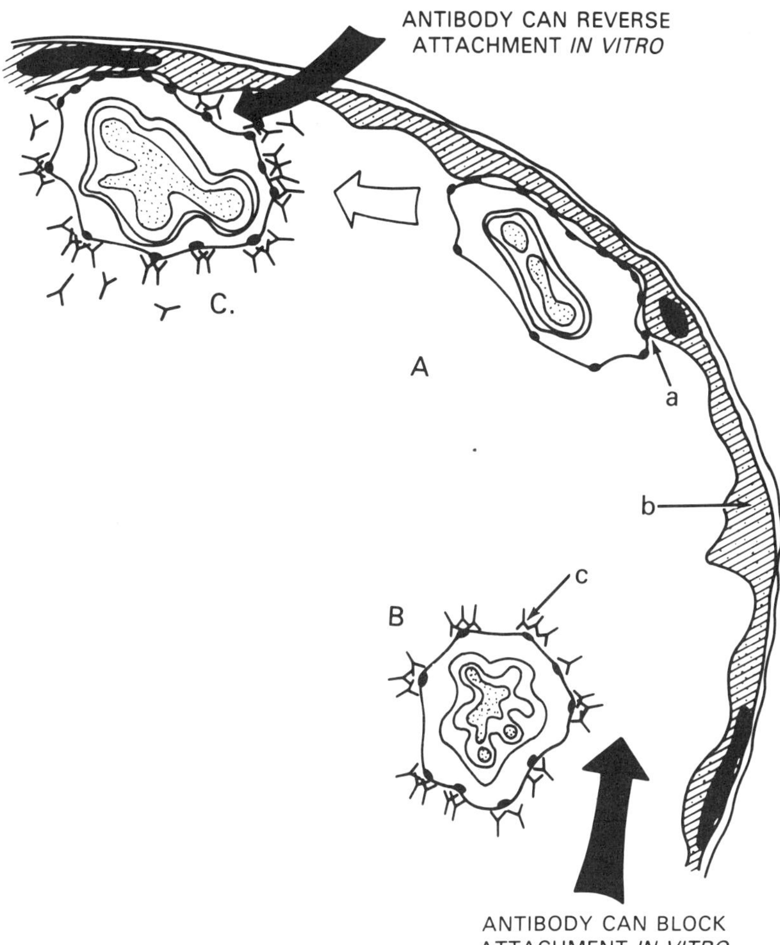

Figure 9. Antibody-dependent blockade of attachment of *P. falciparum*-infected erythrocytes to endothelium. (A) Trophozoite- and schizont-infected erythrocytes attach via knoblike protrusions on the surface membrane (a) to endothelial cells (b) *in vivo* (Luse and Miller, 1971) and *in vitro* (Udeinya *et al.*, 1981). (B) Preincubation of infected erythrocytes with appropriate anti-malarial antibody (c) specifically blocks attachment of infected erythrocytes to endothelial cells (Udeinya *et al.*, 1981; I. J. Udeinya and L. H. Miller, unpublished data, 1983) or to melanoma cells (I. J. Udeinya, J. H. Leech, and L. H. Miller, unpublished data, 1983) which also specificially bind *P. falciparum*-infected erythrocytes (Schmidt *et al.*, 1982). This schematic drawing shows antibody bound to the knobs, thereby preventing their interaction with the surface of endothelial cells. (C) If appropriate anti-malarial antibody is added *after* infected erythrocytes have attached to endothelial or melanoma cells (i.e., A), the infected cells are released (I. J. Udeinya, L. H. Miller, and I. A. McGregor, unpublished observations, 1983). This reversal of attachment by antibody suggests that the individual receptor–ligand interactions at the points of attachment are continuously broken and reformed.

anoma) and cause displacement of the parasitized cells. (Fig. 9C). The latter result suggests that the binding interaction between knob components on parasitized erythrocytes and the surface of the cells to which they attach is a relatively low-affinity interaction with continuous breakage and reforming of the cell–cell contact such that antibody of higher affinity added subsequently can bind to the surface of the parasitized erythrocyte at the knob and disrupt the interaction. These experiments suggest that the inhibition on reversal of binding by immune sera directly reflects the degree of *in vivo* protection acquired against the same *P. falciparum* strain. If antibody were to displace parasitized cells from sites of endothelial attachment *in vivo*, the released cells might be subject to a battery of specific and nonspecific immune defense mechanisms localized in the spleen that would result in parasite destruction. Since these parasitized cells would be coated in antibody, opsonization and phagocytosis by splenic macrophages might be a major parasiticidal mechanism.

The specificity of this blocking action of anti-malarial antibody (Fig. 9) with respect to components on the parasitized erythrocyte membrane is of great interest. Preabsorption of the immune *Aotus* sera with uninfected erythrocytes had no effect on the ability of these sera to reverse or inhibit binding (I. J. Udeinya and L. H. Miller, unpublished data, 1983). By contrast, absorption with infected erythrocytes caused a significant reduction in the ability of these sera to reverse or inhibit binding. Protein A-purified IgG reversed binding to the same degree as serum, while the portion of serum that did not bind to protein A, including IgM, had no effect (I. J. Udeinya and L. H. Miller, unpublished data, 1983). These results show that the IgG fraction of immune sera contains antibody to a parasite-dependent antigenic determinant on the surface of infected erythrocytes, which can block or reverse endothelial binding. Although it is possible that this antigenic determinant constitutes the binding site of the knob component responsible for endothelial attachment, anti-malarial antibody could also conceivably block attachment by reacting with either another region of the binding molecule or with another molecule unrelated functionally to binding that is physically close to the binding molecule.

The detailed mechanism of the antibody-mediated blockade of attachment is especially important for biochemical studies attempting to identify the binding component(s) of knobs on *P. falciparum*-infected erythrocytes. Antisera that block attachment may or may not do so through recognition of the binding molecule. However, this detail is less important when one considers the potential implications of antibody-mediated blockade of endothelial attachment *in vivo*. An antibody to any knob component that exhibits the effect of blocking or reversing endothelial attachment and of displacing mature parasitized erythrocytes into the peripheral circulation might result in similar consequences of parasite destruction. If passively transferred anti-malarial antibody could reverse endothelial attachment *in vivo* as it does *in vitro*, passive transfer might result in the appearance of these parasitized erythrocytes in the peripheral circulation. We would then expect such antibody-coated erythrocytes to be cleared by opsonization and phagocytosis, especially in the spleen.

E. Summary of Knob Function

Many aspects of this discussion on the role of knobs for parasite survival and of knob antigens as potential targets of *in vivo* parasiticidal immune responses are obviously speculative and based on extrapolation of *in vitro* phenomena to the *in vivo* situation. We conclude this section by recapitulating the major points of certainty and uncertainty. It is well established that many malaria parasites exhibit sequestration of erythrocytes infected with mature parasites. The specificity of this sequestration with respect to the tissues and organs of the host and nature of the particular species of malaria parasite is also demonstrated. In the case of the *falciparum* malarias, several independent lines of evidence indicate that knob components are involved in endothelial attachment—electron microscopy of parasitized cells attached *in vivo* and *in vitro* shows that knobs constitute the points of contact to endothelial cells; K^- parasites that fail to express knobs do not attach to endothelial cells; with K^+ parasites the immature parasitized erythrocytes do not express knobs and do not attach, whereas mature parasitized erythrocytes do express knobs and also attach to endothelial cells. Knobs of *P. falciparum*-infected erythrocytes have been shown to express new parasite-dependent antigens on the cell surface. Anti-malarial antibody can block attachment to endothelial cells or reverse attachment of parasitized cells. It remains to be seen what antigen(s) constitute the targets of this blocking antibody activity and whether such antigens include those responsible for endothelial attachment. The simplest working hypothesis to describe these results with the *falciparum* malarias is to postulate that a new membrane component at the membrane of knobs mediates attachment to endothelial cells. The biologic observations with *P. malariae* and *P. brasilianum*, which express knobs on erythrocytes infected with asexual parasites or gametocytes but do not sequester, do not contradict the hypothesis proposed for *P. falciparum*, a different species of malaria parasite. Instead, we conclude that the presence of the morphologic structure identified as a knob may or may not be associated with the capacity of the parasitized erythrocyte to attach to endothelial cells.

We have suggested that knobs may fulfill several functions and consist of multiple new membrane components, some of which might confer the knob structure, others of which will provide different functions. According to this hypothesis, different malaria species that have knobs might have qualitatively and/or quantitatively different molecular compositions at the knob that confer differences in biologic behavior *in vivo* (i.e., sequestration or not).

An extension of this rationale accounts for the individual tissue distributions of sequestration observed with other malaria parasites that do not possess knobs, such as *P. berghei* and *P. knowlesi.* In these cases, the components responsible for knob structure are reduced or absent, while other components that can mediate endothelial attachment are present. These speculations are amenable to verification as we continue to learn more of the biochemical and functional nature of new or changed membrane constituents on malaria-infected erythrocytes. For the *falciparum* malarias, we have suggested that one of the roles of knobs for this species is to provide endothelial attachment and thereby avoid passage of mature parasitized erythrocytes through the spleen. Specific and nonspecific immune responses against parasitized erythrocytes localized in spleen would therefore be evaded via sequestration. *In vitro* experiments demonstrating that antibody can block or reverse endothelial cell attachment suggest that the humoral arm of immunity might overcome the advantage gained by *falciparum* malarias by sequestration in an immunologically priviledged site (i.e., endothelium distant from the spleen). If antibody to knob components can reverse or block endothelial attachment *in vivo*, one would predict that the released antibody-coated parasitized erythrocytes should be destroyed. Such antibody might therefore constitute a part of host-protective immune responses.

VIII. CONCLUDING REMARKS

We have reviewed the evidence for diverse changes in the membrane morphology and antigenicity of the outer membrane of malaria-infected erythrocytes. Our own results with *P. knowlesi*-infected cells show that the intraerythrocytic malaria parasite has the capacity to inject products of its protein synthesis into this membrane. We expect that many, if not all, of the new antigenic properties of the surface membrane of parasitized erythrocytes will be found to be the result of insertion of malarial proteins in a similar manner. The most fundamental biologic question that we are attempting to answer regarding this membrane is the reason for the appearance of new antigens on infected cells. The immature ring-stage malaria parasite is presumably separated from immune recognition mechanisms by the physical presence of the host erythrocyte membrane, the erythrocyte cytoplasm, and the parasitophorous vacuole membrane. Why does the mature trophozoite express new antigens on the host cell membrane, which appear to increase in density as the schizont matures, such that the parasite advertises its presence to host-recognition mechanisms?

We have suggested that the intracellular parasite might have an obligatory requirement to alter the transport properties of the outer erythrocyte membrane such that it can obtain essential metabolites from plasma. New transport proteins might be inserted to fulfill this role and thereby create new membrane antigens.

Second, from the capacity of erythrocytes infected with *P. falciparum* trophozoites and schizonts to sequester via knob-mediated attachment to endothelial cells lining the blood vessels, we have suggested that new antigens in the knob structure might be required to bind to endothelial cells. This property enables these parasitized erythrocytes to avoid passage through the spleen and might therefore constitute a mechanism for evasion of localized splenic immune responses. The presence of knobs, and presumably new antigens, on the surface of erythrocytes infected with other species of malaria parasites that do not sequester leads us to suggest that knobs might possess diverse functions for parasite survival and that different malaria species exhibit these functions to different extents.

The capacity of cloned *P. knowlesi* parasites to express antigenically and structurally different variant (SICA) antigens on the surface of parasitized erythrocytes permits this parasite to evade variant-specific immune responses. This property of antigenic variation does not explain, however, why the variant antigen is expressed. New insight into the role of the SICA antigen has come from comparison of the *in vivo* properties of SICA[+] parasites that express variant antigen and SICA[-] parasites that fail to express variant antigen on the erythrocyte surface. SICA[-] parasites appear to be less virulent in nonsplenectomized monkeys than are SICA[+] parasites. Initial studies on challenging animals with SICA[+] parasites after drug-assisted cure of an initial SICA[-] infection indicate that potent parasiticidal responses against SICA[+] parasites had been evoked. In contrast, repeated infections with SICA[+] parasites appear to induce a much less effective immune response. We have concluded that variant-transcending parasiticidal immune responses are poorly elicited by a SICA[+] infection, while they appear to be better elicited by a SICA[-] infection. The only phenotypic difference between these parasites that we have identified to date is the presence or absence of the SICA antigen on the cell surface. If there are no other phenotypic differences, our results would imply that the SICA antigen itself exerts an immunosuppressive effect on the development of variant-transcending immunity to *P. knowlesi.* The molecular basis for the different *in vivo* properties of SICA[-] and SICA[+] parasites is obviously of great interest in resolving this question.

Our methodologic approach to answering the questions and speculations raised by the results reviewed may be summarized as follows:

1. Define the biologic properties of different isolates of malaria parasites both *in vivo* and *in vitro. In vitro* properties of interest include virulence, immunogenicity on infection or immunization and subsequent challenge, and degree and organ distribution of sequestration. *In vitro* properties of interest include agglutinability or surface-membrane fluorescence of infected erythrocytes with various antisera, assays for altered functional properties of the erythrocyte membrane such as membrane transport, the ability of infected cells to bind to endothelial cells or the amelanotic melanoma cell line (and the capacity of various sera to block or reverse this attachment), and the nature of morphologic alterations in the outer membrane of infected erythrocytes identified by electron microscopy.

2. Attempt wherever possible to obtain clones of parasites that exhibit defined and stable phenotypic properties of the erythrocyte membrane of infected cells as defined in the first step. It has been possible with *P. knowlesi* to obtain cloned parasites of different phenotypes derived one from the other by selection/modulation *in vivo*. These parasites are genotypically very similar, if not identical (e.g., cloned antigenic variants of the SICA antigen derived from an original clone after antigenic variation *in vivo*; cloned SICA[-] parasites derived from a cloned SICA[+] parasite by passage in splenectomized monkeys). With genotypically similar or identical parasites, it becomes feasible to ascribe differences in biologic phenotype to a relatively small number of differences in phenotype that can be defined by biochemical or immunochemical analysis.
3. Perform biochemical, immunochemical, and molecular biologic analyses of the molecular and genetic basis for the observed differences in biologic phenotype of the different isolates/clones. Molecular probes critical in this analysis will include antisera defined in biologic assays (i.e., agglutination, immunoelectron microscopy, blockade/reversal of endothelial attachment) and monoclonal antibodies prepared against the molecules suspected to account for the phenotype of interest.
4. Demonstrate directly that the parasite phenotype is determined by the molecule(s) identified in steps 1-3. For the SICA antigen, this entails the demonstration that the antigenically variant proteins are on the cell surface, as has been demonstrated, and the eventual production of monoclonal antibodies to several variant molecules to explore in detail their subcellular location and antigenic structure. Molecular biologic analysis of the genome of cloned parasites should confirm that *P. knowlesi* has the genetic capacity to express a repertoire of variant molecules. For the molecules that mediate attachment of infected erythrocytes to endothelial cells, we would hope either to block attachment in a competitive binding assay using purified membrane components from the erythrocyte surface or to produce a monoclonal antibody that will specifically block or reverse attachment.

This approach should allow us to explore the *in vivo* roles of new parasite components in the erythrocyte membrane for parasite survival and as potential targets of host-immune responses.

ACKNOWLEDGMENTS

We are grateful to several colleagues in this laboratory for helpful discussions, and especially to Dr. L. H. Miller and Dr. I. J. Udeinya for allowing us to include some of their unpublished results. We thank Mrs. Wilma Davis for typing and for editorial assistance.

IX. REFERENCES

Abele, D. C., Toloie, J. E., Hill, G. J., Contacos, P. G., and Evans, C. B., 1965, Alterations in serum proteins and IgG antibody production during the course of induced malarial infections in man, *Am. J. Trop. Med. Hyg.* **14**:191.

Aikawa, M., Rabbege, J. R., and Wellde, B. T., 1972, Junctional apparatus in erythrocytes infected with malarial parasites, *Z. Zellforsch. Mikrosk. Anat.* **124**:72.

Aikawa, M., Miller, L. H., and Rabbege, J., 1975, Caveola-vesicle complexes in the plasmalemma of erythrocytes infected by *Plasmodium vivax* and *P. cynomolgi, Am. J. Pathol.* **79**:285.

Aikawa, M., Hseieh, C-L., and Miller, L. H., 1977, Ultrastructural changes of the erythrocyte membrane in Ovale-type malarial parasites, *J. Parasitol.* **63**:152.

Alger, N. E., 1963, Distribution of schizonts of *Plasmodium berghei* in tissues of rats, mice, and hamsters, *J. Protozool.* **10**:6.

Allison, A. C., 1981, Cell-mediated immune responses in malaria, in: *Immunological Methods in Malariology* (Report and working papers of a meeting under the auspices of the Mérieux Foundation/WHO, Lyons, France, Sept. 9–12, pp. 1–14.

Ash, J. E., and Spitz, S., 1945, *Pathology of Tropical Diseases*, W. B. Saunders, Philadelphia and London.

Atkinson, J. P., Glew, R. H., Neva, F. A., and Frank, M. M., 1975, Serum complement and immunity in experimental simian malaria. II. Preferential activation of early components and failure of depletion of late components to inhibit protective immunity, *J. Infect. Dis.* **131**:26.

Barnwell, J. W., Howard, R. J., and Miller, L. H., 1982*a*, Altered expression of *Plasmodium knowlesi* variant antigen on the erythrocyte membrane in splenectomized rhesus monkeys, *J. Immunol.* **128**:224.

Barnwell, J. W., Howard, R. J., and Miller, L. H., 1982*b*, The influence of the spleen on the expression of surface antigens on parasitized erythrocytes, in: *Malaria and the Red Cell*, CIBA Foundation Symposium No. 94, pp. 117-136.

Barnwell, J. W., Howard, R. J., Coon, H. G., and Miller, L. H., 1983, Splenic requirement for antigenic variation and expression of the variant antigen on the erythrocyte membrane in cloned *Plasmodium knowlesi* malaria, *Infect. Immun.* **40**:985.

Bignami, A., and Bastianelli, G., 1890, *Riforma Medica* **6**:1334.

Biozzi, G., Halpern, B. N., Benacerraf, B., and Stiffel, C., 1975, in: *Physiopathology of the Reticuloendothelial System* (symposium), p. 204, Blackwell Scientific, Oxford.

Branton, M. B., 1978, Studies of clonal populations of NK65 *Plasmodium berghei*, Ph.D. thesis, University of Illinois at Urbana, Champaign, p. 167.

Bray, R. S., and Sinden, R. E., 1979, The sequestration of *Plasmodium falciparum* infected erythrocytes in the placenta, *Trans. R. Soc. Trop. Med. Hyg.* **73**:716.

Brookes, C. B., and Kreier, J. P., 1978, Role of surface coat on *in vitro* attachment and phagocytosis of *Plasmodium berghei* by peritoneal macrophages, *Infect. Immun.* **20**:927.

Brown, G. V., Anders, R. F., Mitchell, G. F., and Heywood, P. F., 1982, Target antigens of purified human immunoglobulins which inhibit growth of *Plasmodium falciparum in vitro, Nature (Lond.)* **297**:591.

Brown, I. N., 1969, Immunological aspects of malaria infection, *Adv. Immunol.* **11**:267.

Brown, I. N., Brown, K. N., and Hills, L. A., 1968, Immunity to malaria: The antibody response to antigenic variation by *Plasmodium knowlesi*, *Immunology* **14**:127.

Brown, J., and Smalley, M. E., 1981, Inhibition of the *in vitro* growth of *Plasmodium falciparum* by human polymorphonuclear neutrophils, *Clin. Exp. Immunol.* **46**:106.

Brown, K. N., 1971, Protective immunity to malaria provides a model for the survival of cells on immunologically hostile environment, *Nature (Lond.)* **230**:163.

Brown, K. N., 1973, Antibody induced variation in malaria parasites, *Nature (Lond.)* **242**:49.

Brown, K. N., 1976, Resistance to malaria, in: *Immunology of Parasitic Infections* (S. Cohen and E. Sadun, eds.) p. 268, Blackwell Scientific, Oxford.

Brown, K. N., 1977, Antigenic variation in malaria, in: *Immunity to Blood Parasites of Animals and Man* (L. H. Miller, J. A. Pino, and J. J. McKelvey Jr., eds.), pp. 5–25, Plenum Press, New York.

Brown, K. N., and Brown, I. N., 1965, Immunity to malaria: Antigenic variation in chronic infections of *Plasmodium knowlesi*, *Nature (Lond.)* **208**:1286.

Brown, K. N., and Hills, L. A., 1974, Antigenic variation and immunity to *Plasmodium knowlesi*: Antibodies which induce antigenic variation and antibodies which destroy parasites, *Trans. R. Soc. Trop. Med. Hyg.* **68**:139.

Brown, K. N., Brown, I. N., and Hills, L. A., 1970*a*, Immunity to malaria. I. Protection against *Plasmodium knowlesi* shown by monkeys sensitized with drug-suppressed infections or by dead parasites in Freund's Adjuvant, *Exp. Parasitol.* **28**:301.

Brown, K. N., Brown, I. N., Trigg, P. I., Phillips, R. S., and Hills, L. A., 1970*b*, Immunity to malaria. II. Serological response of monkeys sensitized by drug-suppressed infection or by dead parasitized cells in Freund's Complete Adjuvant, *Exp. Parasitol.* **28**:318.

Butcher, G. A., and Cohen, S., 1972, Antigenic variation and protective immunity in *Plasmodium knowlesi* malaria, *Immunology* **23**:503.

Cadigan, F. C., Jr., and Chaicumpa, V., 1969, *Plasmodium falciparum* in the white-handed gibbon: Protection afforded by previous infection with homologous and heterologous strains obtained in Thailand, *Mil. Med.* **134**:1135.

Callow, L. L., 1977, Vaccination against Bovine babesiosis, in: *Immunity to Blood Parasites of Animals and Man* (L. H. Miller, J. A. Pino, and J. J. McKelvey Jr., eds.), pp. 121–150, Plenum Press, New York.

Cannon, P. R., and W. H. Taliaferro, 1931, Acquired immunity in avian malaria. III. Cellular reactions in infection and superinfection, *J. Prevent. Med.* **5**:37.

Carter, R., 1970, Enzyme variation in *Plasmodium berghei*, *Trans. R. Soc. Trop. Med. Hyg.* **65**:586.

Carter, R., and Voller, A., 1975, The distribution of enzyme variation in populations of *Plasmodium falciparum* in Africa, *Trans. R. Soc. Trop. Med. Hyg.* **69**:371.

Celada, A., Cruchaud, A., and Perrin, L. H., 1982, Opsonic activity of human immune serum on *in vitro* phagocytosis of *Plasmodium falciparum* infected red blood cells by monocytes, *Clin. Exp. Immunol.* **47**:635.

Chow, J. S., and Kreier, J. P., 1972, *Plasmodium berghei*: Adherence and phagocytosis by rat macrophages *in vitro*, *Exp. Parasitol.* **31**:13.

Chulay, J. D., Haynes, J. D., and Diggs, C. L., 1981, Inhibition of *in vitro* growth of *Plasmodium falciparum* by immune monkey serum, *J. Infect. Dis.* **144**:270.

Clark, H. C., 1939, The first twelve months of infancy as a test for the community incidence of initial attacks of malaria, *South. Med. J.* **30**:348.

Clark, H. C., and Tomlinson, W. J., 1949, in: *Pathologic Anatomy of Malaria* (M. F. Boyd, ed.), 874 pp. W. B. Saunders, Philadelphia.

Clayton, C. E., Sacks, D. L., Ogilvie, B. M., and Askomas, A., 1979, Membrane fractions of trypanosomes mimic the immunosuppressive and mitogenic effects of living parasites on the host, *Parasite Immunol.* **1**:241.

Cohen, S., 1976, Survival of parasites in the immunized host, in: *Immunology of Parasitic Infections* (S. Cohen and E. H. Sadun, eds.), pp. 35–46, Blackwell Scientific, Oxford.

Cohen, S., 1979, Immunity to malaria, *Proc. R. Soc. Lond. B* **203**:323.

Cohen, S., and Butcher, G. A., 1970, Properties of protective malarial antibody, *Immunology* **19**:369.

Cohen, S., and Butcher, G. A., 1971, Serum antibody in acquired malarial immunity, *Trans. R. Soc. Trop. Med. Hyg.* **65**:125.

Cohen, S., and Mitchell, G. H., 1978, Prospects for immunization against malaria, *Curr. Top. Microbiol. Immunol.* **80**:97.

Cohen, S., McGregor, I. A., and Carrington, S. C., 1961, Gamma globulin and acquired immunity to human malaria, *Nature (Lond.)* **192**:733.

Cohen, S., Butcher, G. A., and Crandall, R. B., 1969, Action of malarial antibody *in vitro*, *Nature (Lond.)* **223**:368.

Cox, H. W., 1957, Observations on induced chronic *Plasmodium berghei* infections in white mice, *J. Immunol.* **79**:450.

Cox, H. W., 1959, A study of relapse *Plasmodium berghei* infections isolated from white mice, *J. Immunol.* **82**:209.

Cox, H. W., 1962, The behaviour of *Plasmodium berghei* strains isolated from relapsed infections of white mice, *J. Protozool.* **9**:114.

David, P.H., Hommel, M., and Oligino, L. D., 1982, *Plasmodium falciparum*: Interactions of infected erythrocytes with ligand-coated agarose beads, *Mol. Biochem. Parasitol.* **4**:195.

Deans, J. A., and Cohen, S., 1979, Localization and chemical characterization of *Plasmodium knowlesi* schizont antigens, *Bull WHO* **57**(Suppl. 1):93.

Deans, J. A., Dennis, E. D., and Cohen, S., 1978, Antigenic analysis of sequential erythrocytic stages of *Plasmodium knowlesi*, *Parasitology* **77**:333.

Desowitz, R. S., and Barnwell, J. W., 1976, *Plasmodium berghei*: Deep vascular sequestration of young forms in the heart and kidney of the white rat, *Ann. Trop. Med. Parasitol.* **70**:475.

Desowitz, R. S., and Miller, L. H., 1982, A perspective on malaria vaccines, *Bull. WHO* **58**:897.

Desowitz, R. S., Miller, L. H., Buchannan, R. D., and Permpanich, B., 1969, The sites of deep vascular schizogony in *Plasmodium coatneyi* malaria, *Trans. R. Soc. Trop. Med. Hyg.* **63**:198.

Diggs, C. L. and Osler, A. G., 1975, Humoral immunity in rodent malaria, III: Studies on the site of antibody action, *J. Immunol.* **114**:1243.

Diggs, C. L., Wellde, B. T., Andersen, J. S., Weher, R. M., and Rodriquez, E. Jr., 1972, The protective effect of African human immunoglobulin G in *Aotus trivirgatus* infected with Asian *Plasmodium falciparum*, *Proc. Helminthol. Soc. Wash.* **39**:449.

Diggs, C. L., and Osler, A. G., 1975, Humoral immunity in rodent malaria, III: Studies on the site of antibody action, *J. Immun.* **114**:1243.

Dvorak, J. A., Miller, L. H., Shiroishi, T., and Whitehouse, W. C., 1975, Invasion of erythrocytes by malaria merozoites, *Science* **187**:748.

Eaton, M. D., 1938, The agglutination of *Plasmodium knowlesi* by immune serum, *J. Exp. Med.* **67**:857.

Epstein, N., Miller, L. H., Kaushal, D. C., Udeinya, I. J., Rener, J., Howard, R. J., Asofsky, R., Aikawa, M., and Hess, R. L., 1981, Monoclonal antibodies against a specific surface determinant on malarial (*Plasmodium knowlesi*) merozoites block erythrocyte invasion, *J. Immunol.* **127**:212.

Facer, C. A., 1980*a*, Direct Coombs antiglobulin reactions in Gambian children with *Plasmodium falciparum* malaria. II. Specificity of erythrocyte bound IgG, *Clin. Exp. Immunol.* **39**:279.

Facer, C. A., 1980*b*, Direct antiglobulin reactions in Gambian children with *P. falciparum* malaria. III. Expression of IgG subclass determinants and genetic markers and association with anaemia, *Clin. Exp. Immunol.* **41**:81.

Facer, C., Bray, R. S., and Brown, J., 1979, Direct Coombs antiglobulin reactions in Gambian children with *Plasmodium falciparum* malaria I. Incidence and class specificity, *Clin. Exp. Immunol.* **35**:119.

Freeman, R. R., and Parish, C. R., 1981, Polyclonal B-cell activation during rodent malarial infections, *Clin. Exp. Immunol.* **32**:41.

Freeman, R. R., Trejdosiewicz, A. J., and Cross, G. A. M., 1980, Protective monoclonal antibodies recognizing stage-specific merozoite antigens of a rodent malaria parasite, *Nature (Lond.)* **284**:366.

Fremount, H. N., and Miller, L. H., 1975, Deep vascular schizogony in *Plasmodium fragile*: Organ distribution and ultrastructure of erythrocytes adherent to vascular endothelium, *Am. J. Trop. Med. Hyg.* **24**:1.

Garnham, P. C. C., 1966, *Malaria Parasites and Other Hemosporidia*, pp. 287–383, Blackwell Scientific, Oxford.

Glew, R. H., Atkinson, J. P., Frank, M. M., Collins, W. E., and Neva, F. A., 1975, Serum complement and immunity in experimental simian malaria I. Cyclical alterations in C4 related to schizont rupture, *J. Infect. Dis.* **131**:17.

Green, T. J., and Kreier, J. P., 1978, Demonstration of the role of cytophilic antibody in resistance to malaria parasites (*Plasmodium berghei*) in rats, *Infect. Immun.* **19**:138.

Green, T. J., Morhardt, M., Brackett, R. G., and Jacobs, R. L., 1981, Serum inhibition of merozoite dispersal from *Plasmodium falciparum* schizonts: Indicator of immune status, *Infect. Immun.* **31**:1203.

Greenwood, B. M., 1974, Possible role of a B-cell mitogen in hypergammaglobulinaemia in malaria and trypanosomiasis, *Lancet* **1**:435.

Greenwood, B. M., Muller, A. S., and Valkenburg, H. A., 1971*a*, Rheumatoid factor in Nigerian sera, *Clin. Exp. Immunol.* **9**:161.

Greenwood, B. M., Playfair, J. H. L., and Torrigiani, G., 1971*b*, Immunosuppression in murine malaria. I. General characteristics, *Clin. Exp. Immunol.* **8**:467.

Grun, J. L., and Weidanz, W. P., 1981, Immunity to *Plasmodium chabaudi adami* in the B-cell deficient mouse, *Nature (Lond.)* **290**:143.

Gutierrez, Y., Aikawa, M., Freemount, H. N., and Sterling, C. R., 1976, Experimental infection of *Aotus* monkeys with *Plasmodium falciparum*. Light and electron microscopic changes, *Ann. Trop. Med. Parasitol.* **70**:26.

Hamburger, J., and Kreier, J. P., 1976, *Plasmodium berghei*. Use of free blood stage parasites to demonstrate protective humoral activity in the serum of recovered rats, *Exp. Parasitol.* **40**:158.

Holder, A. A., and Freeman, R. R., 1981, Immunization against blood-stage rodent malaria using purified parasite antigens, *Nature (Lond.)* **294**:361.

Hommel, M., and David. P. H., 1981, *Plasmodium knowlesi* variant antigens are found on schizont-infected erythrocytes but not on merozoites, *Infect. Immun.* **33**:275.

Hommel, M., David, P. H., and Oligino, L. D., 1983, Surface alterations of erythrocytes in *Plasmodium falciparum* malaria, *J. Exp. Med.* **157**:1137.

Houba, V., Faulk, W. P., and Malola, Y.G., 1974, Heterophilic antibodies in relation to malarial infection: Population and experimental studies, *Clin. Exp. Immunol.* **18**:89.

Howard, R. J., 1982, Alterations in the surface membrane of red blood cells during malaria, *Immunol. Rev.* **61**:67.

Howard, R. J., Barnwell, J. W., Kao, V., Daniel, W. A., and Aley, S. B., 1982*a*, Radioiodination of new protein antigens on the surface of *Plasmodium knowlesi* schizont-infected erythrocytes, *Mol. Biochem. Parasitol.* **6**:343.

Howard, R. J., Barnwell, J. W., and Kao, V., 1982*b*, Tritiation of protein antigens of *Plasmodium knowlesi* schizont-infected erythrocytes using pyridoxal phosphate-[^{3}H]-sodium borohydride, *Mol. Biochem. Parasitol.* **6**:369.

Howard, R. J., Barnwell, J. W., and Kao, V., 1983, Antigenic variation in *Plasmodium knowlesi* malaria: Identification of the variant antigen on infected erythrocytes, *Proc. Natl. Acad. Sci. USA* **80**:4129.

Hunter, K. W., Winkelstein, J. A., and Simpson, T. W., 1979, Serum opsonic activity in rodent malaria: Functional and immunochemical characteristics *in vitro*, *J. Immunol.* **123**:2582.

Hunter, K. W., Finkelman, F. D., Strickland, G. T., Sayles, P. C., and Scher, I., 1980, Murine malaria: Analysis of erythrocyte surface-bound immunoglobulin by flow microfluorimetry, *J. Immunol.* **125**:169.

Jacobs, R. L., and Warren, M., 1967, Sequestration of schizonts in the deep tissues of mice infected with chloroquine-resistant *Plasmodium berghei*, *Trans. R. Soc. Trop. Med. Hyg.* **61**:273.

Jandl, J. H., and Kaplan, M. E., 1960, The destruction of red cells by antibodies in man. III. Quantitative factors influencing the patterns of hemolysis *in vivo*, *J. Clin. Invest.* **39**:1145.

Jandl, J. H., Jones, A. R., and Castle, W. B., 1957, The destruction of red cells by antibodies in man. I. Observations on the sequestration and lysis of red cells altered by immune mechanisms, *J. Clin. Invest.* **36**:1428.

Jayawardena, A. N., 1981, Immune responses in malaria, in: *Parasitic Diseases*, Vol. 1: *The Immunology*, p. 85, Marcel Dekker, New York.

Jayawardena, A. N., Targett, G. A. T., Leuchans, E., and Davies, A. J. S., 1978, The immunological response of CBA mice to *P. yoelii*. II. The passive transfer to immunity with serum and cells, *Immunology* **33**:157.

Jeffery, G. M., 1966, Epidemiological significance of repeated infections with homologous and heterologous strains and species of *Plasmodium*, *Bull WHO* **35**:873.

Jervis, H. R., Sprinz, H., Johnson, A. J., and Wellde, B. T., 1972, Experimental infection with *Plasmodium falciparum* in *Aotus* monkeys. II: Observations on host pathology, *Am. J. Trop. Med. Hyg.* **21**:272.

Jilly, P., 1969, Anaemia in parturient women, with special reference to malaria infection of the placenta, *Ann. Trop. Med. Parasitol.* **63**:109.

Johnson, J. G., Epstein, N., Shiroishi, T., and Miller, L. H., 1980, Factors affecting the ability of isolated *Plasmodium knowlesi* merozoites to attach to and invade erythrocytes, *Parasitology* **80**:539.

Kilejian, A., 1979, Characterization of a protein correlated with the production of knob-like protrusions on membranes of erythrocytes infected with *Plasmodium falciparum*, *Proc. Natl. Acad. Sci. USA* **76**:4650.

Kilejian, A., 1980, Stage-specific proteins and glycoproteins of *Plasmodium falciparum*: Identification of antigens unique to schizonts and merozoites, *Proc. Natl. Acad. Sci. USA* **77**:3695.

Kilejian, A., 1981, Alterations of human erythrocyte membranes due to infection with *Plasmodium falciparum*, in: *The Biochemistry of Parasites* (G. M. Slutzky, ed.), pp. 67–73, Pergamon Press, Oxford.

Kilejian, A., Abati, A., and Trager, W., 1977, *Plasmodium falciparum* and *Plasmodium coatneyi*: Immunogenicity of "knob-like protrusions" on infected erythrocyte membranes, *Exp. Parasitol.* **42**:157.

Kim, K. J., Taylor, K. W., Evans, C. B., and Asofsky, R., 1980, Radioimmunoassay for detecting antibodies against murine malarial parasite antigens: Monoclonal antibodies recognizing *Plasmodium yoelii* antigens, *J. Immunol.* **125**:2526.

Knowles, G., and Walliker, D., 1980, Variable expression of virulence in the rodent malaria *Plasmodium yoelii yoelii*, *Parasitology* **81**:211.

Kreier, J. P., and Green, T. J., 1980, The vertebrate host's immune response to Plasmodia, in: *Malaria,* Vol. 3: *Immunology and Immunization* (J. P. Kreier, ed.), pp. 111-162, Academic Press, New York.

Langreth, S. G., and Reese, R. T., 1978, Antigenicity of the infected erythrocyte surface in falciparum malaria, *J. Cell Biol.* **79**:89*a*.

Langreth, S. G., and Reese, R. T., 1979, Antigenicity of the infected erythrocyte and merozoite surfaces in falciparum malaria, *J. Exp. Med.* **150**:1241.

Langreth, S. G., Jensen, J. B., Reese, R. T., and Trager, W., 1978, Fine structure of human malaria *in vitro*, *J. Protozool.* **25**:443.

Langreth, S. G., Reese, R. T., Motyl, M. R., and Trager, W., 1979, *Plasmodium falciparum*: Loss of knobs on the infected erythrocyte surface after long term cultivation, *Exp. Parasitol.* **48**:213.

Luse, S.A., and Miller, L. H., 1971, *Plasmodium falciparum* malaria: Ultrastructure of parasitized erythrocytes in cardiac vessels, *Am. J. Trop. Med. Hyg.* **20**:655.

Lustig, H. J., Nussenzweig, V., and Nussenzweig, R. S., 1977, Erythrocyte membrane-associated immunoglobulins during malaria infection of mice, *J. Immunol.* **119**:210.

Lustig-Shear, H., Nussenzweig, R. S., and Bianco, C., 1979, Immune phagocytosis in murine malaria, *J. Exp. Med.* **149**:1288.

Maegraith, B., 1969, Complications of falciparum malaria, *Bull. N.Y. Acad. Med.* **45**:1061.

McBride, J. S., Walliker, D., and Morgan, G., 1982, Antigenic diversity in the human malaria parasite *Plasmodium falciparum, Science* **217**:254.

McGregor, I. A., 1960, Demographic effects of malaria with special reference to the stable malaria of Africa, *W. Afr. Med. J. J. Nigerian Pract.* **9**:260.

McGregor, I. A., 1972, Immunology of malarial infection and its possible consequences, *Br. Med. Bull.* **28**:22.

McGregor, I. A., 1978, Topical aspects of the epidemiology of malaria, *Isr. J. Med. Sci.* **14**:523.

McGregor, I. A., 1981, Immune phenomenon in naturally acquired human malaria, in: *Immunological Methods in Malariology* (report and working papers of a meeting under the auspices of the Mérieux Foundation/WHO Lyons, France, Sept. 9–12), pp. 1–12.

McGregor, I. A., and Williams, K., 1978, Value of the gel precipitation test in monitoring the endemicity of malaria in a rural African village, *Isr. J. Med. Sci.* **14**:697.

McGregor, I. A., Carrington, S. P., and Cohen, S., 1963, Treatment of East African *P. falciparum* malaria with West-African human γ-globulin, *Trans. R. Soc. Trop. Med. Hyg.* **57**:170.

Miller, L. H., 1969, Distribution of mature trophozoites and schizonts of *Plasmodium falciparum* in the organs of *Aotus trivirgatus*, the night monkey, *Am. J. Trop. Med. Hyg.* **18**:860.

Miller, L. H., 1972, The ultrastructure of red cells infected by *Plasmodium falciparum* in man, *Trans. R. Soc. Trop. Med. Hyg.* **66**:459.

Miller, L. H., and Fremount, H. N., 1969, The sites of deep vascular schizogony in chloroquine-resistant *Plasmodium berghei* in mice, *Trans. R. Soc. Trop. Med. Hyg.* **63**:195.

Miller, L. H., Fremount, H. N., and Luse, S. A., 1971*a*, Deep vascular schizogony *of Plasmodium knowlesi* in *Macaca mulatta*, *Am. J. Trop. Med. Hyg.* **20**:816.

Miller, L. H., Usami, S., and Chien, S., 1971*b*, Alteration in the rheologic properties of *Plasmodium knowlesi*-infected red cells. A possible mechanism for capillary obstruction, *J. Clin. Invest.* **50**:1451.

Miller, L. H., Aikawa, M., and Dvorak, J., 1975, Malaria (*Plasmodium knowlesi*) merozoites: Immunity and the surface coat, *J. Immunol.* **114**:1237.

Miller, L. H., Powers, K. G., and Shiroishi, T., 1977, *Plasmodium knowlesi*: Functional immunity and antimerozoite antibodies in rhesus monkeys after repeated infection, *Exp. Parasitol.* **41**:105.

Miller, L. H., McAuliffe, F. M., and Johnson, J. G., 1979, Invasion of erythrocytes by malaria merozoites, *Prog. Clin. Biol. Res.* **30**:497.

Mims, C. A., 1964, Aspects of the pathogenesis of virus diseases, *Bacteriol. Rev.* **28**:30.

Mitchell, G. H., Butcher, G. A., Voller, A., and Cohen, S., 1976, The effect of human immune IgG on the *in vitro* development of *Plasmodium falciparum*, *Parasitology* **72**:149.

Newbold, C. I., Boyle, D. B., Smith, C. C., and Brown, K. N., 1982, Identification of a schizont- and species-specific surface glycoprotein on erythrocytes infected with rodent malarias, *Mol. Biochem. Parasitol.* **5**:45.

Nussenzweig, R. S., Cochran, A. H., and Lustig, H. J., 1978, Immunological responses, in: *Rodent Malaria* (R. Killick-Kendrick and W. Peters, eds.), pp. 248–307, Academic Press, New York.

Perkins, M., 1982, Surface proteins of schizont-infected erythrocytes and merozoites of *Plasmodium falciparum, Mol. Biochem. Parasitol.* **5**:55.

Perrin, L. H., Ramirez, E., Hsiang, E., and Lambert, P. H., 1980, *Plasmodium falciparum*: Characterization of defined antigens by monoclonal antibodies, *Clin. Exp. Immunol.* **41**:91.

Perrin, L. H., Ramirez, E., Lambert, P. H., and Miescher, P. A., 1981, Inhibition of *Plasmodium falciparum* growth in human erythrocytes by monoclonal antibodies, *Nature (Lond.)* **289**:301.

Phillips, R. S., 1970, *Plasmodium berghei*: Passive transfer of immunity by antisera and cells, *Exp. Parasitol.* **27**:479.

Phillips, R. S., and Jones, V. E., 1972, Immunity to *Plasmodium berghei* in rats: Maximum levels of protective antibody activity are associated with eradication of the infection, *Parasitology* **64**:117.

Phillips, R. S., Wolstencroft, R. A., Brown, I. N., Brown, K. N., and Dumonde, D. C., 1970, Immunity to malaria. III. Possible occurrence of a cell-mediated immunity to *Plasmodium knowlesi* in chronically infected and Freund's Complete Adjuvant-sensitized monkeys, *Exp. Parasitol.* **28**:339.

Playfair, J. H. L., 1978, Effective and ineffective immune responses to parasites: Evidence from experimental models, *Curr. Top. Microbiol. Immunol.* **80**:37.

Poels, L. G., van Niekerk, C. C., Franken, M. A. M., and E. H. van Elven, 1977, *Plasmodium berghei*: Selective release of "protective" antigens, *Exp. Parasitol.* **42**:182.

Powell, R. D., McNamara, J. V., and Rieckmann, K. H., 1972, Clinical aspects of acquisition of immunity to Falciparum malaria, *Proc. Helminth. Soc. Wash.* **39**(special issue):51.

Quinn, T. C., and Wyler, D. J., 1979, Intravascular clearance of parasitized erythrocytes in rodent malaria, *J. Clin. Invest.* **63**:1187.

Rank, R. G., and Weidanz, W. P., 1976, Nonsterilizing immunity in avian malaria: An antibody-independent phenomenon, *Proc. Soc. Exp. Biol. Med.* **151**:257.

Reese, R. T., and Motyl, M. R., 1979, The effects of immune serum and purified immunoglobulin from owl monkeys, *J. Immunol.* **123**:1894.

Redmond, W. B., 1941, Immunity to human malaria: Characteristics of immunity, in: *Human Malaria* (F. R. Moulton, ed.), pp. 231–238, American Association for the Advancement of Science, Washington, D.C.

Rigdon, R. H., 1942, A consideration of the mechanism of death in acute *Plasmodium falciparum* infection; report of a case, *Am. J. Hyg.* **36**:269.

Roth, R. L., and Herman, R., 1979, *Plasmodium berghei*: Correlation of *in vitro* erythrophagocytosis with the dynamics of early-onset anemia and reticulocytosis in mice, *Exp. Parasitol.* **47**:169.

Rudzinska, M. A., and Trager, W., 1968, The fine structure of trophozoites and gametocytes in *Plasmodium coatneyi*, *J. Protozool.* **15**:73.

Sacks, D. L., Selkirk, M. E., Ogilvie, B. M., and Askonas, B. A., 1980, Intrinsic immunosuppressive activity of different trypanosome strains varies with parasite virulence, *Nature (Lond.)* **283**:476.

Sacks, D. L., Bancroft, A., Howard Evans, W., and Askonas, B. A., 1982, Incubation of trypanosome-derived mitogenic and immunosuppressive products with peritoneal macrophages allows recovery of biological activities from soluble parasite fractions, *Infect. Immun.* **36**:160.

Sadun, E. H., 1966, Research in malaria, *Mil. Med.* **131**(Suppl.):847.

Sadun, E. H., Hickman, R. L., Wellde, B. T., Moon, A. P., and Udeozo, I. O. K., 1966, Active and passive immunization of chimpanzees infected with West African and Southeast Asian strains of *Plasmodium falciparum*, *Mil. Med.* **131**(Suppl.):1250.

Scheibel, L. W., Ashton, S. H., and Trager, W., 1979, *Plasmodium falciparum*: Microaerophilic requirements in human red blood cells, *Exp. Parasitol.* **47**:410.

Schmidt, J. A., Udeinya, I. J., Leech, J. H., Hay, R. J., Aikawa, M., Barnwell, J., Green, I., and Miller, L. H., 1982, *Plasmodium falciparum* malaria. An amelanotic melanoma cell line bears receptors for the knob ligand on infected erythrocytes, *J. Clin. Invest.* **70**:379.

Schmidt-Ullrich, R., and Wallach, D. F. H., 1978, *Plasmodium knowlesi*-induced antigens in membranes of parasitized rhesus monkey erythrocytes, *Proc. Natl. Acad. Sci. USA* **75**:4949.

Schmidt-Ullrich, R., Wallach, D. F. H., and Lightholder, J., 1979*a*, Two *Plasmodium knowlesi*-specific antigens on the surface of schizont-infected rhesus monkey erythrocytes induce antibody production in immune hosts, *J. Exp. Med.* **150**:86.

Schmidt-Ullrich, R., Wallach, D. F. H., and Lightholder, J., 1979*b*, Fractionation of *Plasmodium knowlesi*-induced antigens of rhesus monkey erythrocyte membranes, *Bull. WHO* **57**(Suppl. 1):115.

Schmidt-Ullrich, R., Miller, L. H., Wallach, D. F. H., Lightholder, J., Powers, K. G., and Gwadz, R. W., 1981, Rhesus monkeys protected against *Plasmodium knowlesi* malaria produce antibodies against a 65,000-M_r *P. knowlesi* glycoprotein at the surface of infected erythrocytes, *Infect. Immun.* **34**:519.

Schnitzer, B., Sodeman, T., Mead, M., and Contacos, P. G., 1972, Pitting function of the spleen in malaria: Ultrastructural observations, *Science* **177**:175.

Shakespeare, P. G., Trigg, P. I., and Tappenden, L., 1979, Some properties of *Plasmodium knowlesi* membranes in the simian malaria parasite, *Ann. Trop. Med. Parasitol.* **73**:333.

Sherman, I. W., 1979, Biochemistry of *Plasmodium* (Malarial parasites), *Microbiol. Rev.* **43**:453.

Shortt, H. E., Pardit, S. R., Menon, K. P., and Swaminath, C., 1938, Absence of effective immunity after cure of protozoal infections, *Indian J. Med. Res.* **25**:763.

Sinton, J. A., 1939, A summary of our present knowledge on the mechanism of immunity in malaria, *J. Malaria Inst. India* **2**:71.

Smith, D. H., and Theakston, R. D. G., 1970, Comments on the ultrastructure of human erythrocytes infected with *Plasmodium malariae*, *Ann. Trop. Med. Parasitol.* **64**:329.

Spitz, S., 1946, The pathology of acute falciparum malaria, *Mil. Surg.* **99**:555.

Sterling, C. R., Aikawa, M., and Nussenzweig, R. S., 1972, Morphological divergence in a mammalian malarial parasite: The fine structure of *Plasmodium brasilianum*, *Proc. Helminth. Soc. Wash.* **39**(special issue):109.

Sterling, C. R., Seed, T., Aikawa, M., and Rabbege, J., 1975, Erythrocyte membrane alteration induced by *Plasmodium simium* in *Siamiri sciureus*: A relation to Schüffners dots, *J. Parasitol.* **61**:177.

Tait, A., 1981, Analysis of protein variation in *Plasmodium falciparum* by two-dimensional gel electrophoresis, *Mol. Biochem. Parasitol.* **2**:205.

Taliaferro, W. H., 1949, Immunity to the malaria infections, in: *Malariology*, Vol. II (M. F. Boyd, ed.), pp. 935–965, W. B. Saunders, Philadelphia.

Taliaferro, W. H., and Cannon, P. R., 1936, The cellular reactions during primary infections and superinfections of *Plasmodium brasilianum* in Panamanian monkeys, *J. Infect. Dis.* **59**:72.

Taliaferro, W. H., and Mulligan, H. W., 1937, The histopathology of malaria with special reference to the function and origin of the macrophages in defence, *Indian Med. Res. Mem.* **29**:1.

Targett, G. A. T., and A. Voller, 1965, Studies on antibody levels during vaccination of rhesus monkeys against *Plasmodium knowlesi*, *Br. Med. J.* **2**:1104.

Tella, A., and Maegraith, B. C., 1965, Physiopathological changes in primary acute blood-transmitted malaria and *Babesia* infections. II. A comparative study of serum protein levels in infected rhesus monkeys, mice and puppies, *Ann. Trop. Med. Parasitol.* **59**:153.

Terry, R. J., 1978, Immunodepression in parasitic infections, in: *Immunity in Parasitic Disease,* Vol. 72, p. 161, *INSERM* Paris.

Tosta, C. E., and Wedderburn, N., 1980, Immune phagocytosis of *Plasmodium yoelii*-infected erythrocytes by macrophages and eosinophils, *Clin. Exp. Immunol.* **42**:114.

Trager, W., Rudzinska, M. A., and Bradbury, P. C., 1966, The fine structure of *Plasmodium falciparum* and its host erythrocytes in natural malaria infections in man, *Bull. WHO* **35**:883.

Trager, W., Stanley, H. S., Allen, R. D., and Allen, N. S., 1982, Knobs on the surface of erythrocytes infected with *Plasmodium falciparum*: Visualization by video-enhanced, differential interference contrast microscopy, *J. Parasitol.* **68**:332.

Udeinya, I. J., Schmidt, J. A., Aikawa, M., Miller, L. H., and Green, I., 1981, Falciparum malaria-infected erythrocytes specificially bind to cultured human endothelial cells, *Science* **213**:555.

Vincent, H. M., and Wilson, R. J. M., 1980, Malarial antigens on infected erythrocytes, *Trans. R. Soc. Trop. Med. Hyg.* **74**:452.

Voller, A., and Richards, W. H. G., 1970, Immunity to *Plasmodium falciparum* in owl monkeys (*Aotus trivirgatus*), *Z. Tropenmed. Parasitol.* **21**:159.

Voller, A., and Rossan, R. N., 1969*a*, Immunological studies with simian malarias I. Antigenic variants of *Plasmodium cynomolgi bastianellii*, *Trans. R. Soc. Trop. Med. Hyg.* **63**:46.

Voller, A., and Rossan, R. N., 1969*b*, Immunological studies on simmian malaria parasites. IV. Heterologous superinfection of monkeys with chronic *Plasmodium knowlesi* infections, *Trans. R. Soc. Trop. Med. Hyg.* **63**:837.

Voller, A., Richards, W. H. G., Hawkey, C. M., and Ridley, D. S., 1969, Human malaria (*Plasmodium falciparum*) in owl monkeys (*Aotus trivirgatus*), *J. Trop. Med. Hyg.* **72**:153.

Wallach, D. F. H., and Conley, M., 1977, Altered membrane proteins of monkey erythrocytes infected with simian malaria, *J. Mol. Med.* **2**:119.

Weinbaum, F. I., Evans, C. B., and Tigelaar, R. E., 1976, Immunity to *Plasmodium berghei yoelii* in mice I. The course of infection in T cell and B cell deficient mice, *J. Immunol.* **117**:1999.

Wellde, B. T., and Sadun, E. H., 1967, Resistance produced in rats and mice by exposure to irradiated *P. berghei*, *Exp. Parasitol.* **21**:310.

Wery, M., Weyn, J., Timperman, A., and Hendrix, L., 1979, Observations on the virulence and the antigenic characters of cloned and uncloned lines of the Anka isolate of *Plasmodium berghei.* 1. Production of recrudescent parasitemias in immunized mice, *Ann. Soc. Belge. Med. Trop.* **59**:347.

Wilson, R. J. M., 1981, Methods for the biological characterization of malaria parasites:

Characterization of *Plasmodium falciparum* with L, R and S-antigens, *WHO/MAL/* 81.937:1–14.

Wilson, R. J. M., and Phillips, R. S., 1976, Method to test inhibitory antibodies in human sera to wild populations of *Plasmodium falciparum*, *Nature (Lond.)* **263**:132.

Woodruff, A. W., Ziegler, J. L., Hathaway, A., and Gwata, T., 1973, Anemia in African trypanosomiasis and "big spleen disease" in Uganda, *Trans. R. Soc. Trop. Med. Hyg.* **67**:329.

World Health Organization Technical Report Series, 1975, *Developments in Malaria Immunology*, No. 579, p. 16.

Wunderlich, F., Stübig, H., and Könígk, E., 1982, Intraerythrocytic development of *Plasmodium knowlesi*: Structure, Temperature-and Ca^{2+}-response of the host and parasite membranes, *J. Protozool.* **29**:49.

Zuckerman, A., 1945, *In vitro* opsonic tests with *Plasmodium gallinaceum* and *Plasmodium lophurae*, *J. Infect. Dis.* **77**:28.

Zuckerman, A., 1977, Current status of the immunology of blood and tissue protozoa. II, *Plasmodium*, *Exp. Parasitol.* **42**:374.

Chapter 6

Murine T-Cell Responses to Protozoan and Metazoan Parasites: Functional Analysis of T-Cell Lines and Clones Specific for Leishmania tropica and Schistosoma mansoni

J. A. Louis, G. Lima, J. Pestel, and R. Titus

WHO Immunology Research and Training Centre
Institute of Biochemistry
University of Lausanne
CH-1066 Epalinges, Switzerland

and

H. D. Engers

Department of Immunology
Swiss Institute for Experimental Cancer Research
CH-1066 Epalinges, Switzerland

I. *LEISHMANIA TROPICA* MAJOR

Leishmanias are protozoan parasites transmitted to mammalian hosts by the bite of sandflies. After injection into the skin, the flagellate forms (promastigotes) are taken up by macrophages, where they transform into the aflagellate form (amastigotes). Cutaneous leishmaniasis is caused by parasites belonging to the *L. tropica* (Old World), *L. mexicana* and *L. brasiliensis* complex (New World) Mucocutaneous forms of the disease usually result from infection with *L. brasiliensis* and visceral leishmaniasis, commonly named kala-azar, is caused by *L. donovani*. The clinical manifestations observed in patients infected with these parasites form a spectrum of diseases similar to that described in leprosy. The types of pathologic changes observed in infected human subjects appear to be dependent on factors related to both the parasite and the host response. Furthermore, observations that the development of parasite-specific delayed-type hypersensitivity (DTH) responses are often concommitant with the resolution of cutaneous human lesions point to a major role played by cell-mediated immunity in the healing of cutaneous lesions (Alder, 1964). The absence of specific DTH reactions during the active phase of visceral leishmaniasis support this hypothesis.

Experimental infection of inbred mice with *L. tropica* results in the development of lesions very similar to those found in human leishmaniasis and depending on their genetic background, the entire spectrum of clinical manifestations is observed (Handman *et al.*, 1979; Behin *et al.*, 1979; Nasseri and Modabber, 1979). Experiments performed in the murine model of leishmaniasis (*L. tropica*) also indicate that cell-mediated immunity is essential for the healing of cutaneous lesions and for the development of protective immunity to reinfection. An elegant series of experiments performed in Mitchell's laboratory in Australia have clearly established that T cells are essential in the development of immunity to *L. tropica* in mice. In brief, Mitchell and co-workers have demonstrated that athymic nu/nu mice exhibit a severe generalized disease after infection with *L. tropica* and that transfer of syngeneic T lymphocytes confers resistance to infection in these highly susceptible mice (Handman *et al.*, 1979; Mitchell *et al.*, 1980).

Balb/c mice have been found by many investigators to be extremely susceptible to *L. tropica* infection, i.e., these mice usually fail to heal cutaneous lesions that progress inexorably. This experimental model of infection is now being extensively studied in an effort to identify possible defect(s) responsible for the exquisite susceptibility to *L. tropica* exhibited by mice of this strain. It has been suggested that the establishment of *L. tropica* in macrophages from Balb/c mice is better than in macrophages from resistant mice, such as NZB (Handman *et al.*, 1979). This observation relates the susceptibility to *L. tropica* infection to the permissiveness of macrophages in supporting multiplication of *L. tropica* organisms *in vitro*. Recent results also indicate that the susceptibility exhibited by Balb/c mice is correlated with specific suppression of DTH reactivity to parasite antigens during the course of experimental infection (Howard *et al.*, 1980). This impairment of parasite-specific DTH responses in Balb/c mice was shown to result from the activity of antigen-specific suppressor T cells . Interestingly, the generation of parasite-specific suppressor T cells was abrogated in Balb/c mice sublethally irradiated with 550 rad before infection (Howard *et al.*, 1981). The suppressive activity of spleen cells of *L. tropica*-infected Balb/c mice was limited to the induction phases of specific DTH responses, since no suppression of the effector arm of the response was observed. Furthermore, cells mediating suppression were characterized as being T lymphocytes expressing the Lyt $1^{+}2^{-}$ phenotype (Liew *et al.*, 1982). As has already been postulated, the large number of *Leishmania* antigens resulting from the permissiveness of Balb/c macrophages could account for the preferential triggering of suppressor T cells in these mice. The induction of suppressive T cells in Balb/c mice could also be the result of a selective defect of infected Balb/c macrophages in their antigen-presenting function. Observations showing a reduced expression of H-2 antigens on infected macrophages furnish experimental evidence for this contention (Handman *et al.*, 1978). This hypothesis is also supported by recent observations of Gorczynski and Mac Rae (1982), which showed that, although adherent skin macrophages from susceptible Balb/c mice were functionally equivalent to macrophages from resistant mice in the presentation of degraded *L. tropica* antigens, once infected with living parasites Balb/c macrophages were not able to trigger immune T cells. As suggested by Gorczynski, (1982), difference in the expression of antigen on parasitized resistant and susceptible macrophage could account for the differential triggering of T-cell subsets. Taken together, these observations indicate that T cells play a crucial role in determining the outcome of the disease, at least in experimental models of infection in rodents. In this experimental system, although T cells appear instrumental in the destruction of parasites inside macrophages (Louis *et al.*, 1982), they are capable of suppressing other anti-*Leishmania* T-cell responses beneficial to the host (Howard *et al.*, 1980).

The precise mechanism(s) by which T cells are involved in the elimination of parasites is not yet fully understood. It is possible that DTH triggered by *L. tropica*-specific T cells represents the major effector pathway by which parasites are eliminated. Specific T cells might also induce the activation of macrophages infected with *Leishmania*, resulting in the killing of intracellular parasites (Mauel *et al.*, 1981; Louis *et al.*, 1982). Sensitized T lymphocytes could also be cytolytic for *Leishmania*-infected macrophages (Bryceson *et al.*, 1970) or could influence the quality and magnitude of the anti-*Leishmania* antibody response.

In an attempt to correlate healing of lesions with a particular T-cell function, we have devised methods by which to measure functionally specific T-cell responses to *Leishmania* antigens in mice. Draining lymph node cells from mice primed with *L. tropica* antigens were capable of (1) producing specific proliferative responses upon challenge *in vitro* with parasite antigens (Louis *et al.*, 1979); (2) providing specific helper activity for an *in vitro* antibody response in a hapten-carrier (DNP-*L. tropica*) system; (3) transferring antigen-specific DTH responses to normal mice (Louis *et al.*, 1982); and (4) secreting macrophage activating factor (MAF) after *in vitro* stimulation with parasite antigens (R. Titus and J. A. Louis, submitted). One aim of this review is to present the results of experiments recently initiated in our laboratory in an attempt to correlate susceptibility to infection with a defect in a particular parasite-specific T-cell function. Furthermore, in order to assess directly the role of T cells in the resolution of cutaneous lesions induced by *L. tropica*, we have generated homogeneous populations and clones of T lymphocytes specific for *L. tropica* antigens. Studies pertaining to the functional characteristics of these T-cell populations and clones are presented.

A. Description of Methods Used for the Functional Analysis of Specific T-Cell Responses in Mice Primed with *L. tropica* Antigens

1. Specific T-Lymphocyte Proliferative Response in Vitro

Previous work from this laboratory has documented the presence in lymph node (LN) cells from mice primed subcutaneously (s.c.) with *L. tropica* antigen emulsified in complete Freund's adjuvant (CFA) of cells capable of a proliferative response upon challenge *in vitro* with parasites (Louis *et al.*, 1979). This assay was adapted from a method described for the measurement of T-lymphocyte activation by protein antigens *in vitro* (Corradin *et al.*, 1977). The secondary *in vitro* proliferative response to *L. tropica* was found to be totally specific for *L. tropica*, and it was shown that T cells were responsible for the observed proliferation.

2. Helper Activity of L. tropica-Primed LN Cells

The helper function of LN cells obtained from mice primed s.c. with *L. tropica* antigens emulsified in CFA was tested in a secondary *in vitro* anti-trinitrophenyl (TNP) antibody response after challenge with TNP coupled *L. tropica* parasites (TNP-*L. tropica*). In brief, various numbers of primed LN cells were mixed with a constant number of spleen cells from mice primed to TNP-KLH in the presence of TNP-*Leishmania*. Six days later, the number of anti-TNP plaque-forming cells (PFCs) was enumerated by the slide modification (Golub *et al.*, 1968) of the Jerne plaque assay, using TNP-sheep red blood cells (Rittenberg and Pratt, 1969) as indicator cells. Results in Table I demonstrate that LN cells from mice primed with *L. tropica* are capable of providing helper activity in a secondary anti-TNP response. This effect was demonstrated to be carrier specific and T-cell mediated.

3. Delayed-Type Hypersensitivity Response in Mice Primed to L. tropica

Mice were primed by a s.c. injection of *L. tropica* antigens emulsified in CFA and 10 days later were challenged with *L. tropica* in the footpad. The increase in footpad thickness was measured 18–24 hr later. The observed swelling response was specific for *Leishmania* antigen and was T-cell mediated. Upon histologic examination, the classic hallmarks of DTH were observed.

Table I. Ability of LN from *L. tropica*-Primed Mice to Provide Specific Helper Capacity for *L. tropica* in a Secondary Anti-TNP PFC response *in Vitro*

LN primed to[a]	Antigen *in vitro*[b]	TNP-specific PFC response/10^6 spleen cells plated[c]
L. tropica	TNP–*S. mansoni*	0
L. tropica	TNP–*L. tropica*	657
S. mansoni	TNP–*L. tropica*	0
S. mansoni	TNP–*S. mansoni*	845

[a] Balb/c mice were primed s.c. at the base of the tail with 5×10^6 frozen and thawed *L. tropica* emulsified in CFA or 100 μg *Schistosoma mansoni* (*S. mansoni*) antigen emulsified in CFA as indicated.
[b] One week after priming, 5×10^5 draining LNC were mixed *in vitro* with 5×10^5 TNP-primed spleen cells and the cultures were stimulated with either 5×10^4 TNP-coupled *L. tropica* or 75 μg TNP-coupled *S. mansoni* antigen as indicated.
[c] Six days after the cultures were initiated, the number of TNP specific PFCs was enumerated using the Jerne plaque assay described in Section I.A. The number of TNP-specific PFCs was calculated by subtracting the number of PFCs obtained using uncoupled sheep red blood cells (in this experiment, 43) from that obtained using TNP-coupled sheep red blood cells.

B. Influence of Infection with *L. tropica* on the Development of Specific T-Cell Responses in Mice

It has been previously reported that whereas parasite-specific DTH reactivity developed in CBA mice infected with *L. tropica*, this response decreased gradually in highly susceptible infected Balb/c mice. This observation indicates that in Balb/c mice at least one specific T-cell function, i.e., DTH, is impaired as a result of infection with *L. tropica* (Howard *et al.*, 1980).

Using the methodology described in Section I. A, we have further investigated the influence of infection with *L. tropica* on the development of various specific T-cell responses in susceptible and resistant mice after challenge with an immunogenic form of antigen. According to the protocol depicted in Fig. 1, groups of Balb/c (susceptible) and CBA (resistant) mice were infected with 20×10^6 *L. tropica* promastigotes administered s.c. in the shaved rump. At various times thereafter, a frozen and thawed preparation of *L. tropica* emulsified in CFA was injected s.c. at the base of the tail. Similarly immunized normal mice served as controls. Seven days after immunogenic challenge, draining LN cells were assessed for their ability to proliferate to, and provide help for,

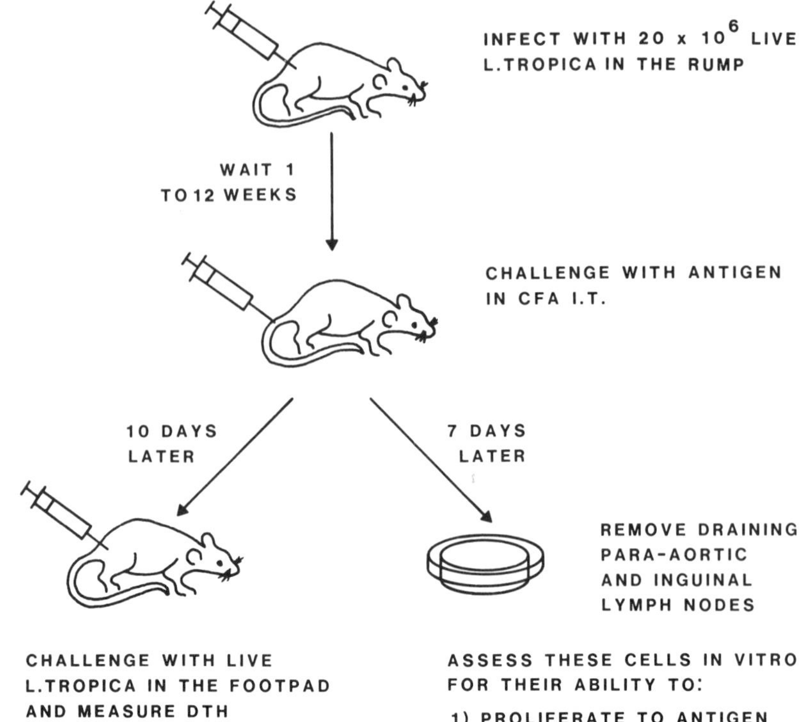

Figure 1. Protocol used to study the effect of infection with *L. tropica* on the development of various specific T-cell responses after immugenic challenge.

specific *Leishmania* antigens *in vitro*. In addition, similarly treated animals, at 10 days after immunogenic challenge, were injected with *L. tropica* parasites in the footpad; the intensity of DTH responses was assessed 18–24 hr postinjection by measuring the increase in footpad thickness.

After priming *L. tropica* antigens emulsified in CFA, draining LN cells from both Balb/c and CBA mice exhibited (1) a substantial proliferative response *in vitro* after challenge with the parasites, and (2) helper activity for an anti-TNP PFC response to challenge with TNP-*L. tropica* (Tables II and III). Mice from both strains also displayed significant DTH reactivity 10 days after priming with *L. tropica* in CFA. These observations support the notion that lymphoid tissues from susceptible Balb/c mice contain a number of T-cell percursors capable of recognizing *Leishmania* antigens similar to that seen in resistant CBA mice. This contention is further supported by recent work in this laboratory showing that the frequency of T-cell precursors capable of mounting specific proliferative responses to *Leishmania* antigens is similar in immunized Balb/c and CBA mice.

Table II. Effect of Infection on the Capacity of LN Cells from Primed Mice to Mount a Specific Proliferative Response *in Vitro*

Antigen[a] or mitogen	Response: Incorporation of [^{3}H]-TdR, CPM (±SE)[b]			
	Balb/c	Balb/c infected	CBA	CBA-infected
2 × 10^5 *L. tropica* per ml culture	116,398 (8,134)	14,635 (2,455)	140,238 (9,800)	113,489 (933)
5 × 10^5 *L. tropica* per ml culture	92,288 (5,310)	14,068 (1,156)	111,190 (6,397)	81,304 (7,898)
10 × 10^5 *L. tropica* per ml culture	64,943 (1,421)	12,186 (962)	78,245 (1,712)	58,408 (6,094)
Con A	27,458 (1,717)	32,547 (1,694)	118,457 (1,759)	150,580 (12,664)

[a] *L. tropica* used in culture as a frozen and thawed preparation; concanavalin A (con A, Pharmacia, Sweden) used at 1 μg/ml of culture.
[b] Groups of Balb/c and CBA mice were infected with 20 × 10^6 *L. tropica* s.c. in the rump and 5 weeks later were challenged, along with groups of appropriate control mice, s.c. at the base of the tail with 5 × 10^5 frozen and thawed *L. tropica* emulsified in CFA. One week later, cells derived from the draining lymph nodes were assessed *in vitro* for proliferation to specific antigen and to mitogen. Data shown are for 4 × 10^5 LNC/well and are expressed as the average incorporation of [^{3}H]-TdR in CPM ±SE.

In contrast to these results, seen in normal mice, although LN cells from infected CBA mice challenged with *L. tropica* antigens in CFA were capable of a substantial proliferative response *in vitro*, LN cells from infected Balb/c mice did not proliferate significantly upon *in vitro* challenge with *Leishmania* antigens (Table II). However, the specific helper capacity of LN cells was not impaired as a result of infection in either strain of mouse (Table III).

Table III. Effect of Infection on the Capacity of LN Cells from Primed Mice to Provide Specific Helper Activity in an Anti-TNP *Leishmania* Response

Number of LNC[a] tested for helper activity (× 10^5)	PFC response/10^6 spleen cells plated[b]			
	Balb/c	Balb/c infected	CBA	CBA-infected
5	1520	2567	1856	3524
1	893	1500	733	2400
0.5	467	277	397	1127
0.1	20	180	77	467

[a] LN cells obtained as described in footnote *b*, Table II.
[b] For technique, see Table I.

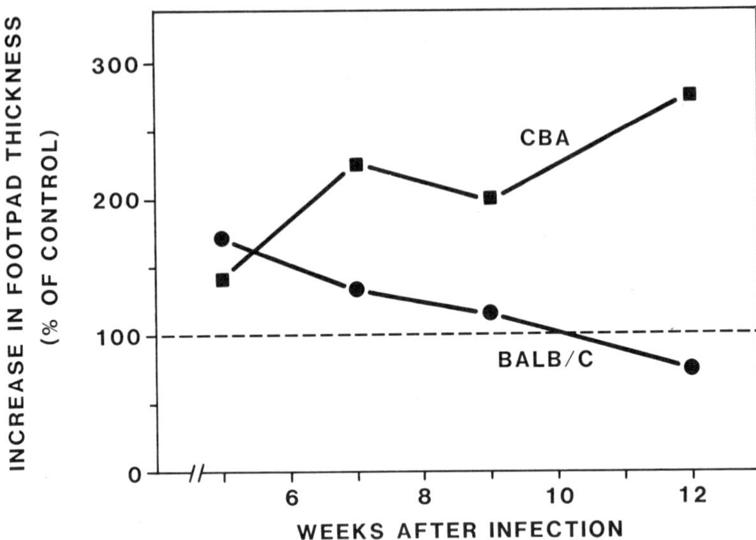

Figure 2. Effect of infection with *L. tropica* on the ability of an immunogenic challenge to sensitize mice for parasite-specific DTH. Groups of CBA and Balb/c mice were infected s.c. with *L. tropica* and at various times thereafter were challenged s.c. at the base of the tail with *L. tropica* antigens in CFA; 1 week later, *L. tropica* was injected into the footpad and the net increase in footpad thickness was recorded 18 hr later. Results are expressed as the percentage of the swelling response seen in similarly immunized uninfected mice.

Finally, whereas the ability to sensitize for specific DTH reactivity increased over time in infected CBA mice, it steadily decreased in infected Balb/c mice (Fig. 2). The latter result confirms and extends observations that indicated that the specific DTH reactivity that evolves as a result of infection is only transient in Balb/c mice (Howard *et al.*, 1980).

Taken together, this series of experiments indicates that a state of immunologic unresponsiveness to *Leishmania* antigens develops in infected Balb/c mice. This unresponsive state selectively operates on T-cell subsets involved in proliferative and DTH responses to *L. tropica* but does not interfere with the activity of specific helper T cells. The mechanism by which this phenomenon evolves is unknown but could be caused by the activities of suppressor T cells or by feedback inhibition by specific antibody, or both.

C. *Leishmania*-Specific T-Cell Lines

1. Maintenance in Vitro of T-Cell Populations Specific for Leishmania tropica Antigens

Blast cells from cultures of primed LN cells responding to *L. tropica* were isolated on Percoll density gradients (Kurnick *et al.*, 1979). These blasts were subsequently maintained in the absence of parasite antigens in medium containing supernatants of mixed leukocyte cultures as a source of interleukin-2 (IL-2) (Ryser *et al.*, 1978). After removal from IL-2-containing medium, these blasts could be restimulated by *L. tropica*, provided that additional macrophages were added to the culture system (Louis *et al.*, 1981). The antigen-induced proliferative response of these blasts was strictly specific, i.e., antigens unrelated to the parasites failed to provoke proliferative responses. In our experience, the blasts isolated from *L. tropica*-primed LN cells could be repeatedly restimulated by parasites *in vitro* without losing their capacity to exhibit a specific proliferative response. After the cycle of isolation from responding LN cells on Percoll, maintenance in IL-2-containing medium, specific restimulation by parasite antigens, maintenance in IL-2-containing medium, and so on, blast cells specific for *L. tropica* have now been maintained for several months. Blast cells obtained and maintained according to this protocol were T cells, since they expressed the Thy 1.2 and Lyt $1^{+}2^{-}$ surface phenotypes. Furthermore, these specific T-cell populations were examined for possible MHC restriction in the interactions between antigen-specific T cells and parasite antigen presenting cells. Using congenic recombinant strains of mice, it was possible to show that *Leishmania*-specific B10.AQR T blasts could be specifically restimulated in the presence of B10.AQR or B10.A(4R) irradiated spleen cells as a source of antigen-presenting cells, but not in the presence of B10.T (6R) or B10.A (5R)-irradiated spleen cells (Fig. 3). These data indicate that *Leishmania*-specific T cells can recognize *Leishmania* antigens only in the context of I-A identical antigen-presenting cells (Louis *et al.*, 1981). Further experiments have shown that F_1 T blasts could be positively selected to respond to *Leishmania* antigen within the context of only one parental haplotype antigen-presenting cell. These data confirm observations and extend to a protozoan parasite antigenic system demonstrating two distinct T-cell populations within a specific F_1 T cells, each recognizing the antigen on presenting cells from only one parental haplotype (Paul *et al.*, 1977; Thomas and Shevach, 1978; Ziegler and Unanue, 1979).

2. *Functional Analysis of Leishmania-Specific T-Cell Populations Maintained in Vitro*

The helper activity of the *L. tropica*-specific T-cell populations maintained *in vitro* has been investigated in a secondary anti-hapten antibody response *in vitro* in a manner similar to that described for LN cells in Section I.A. In brief, cultures containing pure B-lymphocyte populations obtained from spleen cells of mice primed with TNP-KLH and *L. tropica*-specific T blasts maintained *in vitro* were challenged with TNP-*L. tropica*. Five days later, the number of anti-TNP-PFC per culture was enumerated. It was observed that challenge with TNP-*L. tropica* induced high secondary anti-TNP IgG PFC responses in these cultures. The magnitude of the PFC response was found to be dependent on the number of TNP-*L. tropica* and *Leishmania*-specific T cells added (Zubler and

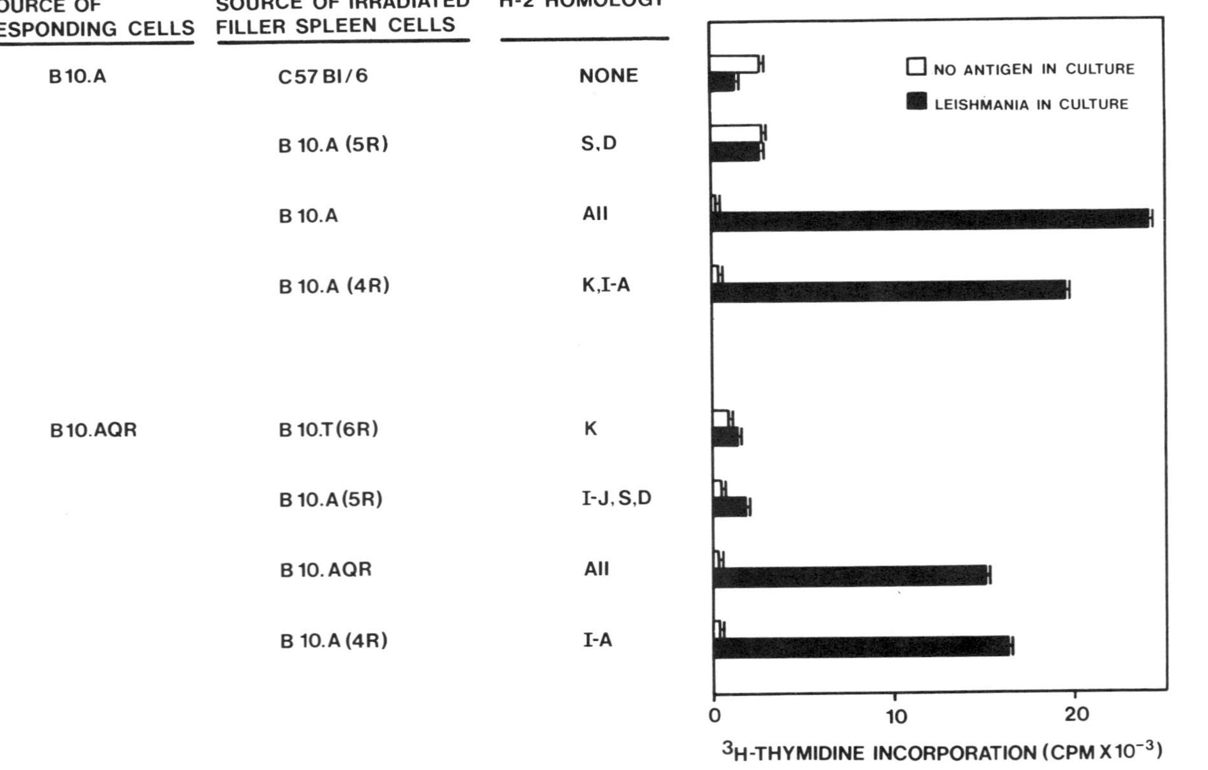

Figure 3. Antigen-specific proliferative response of B10.A and B10.AQR *Leishmania*-immune T blasts as a function of the haplotype of added spleen cells. Parasite-specific blasts derived from *Leishmania*-immune B10.A or B10.AQR mice were maintained in IL-2 for 5 days. These blasts were then restimulated with 10^6 *L. tropica* in the presence of 5×10^6-irradiated normal spleen cells obtained from mice of the indicated strains. After 5 days of culture, proliferation was assessed by measuring [^{3}H]-TdR incorporation with a 16-hr pulse. (From Louis *et al.*, 1981.)

Louis, 1981). Furthermore, the carrier specificity of the response and the necessity of linkage between the hapten and the carrier (*L. tropica*) for the induction of a secondary anti-hapten antibody response in this system was demonstrated experimentally (Zubler and Louis, 1981). Using a culture system in which antigen-presenting cells were H-2 identical with the *L. tropica*-specific T blasts, it was possible to demonstrate the necessity for H-2 identity between TNP-primed B cells and *L. tropica*-specific T cells for an optimal anti-TNP responses in this experimental system. These results are in agreement with observations made in several experimental systems (Katz *et al.*, 1973; Kappler and Marrack, 1978).

It is accepted that activation of macrophages leads to their ability to destroy intracellular *Leishmania* (Mauel *et al.*, 1978). Antigen or polyclonal activation of T lymphocytes leads to the production of lymphokines capable of activating macrophages resulting in intracellular destruction of *Leishmania* (Buchmüller and Mauel, 1979; Nacy *et al.*, 1981). We therefore studied the ability of *in vitro*-maintained *Leishmania*-specific T blasts to activate macrophages parasitized with *Leishmania* (Louis *et al.*, 1982). Peritoneal exudate macrophages parasitized with *Leishmania* were incubated with parasite-specific T-cell blasts; at various times thereafter, the number of surviving intracellular *L. tropica* was recorded. It was observed that *Leishmania* parasites disappeared readily from macrophages in cultures containing *L. tropica*-specific T cells. This effect was not observed in the presence of T blasts specific for an antigen unrelated to *Leishmania*, indicating that in this system macrophage activation did require restimulation of specific T lymphocytes by *L. tropica* antigens and was not the consequence of active secretion of lymphokines by blast T cells maintained *in vitro*. At this point, it is worth mentioning that, using a short term ^{51}Cr-release assay, these *L. tropica*-specific T blasts were not found to be directly cytolytic for parasitized syngeneic macrophages (Louis *et al.*, 1982).

Since it is generally believed that DTH responses are extremely important in host destruction of cells bearing foreign antigens (Liew and Simpson, 1980; Loveland *et al.*, 1981), the ability of *L. tropica*-specific T-cell lines to transfer specific DTH responses to normal mice was investigated. Normal mice injected s.c. with as few as 10^5 antigen-specific T blasts and *Leishmania* antigens exhibited specific DTH responses, measured by the increase in footpad thickness 18-24 hr later. Furthermore, a transfer of DTH reactivity to normal mice was also observed after intravenous (i.v.) injection of *Leishmania*-specific T-cell lines (Fig. 4). Under these two experimental conditions, a requirement for identity at the level of H-2 between T blasts and recipient mice was necessary for the transfer of specific DTH responses. Preliminary experiments indicate that the restricting element was located in the I region of the H-2 complex. The possibility that these functional activities are mediated by different T-cell subsets present in the parasite-specific T-cell lines is discussed in Section I.D., which describes *Leishmania*-specific T-cell clones.

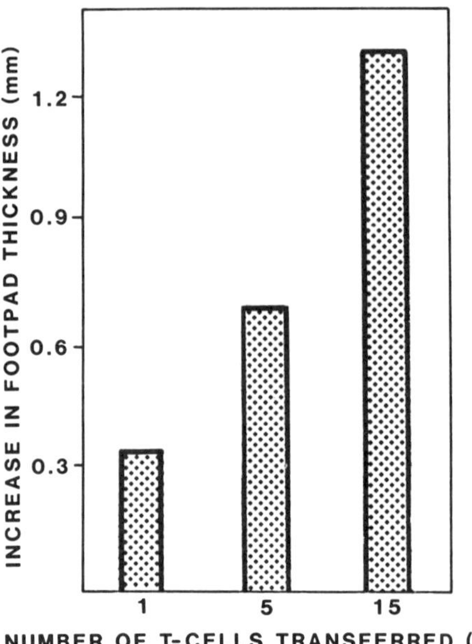

Figure 4. Transfer of delayed-type hypersensitivity to normal mice by i.v. administration of *L. tropica* specific T cells. The indicated numbers of parasite-specific T blasts were injected i.v. into normal mice. Four days later, *L. tropica* antigen(s) was injected into the footpads and the net increase in footpad thickness was measured 24 hr later; the values for mice receiving antigen alone (0.05 mm) have been subtracted.

3. *In Vivo Effect of Leishmania-Specific T-Cell Lines*

Inasmuch as *Leishmania*-specific T-cell lines were shown to be capable of performing a variety of specific immunologic functions both *in vitro* (proliferation, helper activity, and macrophage activation) and *in vivo* (transfer of specific DTH), it was of interest to study the effect that these propagated T cells might have on the course of the disease.

In the first experiment, groups of CBA (resistant) and Balb/c (susceptible) mice were injected i.v. with 10^7 *L. tropica*-specific *in vitro* propagated T cells. After 2 hr, 10^7 *L. tropica* promastigotes were injected in the footpad and the degree of footpad swelling (Weintraub and Weinbaum, 1977) was measured at regular intervals thereafter. Control mice received parasites alone. Results in Fig. 5 indicate that the *L. tropica*-induced lesions were substantially larger in mice receiving *L. tropica* specific T-cell lines. These unexpected results were confirmed in a second experiment using Balb/c mice. These animals were injected intravenously with 10^7 *L. tropica*-specific blasts and 2 hr later infected s.c. on the shaved rump with 10^7 *L. tropica*. As shown in Fig. 5, a similar enhancing effect by specific T cells on the size of cutaneous lesions was observed. Since it has been recently reported that Balb/c mice irradiated (550 rad) before infection were able to heal *L. tropica*-induced lesions (Howard *et al.*, 1981), the effect of parasite-specific T-cell lines was tested in this experimental system. Control Balb/c mice were irradiated (550 rad) and 24 hr later were infected s.c. on the rump with 10^7 *L. tropica* promastigotes. As reported by Howard

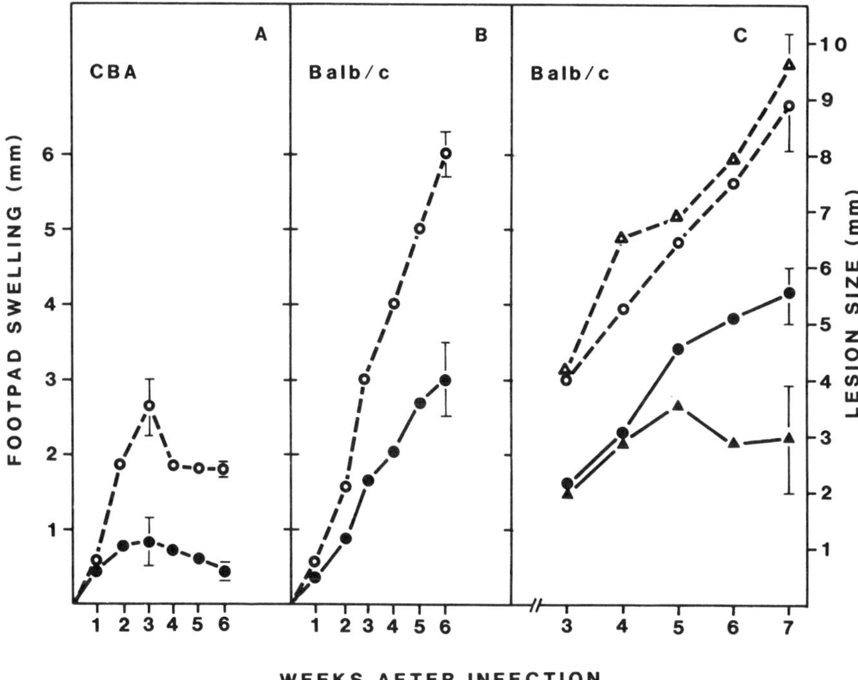

Figure 5. Effect of *L. tropica*-specific T-cell blasts on the development of lesions produced by the parasite. Groups of CBA mice (A) and Balb/c mice (B) were injected i.v. with 10^7 *L. tropica*-specific T blasts and 2 hours later with 10^7 *L. tropica* promastigotes injected in the footpad. At weekly intervals, footpad thickness was measured (○---○). Control mice received only *L. tropica* (●——●). (C) Groups of either normal (○---○) or irradiated (550 rad) (△---△) Balb/c mice were given i.v. 10^7 *L. tropica*-specific T blasts and 24 hr later were infected s.c. on the rump with 10^7 *L. tropica* promastigotes. Normal (●——●) or irradiated (▲——▲) control mice received *L. tropica* alone. At weekly intervals after infection, the mean lesion diameter was recorded.

et al. (1981), the lesions observed in these Balb/c mice were significantly less severe than those seen in nonirradiated infected controls. In sharp contrast, irradiated Balb/c mice receiving *L. tropica*-specific T cells exhibited, after infection, severe, progressive lesions characteristic of cutaneous leishmaniasis in mice of this strain (Fig. 5). The specificity of the lesion-enhancing effect of *L. tropica*-specific T cells was studied in irradiated Balb/c mice receiving ovalbumin-specific T cells before infection with *L. tropica*. In these mice, lesion size was equivalent to that seen in control mice. (G. Lima, R. Titus, H. Engers, and J. A. Louis, in preparation, 1983).

It should be emphasized that the enhancing effect of *L. tropica*-specific T cells on lesion size has been observed over an admittedly short period of time and that it remains to be established as to whether this effect is the result of an increased load of parasites or of extensive tissue necrosis induced by parasite-specific T cells, or both.

Interestingly, preliminary experiments indicate that, by 14 days postinfection, mice injected with *L. tropica*-specific T-cell lines exhibited in their serum substantial levels of antibodies directed against glycoproteins of *L. tropica* promastigotes (Nguyen *et al.*, 1982). In contrast, very low levels of *Leishmania*-specific antibodies were observed, at this time, in the serum of control infected mice (Fig. 6).

Clearly, further investigations are required in order to determine the mechanism(s) by which T-cell lines specific for *Leishmania* antigen have an exacerbative effect on lesions induced by these parasites. In these studies, an eventual role for T-cell-dependent antibodies should be considered.

D. *Leishmania*-Specific T-Cell Clones

The results described thus far indicate that *Leishmania*-specific T cells resulting from stimulation *in vitro* of primed LN cells with *Leishmania* antigens could, after propagation in IL-2-containing medium, perform several immunologic functions, such as (1) helper activity for a secondary anti-hapten response *in vitro*, (2) activation of parasitized macrophages resulting in parasite destruction; and (3) transfer of specific DTH reactions to normal recipient mice. It was therefore important to determine whether these various functions were mediated by different T-cell subsets; for this purpose, *Leishmania*-specific T-cell clones were derived and analyzed functionally.

Inasmuch as the frequency of parasite-specific proliferative and helper T cells in populations isolated from primed LNC cultures responding to *L. tropica* and maintained for 3 days in IL-2-containing medium was high, i.e., on the order of 1:25 (Louis *et al.*, 1982), these populations were used to derive *L. tropica*-specific T-cell clones either under limiting dilution conditions or by micromanipulation (Louis *et al.*, 1982). In brief, limiting numbers of T cells (0.5 cell/microtiter well) were cultured together with irradiated spleen cells (as a source of filler cells), *L. tropica* antigen, and IL-2. Microcultures, where growth was observed, were subcultured 10 days later in a larger volume together with fresh filler spleen cells, parasite antigens, and IL-2. Clones were maintained by reculturing them approximately every 2 weeks. In other cloning experiments, individual T lymphocytes were micromanipulated into microtiter wells. Clones obtained by these two approaches could be maintained for 3 months in continuous culture.

Characterization of several cloned T-cell lines has revealed that (1) they all displayed the Thy 1.2, Lyt $1^{+}2^{-}$ phenotype; (2) after removal from IL-2-conditioned medium, they exhibited a specific proliferation upon challenge with the parasites *in vitro*; and (3) they possessed helper activity as tested in a secondary anti-hapten antibody response *in vitro*. As for the original T-cell populations, T-cell clones recognized parasite antigens within the context of H-2-identical

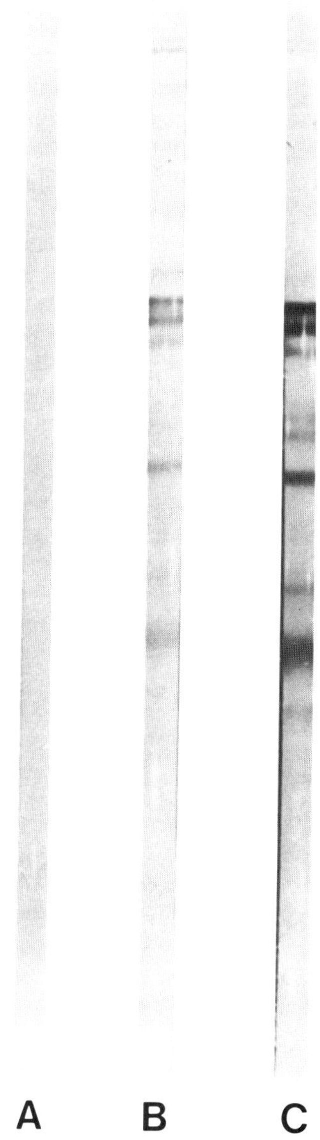

Figure 6. Anti-*Leishmania* humoral response in CBA mice injected with 10^7 *L. tropica*-specific T-cell blasts before infection with *L. tropica*. The glycoproteins of *L. tropica* promastigotes were separated by SDS-PAGE, transferred into nitrocellulose paper, and used as antigens in an enzyme-linked immunoassay to determine the presence of antibodies directed against each of them in the serum of either normal mice (A), infected mice 14 days after infection (B), and T blast transferred and infected mice (14 days after treatment) (C).

macrophages and their helper activity was also restricted to B cells compatible at the I-A region of the MHC.

All the clones tested were also found to transfer specific DTH responses after transfer to unimmunized mice and to activate specifically *Leishmania*-infected macrophages. These results confirm, in a parasite antigen system, previous observations which demonstrated that a given T-cell clone can perform a variety of immunologic functions (Bianchi *et al.*, 1981; Lin and Askonas, 1981).

E. Concluding Remarks

It has previously been shown that Lyt 1^+2^- lymphocytes from CBA (resistant) mice recovering from cutaneous lesions induced by *L. tropica* are capable of conferring protective immunity upon transfer in normal mice (Liew *et al.*, 1982). In contrast, Lyt 1^+2^- splenic lymphocytes from infected susceptible Balb/c mice were able to inhibit the induction of *Leishmania*-specific DTH reactions in syngeneic recipient Balb/c mice (Howard *et al.*, 1981; Liew *et al.*, 1982). Furthermore, results of experiments with nu/nu Balb/c athymic mice indicated that depending on the number of cells injected, Lyt 1^+2^- cells from normal mice either conferred some resistance or had a suppressive effect (Mitchell *et al.*, 1981). The results presented here indicate that parasite-specific Lyt 1^+2^- T-cell lines obtained from both CBA and Balb/c mice can aggravate the cutaneous lesions induced by *L. tropica* infection in these two strains of mice. It is puzzling that these same T-cell lines have been shown to perform a variety of immunologic functions, some of which are assumed to be beneficial for the host. For example, these *Leishmania*-specific T-cell lines were able to transfer specific DTH responses even in infected Balb/c recipients. The mechanism(s) responsible for this aggravation of lesions by specific T cells is unknown, and it remains to be determined whether it is accompanied by excessive parasite growth. Inasmuch as the lesion promoting T cells could represent a particular T-cell subset, we are now investigating the effect of several *L. tropica*-specific T-cell clones on the course of cutaneous lesions.

II. *SCHISTOSOMA MANSONI*

Schistosomiasis is a helminthic disease that afflicts ~ 300 million people in tropical and subtropical regions. Three closely related species, *Schistosoma mansoni*, *S. haematobium*, and *S. japonicum*, are capable of infecting primates and human subjects. During its life cycle, this metazoan parasite requires a specific snail as intermediate host. The infective larvae (cercariae) penetrate through the host skin; shistosomulae, after passage to the lungs develop in the portal vein. From the liver, worm pairs migrate to the small mesenteric or pelvic veins, or both. Large quantities of eggs are produced that invade the appropriate intermediate host (snail), after their elimination from the host. Macrophages and eosinophils have now been recognized as potent effector cells in the destruction of schistosomulae *in vitro* (Butterworth *et al.*, 1980; Capron *et al.*, 1982); however, their parasiticidal activity requires the presence of specific antibody, i.e., IgG and IgE antibodies for eosinophils and IgE antibodies for macrophages (Joseph, 1982; Dessaint *et al.*, 1982).

Although the absolute inefficiency of T cells in exerting a direct cytolytic activity on schistosoma larvae has been documented (Butterworth *et al.*, 1979), the importance of other subsets of parasite-specific T lymphocytes in the elimination of schistosomulae is not yet fully appreciated. It is likely that helper T cells specific for schistosoma antigens dictate the quality and specificity of the anti-schistosome antibody B-cell response. Using hapten-carrier conjugates made of TNP-coated schistosomulae as immunogen, helper T-cell activity has indeed been demonstrated in the spleens of rats and mice infected with *S. mansoni* (Ramalho-Pinto *et al.*, 1976). Furthermore, parasite-specific DTH reactions (immunologic responses mediated through specific T cells) are associated with the release of soluble eggs antigens and granuloma formation observed during the course of schistosomiasis appears to be T-cell dependent (Chensue *et al.*, 1981; Doughty and Phillips, 1982).

To permit analysis of the role of T cells in schistosomiasis, we devised methods similar to those described in the *L. tropica* system by which to obtain homogeneous populations and clones of T lymphocytes specific for schistosoma antigens. Some of the characteristics of these T cells are summarized in this review.

A. Murine T-Lymphocyte-Dependent Proliferative Responses to Schistosoma Antigens: Characteristics and Specificity

Inguinal and periaortic LN cells from mice primed by a s.c. injection at the base of the tail of soluble adult worm (*S. mansoni*) extracts (Ag Sm) (Capron *et al.*, 1968) were capable of extensive proliferative responses *in vitro* after challenge with the same parasite antigen. Although priming of mice with only 1 μg Ag Sm (1 μg dry product being equivalent to ± 0.5 μg protein) resulted in significant proliferation after challenge of LN cells *in vitro*, sensitization with 100 μg of antigen consistently resulted in maximal proliferative responses. The data presented in Fig. 7 indicate that, after priming with 100 μg Ag Sm, optimal proliferative responses *in vitro* were also obtained after stimulation of cultures with 100 μg antigen.

Specificity analysis has demonstrated that (1) host antigens (hamster) acquired by adult worms used to prepare Ag Sm were not responsible for the observed proliferative response, (2) protozoan parasite (*L. tropica*) antigens and extracts from other trematodes (*Fasciola hepatica*) did not induce proliferation of Ag Sm-primed LN cells, and (3) Ag Sm. and *S. mansoni* antigens obtained by other procedures displayed extensive cross-reactivity when tested in this LN-proliferative assay (J. Pestel and J. A. Louis, unpublished observations, 1983). Since incubation products of living adult schistosomes (I.P. Sm) displayed complete antigenic cross-reactivity with Ag Sm, I.P. Sm was used as immunogen in several experiments. Finally, the T-cell dependence of the observed proliferative response has been established (J. Pestel and J. A. Louis, unpublished observations, 1983).

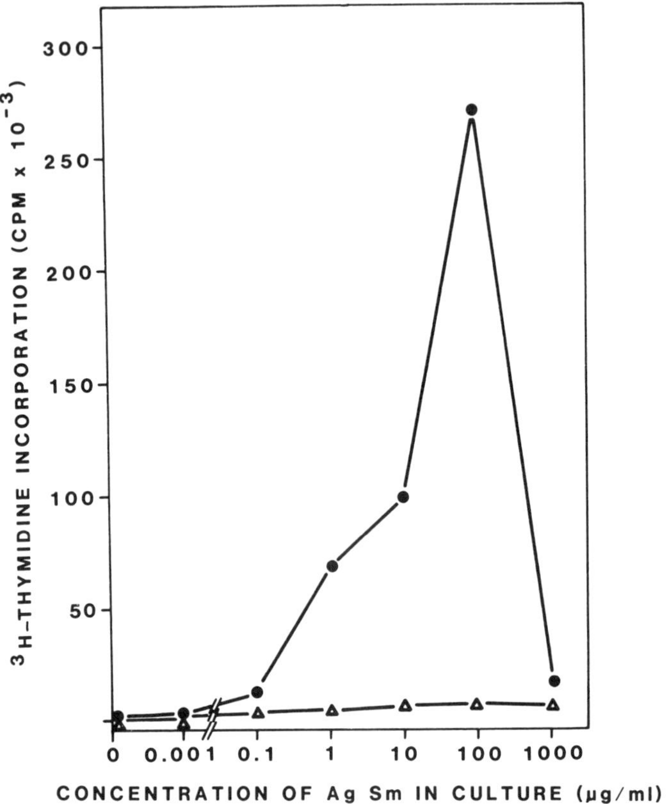

Figure 7. Dose dependence of the parasite antigen-induced proliferation of primed LN cells *in vitro*. LN cells from Balb/c mice injected with 100 µg Ag Sm (soluble extract of adult worms) emulsified in CFA (●——●) or with CFA alone (△——△) were cultured with various amounts of Ag Sm. Proliferation was assessed after 5 days of culture by measuring the incorporation of [^{3}H]thymidine using a 16-hr pulse.

B. Helper Activity of LN Cells from Mice Primed with Schistosoma Antigens

LN cells from mice primed with incubation products of living adult schistosomes (I.P. Sm) were cultured with spleen cells obtained from mice primed with TNP-KLH in the presence of trinitrophenylated I.P. Sm (TNP-I.P. Sm). Five days later, the number of anti-TNP PFC per culture was recorded as described above. TNP-I.P. Sm triggered a secondary anti-TNP response in these cultures. The inability to induce a significant anti-TNP response in similar cultures challenged with TNP-*L. tropica* or in cultures containing *L. tropica*-specific LN cells challenged with TNP-I.P. Sm demonstrates the carrier specificity of the observed response.

Table IV. Incubation Products of *S. mansoni* Adult Worms (I.P. Sm)-Induced Restimulation of T-Cell Blasts after Propagation in IL-2-Containing Medium [a]

Antigen added to cultures	Response: Incorporation of [^{3}H]-TdR, CPM (± SD)	
	OVA-induced blasts	I.P. Sm-induced blasts
OVA	12,524 (1052)	246 (120)
I.P. Sm	382 (110)	11,296 (834)
None	405 (105)	410 (122)

[a] I.P. Sm and OVA-induced LNC were cultured for 5 days, respectively, with I.P. Sm and OVA (100 μg/ml), the resulting blasts, after Percoll gradient, were further incubated for few days in medium containing IL-2. After removal of IL-2, washed blasts cells (10^4/well) were cultured with 10^6 syngeneic filler irradiated (2000 rad) spleen cells (as source of macrophages) in presence of I.P. Sm (100 μg/ml). Four days later, proliferation was assessed by measuring [^{3}H]thymidine incorporation with a 16-hr pulse (triplicate cultures). Data are expressed as the average incorporation of [^{3}H]-TdR in CPM ± standard deviation (SD).

C. Schistosome-Specific T-Cell Lines

Homogeneous populations of T cells specific for I.P. Sm were obtained according to procedures described in Section I. In brief, primed LN cells were challenged with parasite antigens; after 5 days, blast cells were isolated on Percoll density gradients. The blast cells were subsequently propagated in antigen-free IL-2-containing medium for a few days. After this time, more than 95% of cells expressed the Thy 1.2, Lyt 1^+2^- phenotype. Furthermore, after removal from IL-2-containing medium, these T cells could be restimulated only by specific antigen and in the presence of irradiated syngeneic spleen cells as a source of antigen-presenting cells (Table IV). Maintenance *in vitro* of these I.P. Sm-specific T cells was achieved by cycles of specific restimulation in the presence of syngeneic macrophages followed by cultivation in IL-2-containing medium. Preliminary experiments aimed at determining the frequency of I.P. Sm specific T cells in various lymphocyte populations were carried out using limiting dilution techniques. Results obtained clearly indicate that, compared with primed LN cells, continuously propagated T blasts were significantly enriched in parasite-specific T-cell precursors. Using a protocol similar to that described, these continuously propagated T cells were also demonstrated to provide helper activity for secondary anti-TNP responses in cultures challenged with TNP–I.P. Sm.

D. Cloned T Cells Specific for Schistosoma Antigens

T-cell clones specific for schistosoma antigens were derived from either cultures of primed LN cells responding to I.P. Sm or from the continuous T-cell

Table V. Proliferative Response and Helper Activity of I.P. *S. mansoni*-Specific T-Cell Clones

Clone	Proliferation[a] (Incorporation of [^{3}H]-TdR, CPM, ± SD)	Helper activity[b] (TNP-specific PFC response/10^6 spleen cells plated
HA	2,650 (230)	680
HB	4,400 (1300)	100
HG	5,861 (723)	270
HI	7,731 (2522)	540
HP	12,452 (775)	1100
HT	13,618 (1039)	450

[a] Cloned T cells (10^4) were challenged with 10 μg I.P. Sm in the presence of 10^6 irradiated syngeneic spleen cells. After four days, proliferation was assessed using a 16-hr pulse with [^{3}H]thymidine (triplicate cultures).
[b] Cloned T cells (10^4), 5×10^5 TNP-KLH-primed spleen cells were cultured in the presence of 60 ng TNP–I.P. Sm. Five days later, indirect anti-TNP PFCs were determined. Background values were subtracted.

lines described. The first cloning procedure consisted of culturing limiting numbers (1 cell/well) of responding cells in the presence of IL-2, I.P. Sm (50 μg/ml), and irradiated syngeneic spleen cells. In another series of experiments, individual responding cells were micromanipulated into microtiter wells also containing the specific antigen, IL-2, and irradiated filler spleen cells. Results obtained so far indicate that these T-cell clones could be maintained for up to 8 weeks by reculturing at 2-week intervals. After removal from IL-2-containing medium, all six clones obtained exhibited, with a variable intensity, specific proliferative responses after challenge with I.P. Sm in the presence of irradiated filler cells (Table V). Furthermore, the six clones were able to provide antigen-specific helper activity in a secondary *in vitro* anti-hapten antibody response (Table V). Interestingly, no correlation between the proliferative capacity and the helper activity of individual T-cell clones was observed.

E. Concluding Remarks

It is hoped that the schistosome-specific T-cell populations and clones described will provide a tool by which to analyze directly the role of T cells in the course of experimental schistosomaisis.

These T-cell lines and clones might be of interest in understanding the mechanism by which recently described schistosome-derived factors (Capron and Dessaint, 1982) inhibit the host-immune response. Since it has been reported recently that activated macrophages are able to kill schistosomulae *in vitro* (Sher *et al.*, 1982), it would be of interest to determine whether these parasite-specific T-cell lines and clones could, after recognition of specific antigen, release lymphokines capable of activating macrophages.

The possibility that individual T-cell clones could provide selective help for specific IgE responses merits further investigation.

Finally, the possible identification of cloned T-cell lines with helper activity for an antischistosome IgE B-cell response could provide a powerful system for the analysis of the regulatory mechanisms influencing this response.

ACKNOWLEDGMENTS

We would like to thank Dr. A. Capron, Dr. H. Isliker, and Dr. P.-H. Lambert for their constant support and Dr. T. Brunner, Dr. J.-C. Cerottini, and Dr. J. Mauel for valuable discussions. This work was supported by grants from the Swiss National Science Foundation and the UNDP/World Bank/WHO Special Programme for Research in Tropical Diseases.

III. REFERENCES

Adler, S., 1964, in: *Advances in Parasitology* (B. Dawes, ed.), vol. 2, pp. 35–96, Academic Press, London.

Behin, R., Mauel, J., and Sordat, B., 1979, *Leishmania tropica*: Pathogenicity and *in vitro* macrophage function in strains of inbred mice, *Exp. Parasitol.* **48**:81.

Bianchi, A. T. J., Hooijkaas H., Benner, R., Tees, R., Nordin, A. A., and Schreier, M. H., 1981, Clones of helper T cells mediate antigen-specific, H2 restricted DTH, *Nature* **290**:62.

Bryceson, A. D. M., Bray, R. S., Wolstencroft, R. A., and Dumonde, D. C., 1970, Immunity in cutaneous leishmaniasis of the guinea-pig, *Clin. Exp. Immunol.* **7**:301.

Buchmüller, Y., and Mauel, J., 1979, Studies on the mechanisms of macrophage activation. II. Parasite destruction in macrophages activated by supernates from concanavalin A-stimulated lymphocytes, *J. Exp. Med.* **150**:359.

Butterworth, A. E., Vadas, M. A., Martz, E., and Sher, A., 1979, Cytolytic T lymphocytes recognize alloantigens on schistosomula of *Schistosoma mansoni*, but fail to induce damage, *J. Immunol.* **122**:1314.

Butterworth, A. E., Vadas, M. A., and David, J. R., 1980, Mechanisms of eosinophil mediated helminthotoxicity, in: *The Eosinophil in Health and Disease* (A. A. F. Mahmoud and F. F. Austen, eds.), pp. 253–272, Grune & Stratton, New York.

Capron, A., and Dessaint, J. P., 1982, Schistosomes as potential source of immunopharmacological agents, in: *Clinics in Immunology and Allergy* (A. Capron, ed.), vol. 2, pp. 613–619, W.B. Saunders, Philadelphia.

Capron, A., Biguet, J., Vernes, A., and Afchain, D., 1968, Structures antigéniques des helminthes. Aspects immunologiques des relations hôte-parasite, *Pathol. Biol.* **16**:121.

Capron, A., Dessaint, J. P., Haque, A., and Capron, M., 1982, Antibody dependent cell-mediated cytotoxicity against parasites, in: *Immunity and Concomitant Immunity in Infectious Diseases* (G. Kallos, ed.), pp. 234–267, S. Karger, Basel.

Chensue, S. W., Wellhausen, S. R., and Boros, D. L., 1981, Modulation of granulomatous hypersensitivity II. Participation of Lyl$^+$ and Ly 2$^+$ T lymphocytes in the suppression of granuloma formation and lymphokine production in *Schistosoma mansoni* infected mice, *J. Immunol.* **127**:363.

Corradin, G., Etlinger, H. M., and Chiller, J. M., 1977, Lymphocyte specificity to protein antigens. I. Characterization of the antigen-induced *in vitro* T cell-dependent proliferative response with lymph node cells from primed mice, *J. Immunol.* **119**:1048.

Dessaint, J. P., Capron, A., Joseph, M., Auriault, C., and Pestel, J., 1982, Macrophage-mediated IgE antibody-dependent cellular cytotoxicity to helminth parasites and IgE dependent macrophage activation, in: *Macrophage-Mediated Antibody-Dependent Cellular Cytotoxicity* (H. S. Koren, ed.), Dekker, New York (in press).

Doughty, B. L., and Phillips, S. M., 1982, Delayed hypersensitivity granuloma formation and modulation around *Schistosoma mansoni* eggs *in vitro*. II. Regulatory T cell subset, *J. Immunol.* **128**:37.

Golub, E. S., Mishell, R. I., Weigle, W. O., and Dutton, R. W., 1968, A modification of the hemolytic plaque assay for use with protein antigens, *J. Immunol.* **100**:133.

Gorczynski, R. M., 1982, Nature of resistance to leishmaniasis in experimental rodents, *Dev. Comp. Immunol.* **6**:199.

Gorczynski, R. M., and MacRae, S., 1982, Analysis of subpopulation of glass-adherent mouse skin cells controlling resistance/susceptibility to infection with *Leishmania tropica*, and correlation with the development of independent proliferative signals to Lyt 1$^+$/Lyt 2$^+$ lymphocytes, *Cell. Immunol.* **67**:74.

Handman, E., Ceredig, R., and Mitchell, G. F., 1979, Murine cutaneous leishmaniasis: Disease patterns in intact and nude mice of various genotypes and examination of some differences between normal and infected macrophages, *Aust. J. Exp. Biol. Med. Sci.* **57**:9.

Howard, J. G., Hale, C., and Liew, F. Y., 1980, Immunological regulation of experimental cutaneous leishmaniasis. III. Nature and significance of specific suppression of cell-mediated immunity in mice highly susceptible to *Leishmania tropica*, *J. Exp. Med.* **152**:594.

Howard, J. G., Hale, C., and Liew, F. Y., 1981, Immunological regulation of experimental cutaneous leishmaniasis. IV. Prophylactic effect of sublethal irradiation as a result of abrogation of suppressor T cell generation in mice genetically susceptible to *Leishmania tropica*, *J. Exp. Med.* **153**:557.

Joseph, M., 1982, Phagocytes against helminths, in: *Clinics in Immunology and Allergy* (A., Capron, ed.), vol. 2, pp. 567–596, W. B. Saunders, Philadelphia.

Kappler, J. W., and Marrack, P., 1977, The role of H-2 linked genes in helper T-cell function. I. *In vitro* expression in B cells of immune response genes controlling helper T cell activity, *J. Exp. Med.* **146**:1977.

Katz, D. H., Hamaoka, T., and Benacerraf, B., 1973, Cell interactions between histoincompatible T and B lymphocytes. II. Failure of physiologic cooperative interactions between T and B lymphocytes from allogeneic donor strains in humoral response to hapten–protein conjugates, *J. Exp. Med.* **137**:1045.

Kurnick, J. T., Gronvik, K. O., Kimura, A. K., Linblon, J. B., Skoog, V. T., Sjosberg, O., and Wigzell, H., 1979, Long term growth *in vitro* of human T cell blasts with maintenance for specificity and function, *J. Immunol.* **122**:1255.

Liew, F. Y., and Simpson, E., 1980, Delayed-type hypersensitivity responses to H-Y: Characterization and maping of Ir genes, *Immunogenetics* **11**:255.

Liew, F. Y., Hale, C., and Howard, J. C., 1982, Immunological regulation of experimental cutaneous leishmaniasis. V. Characterization of effector and specific suppressor T cells, *J. Immunol.* **128**:1917.

Lin, L. Y. and Askonas, B. A., 1980, Cross-reactivity for different type A influenza viruses of a cloned T-killer cell line, *Nature (Lond.)* **288:**164.

Louis, J. A., Moedder, E., Behin, R., and Engers, H. D., 1979, Recognition of protozoan parasite antigens by murine T lymphocytes. I. Induction of specific T lymphocyte-dependent proliferative response to *Leishmania tropica*, *Eur. J. Immunol.* **9:**841.

Louis, J. A., Moedder, E., MacDonald, H. R., and Engers, H. D., 1981, Recognition of protozoan parasites by murine T lymphocytes. II. Role of the H-2 gene complex in interactions between antigen-presenting macrophages and *Leishmania*-immune T lymphocytes, *J. Immunol.* **126:**1661.

Louis, J. A., Zubler, R. H., Coutinho, S-G., Lima, B., Behin, R., Mauel, J., and Engers, H. D., 1982, The *in vitro* generation and functional analysis of murine T cell populations and clones specific for a protozoan parasite, *Leishmania tropica*, *Immunol. Rev.* **61:**215.

Loveland, B. E., Hogarth, P. M., Ceredig, R., and McKenzie, I. F. C., 1981, Cells mediating graft rejection in the mouse. I. Lyt-1 cells mediate skin graft rejection, *J. Exp. Med.* **153:**1044.

Mauel, J., Buchmüller, Y., and Behin, R., 1978, Studies on the mechanisms of macrophage activation. I. Destruction of intracellular *Leishmania enriettii* in macrophages activated by cocultivation with stimulated lymphocytes, *J. Exp. Med.* **148:**393.

Mauel, J., Behin, R., and Louis, J. A., 1981, *Leishmania enriettii*: Immune induction of macrophage activation in an experimental model of immunoprophylaxis in the mouse, *Exp. Parasitol.* **52:**331.

Mitchell, G. F., Curtis, J. M., Handman, E., and MacKenzie, I. F., 1980, Cutaneous leishmaniasis in mice: Disease patterns in reconstituted nude mice of several genotypes infected with *Leishmania tropica*, *Aust. J. Exp. Biol. Med. Sci.* **58:**521.

Mitchell, G. F., Curtis, J. M., Scollay, R. G., and Handman, E., 1981, Resistance and abrogation of resistance to cutaneous leishmaniasis in reconstituted Balb/c nude mice, *Aust. J. Exp. Biol. Med. Sci.* **59:**539.

Nacy, C. A., Meltzer, M. S., Leonard, E. J., and Wyler, D. J., 1981, Intracellular replication and lymphokine-induced destruction of *Leishmania tropica* in C3H/HeN mouse macrophages, *J. Immunol.* **127:**2381.

Nasseri, M., and Modabber, F. Z., 1979, Generalized infection and lack of delayed hypersensitivity in Balb/c mice infected with *Leishmania tropica* major, *Infect. Immun.* **26:**611.

Nguyen, T. V., Mauel, J., Etges, R., and Bordier, C., 1982, Identification of immunogenic proteins during the infection of different strains of mice by *Leishmania tropica*, *Fed. Proc.* **41:**582.

Paul, W. E., Shevach, E. M., Pickeral, S., Thomas, D. W., and Rosenthal, A. S., 1977, Independent populations of primed F_1 guinea-pig T lymphocytes respond to antigen pulsed parental peritoneal exudate cells, *J. Exp. Med.* **145:**618.

Ramalho-Pinto, F. J., Goldring, O. L., Playfair, J. H. L., and Smithers, S. R., 1976, T cell response to surface components of *Schistosoma mansoni*, in: *Biochemistry of Parasites and Host-Parasite Relationships* (H. Van den Bossche, ed.), pp. 291–298, North-Holland, Amsterdam.

Rittenberg, M. B., and Pratt, K. L., 1969, Anti-TNP plaque assay: Primary response of Balb/c mice to soluble and particulate immunogen, *Proc. Soc. Exp. Biol. Med.* **132:**575.

Ryser, J. E., Cerottini, J. C., and Brunner, K. T., 1978, Generation of cytolytic T lymphocytes *in vitro.* IX. Induction of secondary CTL responses in primary long term MLC by supernatants from secondary MLC, *J. Immunol.* **120:**370.

Sher, A., Jaimes, S., Simpson, A. J. G., Lazdins, J. K., and Meltzer, M. S., 1982, Macrophages as effector cells of protective immunity in murine schistosomiasis. III. Loss of susceptibility to macrophage-mediated killing during maturation of *S. mansoni* schistosomula from the skin to the lung stage, *J. Immunol.* **128:**1876.

Thomas, D. W., and Shevach, E. M., 1978, Nature of the antigenic complex recognized by T lymphocytes. V. Genetic predisposition of independent F_1 T cell subpopulations responsive to antigen pulsed parental macrophages, *J. Immunol.* **120:**638.

Weintraub, J., and Weinbaum, F. I., 1977, The effect of BCG on experimental cutaneous leishmaniasis in mice, *J. Immunol.* **118:**2288.

Ziegler, K., and Unanue, E. R., 1979, The specific binding of *Listeria* monocytogenes-immune T lymphocytes to macrophages. I. Quantitation and role of H-2 gene products, *J. Exp. Med.* **150:**1143.

Zubler, R. H., and Louis, J. A., 1981, Clonal assay for T helper-cells (T_n) and conditions for H-2 restricted linked cooperation between T_h and hapten-primed B cells. Application to the quantitation of hemocyanin or *Leishmania tropica*-specific T_h in primed mice, *J. Immunol.* **127:**1924.

Chapter 7

Immunobiology of African Trypanosomiasis

Georges E. Roelants and Margaret Pinder

Centre de Recherches sur les Trypanosomoses Animales (C.R.T.A.)
B. P. 454
Bobo-Dioulasso, Upper Volta, West Africa

I. INTRODUCTION

African trypanosomes are flagellated *Protozoa* that include the causative agents of important human and animal diseases (Table I) and are fascinating biologically.

The interest of African trypanosomes to immunologists is multiple. Each species comes in a multitude of different antigenic types owing to the expression of different surface-coat glycoproteins. These glycoproteins have homologous C-terminal and variable N-terminal portions and can be classified in subgroups similar to the isotypes of immunoglobulins. There is intense rearrangement of genes coding for these glycoproteins in the trypanosome genome. Epitopes on these glycoproteins are highly immunogenic, but antigenic variation often permits the trypanosome to escape the specific antibody response while antigen-antibody complexes may have pathogenic effects. Moreover, trypanosomes induce polyclonal activation of B lymphocytes that may lead to exhaustion of their potential. This activation and the intense anemia seen in the disease unbalance stem cell maturation. Finally, several immunosuppressive mechanisms are activated involving suppressor T cells and macrophages.

It is enthralling to see that most of the mechanisms unraveled by basic immunologists concerning the ontogeny, effector, and regulatory aspects of the lymphoid system are not only active, but operating concomitantly in trypanosomal infections. The outcome depends in part on the balance between the various immunoregulatory circuits and is very different in various host-parasite combinations. Besides, some host species are nonpermissive for certain trypanosomes without intervention of an acquired immune response. There is also great variation in the susceptibility of permissive hosts.

Table I. Classification of the Genus *Trypanosoma*, Section *Salivaria*[a]

Subgenus	Species	Subspecies	Important hosts	Pathogenicity
Duttonella	*vivax*	–	*Bovidae*	+
Nannomonas	*congolense*	–	*Bovidae*	+
Trypanozoon	*brucei*	*brucei*	*Bovidae*	±
		gambiense	Humans	+
		rhodesiense	Humans	+
	evansi[b]	–	Camel, horse, dog	+
	equiperdum[b]	–	Horse, donkey	+

[a]Two other *Trypanosomae* found in this review are from the genus *Stercoraria* subgenus *Herpetosoma*: the *lewisi* species found in rats and the *musculi* species in mice. They are not pathogenic.
[b]These two species are not cyclically transmissible by tsetse fly. *T. evansi* is transmitted mechanically by tabanids. *T. equiperdum* transmission is venereal.

These various aspects are considered in turn; they obviously do not constitute a comprehensive overview of African trypanosomes and trypanosomiasis. For further information, see Mulligan (1970) and the relevant chapters in Lumsden and Evans (1979) and Roelants and Williams (1982).

II. TRYPANOSOME ANTIGENS

In keeping with the bias of this review, which bears on immunologic aspects of trypanosomiasis, we divide the trypanosome antigens into those that are exposed on the surface of live trypanosomes and can interact with the immune system and those that are buried.

A. Variant-Surface Glycoprotein

One of the most well-known facets of African trypanosomes is their ability to change their antigenic character in the mammalian host. In the bloodstream, this characteristic leads to successive waves of parasites resulting from the sequential emergence of new variants and their destruction by the immune system. Thus antigenic variation constitutes a particularly efficient means of escaping host-immune defenses.

An in-depth coverage of the state of the art was published by Doyle (1977), and several reviews have appeared since (Vickerman, 1978; Cross *et al.*, 1980; Roelants and Williams, 1982). We cover only the most salient points or more recent advances.

1. Biochemistry

The surface antigenic character of a given live trypanosome in the bloodstream is entirely carried by a unique, variant-surface glycoprotein (VSG), which forms a dense outer coat 12-15 μm thick surrounding the whole surface of the parasite, including the flagellum (Vickerman, 1969; Vickerman and Luckins, 1969; Allsop *et al.*, 1971). It is the expression of new VSG with different antigenic characters that is responsible for the phenomenon of antigenic variation.

Work done on VSGs of *Trypanosoma* has shown that they consist of a single polypeptide chain having a rather uniform apparent molecular weight of 60,000-65,000, but widely different isoelectric points (Cross, 1975). They contain about 600 amino acid and 20 monosaccharide residues that usually appear to be segregated preferentially toward the C-terminal end of the polypeptide (Johnson and Cross, 1977, 1979). These general characteristics seem to hold true with minor variations for *T. congolense* (Rovis *et al.*, 1978; Onodera *et al.*, 1981) and *T. b. rhodesiense* (Olenick *et al.*, 1981).

Sequences of the N-terminal 20–35 amino acids of four *T. b. brucei* (Bridgen *et al.*, 1976) and two *T. congolense* (Onodera *et al.*, 1981) VSGs show considerable differences. Using cDNA derived amino acid sequences from *T. b. brucei*, VSG sequence homologies were detected near the C-terminal extending 75–105 amino acids into the protein (Matthyssens *et al.*, 1981; Rice-Ficht *et al.*, 1981). Moreover, analyses of nine such C-terminal amino acid sequences demonstrated the existence of at least two subsets (Rice-Ficht *et al.*, 1981). This finding suggests that VSG can be classified into subsets, on the basis of C-terminal homology, somewhat similar to the isotype classification of immunoglobulins. The closeness of sequences within a subset and the number of different subsets await more sequence data. Homologies were also found at the 3′-end region of mRNA coding for different VSGs of *T. b. brucei* (Majumder *et al.*, 1981).

The findings that the carbohydrate moiety are concentrated on the C-terminal end of the molecule and that concanavalin A (con A) binds to live trypanosomes only after treatment with trypsin, led to the conclusion that the C-terminus was close to the plasma membrane and hidden, while the N-terminus was outermost and exposed to the environment (Cross, 1979*a*). More direct studies have to be done, however, on the projection of VSGs to determine to which of the two main classes of surface α-helical proteins they belong—the antiparallel single-chain or the multiple-chain parallel molecules (Cohen and Phillips, 1981).

As stated at the beginning of this section, immunologists should keep in mind that the coat is very dense and that most of the VSG molecule on a live trypanosome is not available for interaction with the immune system. Monoclonal antibodies to VSG are very useful not only as exquisite reagents for the characterization of VSG (Pearson *et al.*, 1980, 1981; Roelants and Pinder, 1981;

Lyon *et al.*, 1981) but also for mapping, on a given VSG, both the exposed and buried determinants. Indeed, Pearson *et al.* (1981) found that of 10 different monoclonal antibodies obtained against two VSGs of *T. b. brucei* only two recognized exposed determinants on intact organisms, while the others bound only to acetone-fixed trypanosomes. It is thus possible that the intact organism presents very few different antigenic determinants to the environment.

2. *Repertoire*

Exposed varying epitopes on the VSG are the molecular basis of antigenic variation and thus one of the most important mechanisms of escape from the immune response. A crucial question is: how broad is the repertoire of VSG?, i.e., is there any hope that the immune system will cope?

Several questions should be asked. How broad is the total possible repertoire? Which portion of this repertoire is generally used by a trypanosome? Are some VSGs more frequently expressed than others? Are VSGs expressed in a defined order? What happens after cyclical passage through the fly?

The total number of different antigenic variants is unknown, but lies certainly in the thousands. The greatest number so far recorded as derived from a single clone is 101 during experimental infection of rabbits with *T. equiperdum* (Capbern *et al.*, 1977). This number is not the limit, however, and might reflect the exhaustion of the investigators rather than of the trypanosome.

The total number of possible variants obtainable in experimental infection may be less interesting that the number usually expressed during infection, particularly those found soon after fly bite. Clearly, if this number is limited, or if certain variants always appear before others, the possibility of vaccination may be entertained; if, on the contrary, hundreds of antigenic variants can appear at random, the task would be hopeless. There are differences of opinion in the interpretation of available data. It appears to us that there is no definite ordered sequence of appearance of VSGs on the trypanosome surface, although there may be a statistical chance for some to appear before others (Kosinski, 1980).

One of the important concepts on this subject has been introduced and developed by Gray (1965, 1966). From the analysis of antisera obtained after experimental infection in rabbits, he concluded that certain antigenic variants appear with a high frequency. They were called predominant antigens. Moreover, Gray claimed that any bloodstream-form variant reverted to one of a limited number of basic variants after cyclic passage through *Glossina.* One should be cautious in the interpretation of these studies, for they rely on uncloned stocks of trypanosomes expressing mixtures of variant antigens and on antisera to characterize them rather than on cloned organisms and biochemical analysis of VSG. Moreover, sometimes tests such as enzyme-linked immunosorbent assay (ELISA) are used that also detect antibody directed against irrelevant epitopes (Gray and Luckins, 1980).

Other studies have been made on the possibility of reversion to a basic type of variant antigen after passage of bloodstream forms through the fly. These have been reviewed in great detail by Doyle (1977). The most one can conclude is that experimental proof of consistent reversion to basic types is lacking. Moreover, no observation of basic types has been made on infected *Glossina* or mammalian hosts in field situations. If there is a reversion to the expression of a basic VSG after fly passage, one would expect that animals kept exposed to tsetse flies in the same area for long periods of time would develop immunity to trypanosome infection. Soltys (1955) apparently shows that *Zebus* treated prophylactically with trypanocides in an area of mild challenge with *T. congolense*-infected *G. austeni* could survive without treatment after a period of 2.5 years. In contrast, Wilson *et al.* (1975*a,b*, 1976), using various regimes of prophylaxy in Boran steers maintained in tsetse-infected areas found that no significant resistance developed during 2 years of continuous exposure to the disease in either calves (0-2 years of age) or in breeding herds (2-4 years of age). Gray (1970) studied the antigenic relationships of isolates of *T. b. brucei* collected from a herd of cattle kept in one locality for 5 years and concluded that, apart from one exception, there was little antigenic relationship among the trypanosomes isolated from year to year. Similarly, Dar *et al.* (1973) compared the antigenic types of *T. vivax* isolated in two different areas of West Africa. These workers detected inter- and intra-area cross-reactions and no evidence for a basic antigen specific for one area. It therefore appears unlikely that the number of variants appearing in the blood of tsetse-fly-exposed animals is really restricted, even in a given geographic area, unless the challenge is very mild, making the possibility of immunization remote.

The first form of trypanosomes penetrating the mammalian host is not the bloodstream form, but the metacyclic, which originates in the fly salivary gland. The metacyclic is the only fly stage that has a glycoprotein coat; workers have been interested in analyzing how variable the salivary gland trypanosome coat antigens are. Jenni (1977*a,b*) and Hudson *et al.* (1980), using *T. b. bucei*, and Nantulya *et al.* (1980*a,b*), using *T. congolense*, have concluded that variant-antigen heterogeneity is restricted within the fly salivary gland, but that upon injection into the host, metacyclics differentiate about immediately in bloodstream forms with heterogeneous VSG. Other investigators have reported that there are multiple variable antigenic types in the metacyclic population of *T. b. brucei* (Le Ray *et al.*, 1978; Barry *et al.*, 1979). Whether antigenic diversity already exists within the fly or immediately upon transmission, it appears again that the repertoire of variant types invading the host is large enough to preclude the possibility of practical vaccination. This could have been predicted by the lack of naturally acquired immunity by animals exposed to infected flies in the field.

3. Genomic Arrangements of VSG Genes

If the repertoire of VSG is too large to envision the likelihood of practical vaccination, an understanding of the mechanism of antigenic variation may at least give us a handle to interfere with the process. Following the recent developments in genetic engineering and their pioneering application to trypanosomes by Williams and co-workers (1978, 1979), a large number of publications have recently appeared on the genomic arrangement and expression of VSG genes.

There is a certain discrepancy between the results obtained by different groups working on *T. b. brucei* VSG clones. In a series of experiments, Williams and his co-workers have described considerable genomic rearrangement around the VSG genes, especially beyond the 3′-untranslated region of the gene. They also found two to four gene copies, which would suggest a diploid organism. Neither the number of genes nor their mode of arrangement is correlated with gene expression (Williams *et al.*, 1979, 1980, 1981; Marcu and Williams, 1981). In other cases, two VSG gene copies were found only in organisms expressing that particular VSG–one basic copy and one expression-linked copy (ELC). Besides the ELC, more complex rearrangements were also found to occur (Hoeijmakers *et al.*, 1980*a,b*; Borst *et al.*, 1980*a,b*; 1981; Boothroyd *et al.*, 1980, 1981; Agabian *et al.*, 1980; Pays *et al.*, 1981*a*). Recent evidence seem to show that the ELC of the VSG gene is the one transcribed (Pays *et al.*, 1981*b*). The relative importance of the two main types of rearrangements and the control of gene expression are under intense study. It is clear, however, from the data already available that the control is at the level of transcription and that the two different types of rearrangement do not correspond to the two different subsets of 3′-end sequence homologies described by Rice-Ficht *et al.* (1981). To make the story even more complicated, studies on eight polymorphic enzymes (Gibson *et al.*, 1980) have provided very strong evidence for diploidy and mating in trypanosomes (Tait, 1980).

B. Internal Antigenic Determinants

The variant glycoproteins have attracted most of the attention in this field of study, but it is evident that many other structures have antigenic determinants common to all members of a species, or even to all species of trypanosomes. Since they are not available for interaction with the immune system *in situ*, these structures have no relevance for the elicitation of a protective immune response. They may certainly obscure the detection of the relevant anti-VSG response when improper methods are used, such as ELISA and immunofluorescence on acetone-fixed cells. Antibodies against shared antigenic determinants are, however, of considerable diagnostic value when using these techniques (Voller *et al.*, 1975; Luckins, 1977; Vervoort *et al.*, 1978; Magnus *et al.*, 1978; Roffi *et al.*, 1978; Mangenot *et al.*, 1979; reviewed in Voller and De Savigny, 1981).

It is noteworthy that the epitopes present on the buried C-proximal end of the VSG itself have to be classified within the internal antigenic determinant group, as they are not available to the immune system *in situ* and present a certain degree of homology. VSG glycopeptides have been described by Barbet and McGuire (1978) that elicit in rabbits antibody cross-reacting with various VGS when tested by radioimmunoassay. They were shown to be attached at, or very close to, the C-terminus, to be variable in size and oligosaccharide composition, and to be of little relevance to the antigenicity of native antigen *in situ* (Cross, 1979*b*; Holder and Cross, 1981).

Common antigenic determinants may play a critical role in the pathogeny of the disease by triggering an immune response after the destruction of the parasite by VSG specific antibodies and by forming immune complexes (Section IV).

III. PROTECTIVE IMMUNE RESPONSE

A. Antibody

1. Protection by Infection and Cure

It has been amply demonstrated that domestic animals and laboratory rodents can be rendered resistant to a subsequent challenge of homologous trypanosomes, by either needle or fly infection, followed by drug- or self-cure, or by inoculation of irradiated trypanosomes (Laveran and Mesnil, 1902; Ehrlich, *et al.*, 1909; Browning and Gulbransen, 1936; Hornby, 1941; Fulton and Lourie, 1946; Wiesenhutter, 1970; Wellde *et al.*, 1975; Nantulya *et al.*, 1980*b*; Wellde *et al.*, 1981). All these workers and many others in this field have shown that such resistance is only effective against homologous trypanosomes and confers no protection against challenge by antigenically different trypanosomes, whether derived from the same or a different stock. Few investigators have attempted to immunize with more than one stock at a time. Scott *et al.*(1978) tried to immunize cattle with an 11-stabilate cocktail, but all the animals succumbed to challenge with 9 of the 11 stabilates 40 days later, however a massive challenge was used, i.e., 10^8-10^9 trypanosomes of each stabilate. It is therefore not surprising that exposure of cattle to natural challenge and periodic treatment with trypanocides does not lead to resistance, except in one instance in which the challenge was limited to one species of trypanosome and vector (Section 11.A.2). Only in the case of the stercorarian trypanosome *T. musculi* does immunity appear to be effective against different stocks (Viens *et al.*, 1975), but these trypanosomes do not appear to undergo antigenic variation (D'Alesandro, 1972). Furthermore, stercorarian trypanosomes exhibit many other biologic differences from the salivarian ones, and for these reasons antibody-mediated mechanisms dealing with them are not discussed further.

Proctection of laboratory rodents to homologous challenge can be transferred by serum (Mesnil and Brimont, 1909; Levaditi and McIntosh, 1910); the active principle is γ-globulin (Dodin *et al.*, 1962; Takayanagi and Enriquez, 1973). Protection is dependent on the presence of high-titer antibodies and not too large a challenge inoculum (Seed and Gam, 1966); indeed, this is the basis of the test for trypanosomal neutralization of infectivity (Soltys, 1957; Lumsden *et al.*, 1973). It is assumed that only antibodies against exposed determinants of VSG are effective, but this possibility awaits confirmation by using monoclonal antibodies. Surprisingly, passive transfer of antibodies to newborn rats in colostrum does not appear to give protection against homologous *T. b. gambiensie* challenge (Takayanagi *et al.*, 1978); the importance of such transfer in domestic animals needs investigation. Protection can be transferred to *T. b. rhodesiense* in mice by splenic B cells for 10 days after immunization, but T cells are ineffective (Campbell and Phillips, 1976).

2. Primary and Secondary Responses

The lack of in-depth analysis of the anti-VSG response in classes and subclasses of immunoglobulins and their biologic functions is particularly true for the secondary response. Studies of certain aspects have been reported, however.

The primary response of laboratory rodents and domestic cattle to VSG usually reaches a maximum 7–14 days after challenge, then declines slowly, responses being detectable for more than 100 days (Seed *et al.*, 1969; Wilson and Cunningham, 1972; Herbert *et al.*, 1980; Pinder *et al.*, 1983). Reports differ as to the duration of *in vivo* protection, ranging in cattle from 35 days (Wiesenhutter, 1970) to 14 months (Wellde *et al.*, 1973).

Investigators using the trypanosome agglutination test have reported that the primary response of rabbits to *T. b. gambiense* and rats or cattle to *T. b. brucei* is predominantly IgM (Seed *et al.*, 1969; Zahalsky and Weinberg, 1976; Pinder *et al.*, 1982*a*). Differentiation of IgM from IgG was by G-200 column separation or mercaptoethanol sensitivity. Furthermore, Takayanagi and Enriquez (1973) also report that after primary challenge of mice with *T. b. gambiense*, purified IgM is 200-fold more effective in a neutralization of infectivity assay and 20-fold more effective in an agglutination assay than is IgG.

Other investigators report that the primary response of mice and cattle to *T. b. brucei* or mice to *T. b. rhodesiense* is predominantly IgG (Campbell *et al.*, 1978; Nantulya *et al.*, 1979; Sacks and Askonas, 1980; Mansfield *et al.*, 1981). Nantulya and colleagues used a radioimmunoassay with purified VSG; the other three groups used fluorescence on fixed trypanosomes, which they claim is VSG specific. IgM and IgG are differentiated using class-specific reagents. It is possible that such assays are relatively insensitive in detecting IgM (Osler, 1971).

Sacks and Askonas (1980) also highlight the effect of trypanosome-induced immunodepression on the specific anti-VSG response, noting that IgG responses were strongly suppressed early in infections, whereas IgM was still produced. This finding may indicate that suppressor mechanisms act preferentially on T-helper cells, since the IgM response to VSG appears to be T-independent and occurs equally in nu/+ and nu/nu mice, whereas only nu/+ are able to mount an IgG response (Campbell *et al.*, 1978; Mansfield *et al.*, 1981). As might be predicted from the results of Takayanagi and Enriquez (1973), immunized nu/nu mice withstand homologous rechallenge (Campbell *et al.*, 1978). In contrast, immunization of B-cell-deficient mice does not induce protection (Campbell *et al.*, 1977). The agglutination response of mice to *T. b. gambiense* is reduced by pretreatment with dextran sulfate or carrageenan, agents toxic to macrophages (Ishizaka *et al.*, 1977) and is therefore presumably macrophage dependent (Oka *et al.*, 1981).

Few reports are available on the secondary response to deliberate reinfection with the same stock or clone. It appears, however, that the kinetics, peak titer of antibody, and duration of detectable activity are similar to those after primary challenge in rats infected with *T. b. brucei* Zahalsky and Weinberg, 1976), in mice inoculated with *T. b. gambiense* homogenate (Oka *et al.*, 1981), or in cattle infected with *T. b. brucei* (Pinder *et al.*, 1983). The secondary response in rats was almost totally mercaptoethanol resistant, whereas we have found that in cattle IgM still predominates in the secondary response, although appreciable IgG is produced.

It has been demonstrated that cattle infected with *T. b. brucei* may show two peaks of antibody to the infecting organisms, presumably because of reappearance of organisms with the same or similar VSG (Nantulya *et al.*, 1979). In these instances, the second peak may be of higher titer than the first. Musoke *et al.* (1981) reported that VSG-specific antibodies in cattle were found in IgM, IgG_1 but not IgG_2 subclasses after both primary and secondary challenge. We have observed both IgG_1 and IgG_2 antibody production (Pinder *et al.*, 1983). (Musoke and colleagues also demonstrated that in the first peak, IgM antibodies are 10-fold more efficient at neuralization than are those of the IgG_1 class, confirming results of previous workers; the reverse was true of the second peak. Few studies have examined the cellular requirements of the secondary response; nevertheless, because it can be enhanced in the case of *T. b. brucei* in mice by pretreatment with carrageenan and this enhancement can be blocked by anti-T-cell serum, it is suggested that a memory pool of T cells has been elicited by priming (Oka *et al.*, 1981).

3. *Trypanosome Elimination*

One of the most studied biologic functions of anti-trypanosome antibodies is the enhancement of phagocytosis. Both attachment and ingestion of parasites of the *brucei* species by rodent or rabbit macrophages *in vitro* are greatly enhanced by homologous antiserum (Mesnil and Brimont, 1909; Lumsden and Herbert, 1967; Takayanagi *et al.*, 1974*a*; Stevens and Moulton, 1978; Cook, 1981). Trypanosome agglutinating antibodies may be responsible for this increase in phagocytosis (Takayanagi *et al.*, 1974*b*), although this activity is only present for 2 or 3 weeks after infection (Tizard and Soltys, 1970; Stevens and Moulton, 1978). Takayanagi *et al.* (1977) also report that trypanosome antigens involved in macrophage binding are exclusively those of the VSG. Contradictory results were obtained by Cook (1981), who reports that using serum from rabbits infected with *T. b. brucei* attachment of trypanosomes to cultured peritoneal cells did not show a correlation with agglutinating antibodies when successive variant subpopulations were used. Cook also reported that such antibodies persisted for several weeks postinfection. These differences may be attributable to the use of undiluted serum and to the high parasite:macrophage ratios employed; perhaps under such conditions cross-reacting determinants or common antigens are important (Barbet and McGuire, 1978; Cross, 1979*b*).

The role of complement and the need for macrophage activation by trypanosomes in the *in vitro* attachment and ingestion is equivocal. Most workers report similar results with immune or normal macrophages, and addition of complement or heat inactivation of the antiserum is without effect. Contradictory results are reported by Stevens and Moulton (1978) working with *T. b. brucei* in the deer mouse (*Peromyscus maniculatus*); these may reflect differences in the class of antibodies present or species differences, or both factors.

A series of studies on the clearance of radiolabeled *T. b. brucei* in mice have clarified these *in vitro* studies by highlighting the mechanisms that eliminate trypanosomes *in vivo* (Holmes *et al.*, 1979; Macaskil *et al.*, 1980). In nonimmune animals, 60 min postinjection, 70% of trypanosomes remain in the circulation, whereas in immune animals 60% of the inoculum is taken up by the liver, and less than 3% remains in the circulation. This clearance from the circulation is antibody-dependent; immune serum passive transfer can result in as high levels of clearance as seen in actively immunized animals. Unfortunately, no investigation of VSG specificity was made. It was demonstrated, however, that *in vivo* activation of macrophages by trypanosomes in the absence of antibody did not lead to increased clearance and that the nonspecific macrophage stimulants bacillus Calmette-Guerin (BCG) and *Clostridium parvum* only marginally increased clearance values. Therefore, *in vivo* macrophage activation does not appear to be a prerequisite for clearance.

The importance of complement components in the *in vivo* control of parasitemia is equivocal. Faghihi Shirazi *et al.* (1980) demonstrated that C3 depletion does not alter the parasitemic profile of *T. b. brucei* in mice; successive waves of parasitemia appear to be controlled by IgM (Sacks and Askonas, 1980), and mouse IgM is a very inefficient activator of C3 (Klaus *et al.*, 1979). The *in vivo* relevance of complement-mediated trypanosome lysis so readily demonstrable *in vitro* microscopically (Lumsden *et al.*, 1973) or metabolically (Desowitz, 1959; Diggs *et al.*, 1976) is therefore questionable. Macaskil *et al.* (1980) were able to demonstrate reduced trypanosome clearance in C3-depleted mice. It is possible that the residual activity without C3 is sufficient to control parasitemia in a normal infection.

4. Concluding Remarks

The most important factors in controlling parasitemia in mice appear to be anti-VSG IgM and phagocytosis by macrophages; anti-VSG IgG antibody ceases to be produced early in infection as a result of immunosuppression. The bovine immunosuppression is less profound, and probably both IgM and IgG anti-VSG antibodies continue to be produced. Mice almost always die from the infection; if drug-cured, however, they survive a subsequent homologous challenge, highlighting the importance of a secondary response. Possibly, secondary responses against reoccurring VSG is a major contributing factor to the survival of the bovine host.

B. Cell-Mediated Immunity

The cell-mediated mechanisms directed specifically against trypanosome determinants and their eventual relevance for protection are less well characterized than is the humoral response.

Tizard and Soltys (1971) found that rabbits infected with *T. b. brucei* or with *T. b. rhodesiense* showed signs of both immediate (Arthus type) and delayed type hypersensitivity (DTH). Skin thickness and histologic examination were the two criteria used. Peak Arthus reaction occurred at 2 weeks postinfection and peak DTH at 4 weeks. The reactions were not strain specific and no effort was made to show that they were trypanosome-specific. In rabbits infected with *T. congolense* and drug cured 6 weeks later, immediate hypersensitivity of the Arthus type was found upon intradermal challenge with sonically lysed homologous trypanosomes 2–10 weeks postinfection (Mansfield and Kreier, 1972*a*). There was no evidence for DTH nor did macrophage inhibition factor (MIF) production occur *in vitro*.

Working in mice with *T. b. rhodesiense*, Finerty *et al.* (1978) reported evidence for DTH, but the specificity of the results for trypanosomes was not demonstrated.

Gasbarre *et al.* (1980) studied cell-mediated immunity to *T. b. brucei* in A/J mice using a T-lymphocyte-dependent proliferation response *in vitro*. The subcutaneous priming was effective only with complete Freund's adjuvant (CFA) and live trypanosomes. The reaction was expressed by inguinal and periaortic lymph nodes cells *in vitro* in the presence of homologous live trypanosomes. Absence of stimulation by fetal calf serum served as a specificity control, but no heterologous trypanosome was used. Stimulation was strong, with indexes around 100, 5, and 10 days after priming, variable at day 15, and low to null afterward, because of an unresponsiveness that appeared to affect both T lymphocytes and macrophages (Gasbarre *et al.*, 1981*a*). Since injection of CFA was compulsatory for priming, it is difficult to assess the importance of this *in vitro* T-cell reactivity in natural infections.

An *in vitro* proliferative response to sonicated trypanosome was also analyzed in peripheral blood leukocytes (PBLs) of Boran and Hereford steers exposed to a clone of *T. congolense* (Emery *et al.*, 1980*a*). PBLs from animals infected for 7 weeks, treated with Berenil, and tested 8 weeks later showed a moderate (stimulation indexes of 10) *in vitro* proliferative response, but PBLs from infected animals that were not cured did not react. A weak reaction was also seen 2 weeks after priming with formalin-fixed trypanosomes emulsified in CFA. Neither nonviable nonfixed trypanosomes in CFA nor isolated VSGs primed these cattle.

It appears to us that, at this stage, certain means have been developed to test the activation of trypanosome-specific T cells, but that the importance of these cells in conferring protective immunity and their specificity in terms of species and antigenic types of trypanosomes is unknown. It should be emphasized that in immune cell transfer expreiments T cells alone cannot confer protection (Campbell and Phillips, 1976).

Another cellular reaction is the development of a skin reaction after a tsetse bite, starting about 5-7 days after inoculation of trypanosome by *Glossina* and peaking at about 9-12 days. The histologic appearance of this chancre, as it has been termed, has been described by different investigators as that of a polymorphononuclear or macrophage and lymphocyte infiltration, or mixtures of those cells (reviewed in Emery *et al.*, 1980*b*; Emery and Moloo, 1981). There is some indication that cattle do not develop a chancre upon homologous rechallenge (Emery *et al.*, 1980*b*), and there is evidence that animals of low susceptibility to trypanosomiasis (Section VI) develop smaller chancres when bitten by an infected fly (G. A. Kol *et al.*, in preparation).

IV. DELETERIOUS EFFECTS OF THE IMMUNE RESPONSE

The host antibody response to VSG as well as other trypanosome antigens during the successive waves of parasitemia is often of high titer and might have a pathogenic effect.

Low complement levels are a feature of trypanosomiasis in humans and in experimental animals and have been attributed, at least in part, to immune-complex formation (Greenwood and Whittle, 1976; Nielsen and Sheppard, 1977; Fruit *et al.*, 1977). These complexes may be either circulating (Fruit *et al.*, 1977; Lindsley *et al.*, 1981; Rickman *et al.*, 1981) or deposited, causing damage to kidneys (Nagle *et al.*, 1974; Van Marck and Vervoort, 1980) and other tissues (Lambert and Galvao-Castro, 1977; Galvao-Castro *et al.*, 1978). These last workers have studied in detail the relationship between the host-immune response and tissue lesions, especially those in striated muscle, associated with *T. b. brucei* infection of normal and immune-deficient mice. Outbred OF_1 and Balb/c mice infected with 10^4 *T. b. brucei* (T227) mounted a good immune response against the VSG, as detected by agglutination and immunofluorescence. These mice had large amounts of circulating immune complexes and deposits in tissue lesions containing IgM, IgG, and VSG. Newborn mice showed high parasitemia, but no antibody and no inflammation foci. Mice irradiated with 400 rad also had high parasitemia, low levels of antibody, and only occasional inflammatory foci. Congenitally athymic nu/nu mice behaved the same as irradiated mice. Lesions were enhanced in nu/nu mice receiving either anti-trypanosome antibody, spleen cells, or T lymphocytes from normal mice. The authors concluded that striated muscle lesion in the acute phase of experimental trypanosomiasis is dependent on the immune response of the host against parasite antigen and involves local formation of immune complexes. This conclusion is supported by the finding of Gasbare *et al.* (1981*b*) comparing CBA and CBA/N mice infected with *T. b. brucei*. CBA/N are deficient in B-cell maturation and antibody production (Scher, 1981); nevertheless, they survive longer after trypanosome infection, and the only difference the authors could detect between the two strains was at the level of immune complexes, which are more abundant in CBA mice. In contrast, Biozzi low-antibody-responder mice (Mouton *et al.*, 1981) show 10-fold higher parasitemia and increased mortality with *T. congolense* Dinderesso stock as compared with high-antibody responders (Pinder, 1983).

It has been suggested that autoimmunity may contribute to the pathogenicity of African trypanosomiasis; indeed, autoantibodies have been found against a rabbit liver antigen (Seed and Gam, 1967). immunoglobulin (Klein *et al.*, 1970), liver, brain, kidney, and heart tissues (MacKenzie and Boreham, 1971; Mansfield and Kreier, 1972*b*), spleen cells (Mitry-Martin *et al.*, 1979), thymocytes (Kobayakawa *et al.*, 1979), and erythrocytes (Kobayakawa *et al.*, 1979; Rickman *et al.*, 1981). However, their presence is variable and depends on approporiate host–parasite combinations (e.g., Klein *et al.*, 1970), and transfer of such antibodies does not induce pathology or death in the recipient animal (Mansfield and Kreier, 1972*b*). No evidence was obtained for cell-mediated autoimmunity (Mansfield and Kreier, 1972*b*). It appears that most investigators are of the opinion that although autoimmunity may occur in African trypanosomiasis, it does not play an important role in the pathogeny of the disease.

V. EFFECTS OF TRYPANOSOME INFECTION ON THE LYMPHOID SYSTEM

Three major effects of trypanosome infection on the lymphoid system have been well demonstrated: hypergammaglobulinemia, polyclonal activation, and immunodepression.

A. Hypergammaglobulinemia and Polyclonal Activation

Hypergammaglobulinemia involving IgM, and to a lesser extent IgG, has been consistently found in trypanosome infection of cattle (Gidel, 1962; Luckins, 1972; Clarkson and Penhale, 1973), monkeys (Houba *et al.*, 1969), humans (Greenwood, 1974*a*; Mattern *et al.*, 1978), mice (Hudson *et al.*, 1976), and sheep (MacKenzie *et al.*, 1979). In humans it is used as an important diagnostic test for sleeping sickness; in addition, elevated IgE levels are found (Herbert *et al.*, 1980).

These increased levels of immunoglobulin consist of antibodies directed against both trypanosome and unrelated antigens, including autoantibodies. The balance between a specific response and polyclonal B-lymphocyte activation appears to vary from observation to observation. For instance, Corsini *et al.* (1977) have reported than less than 10% of the increased immunoglobulin produced in mice infected with *T. b. brucei* was trypanosome specific, while Musoke *et al.* (1981) claim that all excess IgM and IgG_1 appearing in *T. b. brucei*-infected cattle could be absorbed by a mixture of variant-antigen types of trypanosomes.

Hypergammaglobulinemia is accompanied by extensive proliferation of T, B, and null cells in mice spleen and bone marrow after infection with *T. congolense* (Morrison *et al.*, 1978) and *T. b. brucei* (Mayor-Withey *et al.*, 1978). Increases in the number and size of lymphocytic follicules and germinal centers and the appearance of large numbers of plasma cells in lymph node and spleen have been reported in goats and cattle infected with *T. congolense* or *T. b. brucei* (reviewed in M. Murray *et al.*, 1980) and *T. vivax* (Masake, 1980; Masake and Morrison, 1981).

B. Immunodepression

Considerable evidence has been accumulated showing that trypanosome-infected mice have impaired antibody responses to a large variety of soluble and particulate antigens. Lymphocytes from such mice give a lower response to B- and T-cell mitogens or allogeneic cells *in vitro* and do not generate cytotoxic

T cells; infected mice also present a delay in allogeneic skin graft rejection (reviewed in Roelants and Williams, 1982). Reports have also appeared on immune depression in cattle, but the phenomenon was less profound and less consistent (Holmes *et al.*, 1974; Scott *et al.*, 1977; Sollod and Frank, 1979; Rurangirwa *et al.*, 1978, 1979, 1980*a*, *b*; Masake *et al.*, 1981); the same is true in man (Greenwood *et al.*, 1973; Greenwood, 1974*b*).

Several groups have studied the mechanism of immunosuppression in murine experimental trypanosomiasis. Since the mouse lymphoid system is by far the best known, its study is particularly interesting in understanding trypanosome–lymphoid system interaction. One must keep in mind, however, that the mechanism unraveled may by fully operative only in mice, and even then only in the spleen. It is simpler to present the work of each group separately, since each has it own idiosyncrasies.

1. The Yale Group

The Yale group works with spleen cells of male, 8–12-week-old CBA/CaJ mice (and sometimes Balb/c and C57B1/6) infected with a stock of *T. b. brucei* (S 42), giving a prepatent period of 3 days and proving fatal in 30–35 days. These workers first demonstrated a progressive decrease in the response of spleen cells from infected mice to concanavalin A (con A), phyto-hemagglutinin (PHA) and lipopolysaccharide (LPS) (Jayawardena and Waksman, 1977). Removal of glass-adherent cells from this population restored LPS stimulation to normal and PHA stimulation to half-normal, but had little effect on con A stimulation. Spleen cells of infected mice suppressed the mitogen response of cells from uninfected mice from 11 days postinfection. This suppressive activity was absent from the spleen of infected athymic nude mice, but was insensitive to treatment with anti-Thy 1 serum and complement.

This group then studied the plaque-forming cell (PFC) response in the spleen of infected mice challenged with sheep red blood cells (SRBC) or DNP–Ficoll, a T-dependent and a T-independent antigen, respectively (Eardley and Jayawardena, 1977; Jayawardena *et al.*, 1978). They found an early phase at $\leqslant 5$ days of enhancement of the response to SRBC, but not to DNP–Ficoll. After day 6, infected spleen cells actively suppressed the PFC response of normal spleen cells to both SRBC and DNP–Ficoll. Both early enhancing and late suppressive activities were sensitive to treatment with anti-Thy 1 and complement, in contrast to the mitogen system, and mediated by nylon-wool-purified, i.e., nonadherent, cells. The suppressor cells were shown to be $Ly1^+, 2.3^+$ and did not appear in animals infected 20 weeks after adult thymectomy. The enhancing cells were of the $Ly1^+,2.3^-$ phenotype.

The Yale group concludes that immunodepression in trypanosome-infected mice is mediated primarily by a potent immunosuppressor T cell. Since glass-adherent cells also have suppressor activity, these workers hypothesize that the suppressor T cell stimulated by trypanosomes may release factors having an affinity for macrophages that become suppressors for T- and B-responses.

2. The Mill Hill Group

Most of the experiments performed by the Mill Hill group involved spleen cells of 4–8-month-old CBA/H and (CBA/H × C57B1/6) F_1 mice bred under specific pathogen-free conditions and infected with a clone (NIM2) of a stock of *T. b. brucei* (S42), giving a prepatent period of 4 days and proving fatal in 19–20 days in the CBA/H and in 20–30 days in (CBA/H × C57B1/6) F_1 mice.

These workers find that *T. b. brucei*-infected mice exhibit a lower response of spleen cells to LPS and to PHA (Corsini *et al.*, 1977) as well as a suppression of the functions of T-helper cells, T-dependent and -independent B-memory cells, and mixed lymphocyte reaction (MLR) responding cells (Askonas *et al.*, 1979). The LPS response was partially restored by treatment with anti-Thy 1 serum and C, in contrast to the Yale group finding, and macrophages isolated from suppressed populations could suppress the LPS response of normal spleen cells (Corsini *et al.*, 1977). Again, in contrast to the Yale group, suppressor cells were found to adhere to nylon-wool, and nude mice spleen cells were also unresponsive to LPS, although the inhibition occurred later than in normal mice (Askonas *et al.*, 1979). Unfortunately, nude mouse spleens were not analyzed for the presence of T cells after *T. b. brucei* infection, nor was an active suppressor assay performed, as by Jayawardena and Waksman (1977). Askonas *et al.* (1975) nevertheless concluded that immune depression was caused by suppressor T cells early in infection but that later depression was independent of T cells and probably attributable to intrinsic unresponsiveness of B and T cells. Furthermore, analysis of spleen colony-forming cells in the bone marrow and the restoration of immune function after Berenil treatment led Clayton *et al.* (1980) to conclude that lymphocyte progenitors are also affected by trypanosomes, but that some of the early bone marrow stem cells escape destruction or inactivation and rapidly repopulate peripheral lymphoid organs upon removal of the parasite.

Microsomal membrane fractions from *T. b. brucei* appear able to mimic the immunosuppression of a primary response to SRBCs and the mitogenic effect of living trypanosomes in (CBA/H × C57B1/6) F_1 mice (Clayton *et al.*, 1979*a*; Sacks *et al.*, 1980). Human peripheral blood lymphocytes could be stimulated by a similar fraction extracted from *T. b. rhodesiense* (Selkirk *et al.*, 1981). Suppression of responses to mitogen, MLR, and other functions examined during infection, i.e., all T-cell functions, were not tested.

Although the first article in this series (Corsini *et al.*, 1977) showed that peritoneal exudate cells of infected mice could suppress the LPS response of normal spleen cells, the possible role of the macrophage was not highlighted in many subsequent papers. Recently, Grosskinsky and Askonas (1981) reinvestigated the role of the macrophage in immunosuppression, always using the NIM2 clone of *T. b. brucei* (S42) in (CBA/.H × C57Bl/6) F_1 mice. They analyzed the effect of injection of peritoneal exudate cells containing engulfed trypanosomes on the primary PFC response to SRBC of syngeneic recipient mice. If SRBC were injected at the same time as macrophage-containing trypanosomes, a slight increase (2.4- to 5-fold) of IgG– but not of IgM-PFC was seen. SRBC injected 4 days after transfer showed a 2.5-fold IgM-PFC decrease in one experiments. IgG-PFC in control mice varied, however, by more than 7-fold in three separate experiments; it is difficult to agree with these investigators' conclusion that macrophages are key cells in mediating parasite-induced immune dysfunction. The effect of macrophages containing trypanosomes on other immune responses (especially T) or the appearance of suppressor T cells in the spleen of recipient mice was not investigated.

The bulk of the evidence of this series of experiments points again to involvement of T-suppressor cells and macrophages, as was found by the Yale group, but the direction and sequence of interactions between the cells involved is not clear.

3. The Louisville Group

The experiments performed by the Louisville group were mainly done with spleen and lymph node cells of male C57B1/6 mice weighing 18–21 g infected with a stock (EATRO 1886) of *T. b. rhodesiense*, which gave a prepatent period of 3 days and a mean time to death of 38 days.

These investigators first examined the effect of infection on the *in vivo* response of mice to T-independent (SIII, PVP) and T-dependent (SRBC, TNP-KLH) antigens measured by the enumeration of direct (IgM) PFC (Mansfield and Bagarsa, 1978). The ratio of PFC to the T-independent antigens was enhanced three- to fourfold throughout most of the infection until just before death. This finding is in contrast to the immunosuppression found for another T-independent antigen (DNP-Ficoll) by the Yale group. PFC to both T-dependent antigens were completely abrogated after day 5 of infection, in contrast to the results of Grosskinsky and Askonas (1981), who found little or no suppression of direct PFC to SRBC. Carrier priming with KLH before infection completely abrogated the suppression observed to primary challenge with TNP-KLH. From these results, Mansfield and Bagarsa concluded that intrinsic B-cell function remains intact throughout infection, that B-cell failure shortly precedes death, and that unresponsiveness is associated with the function of T-helper cells. T-cell responsiveness was not evaluated, nor were indirect (IgG) PFC.

Spleen immunosuppression was investigated using a primary *in vitro* antibody response to SRBC with enumeration of direct PFC after 5 days of culture

(Wellhausen and Mansfield, 1979). Spleen cells of *T. b. rhodesiense*-infected mice failed to generate a primary *in vitro* PFC response to SRBC and suppressed this capacity in spleen cells from uninfected syngeneic mice. The suppressor activity was associated with cells able to adhere to nylon-wool and to plastic and was insensitive to treatment with anti-Thy 1 serum and complement. Full suppression was obtained using unfractionated spleen cells 3 days after infection and only 1 day after infection when using adherent cells. The ratio of adherent cells to B cells in the responding population had to be roughly 1:1 in order to obtain complete suppression, but some suppression was already detected at a ratio of 1:10. Lymph node cells from 20-day-infected mice were not suppressed, nor could they or thymocytes suppress the response of normal spleen cells. However, in a further publication, lymph node lymphocytes appeared to be suppressed in their primary response to SRBC within 20 days of infection, but did not suppress the response of normal lymph node cells (Wellhausen and Mansfield, 1980*a*). Unfortunately, suppressor activity of spleen cells from infected animals on these lymph node cells was not tested.

Suppression of target cells seems to require cell-to-cell contact, for it is not operative when effector and target cells are separated by a cell-impermeable membrane or when supernatants of infected spleen cells are used (Wellhausen and Mansfield, 1980*b*). The suppressor population did not lyse target cells. Suppressor activity was not blocked by irradiation at 1500 rads nor by treatment with mitomycin C, but was inhibited by exposure of spleen cells to silica particles. Wellhausen and Mansfield (1980*b*) therefore conclude that the suppressor activity in African trypanosomiasis of the mouse is attributable to a suppressor macrophage. They reinforced the conclusion that suppression is T independent by showing that spleen cells from infected congenitally athymic nude mice do not respond to LPS and suppress the LPS response of spleen cells from uninfected nude mice (Mansfield *et al.*, 1981). Unfortunately, they did not look for T cells in nude mouse spleens after infection. These results are consistent with those of Clayton *et al.* (1979*b*) and at variance with those of Jayawardena and Waksman (1977).

However, Wellhausen and Mansfield (1980*b*) also showed that indomethacin, which inhibits the activity of T-independent macrophage-mediated suppression (Kempt *et al.*, 1980), failed to restore responsiveness to infected spleen cells. This finding suggests that the phenomenon is not restricted to macrophages alone.

4. The Victoria/Bobo-Dioulasso (ex Nairobi) Group

Unlike the results presented for the Yale, Mill Hill, and Louisville groups, which were based mainly on B-cell assays, the series of experiments conducted by the Victoria/Bobo-Dioulasso group focused essentially on absence of reactivity or active suppression of T-cell function in a stock of *T. congolense* (5E-12) known to cause subacute infections of different lengths in various mouse strains (Morrison *et al.*, 1978).

This group first showed that spleen cells from infected 6-8 week old CBA mice did not respond *in vitro* to con A, LPS, or allogeneic mixtures of cells differing at major (H-2) or minor (Mls) lymphocyte-stimulation loci. In addition, cytotoxic lymphocytes were not generated in allogeneic spleen cell mixtures (Pearson *et al.*, 1978). Extensive dose- and time-curve analysis showed that the reduced stimulation occurs over a wide range of conditions and is not merely caused by a shift in either mitogen concentration or in time required to trigger DNA synthesis. T cells are reduced in the spleen of *T. congolense*-infected mice, but limiting dilution analysis of MLR-responding cells showed that this did not account for the lack of stimulation of spleen cells from infected mice (Roelants *et al.*, 1979*a*). Moreover, spleen cells from infected mice did not stimulate allogeneic cells in MLR. This was probably due to the presence of a suppressor cell in the stimulator cell suspension, for purified Ia^+ B cells from infected mice had the same stimulatory capacity as, or an even higher capacity than, those from uninfected mice (Roelants *et al.*, 1979*a*).

In cultures mixing syngeneic cells from uninfected and infected mice, it was found that the latter inhibited the stimulation of the former by LPS, con A, or MLR. The responder populations used were from C3H and C57B1/6 strains (Pearson *et al.*, 1979). The effect was only partially alleviated by treating the suppressor population with mitomycin C. To obtain a suppressive effect on MLR, infected spleen cells had to be added at the onset, or at the latest at 48 hr of culture. The suppressor cells did not simply inhibit the uptake on consume or dilute the thymidine analogue, [5-^{125}I]-2-deoxyuridine, used to label the cells during the last 18 hr (78-96 hr) of the culture (Pearson *et al.*, 1979). Suppression of the response of 500,000 spleen cells in microplate wells could be inhibited by suppressor populations containing as few as 2000 T cells, 2000 macrophages, and 300 B cells.

Attempts to identify the cell(s) responsible for suppression did not give straightforward results, because various cell-enrichment or -depletion procedures, although giving the results expected from the literature with spleen cells from uninfected mice, did not operate with cells from infected mice (Pearson *et al.*, 1979). Thus, macrophages were divided in adherent and nonadherent populations. Most T cells were not killed by treatment with anti-Thy 1 serum and C.. All T blasts appeared in the effluent of nylon column, but the vast majority of small T cells were nylon bound. Finally, iron carbonyl treatment removed all macrophages, but all T cells as well. The bulk of the data appears to show that suppressor cells in these experiments are a subpopulation of Thy 1^+ cells bearing a low density of this marker. Macrophages might play a secondary role, although in some experiments strong suppression was obtained in the absence of macrophages from the suppressor population. Unfortunately, the number of macrophages in the potentially responding population after several days of culture was not ascertained.

In further experiments on infected CBA and C57B1/6 spleen cells exposed to con A, LPS, or MLR or used for active suppression of the same reactivities of normal spleen cells, it was found that the early kinetics of unresponsiveness and suppression were identical and paralleled the appearance of parasites in the peripheral blood (Roelants *et al.*, 1979*b*). Likewise, these reactions were restored by treatment with the trypanocidal drug Berenil. This was accompanied by a gradual return to normal spleen cell composition and architecture with rapid repopulation of the white pulp and widespread formation of germinal centers (Roelants *et al.*, 1979*b*).

Various inbred strains of mice show a rank order in their susceptibility to the clone 13E-1 of the stock of *T. congolense* used, which is more easily expressed by their mean time to death: A/J, 35 days; C3H, 43 days; AKR, 73 days; and C57Bl/10, 109 days. The immunodepression in the spleen cells of infected mice for LPS, con A, and MLR as well as that of the active suppressor activity followed the same rank order in its kinetics and degree of suppression (Pearson *et al.*, 1982).

Finally, it was shown, in agreement with Wellhausen and Mansfield (1979), that all the assays used in this series of experiment did not detect unresponsiveness or suppressive activities in lymph node, PBL, thymus, or bone marrow of *T. congolense*-infected mice (Kar *et al.*, 1981). Only LPS responses were depressed in lymph node cells within 22 days' infection (spleen cell responsiveness was more than 90% suppressed within 9 days in a parallel experiment). It was also shown that lymph node cell reactivities could be suppressed by spleen cells from infected mice and thus that those cells were not intrinsically resistant to suppression. One explanation is that *T. congolense*, which does not invades tissues and multiplies in lymph nodes, does not depress or generate it does so in the spleen, which is more vascularized. This would imply that the spleen suppressor cells are not recirculating. However, *T. b. brucei*, which invades tissures and multiplies in lymph nodes, does not depress or generate suppressor activity in these anatomic locations either (Kar *et al.*, 1982). It appears that lymph nodes lack the precursor or the combination of cells needed to generate suppressor activity under the influence of African trypanosomes.

5. Concluding Remarks

The results presented are sometimes contradictory and have led investigators to different conclusions. We believe that this is attributable to the idiosyncrasies of the various *in vitro* systems used favoring one or another aspect of the suppressor phenomenon, rather than to radically different suppressor systems functioning in different strains of mice or for different antigens and activation assays. In other words, despite discrepancies, we believe that everybody is right, i.e., that both suppressor T cells and macrophages are involved. It is positively vexing that after years of work the precise sequence of events leading to immunosuppression in mice—let alone humans or cattle—remains to be unraveled.

VI. SUSCEPTIBILITY TO TRYPANOSOMIASIS

Trypanosomes, like most parasites, have a limited host range; furthermore, the level of parasitemia and severity of disease vary considerably among the different species of the host range and also, to a more limited extent, among different strains or individuals of the same species. The mechanisms responsible for this spectrum of sensitivity can be broadly divided into two types, *viz.* innate and immune. Innate mechanisms are obviously of importance in host-parasite combinations in which trypanosomes fail to grow even when injected in large numbers (so-called nonpermissive or heterologous hosts). That such hosts exist is well documented, but the basis of nonpermissiveness is not well characterized. However, since there is such a marked contrast between the efficiency of nonpermissive hosts and the inefficient or misdirected immune responses in susceptible hosts, we will discuss possible mechanisms. The existence of variation of susceptibility within the permissive host range of a given trypanosome is not so clear cut, and the evidence for its existence in cattle and mice is presented.

A. Nonpermissive Hosts

1. Toxic Serum Factors

Certain hosts appear to be nonpermissive because of factors present in their sera. The best studied example is the resistance of humans to infection by *T. b. brucei*. As early as 1902, Laveran and Mesnil reported that mice could be cured from infection with *T. b. brucei* by inoculation of small amounts of normal human serum and concluded that this property of the serum is associated with human resistance to this trypanosome. Later workers have demonstrated that human serum is capable of lysing trypanosomes *in vitro* (Yorke *et al.*, 1930). Two different factors may be involved, one active *in vivo* and one *in vitro* (Hawking *et al.*, 1973). Rifkin (1978*a,b*) studied the *in vitro* reaction in some detail and reported that lysis does not occur at temperatures below 25°C, that one of the first measurable effects of human serum is the loss of normal permeability properties of the trypanosome outer membrane, and that the active principle appears to be high-density lipoprotein (HDL). Serum from patients with Tangier disease, characterized by a severe deficiency of HDL, lack trypanocidal activity. Hawking *et al.* (1973) had reported the principle to be an α_2-macroglobulin but agrees that with the techniques used these were contaminated by α-lipoproteins (Hawking, 1979). Subsequent workers have demonstrated that Ca^{2+} is essential for this reaction and furthermore that α_2-macroglobulin can act as a protein ligand for Ca^{2+} (D'Hondt *et al.*, 1979); thus fractions containing both HDL and α_2-macroglobulin are most active. Since D'Hondt *et al.* also found that D-glucose (but not glycerol) can block trypanolysis, they suggest that HDL binds to the trypanosome membrane by the carrier site for D-glucose, and Ca^{2+} might act by binding to the phospholipid moiety of HDL. Subsequent events involve the alteration of membrane structure, possibly by removing lipid components (Rifkin, 1978*a*; D'Hondt *et al.*, 1979).

It is interesting that the binding of HDL does not appear to be affected by the variant glycoprotein, since antigenically similar trypanosomes can be sensitive or resistant, although the proportion of resistant organisms naturally arising in a nonhuman host for each antigenic variant can be widely different (Van Meirvenne *et al.*, 1976).

Another example of nonpermissiveness resulting from naturally occurring toxic serum factors is the resistance of the cotton rat, *Sigmodon hispidus*, to a rodent-adapted strain of *T. vivax*. Normal cotton rat serum agglutinates and subsequently lyses *T. vivax*; the nature of the active principle is not known but does not appear to be an immunoglobulin (Terry, 1976).

Yet another example was briefly mentioned by Albright and Albright (1981): serum from rats (CD line) is normally toxic to the murine parasite *T. musculi*. In cultures containing rat serum and mouse cells, a few surviving parasites were often able to establish a colony resistant to any residual effects of the serum present in the medium. Thus serum-resistant and -nonresistant organisms can be selected in the case of both *T. b. brucei* and *T. musculi.*

2. *Species Serum Dependency*

Serum from other nonpermissive hosts does not appear to be directly toxic to the trypanosome, yet trypanosomes fail to grow in such hosts. In certain instances in which it has been studied, such resistance can be overcome by serum supplements from susceptible hosts.

Most natural isolates of *T. vivax* from cattle, goats, or sheep cause only transient infections in laboratory rodents and usually fail to grow on the second rodent-to-rodent passage. Desowitz and Watson (1952), working with *T. vivax* from sheep, reasoned that the primary inoculation of sheep blood in some way supported the infection and inoculated their subpassaged rats with fresh sheep blood 24 hr after the inoculation of trypanosomes. Heavy parasitemias lasting some days resulted. The active principle was present in the serum and sera from other hosts normally susceptible to *T. vivax* could also be used. During the course of this work, a line of *T. vivax* developed that was able to grow in rats without serum supplementation (Desowitz and Watson, 1953). Furthermore, Leeflang *et al.* (1976) demonstrated that certain (3 of 75) bovine isolates of *T. vivax*, notably those obtained from early natural infection, were capable of serial maintenance in mice for an indefinite period of time without serum supplementation. It therefore appears that *T. vivax* can occur in a rodent-infective form in nature.

Similar investigations have been conducted with the rat trypanosome *T. lewisi*. This trypanosome does not normally grow in the mouse, but daily inoculations of rat serum enable it to multiply for 1-2 weeks, especially in starved animals (Lincicombe, 1958). A mouse-adapted strain of *T. lewisi* capable of growing without serum supplementation was not obtained even after 300 passages. The minimal volume required for supplementation was 0.01-0.05 ml (Lincicombe and Francis, 1961); similar results were obtained for *T. vivax* in rats (Desowitz and Watson, 1952). Greenblatt and co-workers have experienced difficulties in characterizing the active principle involved, first reporting it to be a γ-globulin, then amending this to the slow β-globulins (although the preparation was heterogeneous and contained siderophilin and lipoproteins) and recently stating that transferrin fractions are active (Greenblatt and Lincicombe, 1966; Greenblatt *et al.*, 1969; Greenblatt, 1975). Possibly, several macromolecules might be active, or small molecules might be bound to them.

Several mechanisms have been suggested to explain the action of serum supplementation, including, (1) serum provides essential nutritional requirements lacking in the host, and (2) proteins in the serum block the immune defenses of the nonpermissive host in either a specific or nonspecific manner. It seems unlikely that nutritional factors play an important role in supplementation, since *T. lewisi* can grow in diffusion chambers in the mouse (Greenblatt and Shelton, 1968), and Albright and Albright (1981) report that *T. lewisi* grows *in vitro* in mouse spleen cell cultures supplemented with mouse serum as well as, or better than, in homologous rat cultures.

The possibility that serum blocks specific anti-trypanosome immune mechanisms was first suggested by Desowitz and Watson (1953), who reported that sheep antibody had little lytic effect on freshly isolated *T. vivax*, but could lyse these trypanosomes after washing. These workers concluded that proteins bind to trypanosome surface and hinder attachment of specific antibodies. Other investigators have detected host proteins on the surface of trypanosomes; Ketteridge (1970) reported host α- and β-serum globulins or antigenically similar molecules on the mouse-adapted *T. vivax*, and D'Alesandro (1972) showed that non-antibody rat serum proteins bind avidly to the surface of *T. lewisi*. However, certain nonpermissive hosts eliminate parasites before acquired antibody production, apparently making blockade of acquired immune responses unnecessary. Moreover, it is difficult to envisage how 10–15 microliters of serum could effectively cover trypanosome surface determinants.

3. Trypanosome Clearance

Few studies have been done on the mechanisms by which trypanosomes are eliminated in nonpermissive hosts, except for those of Albright and Albright (1981). These workers found that when *T. lewisi* was inoculated in large numbers into various mouse strains, parasite elimination occurred in all strains in a similar manner. There was a lag period of about 7 hr of constant parasitemia, followed by rapid clearance (half-times of 55–130 min in the various strains). Inoculation of rat serum 1 hr before and 1 hr after the parasites greatly increased the lag period preceding a still rapid elimination. Furthermore, in those mice also treated with silica dust, extensive *T. lewisi* growth occurred, leading to the death of a number of infected mice. In agreement with other workers, they also found daily inoculations of rat serum (without silica dust) ensured maintenance in mice of prolonged *T. lewisi* infections.

Microscopic examination of host leukocytes and parasites during the phase of elimination showed that trypanosomes frequently bound to granulocytes. These observations led the authors to conclude that mice eliminate *T. lewisi* by granulocyte-mediated cytotoxicity involving naturally occurring antibodies. They suggest that immune complexes present in rat serum may bind to granulocytes, perhaps by the Fc receptor, blocking their uptake of the trypanosomes. They gave no evidence, however, for either the existence of natural antibodies in the mouse or the involvement of immune complexes.

We would like to speculate that many of these observations on serum toxicity or supplementation could have a similar molecular basis at the level of the plasma membrane. Human serum HDL appears to be able to destabilize the membrane of *T. b. brucei*. Is it possible that serum from other nonpermissive hosts contains macromolecules that, although not causing sufficient membrane destabilization to lead to lysis, damages the trypanosomes in such a way as to then become phagocytosed? Possibly serum supplementation is effective because it blocks this destabilization. Studies on the chemical nature of the active substances involved in these reactions and their mode of interaction with the trypanosome surface are obviously necessary to substantiate or refute such a possibility. That such mechanisms may be operating in cases in which animals control parasitemia at low levels also requires investigation.

B. Variations of Susceptibility in Permissive Hosts

There is considerable circumstantial evidence that certain breeds of livestock are less susceptible to trypanosomiasis than others. Since the possibility of an effective vaccine against trypanosomiasis appears remote, and other control methods, chemotherapy, and tsetse fly eradication are insufficient, there has been increased interest in such livestock. Evidence for the existence of variable susceptibility in cattle, and in strains of mice that can be used as a model, is discussed in this section.

1. Cattle

Since cattle were first introduced into Africa, herdsmen in tsetse fly-infested areas have selected those animals most able to resist trypanosomiasis. Early explorers noted that certain humpless West African taurine (*Bos taurus*) breeds (e.g., ndama, muturu, baoulé) were able to survive and reproduce in tsetse fly-infested areas in which zebu (*Bos indicus*) cattle died (Pierre, 1906). A recent survey conducted by ILCA (1979) shows that within the tsetse belt taurines predominate in the southwestern countries (e.g., Gambia, Guinea, Ghana, Benin), whereas further East (e.g., Nigeria, Central African Republic) the national herds are almost exclusively zebu. Countries bisected by the tsetse fly boundary have predominantly zebu herds (e.g., Upper Volta, Senegal), but many humpless cattle are found in the southern areas. Such a distribution could reflect the policy that zebu cattle are preferentially kept in large numbers wherever conditions are favorable for cattle raising; in other areas, particularly regions with a heavy infestation of tsetse flies, the humpless cattle are kept, but the numbers of cattle involved are lower than for zebus. To what extent is this distribution the result of trypanosomiasis?

Several experimental approaches have been used to define differing susceptibility of cattle to trypanosomiasis, including infection by syringe, by infected captured flies, and by exposure to natural challenge.

a. Fly Infection. Experiments using natural challenge tend to demonstrate the greatest difference between breeds.

The earliest report is that of Chandler (1952), working in Nigeria. He exposed a herd of 12 zebus, 8 ndamas, and 12 ndama–zebu cross-breds once weekly to a moderate level of riverine tsetse. The ndamas had previous exposure to trypanosomes, whereas the zebus and cross-breds probably had not (stated in Roberts and Gray, 1973). The results obtained show a striking difference between the breeds. All 12 zebus became infected, and 9 of them died from trypanosomiasis. All 12 ndama–zebu cross-breds also became infected, and 4 died of trypanosomiasis. Only four of the eight ndamas showed parasitemia, and then at low levels, and all survived. After 1 year, Chandler then moved five of these ndamas and six of the cross-breds (three from the previous experiment and three others) to an area of heavy challenge with *G. morsitans.* All the ndamas survived, never showing parasitemia, whereas all the cross-breds became infected and four died from trypanosomiasis. Chandler also states that in another experiment in the same region, "10 healthy zebus succumbed to trypanosomiasis within a period of six months."

Another small-scale investigation was conducted by Van Hoeve (1971), also in Nigeria, using 6 ndamas, 6 muturus, and 10 zebus. Some of the zebus and muturus, but not the ndamas, had a history of trypanosomiasis. When exposed to light tsetse fly challenge for 1 year, all the zebus became parasitemic but only 1 died of trypanosomiasis, and only 1 ndama showed an infection and all survived. When the same ndamas and muturus were exposed to heavy fly challenge, all the animals became parasitemic, but none showed clinical signs of trypanosomiasis.

P. K. Murray and M. Murray and colleagues, working in Gambia, report two investigations using natural fly challenge. In the first investigation, zebus and ndamas that had received needle challenge with either *T. b. brucei* 5 months previously or *T. congolense* 2 months previously, or both, were inadvertently exposed to *G. palpalis* challenge (P. K. Murray *et al.*, 1979*a,b*). All the zebus became reinfected, 21 of 31 died, and those that survived were no longer parasitemic. Few of the ndamas showed parasitemia after fly challenge, and none died out of a total of 37. A later study exposed cattle to a high challenge by *G. m. submorsitans* (M. Murray *et al.*, 1981). This investigation used 10 zebus not previously exposed to trypanosomes of which 8 were pregnant and effectively 8 ndamas obtained from the same ranch on which the above *G. palpalis* infestation occurred, of which three were pregnant at the beginning of the investigation. The zebus became heavily parasitemic, with the exception of two animals (#1 and #10) and all died, although #1 and #10 survived longer. The ndamas showed infrequent parasitemia, and only three died. Pregnancey and the stress of lactation clearly influence survival, since the three ndamas that died were lactating and appear to be the three that were pregnant at the start of the investigation. The zebus may have succumbed more rapidly, because of pregnancy; unfortunately, these workers do not indicate whether the two Zebus that survived longer were the two nonpregnant animals.

Another study conducted by Touré *et al.* (1978), in Senegal, gave somewhat different results. In this study animals were exposed to heavy challenge by *G. m. submorsitans* and *G. palpalis*; after 8 months 13 of 16 zebus, two of four cross-breds, and 14 of 24 ndamas had died of trypanosomiasis. Parasitemia was not tabulated in this report, but it is stated that high levels of parasites were found in zebus and 15 of the ndamas, but that the other 9 ndamas never showed parasitemia and survived in good condition.

Finally, we have exposed 10 baoulés and 10 zebus to heavy natural challenge by *G. m. submorsitans* and *G. palpalis* (Roelants *et al.*, 1983). All zebus died 7-12 weeks after exposure. Five baoulés also died during that time. Five other baoulés had transient or no parasitemia, no anemia, and remained in good health for 27 weeks of exposure. The observation was repeated with 10 baoulés: six died and four resisted. Seven baoulés/ndamas crossed from a herd maintained in the same area for 20 years showed neither parasitemia nor anemia in 27 successive weekly examinations.

These rather crude studies demonstrate that more taurine cattle survive natural challenge by tsetse flies than do zebus. A few studies have examined the mechanism of this resistance using more controlled methods of infection. In the earliest published report using artificial infection, Chandler (1958) attempted to compare the resistance of ndamas previously exposed to tsetse fly challenge to that of unexposed ndamas and zebus. There is some evidence indicating the exposed animals were more resistant, and serologic data suggest this was an immune phenomenon.

Stephen (1966) and Roberts and Gray (1973) fed wild-caught *G. m. submorsitans* on cattle that had no previous exposure to trypanosomiasis. Stephen fed flies once a month on three ndamas and three zebus for some years. A trypanocidal drug was administered when necessary, but only during the first year. All cattle became infected and all required drug treatment except one ndama that had lower parasitemia than the others. All three zebus and one ndama died of trypanosomiasis, and the two survivors were in poor condition. Roberts and Gray first challenged two cattle of each breed (ndama, muturu, zebu) with flies six times and did not treat them with trypanocidal drugs. These animals had similar levels of parasitemia for the first 3 months, but subsequently parasitemia remained higher in the zebus. Both zebus and one muturu, but neither ndama, died of trypanosomiasis. Four animals of each breed were challenged twice with captured flies, treated with a trypanocidal drug, and then rechallenged three times with flies. In this experiment no cattle died, levels of parasitemia were similar after primary challenge but after secondary challenge, fewer parasites were detected in ndamas, and the disease was clinically less severe in both taurine breeds. These varying results may have been caused by heavier challenge and concurrent babesial infections in the Stephen's study.

b. Syringe Infection. Two investigations have challenged taurines and zebus using rodent-passaged trypanosomes inoculated by syringe. P. K. Murray and colleagues (1979*a,b*) infected zebu and ndama cattle with various doses of *T. b. brucei* or *T. congolense* either alone or in combination. The zebus had not been previously exposed to trypanosomes, whereas the ndamas probably had. Animals were monitored for 4 months after *T. b. brucei* infection and for 2 months after *T. congolense* infection, before the herd came inadvertently under natural fly challenge. Parasitemic profiles were similar in both breeds, with all animals becoming parasitemic and showing clinical signs. Before fly infection, 9 of the 40 zebus died, but all 37 ndamas survived.

Contradictory results were obtained by Guidot and Roelants (1982) using baoulé and zebu cattle born and raised under tsetse challenge. Cattle were infected first with a West African *T. vivax* and 26 weeks later with a *T. congolense*, stock ILRAD 1180, which in East African experiments killed 50% of zebu and boran cattle within 3-5 months. Both West African breeds survived the *T. vivax* and *T. congolense* infections, and few clinical symptoms were apparent. Also, with both infections all taurines became parasitemic, as compared with only a portion of the zebus. After *T. congolense* infection, three of the baoulés showed a severe drop in hematocrit (32 to 13), whereas it was less severe in the zebus.

The differences between these two studies could be attributable to exposure of certain of the animals to antigenic variants present in the artificial challenge, although Guidot and Roelants used a clone of *T. congolense* isolated in East Africa on West African cattle. Ndamas in the experiment of P.K. Murray *et al.* had probably been previously infected, as had all cattle in the Guidot and Roelants experiment. Negative reactions in immunofluorescence do not reliably indicate that animals have never been infected, as cattle have been shown to become negative 12 weeks after cure (Lotzsch and Deindl, 1974). Titers of variant specific antibody are not reported in either study. An alternative explanation for these differences could be that zebus raised for several generations under tsetse challenge may have heightened resistance to syringe infection by a single stock or clone. However, such zebus succumb rapidly to heavy natural fly challenge (Roelants *et al.*, 1983).

c. *Mechanisms*. These are all the data available to us at the time of writing. Two major points can be noted.

First, under conditions of natural fly challenge, considerably more zebus than West African taurines die of trypanosomiasis (Table II). Thus, there is a difference in disease susceptibility under natural conditions. This susceptiblity does vary within the breeds, however, since some taurines (20%) die and a substantial percentage of zebus (25%) survive essentially under light challenge. The notion of trypano-sensitive and trypano-resistant (so-called trypano-toler-

Table II. Summary of Comparative Susceptibility of West African Cattle Breeds to Trypanosome Challenge

Challenge	Number parasitemic/ total		Number dead/ total	
	Taurine	Zebu	Taurine	Zebu
Light to moderate natural fly[a]	8/51	53/53	0/51	31/53
Heavy natural fly[b]	57/78	50/50	28/78	47/50
Captured fly[c]	22/22	11/11	2/22	7/11
Total fly challenges	87/151	114/114	30/151	85/114
	58%	100%	20%	75%
Needle[d]	57/57	55/57	0/57	9/57

[a]Chandler (1952); Van Hoeve, (1971); P. K. Murray *et al.* (1979*a*).
[b]Chandler (1952); Van Hoeve (1971); Touré *et al.* (1978); M. Murray *et al.* (1981); Roelants *et al.* (1982), and unpublished.
[c]Stephen (1966); Roberts and Gray (1973); Akol *et al.*, in preparation.
[d]P. K. Murray *et al.* (1979*a,b*); Guidot and Roelants (1982).

ant) breeds needs to be reconsidered and far more attention devoted to the susceptibility status of herds and individuals within a breed. This implies that the selection criterion for breeding more resistant cattle should be made on the basis of individual resistance itself, and not on morphologic or genetic markers of the breed, which have been described by several investigators (Coulomb *et al.*, 1977; ILCA, 1979; Quéval, 1982*a,b*; Quéval and Petit, 1982; Quéval *et al.*, 1983) precisely what African herdsmen have been doing empirically for centuries.

Second, these experimental data are insufficiently controlled, with cattle having diverse or unknown histories and inadvertent exposure to tsetse flies occurring in at least one experiment. In another, stages of gestation are not comparable and seem to determine the outcome of infection; often the experimental groups are very small, and occasionally results are not clearly presented. Discussion of the mechanisms involved is therefore difficult.

One finding appears to be clear: Under natural fly challenge zebus show heavier parasitemia than do taurines; indeed, in most studies, surviving taurines are seldom detectably parasitized, whereas in other investigations they are parasitized, but less so than zebus. Moreover, it appears that the difference in parasitemia is often the result of a heavier infection of zebus by *T. vivax* (Chandler, 1952; Touré *et al.*, 1978; M. Murray *et al.*, 1981). For instance, Murray *et al.* (1981) found that zebus and ndamas are equally infected with *T. congolense* (6-14% and 9-10%, respectively), but that there is a tremendous difference in the infection with *T. vivax* (53-56% against 2-6%, respectively). Our studies (Roelants *et al.*, 1983) do not confirm these observations; the few fleeting parasitemias found in resistant baoulés and ndamas/baoulés were mainly *T. vivax*. The resistance of taurines against fly-transmissible *T. vivax* needs more detailed investigation. Clinical symptoms parallel parasitemic load. This has been most clearly demonstrated for the anemia, which is probably the most significant factor in the disease process (Horby, 1921; M. Murray, 1974). In a series of *in vivo* pathologic studies, Dargie *et al.* (1979) demonstrated that the anemia and its underlying processes were broadly related to the parasitemic load. There is thus little evidence to support the contention that taurines, in contrast to zebus, harbor trypanosomes without showing clinical signs of the disease (Pagot, 1974).

Since zebus are more frequently parasitized than are taurines, susceptibility might be controlled, at least in part, at the level of the fly bite. At least two possible mechanisms could be operating, the most obvious being that taurines are less frequently bitten by tsetse flies. Experiments using captured flies demonstrate that if flies are given little choice, flies do feed on taurines, and infection ensues. Infection does influence fly-feeding behavior (Jenni *et al.*, 1980); it is unknown, however, whether flies carrying *T. vivax* behave differently from those infected with *T. congolense*. Carr *et al.* (1974) have reported that zebus found in an area of greater fly challenge had less parasitemia and thicker skins than did animals of the same breed living under conditions of lower challenge. The feeding behavior of variously infected flies on different breeds needs investigation. An alternative explanation for the lower parasitemia of taurines is that they eliminate metacyclic trypanosomes more efficiently. The experiments of Stephen (1966) and Roberts and Gray (1973) using captured flies argues against this possibility.

Mechanisms operating at the level of the fly bite would ensure that taurines do not become parasitized. If these mechanisms are overcome, however, taurines can behave as susceptible animals, i.e., unable to clear the infection, they die (Touré *et al.*, 1978; M. Murray *et al.*, 1981; Roelants *et al.*, 1982). In other instances, they are able to clear the infection more rapidly than zebus (Roberts and Gray, 1973; M. Murray *et al.*, 1979*a*). In these latter cases, acquired immune responses could be important.

Protective immunity against a given antigenic variant depends largely on serum antibody directed aginst exposed antigenic determinants (Section III. A). Thus the possibility that trypano-resistant cattle mount a more effective immune response has often been stated. Desowitz (1959) examined this possibility in a small-scale study using cattle with variable histories. He found that two zebus, two ndamas, and two muturus with little or no previous exposure to trypanosomes showed equally low-titer antibodies after experimental challenge. Three ndamas that had been repeatedly exposed to trypanosomes, both naturally and artificially, showed high titers after similar challenge. The possibility that the latter three ndamas had previously been exposed to the same variant used for challenge was not excluded.

Recently we compared the antibody responses of more susceptible zebu and simmental–ndama cross-breds and less susceptible baoulé and ndama cattle after needle challenge with 10^5 ID_{63} or 10^9-irradiated organisms of a clone or *T. b. brucei* (DiTat 1.1) (Pinder *et al.*, 1983). All animals had been previously exposed to tsetse flies, but before inoculation none had antibodies directed against the clone used. After primary challenge, all animals showed high levels of agglutinating antibodies (Table III), and no differences were found within the groups either in peak titer or kinetics, including duration, of the response. Titers in the complement-mediated lysis test were lower and declined more rapidly than did those in agglutination assay, particularly in animals challenged with irradiated trypanosomes. In all cases, more than 99% of the antibody was IgM, as judged by mercaptoethanol sensitivity and purification by ammonium sulfate precipitation and chromatography. Sera from ndamas and simmental–ndamas taken on day 14 postinfection were also analyzed by a neutralization of infectivity assay, and no significant differences were found. Certain zebus and baoulés were rechallenged with the same clone, and again no differences were found in the secondary response of these two breeds, which was partly IgM and partly IgG_1 and IgG_2. Zebus, baoulés, and ndamas all supported this challenge with *T. b. brucei* with few clinical signs, whereas West African/European cross-bred animals showed severe clinical signs, and one died 18 days postinfection. Clearly, these

Table III. Comparison of Antibody Responses of Cattle of Varying Susceptibility to Trypanosomiasis to Surface Antigenic Determinants of the *T. b. brucei* Clone Ditat 1.1

Breed of cattle *N*	Challenge[a]	% Parasitemic week 1	2	3	Titer at peak response $\log_3$ mean ± SD Agglutination	C′ lysis
Simmental–ndama (5)	1	6	97	–	9.6 ± 0.9	3.8 ± 0.4
Ndama (5)	1	0	77	–	10.2 ± 0.8	5.2 ± 0.8
Zebu–Djibo[b] (7)	1	6	26	26	10.4 ± 0.9	3.9 ± 0.3
Zebu–Bobo[b] (5)	1	14	11	20	9.1 ± 1.1	3.8 ± 0.4
Baoulé (5)	1	6	20	26	10.2 ± 0.8	3.7 ± 0.3
Zebu–Bobo (4)	1, irr	–	–	–	9.2 ± 1.6	2.5 ± 1.0
Baoulé (3)	1, irr	–	–	–	8.0 ± 1.7	3.0 ± 0
Zebu–Bobo (4)	2	7	21	14	7.3 ± 0.5	2.5 ± 0.6
Baoulé (3)	2	10	5	19	7.3 ± 1.2	2.6 ± 1.4

[a]All animals received 10^5 mouse ID_{63} *T. b. brucei*, except those marked "irr," which received 10^9 irradiated trypanosomes of the same clone.
[b]Zebu-Djibo were purchased and infected in the north of Upper Volta beyond the limit of the tsetse belt. Zebu–Bobo were purchased and infected within the region of Bobo–Dioulasso, i.e., within the tsetse belt.
[c]No parasites seen, because aminals were treated with Berenil or received irradiated trypanosomes.

cross-breds are more susceptible to needle challenge than are any of the pure West African cattle. These results indicate that susceptibility to trypanosomiasis in cattle does not correlate with their ability to respond to trypanosome surface antigens.

Other causes of susceptibility differences that need to be investigated may operate at the level of acquired cellular immune mechanisms (Gasbarre *et al.*, 1980), immunodepression, deleterious immune responses, or the ability of more resistant animals to select for the expression of less virulent organisms (McNeillage and Herbert, 1968). In this regard, work at the C.R.T.A. seems to show that the same 13 dominant clones appear in relapse populations of baoulés and zebus infected with a clone of *T. b. brucei*. The possible role of other breed differences in susceptibility to trypanosomiasis, such as the superior capacity for water conservation in zebus, the greater variation in body temperature of ndamas, and the higher levels of eosinophils in ndamas, have been previously reported, but there is little evidence to indicate precise mechanisms (Coulomb *et al.*, 1977). It should also be considered that all the above pathogenicity studies, with the exception of that of Stephen (1966) and Roberts and Gray (1973), were conducted with taurines born within the tsetse belt, although in some cases the level of exposure to trypanosomes was probably slight (P.K. Murray *et al.*, 1979*a,b*, 1981; Van Hoeve, 1971). In contrast, most studies, except ours, used zebu cattle obtained beyond the tsetse belt whose first exposure to flies and trypanosomes occurred during the investigation. Such zebus were not well adapted to the ecology of the experimental zones, a fact that in itself may explain their higher susceptibility to trypanosomiasis without recourse to other more subtle mechanisms.

In conclusion, it appears that increased resistance in cattle does not correlate with acquired humoral immunity. Since this is the major acquired immune mechanism effective against trypanosomes, we would propose that innate mechanisms are of importance. The most striking finding after exposure of cattle to natural fly challenge is that taurines are less parasitized. It is therefore imperative to investigate the feeding behavior of tsetse flies on these breeds under natural conditions.

2. *Mouse*

The heterogeneity of bovine populations as well as their poorly characterized immune system make analysis of the genetics and the mechanisms of susceptibility difficult in this species. Studies have therefore been made in the mouse, in which large numbers of inbred and congenic strains are available. The major findings on mouse susceptibility as defined by survival time are presented in Table IV.

Morrison and colleagues studied nine inbred strain of mice and found a large array of susceptibility (Morrison *et al.*, 1978; Morrison and Murray, 1979) (Table IV, column 1). Mean times to death varied from 15.8 days for A/J to 163.0 days for C57BL/10 mice. These workers observed a non-Poisson type of distribution within a group, indicating individual differences for mice of identical genetic makeup. The prepatent period was very similar in all groups, and infectivity titrations showed identical titers for A/J and C57BL/6 mice. Higher parasitemia and no remission were found in the four most susceptible strains.

Two other *T. congolense* were used (SE-13 and C-24), which were more virulent. There were fewer differences between the strains, but the same general order of susceptibility was maintained. However, Pearson *et al.* (1982), using a clone (13E-1) derived from 5E-12 after one 8-day passage in A/J mice, found it less virulent for A/J than for C3H mice; C57BL/10 remained the less susceptible strain (Table IV, column 2). Studies on strains congenic for the main histocompatibility complex on a C57BL/10 background showed that mice carrying H-2a, k, or d were slightly more susceptible than were H-2b types (Table IV, column 1). However, most survived for >100 days and were quite less susceptible than the inbred strain with corresponding H-2 types, i.e., A/J, C3H, and Balb/c, re-

Table IV. Survival Time of Trypanosome-Infected Mice (Days)

	T. congolense					*T. b. rhodesiense*	*T. b. brucei*	
Mouse strain	5E.12[a]	13E.1[b]	GVR1[c]	DIND.3–4[d]	DIND.3–5[e]	LouTat 1.1[f]	TREU 667[c]	NIM2[g]
A/J	15.8	43	19.0	14.7	–	36.7	–	–
SWR/J	16.9	–	–	–	–	–	–	–
129/J	22.6	–	–	–	–	–	–	–
DBA/1J	36.3	–	–	–	–	–	–	–
Balb/c	49.5	–	7.0	14.1	13.3	30.4	–	20.5
					–			
C3H	59.0	35	14.9	–	–	18.6	26.8	15.4
CBA/J	–	–	9.0	18.5 (62%)	20.8	33.9	54.5	15.7
				Survive (38%)	–			
AKR/A	81.7	73	–	30.0 (70%)	67.2	–	–	–
				Survive (20%)	–			
C57BL/6J	110.2	–	73.7	Survive	Survive	43.6	53.7	25.6
C57BL/10 SgSn	136.0–163.0	109				57.2	–	–
B10.A/SgSn	113.2	–	–	–	–	53.6	–	–
B10.BR/SgSn	124.6	–	–	–	–	51.3	–	–
B10.D2/n	115.5	–	–	–	–	53.2	–	–
NMRI	–	–	–	15.8 (16%)	11.2 (20%)	–	–	–
				Survive (84%)	Survive (80%)			
CFLP	–	–	14.3	–	–	–	62.8	–
PORTON	–	–	21.8	–	–	–	59.0	–
NZB	–	–	37.3	–	–	–	33.1	–

[a]Morrison *et al.* (1978); Morrison and Murray (1979).
[b]Pearson *et al.* (1982).
[c]Jennings *et al.* (1978).
[d]Libeau and Pinder (1981).
[e]Pinder (1983).
[f]Levine and Mansfield (1981).
[g]Clayton (1978).

spectively. It was concluded that the major histocompatibility complex did not play a primary role in determining susceptibility.

Genetic analysis was done on F_1 and backcrosses among C57BL/6, A/J, and C3H (Morrison and Murray, 1979). The (C57BL/6 × A/J)F_1 had the same survival time as the C57BL/6 parent, and some mice survived even longer (220–250 days). However, (C57BL/6 × C3H)F_1 susceptibility lay in between that of the two parents, and (A/J × C3H)F_1 showed a spread overlapping both parents. Backcrosses of B6A.F_1 to C57BL/6 had the same mean survival time as the F_1 and C57BL/6, but with quite a larger scatter. Finally B6A.F_1 to A/J backcrosses had a large range of susceptibility between that of A/J and C57BL/6. Morrison and Murray (1979) concluded that in some combinations the results might be explained by the involvement of two major genes, but that for other combinations, a multiple gene effect must be involved. It is interesting that in these genetic analysis experiments susceptibility did not correlate with the intensity of the parasitemia.

Jennings *et al.* (1978) found that *T. congolense* VGR1 was virulent for all mouse strains used, including outbreds, with the exception of C57BL/6, and to a lesser degree NZB (Table IV, column 3). The same strains of mice did not show appreciable differences in their susceptibility to a *T. b. brucei* (TREU 667), except for C3H, which was more sensitive (Table IV, column 7). In contrast, Clayton (1978) found that C57BL/6 mice were slightly more resistant to the *T. b. brucei* clone NIM2 than were Balb/c, C3H, and CBA/J mice (Table IV, column 8). Using a clone of *T. b. rhodesiense* (LanTat 1.1), Levine and Mansfield (1981) did not find significant differences in several strains, again with the exception of C3H, which was more susceptible (Table IV, column 6). Congenic strains for the H-2 region did not show any differences.

The mechanisms underlying those differences in susceptibility are poorly understood. M. Murray and Morrison (1979*a*) emphasize the correlation between levels of parasitemia and susceptibility. However, careful analysis of all the inbred or congenic strains, the F_1 and the backcrosses used, as well as results obtained with immunodeficient mice show that this correlation does not always exist and is even sometimes reversed (Morrison *et al.*, 1978; Morrison and Murray, 1979). Moreover, no explanation is offered as to the mechanisms by which parasitemia is lower in certain strains, although W. I. Morrison (personal communication, 1982) has recently observed C57BL/6 mice to produce more anti-VSG antibody against the injected trypanosome than do A/J mice.

Another factor may be that the immune potential of susceptible strains is exhausted more completely by polyclonal activation, which is more pronounced in those strains (Morrison *et al.*, 1978). Pearson *et al.* (1983) showed that the kinetics and degree of immunosuppression of the B- and T-spleen cell responses also parallel the degree of susceptibility of infection. Moreover, suppression is established as soon as parasites appear in the peripheral blood and is not correlated with the level of parasitemia (Roelants *et al.*, 1979*b*; Pearson *et al.*, 1982). For instance, A/J and C57BL/10 have identical peaks of parasitemia (10^8/ml) on day 8, and con A stimulation of spleen lymphocytes is 94% depressed in A/J mice, but only 13% depressed in C57BL/10 mice (Pearson *et al.*, 1982). There is no remission of parasitemia in A/J mice, while C57BL/10 clear the parasites by day 9 or 10. Thus it appears that the degree of immunosuppression induced in various strains is an important factor in determining their degree of susceptibility by limiting the immune response to trypanosome VSG.

Whitelaw *et al.* (1980) used CFLP and C57BL/6 mice in the system described by Jennings *et al.* (1978) (Table IV, column 3). These workers found parasitemia to be higher in CFLP mice, but certain parameters to remain essentially unchanged or equally altered in both strains, i.e., erythrocyte survival, plasma, and erythrocyte volume, blood biochemistry, i.e., levels of lactate deshydrogenase, aspartate, and alanine transaminase, glucose, albumin, electrolytes, urea, phosphate, and Ig. The antibody response to horse erythrocytes was suppressed in both strains by day 7. However, an infectivity neutralization assay done with day 10 postinfection sera on the homologous trypanosome showed a difference: Only 1 of 12 recipients of CFLP sera survived as compared with 9 of 12 recipients of C57BL sera. This finding points again to an involvement of acquired immune responses in susceptibility, a conclusion supported indirectly by the observation that resistance in A/J and C57BL/6 mice to *T. congolense* and *T. b. brucei* can be nonspecifically increased by immunostimulants such as *Corynebacteria parvum* and *Bordetella pertussis* (M. Murray and Morrison, 1979*b*). By contrast, levamisole used under conditions reversing immunosuppression has a deleterious effect on the disease in Balb/c, A/J, CBA, NMRI, and C57BL/6 mice (Libeau and Pinder, 1981).

Most investigators allude to the mouse work as a model for bovine trypanosomiasis. However, serious differences do exist. Bovine trypanosomiasis is often a chronic disease that may last 1 year or more, with parasitemia not usually exceeding 10^6 organisms/ml. Moreover, healthy cattle frequently survive a single challenge. In contrast, most trypanosome isolates studied in mice give high levels of parasitemia, $\geqslant 10^9$ organisms/ml, and cause death within a few days or weeks. The most chronic infection is that reported by Morrison *et al.* (1978) in C57BL/10 mice; nevertheless, all these mice still succumb.

We recently used a stabilate (Dind.3–4) from a stock of *T. congolense* (Dinderesso/80/ CRTA/3) that was close to the primary isolate (five passages). (Libeau and Pinder, 1981; Pinder, 1983). The spectrum of sensitivity of inbred mice to this trypanosome is different from that in other studies (Table IV, column 4). All A/J and Balb/c mice die rapidly. In addition, 70% of AKR, 62% of CBA/J, and 16% of NMRI mice die rapidly, as well, but the remaining mice, as well as all C57BL/6 mice, survive and self-cure. They have been followed for 1 year and 4 months, from the point of onset of death from old age. The prepatent period increases with higher resistance from 6.8 days in Balb/c mice to 18.9 days in C57BL/6 mice. Peaks of parasitemia in Balb/c, A/J, CBA/J, AKR, and the NMRI that would die are ~8.6 log.10/ml, whereas for the NMRI that would survive it is ~7.5, and for the C57BL/6 ~5.7, i.e., 1000 times lower than in the first four strains.

A stabilate made after 11 additional mouse passages was more virulent for the susceptible strains but C57BL/6 were still resistant (Table IV, column 5). Inheritance of resistance, defined as low parasitemia, has been studied using F1, F2 and backcrosses between C57BL/6 and Balb/c. In brief, the results are compatible with resistance being a recessive trait controlled mainly by a single autosomal gene (or gene cluster); in addition sex-associated factors appear to confer higher resistance in females (Pinder, 1983). The mechanism of resistance is under investigation. Toxic or enhancing serum factors do not appear to be involved, C57BL/6 do support growth of these trypanosomes, they do not appear to act as a "heterologous host," and preliminary results suggest C57BL/6 make neutralizing antibodies whilst Balb/c do not. The Dind. 3 system may constitute a model that more closely parallels bovine trypanosomiasis.

VII. CONCLUSIONS

In closing, we offer some remarks on what we believe is known and where we consider efforts should be directed.

The single most important practical point concerning the infected host, is probably that anti-VSG antibody—both of the IgM and of IgG classes—can protect. The epitopes involved and their location on the VSG *in situ* are becoming increasingly well known. Sequence analysis of cDNA will soon give us a good idea of homologous and variable regions and permit classification of VSG in subsets. The repertoire of VSGs expressed in nature—even if some dominant, recurring, or basic types exist—is vast, at the metacyclic stage as well, which appears to preclude practical vaccination. Understanding of the mechanisms of genomic rearrangement of VSG genes may give us a handle on antigenic variation. Several excellent groups working on this possibility are making rapid progress.

Antigenic variation would theoretically preclude elimination of the infection; nevertheless, some animals do self-cure. In cattle this mechanism may involve secondary antibody responses to recurring antigenic types. It is crucial that the mechanisms operating in such hosts receive adequate attention.

Cell-mediated reactions to trypanosomes have been demonstrated, including at the site of fly bite, but their relevance to protective immunity is unknown. Work is being actively pursued on the chancre, and we may soon have more information.

The protective effect of anti-VSG antibody may be counterbalanced by immune-complex-induced lesions. Antibody production can be impaired by immunodepression through a variety of mechanisms, including polyclonal B-cell exhaustion, suppressor T cells, and "angry" macrophages. The precise balance among these events and their sequence in various host-parasite combinations is not yet clearly defined. Investigation of the importance of these mechanisms in natural infections is of high priority.

Some hosts are nonpermissive to certain species or subspecies of trypanosomes, involving several types of toxic or supplementary serum factors that need precise biochemical characterization. The importance of such factors in differences in susceptibility of permissive hosts needs investigation.

Experimental approaches have confirmed that certain cattle are more resistant to African trypanosomiasis. This seems to be a property of individuals rather than of breeds, although resistant animals are present in higher number in some breeds than in others. It appears that higher resistance is not attributable to a more effective antibody response. The role of natural immunity mechanisms and of physiologic factors, both at the fly and the host level, as well as their interaction, requires a lot more work. Before more sophisticated studies are undertaken, it might be worthwhile to investigate whether more resistant animals are simply not bitten as often by tsetse flies.

Work in inbred strains of mice has shown that resistance in genetically inherited in this species, but that there are some individual differences even within an inbred strain. Demonstration of genetic transmission of resistance is lacking in the bovine system, and there are tremendous individual differences within breeds. It is imperative that experimental work in this species be done on individual animals of proven susceptibility status.

Finally, we hope we have demonstrated that the immunobiology of African trypanosomiasis has a number of fascinating facets well worth the time and creative efforts of immunologists.

ACKNOWLEDGMENTS

This work has been supported by the Deutsche Gesellschaft für Technische Zusammenarbeit, Federal Republic of Germany, project PN77.2227.5, and by the Institut d'Elévage et de Médecine Vétérinaire des Pays Tropicaux (IEMVT), France. We thank Mrs. S. Zerbo for outstanding secretarial assistance.

VIII. REFERENCES

Agabian, H., Thompson, L., Milhausen, M., and Stuart, K., 1980, Structural analysis of variant and invariant genes, *Am. J. Trop. Med. Hyg.* **29**:1043–1052.

Albright, J. W., and Albright, J. F., 1981, Basis of the specificity of rodent trypanosomes for their natural host, *Infect. Immun.* **33**:355–363.

Allsopp, B. A., Njogu, A. R., and Humphryes, K. E., 1971, Nature and location of *Trypano-*

soma brucei subgroup exoantigen and the relationship to 4S antigen, *Exp. Parasitol.* **29:** 271–284.

Askonas, B. A., Corsini, A. C., Clayton, C. E., and Ogilvie, B. M., 1979, Functional depletion of T- and B-memory cells and other lymphoid cell subpopulations during trypanosomiasis, *Immunology* **36:**313–321.

Barbet, A. F., and McGuire, T. C., 1978, Crossreacting determinants in variant-specific surface antigens of African trypanosome, *Proc. Natl. Acad. Sci. USA* **75:**1989–1993.

Barry, J. D., Hajduk, S. L., Vickerman, K., and Le Ray, D., 1979, Detection of multiple variable antigen types in metacyclic populations of *Trypanosoma brucei*, *Trans. R. Soc. Trop. Med. Hyg.* **73:**205–208.

Boothroyd, J. C., Cross, G. A. M., Hoeijmakers, J. H. J., and Borst, P., 1980, A variant surface glycoprotein of *Trypanosoma brucei* synthesized with a C-terminal hydrophobic "tail" absent from purified glycoproteins, *Nature* (*Lond.*) **288:**624–626.

Boothroyd, J. C., Paynter, C. A., Cross, G. A. M., Bernards, A., and Borst, P., 1981, Variant surface glycoproteins of *Trypanosoma brucei* are synthesized with cleavable hydrophobic sequences at the carboxy- and aminotermini, *Nucl. Ac. Res.* **9:**4735–4744.

Borst, P., Fase-Fouler, F., Frasch, A. C. C., Hoeijmakers, J. H. J., and Weijers, P. J., 1980*a*, Characterization of DNA from *Trypanosoma brucei* and related trypanosomes by restriction endonuclease digestion, *Mol. Biochem. Parasitol.* **1:**221–246.

Borst, P., Frasch, A. C. C., Bernards, A., Hoeijmakers, J. H. J., Van der Ploeg, L. H. T., and Cross, G. A. M., 1980*b*, The genes for variant antigens in trypanosomes, *Am. J. Trop. Med. Hyg.* **29:**1033–1036.

Borst, P., Frasch, A. C. C., Bernards, A., Van der Ploeg, L. H. T., Hoeijmakers, J. H. J., Annberg, A. C., and Cross, G. A. M., 1981, DNA rearrangements involving the genes for variant antigens in *Trypanosoma brucei*, *Cold Spring Harbor Symp. Quant. Biol.* **45:** 935–944.

Bridgen, P. J., Cross, G. A. M., and Bridgen, J., 1976, N-terminal amino acid sequences of variant specific surface antigens from *Trypanosoma brucei*, *Nature* (*Lond.*) **263:**613–614.

Browning, C. H., and Gulbransen, R., 1936, Immunity following cure of experimental *Trypanosoma brucei* infection by a chemotherapeutic agent, *J. Pathol. Bacteriol.* **43:** 479–486.

Campbell, G. H., and Phillips, S. M., 1976, Adoptive transfer of variant-specific resistance to *Trypanosoma rhodesiense* with B lymphocytes and serum, *Infect. Immun.* **14:**1144–1150.

Campbell, G. H., Esser, K. M., and Weinbaum, F. J., 1977, *Trypanosoma rhodesiense* infection in B cell deficient mice, *Infect. Immun.* **18:**434–438.

Campbell, G. H., Esser, K. M., and Phillips, S. M., 1978, *Trypanosoma rhodesiense* infection in congenitally athymic (nude) mice, *Infect. Immun.* **20:**714–720.

Capbern, A., Giroud, C., Baltz, T., and Mattern, P., 1977, *Trypanosoma equiperdum*: Etude des variations antigéniques au cours de la trypanosomose expérimentale du lapin, *Exp. Parasitol.* **42:**6–13.

Carr, W. R., Macleod, J., Woolf, B., and Spooner, R. L., 1974, A survey of the relationship of genetic markers, tick-infestation level and parasitic diseases in Zebu cattle in Zambia, *Trop. Anim. Health Prod.* **6:**203–214.

Chandler, R. L., 1952, Comparative tolerance of West African N'Dama cattle to trypanosomiasis, *Ann. Trop. Med. Parasit.* **46:**127–134.

Chandler, R. L., 1958, Studies on the tolerance of N'Dama cattle to trypanosomiasis, *J. Comp. Pathol.* **68:**253–260.

Clarkson, M. J., and Penhale, W. J., 1973, Serum protein changes in trypanosomiasis in cattle, *Trans. Soc. Trop. Med. Hyg.* **62:**273.

Clayton, C. E., 1978, *Trypanosoma brucei*: Influence of host strains and parasite antigenic type in infection in mice, *Exp. Parasitol.* **44**:202–208.

Clayton, C. E., Sacks, D. L., Ogilvie, B. M., and Askonas, B. A., 1979*a*, Membrane fractions of trypanosomes mimic the immunosuppressive and mitogenic effects of living parasites on the host, *Parasite Immunol.* **1**:241–249.

Clayton, C. E., Ogilvie, B. M., and Askonas, B. A., 1979*b*, *Trypanosoma brucei* infection in nude mice: B lymphocyte function is suppressed in the absence of T lymphocytes, *Parasite Immunol.* **1**:39–48.

Clayton, C. E., Selkirk, M. E., Corsini, C. A., Ogilvie, B. M., and Askonas, B. A., 1980, Murine trypanosomiasis: Cellular proliferation and functional depletion in the blood, peritoneum and spleen related to changes in bone marrow stem cells, *Infect. Immun.* **28**: 824–831.

Cohen, C., and Phillips, G. N., 1981, Spikes and fimbriae: α-helical proteins from surface projections on microorganisms, *Proc. Natl. Acad. Sci. USA* **78**:5303–5304.

Cook, R. M., 1981, Attachment of *Trypanosoma brucei* to rabbit peritoneal exudate cells, *Int. J. Parasitol.* **11**:149–156.

Corsini, A. C., Clayton, C. E., Askonas, B. A., and Ogilvie, B. M., 1977, Suppressor cells and loss of B-cells potential in mice infected with *Trypanosma brucei*, *Clin. Exp. Immunol.* **29**:122–131.

Coulomb, J. Gruvel, J., Morel, P. C., Perreau, P., Quéval, R., Tibayrenc, R., and Provost, A., 1977, *La Trypanotolérance. Synthèse Bibliographique des Connaissances Actuelles*, Institut d'Elévage et de Médecine Vétérinaire des Pays Tropicaux, Paris.

Cross, G. A. M., 1975, Identification, purification and properties of clone-specific glycoprotein antigens constituting the surface coat of *Trypanosoma brucei*, *Parasitology* **71**: 393–417.

Cross, G. A. M., 1979*a*, Immunochemical aspects of antigenic variation in trypanosomes, *J. Gen. Microbiol.* **113**:1–12.

Cross, G. A. M., 1979*b*, Crossreacting determinants in the C-terminal region of trypanosome variant surface antigens, *Nature (Lond.)* **227**:310–312.

Cross, G. A. M., Holder, A. A., Allen, G., and Boothroyd, J. C., 1980, An introduction to antigenic variation in trypanosomes, *Am. J. Trop. Med. Hyg.* **29**:1027–1032.

D'Alesandro, P. A., 1972, *Trypanosoma lewisi*: Production of exoantigens during infection in the rat, *Exp. Parasitol.* **32**:149–164.

Dar, R. K., Paris, J., and Wilson, A. J., 1973, Serological studies on trypanosomiasis in East Africa. IV. Comparison of antigenic types of *Trypanosoma vivax* group organisms, *Ann. Trop. Med. Parasitol.* **67**:319–329.

Dargie, J. D., Murray, P. K., Murray, M., Grimshaw, W., and McIntyre, W. I. M., 1979, Bovine trypanosomiasis: The red cell kinetics of N'Dama and Zebu cattle infected with *Trypanosoma congolense*, *Parasitology* **78**:271–286.

Desowitz, R. S., 1959, Studies on immunity and host parasite relationships. I. The immunological response of resistant and susceptible breeds of cattle to trypanosomal challenge, *Ann. Trop. Med. Parasitol.* **53**:293–313.

Desowitz, R. S., and Watson, H. C. J., 1952, Studies on *Trypanosoma vivax*. III. Observations on the maintenance of a strain in white rats, *Ann. Trop. Med. Parasitol.* **46**:92–100.

Desowitz, R. S., and Watson, H. C. J., 1953, Studies on *Trypanosoma vivax*. IV. The maintenance of a strain in white rats without sheep-serum supplementation, *Ann. Trop. Med. Parasitol.* **47**:62–67.

D'Hondt, J., Van Meirvenne, N., Moens, L., and Kondo, M., 1979, Ca^{2+} is essential cofactor for trypanocidal activity of normal human serum, *Nature (Lond.)* **282**:613–615.

Diggs, C., Flemmings, B., Dillon, J., Snodgrass, R., Campbell, G., and Esser, K., 1976, Immune serum-mediated cytotoxicity against *Trypanosoma rhodesiense*, *J. Immunol.* **116:** 1005–1009.

Dodin, A., Fromentin, H., and Gleye, M., 1962, Mise en évidence d'un antigène vaccinant dans le plasma de souris expérimentalement infectées par diverses espèces de trypanosomes, *Bull. Soc. Pathol. Exot.* **55**:291–299.

Doyle, J. J., 1977, Antigenic variation in the salivarian trypanosomes, *Adv. Exp. Med. Bio.* **93**:31–63.

Eardley, D. D., and Jayawardena, A. N., 1977, Suppressor cells in mice infected with *Trypanosoma brucei*, *J. Immunol.* **119**:1029–1033.

Ehrlich, P., Rochl, W., and Gulbransen, R., 1909, Ueber serum feste Trypanosomenstaemme, *Z. Immunitaetforsch.* **3**:296–299.

Emery, D. L., and Moloo, S. K., 1981, The dynamics of the cellular reactions elicited in the skin of goats by *Glossina morsitans morsitans* infected with *Trypanosoma* (*Nannomonas*) *congolense* or *T.* (*Duttonella*) *vivax*, *Acta Trop.* **38**:15–28.

Emery, D. L., Wells, P. W., and Tenywa, T., 1980*a*, *Trypanosoma congolense:* specific transformation *in vitro* of leukocytes from infected or immunized cattle, *Exp. Parasitol.* **50**:358–368.

Emery, D. L., Akol, G. W. O., Murray, M., Morrison, W. I., and Moloo, S. K., 1980*b*, The chancre–Early events in the pathogenesis of African trypanosomiasis in domestic livestock, in: *The Host-Invader Interplay* (H. Van den Bossche, ed.), pp. 345–356, Janssen Research Foundation, Elsevier/North-Holland Biomedical Press, Amsterdam.

Faghihi Shirazi, M., Holman, M., Hudson, K. M., Klaus, G. G. B., and Terry, R. J., 1980, Complement (C3) levels and the effect of C3 depletion in infections of *Trypanosoma brucei* in mice, *Parasite Immunol.* **2**:155–161.

Finerty, J. F., Krehl, E. P., and McKevin, R. L., 1978, Delayed-type hypersensitivity in mice immunized with *Trypanosoma rhodesiense* antigens, *Infect. Immun.* **20**:464–467.

Fruit, J., Santoro, F., Afchain, D., Duvallet, G., and Capron, A., 1977, Les immuncomplexes circulants dans la trypanosomiase africaine humaine et expérimentale, *Ann. Soc. Belge Med. Trop.* **57**:257–266.

Fulton, J. D., and Lourie, E. M., 1946, The immunity of mice cured of trypanosome infections, *Ann. Trop. Med. Parasitol.* **40**:1–9.

Galvao-Castro, B., Hochmann, A., and Lambert, P.-H., 1978, The role of the host immune response in the development of tissue lesions associated with African trypanosomiasis in mice, *Clin. Exp. Immunol.* **33**:12–24.

Gasbarre, L. C., Hug, K., and Louis, J. A., 1980, Murine T lymphocyte specificity for African trypanosomes. I. Induction of a T lymphocyte-dependent proliferative response to *Trypanosoma brucei*, *Clin. Exp. Immunol.* **41**:97–106.

Gasbarre, L. C., Hug, K., and Louis, J. A., 1981*a*, Murine T lymphocyte specificity for African trypanosomes. II. Suppression of the lymphocyte proliferative response to *Trypanosoma brucei* by systemic trypanosome infection, *Clin. Exp. Immunol.* **45**:165–172.

Gasbarre, L. C., Finerty, J. F., and Louis, J. A., 1981*b*, Nonspecific immune responses in CBA/N mice infected with *Trypanosoma brucei*, *Parasite Immunol.* **3**:273–282.

Gibson, W. C., Marshall, T. F. de C., and Godfrey, D. C., 1980, Numerical analysis of enzyme polymorphism: A new approach to the epidemiology and taxonomy of trypanosomes of the subgenus *Trypanozoon*, *Adv. Parasitol.* **18**:175–246.

Gidel, R., 1962, Etude électrophorétique quantitative en gélose des protéines sériques de bovins, *Rev. Elev. Méd. Vét. Pays Trop.* **15**:259–263.

Gray, A. R., 1965, Antigenic variation in a strain of *Trypanosma brucei* transmitted by *Glossina morsitans* and *G. palpalis*, *J. Gen. Microbiol.* **41**:195–214.

Gray, A. R., 1966, The antigenic relationship of strains of *Trypanosoma brucei* isolated in Nigeria, *J. Gen. Microbiol.* **44:**263–271.
Gray, A. R., 1970, A study of the antigenic relationships of isolates of *Trypanosoma brucei* collected from a herd of cattle kept in one locality for five years, *J. Gen. Microbiol.* **62:** 301–313.
Gray, A. R., and Luckins, A. G., 1980, Features of epidemiological importance in the development of cyclically transmitted stocks of *Trypanosoma congolense* in vertebrate hosts, *Insect. Sci.* **1:**69–72.
Greenblatt, C. L., 1975, Nutritional and immunological factors in the control of infections with *Trypanosoma lewisi*, *Exp. Parasitol.* **38:**342–349.
Greenblatt, C. L., and Lincicombe, D. R., 1966, Identity of trypanosome growth factors in serum. II. Active globulin components, *Exp. Parasitol.* **19:**139–150.
Greenblatt, C. L., and Sheldon, E., 1968, Mouse serum as an environment for the growth of *Trypanosoma lewisi*, *Exp. Parasitol.* **22:**187–200.
Greenblatt, C. L., Jori, L. A., and Cahnman, H. J., 1969, Chromatographic separation of a rat growth factor required by *Trypanosoma lewisi*, *Exp. Parasitol.* **24:**228–242.
Greenwood, B. M., 1974*a*, Possible role of a B-cell mitogen in hypergammaglobulinaemia in malaria and trypanosomiasis, *Lancet* **16:**435–436.
Greenwood, B. M., 1974*b*, Immunosuppression in patients with malaria and sleeping sickness, in: *Parasites in the Immunised Host: Mechanisms of Survival*, Ciba Foundation Symposium, North-Holland, Amsterdam, **25:**138–146.
Greenwood, B. M., and Whittle, H. C., 1976, Complement activation in patients with Gambian sleeping sickness, *Clin. Exp. Immunol.* **24:**133–138.
Greenwood, B. M., Whittle, H. C., and Molyneux, D. H., 1973, Immunosuppression in Gambian trypanosomiasis, *Trans. R. Soc. Med. Hyg.* **67:**846–850.
Grosskinsky, C. M., and Askonas, B. A., 1981, Macrophages as primary target cells and mediators of immune dysfunction in African trypanosomiasis, *Infect. Immun.* **33:**149–155.
Guidot, G., and Roelants, G. E., 1982, Compared susceptibility of Zebu and Baoulé cattle to seringe infection with *Trypanosoma* (*Duttonella*) *vivax* and *T.* (*Nannomonas*) *congolense*, *Rev. Méd. Vét. Pays Trop.* **35:**233–244.
Hawking, F., 1979, The action of human serum upon *Trypanosoma brucei*, *Protozool. Abst.* **3:**199–206.
Hawking, F., Ramsden, D. B., and Whytock, S., 1973, The trypanocidal action of human serum and of baboon plasma, *Trans. R. Soc. Trop. Med. Hyg.* **71:**427–430.
Herbert, W. J., Parratt, D., Van Meirvenne, N., and Lennox, B., 1980, An accidental laboratory infection with trypanosomes of a defined stock. II. Studies on the serological response of the patient and the identity of the infecting organism, *J. Infect.* **2:**113–124.
Hoeijmakers, J. H. J., Frasch, A. C. C., Bernards, A., Borst, P., and Cross, G. A. M., 1980*a*, Novel expression-linked copies of the genes for variant surface antigens in trypanosomes, *Nature* (*Lond.*) **284:**78–80.
Hoeijmakers, J. H. J., Borst, P., Van den Burg, J., Weissmann, C., and Cross, G. A. M., 1980*b*, The isolation of plasmids containing DNA complementary to messenger RNA for variant surface glycoproteins of *Trypanosoma brucei*, *Gene* **8:**391–417.
Holder, A. A., and Cross, G. A. M., 1981, Glycopeptides from variant surface glycoproteins of *Trypanosoma brucei.* C-terminal location of antigenically cross-reacting carbohydrate moieties, *Mol. Biochem. Parasitol.* **2:**135–150.
Holmes, P. H., Mammo, C., Thomson, A., Knight, P. A., Lucken, R., Murray, P. K., Murray,

M., Jennings, F. W., and Urquhart, G. M., 1974, Immunosuppression in bovine trypanosomiasis, *Vet. Rec.* **95**:86–87.

Holmes, P. H., Macaskill, J. A., Whitelaw, D. D., Jennings, F. W., and Urquhart, G. M., 1979, Immunological clearance of ^{75}Se-labelled *Trypanosoma brucei* in mice. I. Aspects of the radiolabelling technique, *Immunology* **36**:415–420.

Hornby, H. E., 1921, Trypanosomes and trypanosomiasis of cattle, *J. Comp. Pathol.* **34**: 211–240.

Hornby, H. E., 1941, Immunisation against bovine trypanosomiasis, *Trans. Soc. Trop. Med. Hyg.* **35**:165–176.

Houba, V., Brown, K. N., and Allison, A. C., 1969, Heterophile antibodies, M-anti-globulins and immunoglobulins in experimental trypanosomiasis, *Clin. Exp. Immunol.* **4**:113–123.

Hudson, K. M., Byner, C., Freeman, J., and Terry, R. J., 1976, Immunodepression, high IgM levels and evasion of the immune response in murine trypanosomiasis, *Nature (Lond.)* **264**:256–258.

Hudson, K. M., Taylor, A. E. R., and Elce, B. J., 1980, Antigenic changes in *Trypanosoma brucei* on transmission by tsetse fly, *Parasite Immunol.* **2**:57–69.

ILCA, 1979, *Trypanotolerant Livestock in West and Central Africa*, International Livestock Center for Africa, Addis Ababa, Ethiopia.

Ishizaka, S., Otani, S., and Morisawa, S., 1977, Effect of carrageenan on immune response. I. Studies on the macrophage dependency of various antigens after treatment with carrageenan, *J. Immunol.* **18**:1213–1218.

Jayawardena, A. N., and Waksman, B H., 1977, Suppressor cells in experimental trypanosomiasis, *Nature (Lond.)* **265**:539–541.

Jayawardena, A. N., Waksman, B. H., and Eardley, D. D., 1978, Activation of distinct helper and suppressor T cells in experimental trypanosomiasis, *J. Immunol.* **121**:622–628.

Jenni, L., 1977*a*, Comparisons of antigenic types of *Trypanosoma (T.) brucei* strains transmitted by *Glossina m. morsitans*, *Acta Trop.* **34**:35–41.

Jenni, L., 1977*b*, Antigenic variants in cyclically transmitted strains of the *T. brucei* complex, *Ann. Soc. Belge Med. Trop.* **57**:383–386.

Jenni, L., Molyneux, D. H., Livesey, J. L., and Galun, R., 1980, Feeding behavior of tsetse flies infected with salivarian trypanosomes, *Nature (Lond.)* **283**:383–385.

Jennings, F. W., Whitelaw, D. D., Holmes, P. H., and Urquhart, G. M., 1978, The susceptibility of strains of mice to infection with *Trypanosoma congolense*, *Res. Vet. Sci.* **25**:399–400.

Johnson, J. G., and Gross, G. A. M., 1977, Carbohydrate composition of variant-specific surface antigen glycoproteins for *Trypanosoma brucei*, *J. Protozool.* **24**:587–591.

Johnson, J. G., and Gross, G. A. M., 1979, Selective cleavage of variant surface glycoproteins from *Trypanosoma brucei*, *Biochem. J.* **178**:689–697.

Kar, S. K., Roelants, G. E., Mayor-Withey, K. S. and Pearson, T. W., 1981, Immune depression in trypanosome-infected mice. VI. Comparison of immune responses of different lymphoid organs, *Eur. J. Immunol.* **11**:100–105.

Kar, S. K., Murray, M., Roelants, G. E., Logan, D. S., and Morrison, W. I., 1983, Immune depression in trypanosome-infected mice. VII. Comparison of the immune suppression generated in spleen and lymph nodes by *T. brucei* and *T. congolense* in two strains of mice (submitted).

Kempt, J., Louis, D., Nattingly, J., Bennet, J., Higuchi, C., Horowitz, M., and Gershon, R. K., 1980, Suppressor cells *in vitro*: Differential Effects of Indomethacin and related compounds, *J. Immunopharmacol.* **2**:471–489.

Ketteridge, D., 1970, The presence of host serum components on the surface of rodent adapted *Trypanosoma vivax*, *J. Protozool.* **17**(Suppl):24.

Klaus, G. G. B., Pepys, M. B., Kitajima, K., and Askonas, B. A., 1979, Activation of mouse complement by different classes of mouse antibody, *Immunology* **38**:687–695.

Klein, F., Mattern, P., and Korman-Bosch, H. J., 1970, Experimental induction of rheumatoid factor-like substances in animal trypanosomiasis, *Clin. Exp. Immunol.* **7**:851–863.

Kobayakawa, T., Louis, J., Izui, S., and Lambert, P.-H., 1979, Autoimmune response to DNA, red blood cells, and thymocyte antigens in association with polyclonal antibody synthesis during experimental African trypanosomiasis, *J. Immunol.* **122**:296–301.

Kosinski, R. J., 1980 Antigenic variation in trypanosomes: A computer analysis of variant order, *Parasitology* **80**:343–357.

Lambert, P.-H., and Galvao-Castro, B., 1977, Rôle de la réponse immune dans la pathogénie de la trypanosomiase africaine, *Ann. Soc. Belge Méd. Trop.* **57**:267.

Laveran, A., and Mesnil, F., 1902, Recherches sur le traitement et la prévention du Nagana, *Ann. Inst. Pasteur* **16**:785–817.

Leeflang, P., Buys, J., and Blotkamp, J., 1976, Studies on *Trypanosoma vivax* infectivity and serial maintenance of natural bovine isolates in mice, *Int. J. Parasitol.* **6**:413–417.

Le Ray, D., Barry, J. D., and Vickerman, K., 1978, Antigenic heterogeneity of metacyclic forms of *Trypanosoma brucei*, *Nature* (*Lond.*) **273**:300–302.

Levaditi, C., and McIntosh, J., 1910, Le mécanisme de la création des variétés de trypanosomes résistants aux anticorps, *C. R. Soc. Biol.* **66**:49–51.

Levine, R. F., and Mansfield, J. M., 1981, Genetics of resistance to African trypanosomes: Role of the H-2 locus in determining resistance to infection with *Trypanosoma rhodesiense*, *Infect. Immun.* **34**:513–518.

Libeau, G., and Pinder, M., 1981, Effet néfaste du lévamisole sur la trypanosomose expérimentale de la souris, *Rev. Elev. Med. Vet. Pays Trop.* **34**:339–404.

Lincicombe, D. R., 1958, Growth of *Trypanosoma lewisi* in the heterologous mouse host, *Exp. Parasitol.* **7**:1–13.

Lincicombe, D. R., and Francis, E. M., 1961, Quantitative studies on heterologous sera inducing development of *Trypanosoma lewisi* in mice, *Exp. Parasitol.* **11**:68–76.

Lindsley, H. B., Janecek, L. L., Gilmansa, A. M., and Hassanei, K. M., 1981, Detection and composition of immune-complexes in experimental African trypanosomiasis, *Infect. Immun.* **33**:407–414.

Lotzsch, R., and Deindl, G., 1974, Trypanosoma congolense. III. Serological responses of experimentally infected cattle, *Exp. Parasitol.* **36**:27–33.

Luckins, A. G., 1972, Studies on bovine trypanosomiasis. Serum immunoglobulin levels in Zebu cattle exposed to natural infection in East Africa, *B. Vet. J.* **128**:523–528.

Luckins, A. G., 1977, Detection of antibodies in trypanosome infected cattle by means of a microplate linked immunosorbent assay, *Trop. Anim. Health Prod.* **9**:53–62.

Lumsden, W. H. R., and Evans, D. A., 1979, *Biology of the Kinetoplastida*, Academic Press, London.

Lumsden, W. H. R., and Herbert, W. J., 1967, Phagocytosis of trypanosomes by mouse peritoneal macrophages, *Trans. R. Soc. Trop. Med. Hyg.* **61**:142.

Lumsden, W. H. R., Herbert, W. J., and McNeillage, G. J. C., 1973, *Techniques with Trypanosomes*, Churchill Livingstone, Edinburgh.

Lyon, J. A., Pratt, J. M., Travis, R. W., Doctor, B. P., and Olenick, J. G., 1981, Use of monoclonal antibody to immunochemically characterize variant-specific surface coat glycoprotein from *Trypanosoma rhodesiense*, *J. Immunol.* **126**:134–137.

Macaskil, J. A., Holmes, P. H., Whitelaw, D. D., McConnel, I., Jennings, F. W., and Urquhart, G. M., 1980, Immunological clearance of ^{75}Se-labeled *Trypanosoma brucei* in mice. 2. Mechanisms in immune animals, *Immunology* **40**:629–635.

MacKenzie, A. R., and Boreham, P. F. L., 1971, The possible involvement of autoimmunity in *Trypanosoma brucei* infections of the rabbit, *Trans. R. Soc. Trop. Med. Hyg.* **65**: 2350.

MacKenzie, P. K. I., Boyt, W. P., and Nesham, V. W., 1979, Serum immunoglobulin levels in sheep during the course of naturally acquired and experimentally induced trypanosomiasis, *Br. Vet. J.* **135**:178–184.

Magnus, E., Vervoort, T., and Van Meirvenne, N., 1978, A card-agglutination test with stained trypanosomes (C.A.T.T.) for the serological diagnosis of *T. b. gambiense* trypanosomiasis, *Ann. Soc. Belge Med. Trop.* **58**:169–176.

Majumder, H. K., Boothroyd, J.-C., and Weber, H., 1981, Homologous 3′-terminal regions of m RNAs for surface antigens of different antigenic variants of *Trypanosoma brucei*, *Nucl. Ac. Res.* **9**:4745–4754.

Mangenot, M., Chaize, J., Desfontaine, M., Duvallet, G., and Moreau, J.-P., 1979, Intérêt de la technique ELISA pour le dépistage dans les foyers de trypanosomiase humaine africaine. Comparaison avec l'immunofluorescence, *Med. Trop.* **39**:527–530.

Mansfield, J. M., and Bagasra, O., 1978, Lymphocyte function in experimental African trypanosomiasis. I. B cell responses to helper T-cell-independent and -dependent antigens, *J. Immunol.* **120**:759–765.

Mansfield, J. M., and Kreier, J. P., 1972*a*, Tests for antibody- and cell-mediated hypersensitivity to trypanosome antigens in rabbit infected with *Trypanosoma congolense*, *Infect. Immun.* **6**:62–67.

Mansfield, J. M., and Kreier, J. P., 1972*b*, Autoimmunity in experimental *Trypanosoma congolense* infections of rabbits, *Infect. Immun.* **5**:648–656.

Mansfield, J. M., Levine, R. F., Dempsey, W. L., Wellhausen, S. R., and Hansen, C. T., 1981, Lymphocyte function in experimental African trypanosomiasis. IV. Immunosuppression and suppressor cells in the athymic nu/nu mouse, *Cell. Immunol.* **63**:210–215.

Marcu, K. B., and Williams, R. O., 1981, Microbial surface elements: The case of variant surface glycoproteins (VSG) genes of African trypanosomes, in: *Genetic Engineering*, Vol. 3 (J. K. Setlow and A. Hollaender, eds.), pp. 129–155, Plenum, New York.

Masake, R. A., 1980, The pathogenesis of infection with *Trypanosoma vivax* in goats and cattle, *Vet. Rec.* **107**:551–557.

Masake, R. A., and Morrison, W. I., 1981, An evaluation of the structural and functional changes in the lymphoid organs of boran cattle infected with *Trypanosoma vivax*, **42**:1738–1746.

Masake, R. A., Pearson, T. W., Wells, P., and Roelants, G. E., 1981, The *in vitro* response to mitogens of leukocytes from cattle infected with *Trypanosoma congolense*, *Clin. Exp. Immunol.* **43**:583–589.

Mattern, P., Masseyeff, R., Michel, R., and Peretti, P., 1978, Etude immunochimique de la 2-macroglobuline des sérums de malades atteints de trypanosomiase africaine à *T. gambiense*, *Ann. Immunol.* **101**:382–393.

Matthysen, G., Michiels, F., Hamers, R., Pays, E., and Steinert, M., 1981, Two variant surface glycoproteins of *Trypanosoma brucei* have a conserved C-terminus, *Nature (Lond.)* **293**:230–233.

Mayor-Withey, K. S., Clayton, C. E., Roelants, G. E., and Askonas, B. A., 1978, Trypanosomiasis leads to extensive proliferations of B, T and null cells in spleen and bone marrow, *Clin. Exp. Immunol.* **34**:359–363.

McNeillage, G. J. C., and Herbert, W. J., 1968, Infectivity and virulence of *Trypanosoma (Trypanozoon) brucei* for mice. II. Comparison of closely related antigenic types, *J. Comp. Pathol.* **78**:345–349.

Mesnil, E., and Brimont, E., 1909, Propriétés protectrices du sérum des animaux trypanosomés: Races résistantes à ces sérums, *Ann. Inst. Pasteur* **23**:129–154.

Mitry-Martin, L., Hudson, K. M., and Terry, R. J., 1979, Autoantibodies to splenocytes in murine trypanosomiasis, *Trans. R. Soc. Trop. Med. Hyg.* **73**:98.

Morrison, W. I., and Murray, M., 1979, *Trypanosoma congolense*: Inheritance of the susceptibility to infection in inbred strains of mice, *Exp. Parasitol.* **48**:364–374.

Morrison, W. I., Roelants, G. E., Mayor-Withey, K. S., and Murray, M., 1978, Susceptibility of inbred strains of mice to *Trypanosoma congolense:* Correlation with changes in spleen lymphocyte populations, *Clin. Exp. Immunol.* **32**:25–40.

Mouton, M., Stiffel, C., and Biozzi, G., 1981, Genetic regulation of high and low immunoresponsiveness, in: *Immunologic Defects in Laboratory Animals* (M. E. Gershwin and B. Merchant, eds.), Vol. 1, pp. 19–47, Plenum, New York.

Mulligan, H. W., 1970, *The African Trypanosomiasis*, Allen and Unwin, London.

Murray, M., 1974, The pathology of African trypanosomiasis, in: *Progress in Immunology* (L. Brent and J. Holborow, eds.), pp. 181–192, North-Holland, Amsterdam.

Murray, M., and Morrison, W. I., 1979*a*, Parasitemia and host susceptibility to African trypanosomiasis, in: *Pathogenicity of Trypanosomes* (G. Losos and A. Chouinard, eds.), pp. 71–81, International Development Research Centre, Ottawa.

Murray, M., and Morrison, W. I., 1979*b*, Non-specific induction of increased resistance in mice to *Trypanosoma congolense* and *Trypanosoma brucei* by immunostimulants, *Parasitology* **79**:349–366.

Murray, M., Morrison, W. I., Emery, D. L., Akol, G. W. O., Masake, R. A., and Moloo, S. K., 1980, Pathogenesis of trypanosome infections in cattle, in: *Isotope and Radiation Research on Animal Diseases and their Vectors*, pp. 15–30, IAEA, Vienna.

Murray, M., Cliffort, D. J., Gettinby, G., Snow, W. F. and McIntyre, W. I. M., 1981, Susceptibility to African trypanosomiasis of Ndama and Zebu cattle in an area of *Glossina morsitans submorsitans* challenge, *Vet. Rec.* **109**:503–510.

Murray, P. K., Murray, M., Wallace, M., Morrison, W. I., and McIntire, W. I. M., 1979*a*, Trypanosomiasis in Ndama and Zebu cattle. An experimental investigation of susceptibility to *Trypanosoma brucei, T. congolense* and mixed infections, in: *International Scientific Council for Trypanosomiasis Research and Control, Proceedings of the 15th Meeting, The Gambia, Eliza Services, Nairobi.* pp. 470–481.

Murray, P. K., Murray, M., Wallace, M., Morrison, W. I., and McIntire, W. I. M., 1979*b*, Trypanosomiasis in Ndama and Zebu cattle. The influence of the weight of infection on the severity of the disease, in: *International Scientific Council for Trypanosomiasis Research and Control, Proceedings of the 15th Meeting, The Gambia, Eliza Services, Nairobi*, pp. 482–487.

Musoke, A. J., Nantulya, V. M., Barbet, A. E., Kironde, F., and McGuire, T. C., 1981, Bovine immune response to African trypanosomes: Specific antibodies to variant surface glycoproteins of *Trypanosoma brucei, Parasite Immunol.* **3**:97–106.

Nagle, R. B., Ward, P. A., Lindsley, H. B., Sadun, E. H., Johnson, A. J., Berkow, R. E., and Hildebrandt, P. K., 1974, Experimental infections with African trypanosomes. VI. Glomerulonephritis involving the alternate pathway of complement activation. *Am. J. Trop. Med. Hyg.* **23**:15–22.

Nantulya, V. M., Musoke, A. J., Barbet, A. F., and Roelants, G. E., 1979, Evidence for reappearance of *Trypanosoma brucei* variable antigen types in relapse populations, *J. Parasitol.* **65**:673–679.

Nantulya, V. M., Doyle, J. J., and Jenni, L., 1980*a*, Studies on *Trypanosoma* (*Nannomonas*) *congolense.* III. Antigenic variation in three cyclically transmitted stocks, *Parasitology* **80**:123–131.

Nantulya, V. M., Doyle, J. J., and Jenni, L., 1980*b*, Studies on *Trypanosoma* (*Nannomonas*) *congolense.* IV. Experimental immunization of mice against tsetse fly challenge, *Parasitology* **80**:133–137.

Nielsen, K., and Sheppard, J., 1977, Activation of complement by trypanosomes, *Experientia* **33**:769–771.

Oka, M., Osaki, H., Furuya, M., Ito, Y., and Oka, Y., 1981, Agglutination antibody responses to *Trypanosoma gambiense* homogenate in mice treated with dextran sulfate and carrageenan, *Zentralbl. Bakteriol. Parasitenkd. Infectionskr. Hyg. Abt. I.* **250**:173–181.

Olenick, J. G., Travis, R. W., and Garson, S., 1981, *Trypanosoma rhodesiense:* Chemical and immunological characterization of variant-specific surface coat glycoproteins, *Mol. Biochem. Parasitol.* **3**:227–238.

Onodera, M., Rosen, N. L., Lifter, J., Hotez, P. J., Bogucki, M. S., Davis, G., Patton, C. L., Konigsberg, W. H., and Richards, F. F., 1981, *Trypanosoma congolense*: Surface glycoproteins of two early bloodstream variants. II. Purification and partial chemical characterization, *Exp. Parasitol.* **52**:427–439.

Osler, G., 1971, Weight estimates of antibody on antigen binding capacity, in: *Methods Immunol. Immunochem.* **3**:73–85.

Pagot, J., 1974, Les races trypanotolérantes, in: *Control Programme for Trypanosomes and Their Vectors* (*Inst. Elev. Méd. Vét. Pays Trop.*), pp. 235–248, Actes du Colloque, Paris.

Pays, E., Lheureux, M., and Steinert, M., 1981*a*, Analysis of the DNA and RNA changes associated with the expression of isotypic variant-specific antigens of trypanosomes, *Nucl. Ac. Res.* **9**:4225–4238.

Pays, E., Lheureux, M., and Steinert, M., 1981*b*, The expression-linked copy of surface antigen gene in *Trypanosoma* is probably to one transcribed, *Nature* (*Lond.*) **292**: 265–267.

Pearson, T. W., Roelants, G. E., Lundin, L. B., and Mayor-Withey, K. S., 1978, Immune depression in trypanosome-infected mice. I. Depressed T lymphocyte responses, *Eur. J. Immunol.* **8**:723–727.

Pearson, T. W., Roelants, G. E., Pinder, M., Lundin, L. B., and Mayor-Withey, K. S., 1979, Immune depression in trypanosome-infected mice. III. Suppressor cells, *Eur. J. Immunol.* **9**:200–204.

Pearson, T. W., Pinder, M., Roelants, G. E., Kar, J. K., Lundin, L. B., Mayor-Withey, K. S., and Hewett, R. S., 1980, Method for derivation and detection of anti-parasite monoclonal antibodies, *J. Immunol. Methods* **34**:141–154.

Pearson, T. W., Kar, S. K., McGuire, T. C., and Lundin, L. B., 1981, Trypanosome variable surface antigens: Studies using two-dimensional gel electrophoresis and monoclonal antibodies, *J. Immunol.* **126**:823–828.

Pearson, T. W., Morrison, W. I., Roelants, G. E., Lundin, L. B., and Mayor-Withey, K. S., 1983, Immune depression in trypanosome-infected mice. V. Correlation of immune suppression with susceptibility to disease in various inbred strains of mice (submitted).

Pierre, C., 1906, L'élevage dans l'Afrique Occidentale Française. Gouvernement Général de l'A.O.F. (Inspection de l'Agriculture), Paris.

Pinder, M., 1983, *Trypanosoma congolense*: Genetic control of resistence to African trypanosome infection in mice (submitted).

Pinder, M., Libeau, G., Hirsch, W., Tamboura, I., Hauck-Bauer, R., and Roelants, G. E., 1983, Anti-trypanosome-specific immune responses in bovids of differing susceptibility to African trypanosomiasis, *Immunology* (in press).

Quéval, R., 1982*a*, Le défaut érythrocytaire de la G6PD dans les races bovines trypanosensibles et trypanotolérantes de l'ouest africain, *Rev. Elev. Méd. Vét. Pays Trop.* **35**:131–136.

Quéval, R., 1982*b*, Polymorphisme de la transferrine chez les bovins. Répartition et fréquence des alléles dans les populations bovines trypanotolérantes et trypanosensibles de l'ouest africain, *Rev. Elev. Méd. Vét. Pays Trop.* **35**:373–380.

Quéval, R., and Petit, J.-P., 1982, Polymorphisme biochimique de l'hémoglobine de populations bovines trypanosensibles et trypanotolérantes et de leur croisement dans l'ouest africain, *Rev. Elev. Méd. Vét. Pays Trop.* **35**:137–146.

Quéval, R., Poivey, J.-P., and Hoste, C., 1983, Blood grouping of West African cattle breeds (in preparation).

Rice-Ficht, A. C., Chen, K. K., and Donelson, J. E., 1981, Sequence homologies near the C-terminus of the variable surface glycoproteins of *Trypanosoma brucei*, *Nature* (*Lond.*) 294:53–57.

Rickman, W. J., Cox, H. W., and Thoongsu, S., 1981, Interaction of immunoconglutinins and immune complexes in cold auto-hemagglutination associated with African trypanosomiasis, *J. Parasitol.* 67:159–163.

Rifkin, M. R., 1978*a*, Identification of the trypanocidal factor in normal human serum: high density lipoprotein, *Proc. Natl. Acad. Sci. USA* 75:3450–3454.

Rifkin, M. R., 1978*b*, *Trypanosoma brucei*: Some properties of the cytotoxic reaction induced by normal human serum, *Exp. Parasitol.* 46:189–206.

Roberts, C. J., and Gray, A. R., 1973, Studies on trypanosome resistant cattle. II. The effect of trypanosomiasis on N'Dama, Muturu and Zebu cattle, *Trop. Anim. Health Prod.* 5:220–233.

Roelants, G. E., and Pinder, M., 1981, Application of monoclonal antibody technology to parasite immunology, in: *The Immune System* (C. M. Steinberg and I. Lefkovist, eds.), Vol. 2, pp. 446–453. S. Karger, Basel.

Roelants, G. E., and Williams, R. O., 1982, African trypanosomiasis, *Crit. Rev. Trop. Med.* 1:31–75.

Roelants, G. E., Pearson, T. W., Tyrer, H. W., Mayor-Withey, K. S., and Lundin, L. B., 1979*a*, Immune depression in trypanosome-infected mice. II. Characterization of the spleen cells involved, *Eur. J. Immunol.* 9:195–199.

Roelants, G. E., Pearson, T. W., Morrison, W. I., Mayor-Withey, K. S., and Lundin, L. B., 1979*b*, Immune depression in trypanosome infected mice. IV. Kinetics of suppression and alleviation by the trypanocidal drug Berenil, *Clin. Exp. Immunol.* 37:457–469

Roelants, G. E., Tamboura, I., Sidiki, D. B., Bassinga, A., and Pinder, M., 1983. Trypanotolerance: An individual not a breed character, *Acta Tropica* 40:99–104.

Roffi, J., Deranin, F., and Diallo, P. B., 1978, Application d'une méthode immuno-enzymatique (ELISA) au dépistage de la trypanosomiase humaine africaine a *Trypanosoma brucei gambiense*, *Méd. Mal. Infect.* 8:9-14.

Rovis, L., Barbet, A. F., and Williams, R. O., 1978, Characterization of the surface coat of *Trypanosoma congolense*, *Nature* (*Lond.*) 271:654–656.

Rurangirwa, F. R., Tabel, H., Losos, G., Masiga, W. N., and Mwambu, P., 1978, Immunosuppressive effect of *Trypanosoma congolense* and *Trypanosoma vivax* on the secondary immune response of cattle to *Mycoplasma mycoides* subsp mycoides, *Res. Vet. Sci.* 25:395–397.

Rurangirwa, F. R., Tabel, H., Losos, G. J., and Tizard, I. R., 1979, Suppression of antibody response to *Leptospira biflexa* and *Brucella abortus* and recovery from immunosuppression after Berenil treatment, *Infect. Immun.* 26:822–826.

Rurangirwa, R. R., Mushi, E. Z., Tabel, H., Tizard, J. R., and Losos, G. J., 1980*a*, The effect of *Trypanosoma congolense* and *T. vivax* infections on the antibody response of cattle to live rinderpest virus vaccine, *Res. Vet. Sci.* 28:264–266.

Rurangirwa, F. R., Tabel, H., Losos, G. J., and Tizard, I. R., 1980*b*, Immunosuppression in bovine trypanosomiasis. The establishment of "memory" in cattle infected with *T. congolense* and the effect of post infection serum on *in vitro* ^{3}H-thymidine uptake by lymphocytes and on leukocyte migration, *Tropenmed. Parasitol.* 31:105–110.

Sacks, D. L., and Askonas, B. A., 1980, Trypanosome induced suppression of antiparasite responses during experimental African trypanosomiasis, *Eur. J. Immunol.* 10:971–974.

Sacks, D. L., Selkirk, M., Ogilvie, B. M., and Askonas, B. A., 1980, Intrinsic immunosuppressive activity of different trypanosome strains varies with parasite virulence, *Nature (Lond.)* 283:476–478.

Scher, I., 1981, B-lymphocyte development and heterogeneity. Analysis with the immune-defective CBA/N mouse strain, in: Immunologic Defects in Laboratory Animals (M. E. Gershwin and B. Merchant, eds.), Vol. 1, pp. 163–190, Plenum, New York.

Scott, J. M., Pegram, R. G., Holmes, P. H., Pays, T. W., Knight, D. A., Jennings, F. W., and Urquhart, G. M., 1977, Immunosuppression in bovine trypanosomiasis: Field studies in using foot-and-month disease vaccine and clostridial vaccine, *Trop. Anim. Health Prod.* 9:159–165.

Scott, J. M., Holmes, P. H., Jennings, F. W., and Urquhart, G. M., 1978, Attempted protection of Zebu cattle against trypanosomiasis using a multistabilate vaccine, *Res. Vet. Sci.* 25:115–117.

Seed, J. R., and Gam, A. A., 1966, Passive immunity to experimental trypanosomiasis, *J. Parasitol.* 52:1134–1140.

Seed, J. R., and Gam, A. A., 1967, The presence of antibody to a normal rabbit liver antigen in rabbits infected with *Trypanosoma gambiense*, *J. Parasitol.* 53:946–950.

Seed, J. R., Cornille, R. L., Risby, E. L., and Gam, A. A., 1969, The presence of agglutinating antibody in the IgM immunoglobulin fractions of rabbit antiserum during experimental African trypanosomiasis, *Parasitology* 59:283–292.

Selkirk, M. E., Ogilvie, B. M.. and Platts-Mills, T. A. E., 1981, Activation of human peripheral blood lymphocytes by a trypanosome-derived mitogen, *Clin. Exp. Immunol.* 45:615–620.

Sollod, A. E., and Frank, G. H., 1979, Bovine trypanosomiasis effect on the immune response of the infected host, *Am. J. Vet. Res.* 40:658–664.

Soltys, M. A., 1955, Studies on resistance to *T. congolense* developed by Zebu cattle treated prophylactically with Antrycide prosalt in an enzootic area of East Africa, *Ann. Trop. Med. Parasitol.* 49:1–8.

Soltys, M. A., 1957, Immunity in trypanosomiasis. I. Neutralization reactions, *Parasitology* 47:375–389.

Stephen, L. E., 1966, Observations on the resistance of West African N'Dama and Zebu cattle to trypanosomiasis following challenge by wild *Glossina morsitans* from an early age, *Ann. Trop. Med. Parasitol.* 60:230–246.

Stevens, D. R., and Moulton, J. E., 1978, Ultrastructural and immunological aspects of the phagocytosis of *Trypanosoma brucei* by mouse peritoneal macrophages, *Infect. Immun.* 19:972–982.

Tait, A., 1980, Evidence for diploidy and mating in trypanosomes, *Nature (Lond.)* 287: 536–538.

Takayanagi, T., and Enriquez, G. L., 1973, Effects of the IgG and IgM immunoglobulins in *Trypanosoma gambiense* infection in mice, *J. Parasitol.* 59:644–647.

Takayanagi, T., Nakatake, Y., and Enriquez, G. L., 1974*a*, Attachment and ingestion of *Trypanosoma gambiense* to the rat macrophage by specific antiserum, *J. Parasitol.* 60:336–339.

Takayanagi, T., Nakatake, Y., and Enriquez, G. L., 1974*b*, *Trypanosoma gambiense:* Phagocytosis *in vitro*, *Exp. Parasitol.* 36:106–113.

Takayanagi, T., Nakatake, Y., and Kato, H., 1977, *Trypanosoma gambiense:* Antigens involved in phagocytosis, *Exp. Parasitol.* 43:414–422.

Takayanagi, T., Takayanagi, M., Yabu, Y., and Kato, H., 1978, *Trypanosoma gambiense*: Immune responses of neonatal rats receiving antibodies from the female, *Exp. Parasitol.* 44:82–91.

Terry, R. J., 1976, Innate resistance to trypanosome infections, in: *Biology of the Kinetoplastida* (W. H. R. Lumsden and D. A. Evans, eds.), Vol. 1, pp. 477–489, Academic Press, London.

Tizard, I. R., and Soltys, M. A., 1970, Macrophage cytophilic antibodies to *Trypanosoma brucei* in rabbits, *Trans. R. Soc. Trop. Med. Hyg.* **65**:407–408.

Tizard, I. R., and Soltys, M. A., 1971, Cell mediated hypersensitivity in rabbits infected with *Trypanosoma brucei* and *Trypanosoma rhodesiense*, *Infect. Immun.* **4**: 674–677.

Touré, S. M., Gueye, A., Seye, M., Ba, M. A., and Mane, A., 1978, Expérience de pathologie comparée entre bovins Zébus et Ndamas soumis à l'infection naturelle par des trypanosomes pathogènes, *Rev. Elev. Méd. Vét. Pays. Trop.* **31**:293-313.

Van Hoeve, K., 1971, Some observations on the performance of Ndama and Muturu cattle under natural conditions in Northern Nigeria *in Report of the International Scientific Council for Trypanosomiasis Research, Thirteenth Meeting, Lagos, OAU/STRC Publication 105, Elisa Services, Nairobi*, pp. 103-106.

Van Marck, E. A. E., and Vervoort, T., 1980, *Trypanosoma brucei*: Detection of antigen deposits in glomeruli of mice vaccinated with purified variable antigen, *Trans. R. Soc. Med. Hyg.* **74**:666-667.

Van Meirvenne, N., Magnus, E., and Janssens, P. G., 1976, The effect of normal human serum on trypanosomes of distinct antigenic type (ETAT 1 to 12) isolated from a strain of *Trypanosoma brucei rhodesiense, Ann. Soc. Belge Med. Trop.* **56**:55-63.

Vervoort, T., Magnus, E., and Van Meirvenne, N., 1978, Enzyme-linked immunosorbent assay (ELISA) with variable antigens for serodiagnosis of *T. b. gambiense* trypanosomiasis, *Ann. Soc. Belge. Méd. Trop.* **58**:177-183.

Vickerman, K., 1969, On the surface coat and flagellar adhesion in trypanosomes, *J. Cell. Sci.* **5**:163–194.

Vickerman, K., 1978, Antigenic variation in trypanosomes, *Nature (Lond.)* **273**:613–617.

Vickerman, K., and Luckins, A. G., 1969, Localization of variable antigens in the surface coat of *Trypanosoma brucei* using ferritin conjugated antibody, *Nature (Lond.)* **224**: 1125–1126.

Viens, P., Targett, G. A. T., and Lumsden, W. H. R., 1975, The immunological response of CBA mice to *Trypanosoma musculi*: Mechanisms of protective immunity, *Int. J. Parasitol.* **5**:235–239.

Voller, A., and De Savigny, D., 1981, Diagnostic serology of tropical parasitic diseases, *J. Immunol. Methods.* **46**:1–29.

Voller, A., Bidwell, D. E., and Bartlett, A., 1975, A serological study on human. *T. rhodesiense* infections using a microscale ELISA, *Tropenmed. Parasitol.* 26:247-251.

Wellde, B. T., Duxbury, R. E., Sadun, E. H., Lanbehn, H. R., Lotzsch, R., Deindl, G., and Warni, G., 1973, Experimental infections with African trypanosomes. IV. Immunization of cattle with gamma-irradiated *Trypanosoma rhodesiense, Exp. Parasitol.* **34**: 62–68.

Wellde, B. T., Schoenbechler, M. J., Diggs, C. L., Langbehn, H. R., and Sadun, E. H., 1975, *Trypanosoma rhodesiense*: Variant specificity of immunity induced by irradiated parasites, *Exp. Parasitol.* **37**:125–129.

Wellde, B. T., Hockmeyer, W. T., Kovatch, R. M., Bhogal, M. S., and Diggs, C. L., 1981, *Trypanosoma congolense*: Natural and acquired resistance in the bovine, *Exp. Parasitol.* **52**:219–232.

Wellhausen, S. R., and Mansfield, J. M., 1979, Lymphocyte function in experimental African trypanosomiasis. II. Splenic suppressor cell activity, *J. Immunol.* **122**:818–824.

Wellhausen, S. R., and Mansfield, J. D., 1980*a*, Lymphocyte function in experimental African trypanosomiasis. III. Loss of lymph node cell responsiveness, *J. Immunol.* **124**:1183–1186.

Wellhausen, S. R., and Mansfield, J. M., 1980*b*, Characteristics of splenic suppressor cell–

target cell interaction in experimental African trypanosomiasis, *Cell. Immunol.* **54:** 414–424.

Whitelaw, D. D., Macaskill, J. A., Holmes, P. H., Jennings, F. W., and Urquhart, G. M., 1980, Genetic resistance to *Trypanosoma congolense* infections in mice, *Infect. Immun.* **27:**707–713.

Wiesenhutter, Von E., 1970, Immunity experiments in Zebu against a *Trypanosoma congolense* infection, *Berl. Munch. Tierarztl. Eschr.* **83:**218–220.

Williams, R. O., Marcu, K. B., Young, J. R., Rovis, L., and Williams, S. C., 1978, A characterization of m RNA activities and their sequence complexities in *Trypanosoma brucei*: Partial purification and properties of the VSSA m RNA, *Nuc. Ac. Res.* **5:**3171–3182.

Williams, R. O., Young, J. R., and Majiwa, P. A. O., 1979, Genomic rearrangements correlated with antigenic variation in *Trypanosoma brucei*, *Nature* (*Lond.*) **282:**847–849.

Williams, R. O., Young, J. R., Majiwa, P. A. O., Doyle, J. J., and Shapiro, S. Z., 1980, Analysis of variable antigen gene rearrangements in *Trypanosoma brucei*, *Am. J. Hyg. Trop. Med.* **29:**1037–1042.

Williams, R. O., Young, J. R., Majiwa, P. A. O., Doyle, J. J., and Shapiro, S. Z., 1981, Contextural genomic rearrangements of variable antigen genes in *Trypanosoma brucei*, *Cold Spring Harbor Symp. Quant. Biol.* **45:**945–950.

Wilson, A. J., and Cunningham, M. P., 1972, Immunological aspects of bovine trypanosomiasis. I. Immune response of cattle to infection with *Trypanosoma congolense* and the antigenic variation of the infecting organisms, *Exp. Parasitol.* **32:**165–173.

Wilson, A. J., Paris, J., and Dar, F. K., 1975*a*, Maintenance of a herd of breeding cattle in an area of high trypanosome challenge, *Trop. Anim. Health. Prod.* **7:**63–71.

Wilson, A. J., Le Roux, J. G., Paris, J. Davidson, C. R., and Gray, A. R., 1975*b*, Observations on a herd of beef cattle maintained in a tsetse area. I. Assessment of chemotherapy as a method for the control of trypanosomiasis, *Trop. Anim. Health. Prod.* **7:**187–199.

Wilson, A. J., Paris, J., Luckins, A. G., Dar, F. K., and Gray, A. R., 1976, Observations on a herd of beef cattle maintained in a tsetse area. II. Assessment of the development of immunity in association with trypanocidal during treatment, *Trop. Anim. Health. Prod.* **8:**1–12.

Yorke, W., Adams, A. R. D., and Murgatroyd, D. F., 1930, Studies in chemotherapy: The action *in vitro* of normal human serum on the pathogenic trypanosomes and its significance, *Ann. Trop. Med. Parasitol.* **24:**115–163.

Zahalsky, A. C., and Weinberg, R. L., 1976, Immunity to monomorphic *Trypanosoma brucei*: Humoral response, *J. Parasitol.* **62:**15–19.

Chapter 8

Rodent Models of Filariasis

Mario Philipp,* M. J. Worms, and R. M. Maizels

Division of Parasitology
National Institute for Medical Research
London NW7 IAA, England

and

Bridget M. Ogilvie

The Wellcome Trust
London NW1 4LJ, England

I. INTRODUCTION

Filariasis is one of the most challenging problems for immunoparasitology today. This long-term chronic infection in humans causes extensive morbidity but little mortality (Nelson, 1979; Ottesen, 1980; Ogilvie and Mackenzie, 1981). Existing chemotherapeutic regimens suffer from inefficacy against mature parasites and may incur unpleasant side effects; control of the arthropod vectors has not yet proved sufficiently effective to interrupt transmission. For these reasons, immunologic analysis, directed toward both detection and prophylaxis of infection, may play a major role in controlling a disease currently afflicting some 300 million people (Sasa, 1976).

Several features of filariasis contribute to its unusual complexity as a group of diseases. First, it is a spectral disease, and infection may have a wide range of sequelae from chronic skin and eye disease and elephantiasis, to the benign carrier state of microfilaremia, and the possibly allergic condition of tropical eosinophilia with pulmonary involvement (Ottesen *et al.*, 1979). Second, there are at least seven species pathogenic to humans, having remarkably different vectors,

*Present address: New England Bio-Labs, Beverly, Massachusetts 01915.

host tissue sites, and susceptibility to chemotherapy. Third, in addition to problems of vector maintenance, the restriction of definitive host specificity is a severe constraint on studying human filarial parasites in the laboratory.

It is this host specificity that has led to a wide diversity of studies of filariasis. Some pathogenic filariae may only be studied in certain, often scarce, species, e.g., *Onchocerca volvulus* in the chimpanzee (Duke, 1962) or *Wuchereria bancrofti* in the silvered leaf monkey (Palmieri *et al.*, 1980). The disparity not only amongst parasites studied, but also amongst the mammalian hosts that will permit such studies, has hampered our advance toward a generalized understanding of the immunobiology of filariasis.

Rodent models of filariasis should therefore circumvent the problems of supply, variability, and poor characterization of host species. Both filarial parasites naturally found in rodents and those artificially introduced have been studied; a synopsis of such host-parasite combinations is presented in Table I. A number of considerations now make the development of inbred rodent models essential. One key requirement is the analysis of lymphocytic populations in filarial infection; cell transfers and cultures of syngeneic cell subsets will be necessary for such studies. Inbred animals also offer, in addition to a greater homogeneity in responsiveness to antigenic challenge, the possibility of detecting strain-associated differences in antiparasite responses, leading to the identification of genetically defined influences or lesions in the host response.

As the best immunologically characterized nonhuman hosts, rats and mice would be the first choice host for immunoparasitologic studies. Cell markers, antibody classes, and congenic strains are all extensively characterized, and reagents are available for their definition or separation. The literature on the immunology of host species is immense, and a wealth of techniques and information is readily at hand for the researcher in filariasis. The purpose of this chapter is to assess where we now stand in attempts to transfer the knowledge gained from many years' work on filarial parasites in nonrodent hosts, to model systems that should be more amenable to experimental manipulation and intervention.

Host responses to developing larvae and adult worms in different rodent models are analyzed in Section II, followed by a review of the mechanisms controlling microfilarial burdens.

II. RESPONSES OF THE HOST TO DEVELOPING LARVAE AND ADULT WORMS

A. Studies in Susceptible Hosts Infected with Filariae Naturally Parasitic in Rodents

The life cycles of some (10) species of naturally occurring filarial parasites of rodents have been described (Schacher, 1973), but to date only four have been unevenly explored as laboratory models. These are *Litomosoides carinii*, *Dipetalonema viteae*, *Monanema globulosa*, and *Breinlia booliati*, and only the first two species have been studied more extensively.

1. *Responses to Developing Larvae and Adult Worms of L. carinii*

Litomosoides carinii has been maintained cyclically in the laboratory for many years in a natural host (cotton rat) and vector mite (*Ornithonyssus bacoti*) system (Williams and Brown, 1945) and is readily transmissible to a number of other laboratory rodent species. In cotton rats, third-stage infective larvae (L3) introduced into the skin migrate within 6 days to the pleural cavities, in which all subsequent development takes place. The moult to the fourth larval stage occurs about 8–10 days after exposure, and the moult to the adult stage about day 24. Mature adult worms shed microfilariae into the pleural fluid by day 40, and these gain access to the blood, appearing in the peripheral circulation from day 50.

A similar course of development is obtained in multimammate rats (*Mastomys natalensis*) (Pringle and King, 1968; Lämmler *et al.*, 1968) and the jird (gerbil) (*Meriones unguiculatus*) (Zein-Eldin, 1965). In addition, young albino rats of several strains are susceptible to infection (Olson *et al.*, 1954; Olson, 1959; Ramakrishnan *et al.*, 1961), but develop immunity to microfilariae, hence suppression of microfilaremia. This host system has therefore been employed especially in the study of latency (Section III).

a. Resistance to Infection Induced by Antigens of Third-Stage Larvae and Other Stages. The results of research with *L. carinii* point to the concept that immunologic stimuli originating from the parasites of a given stage in development may affect both parasites of that stage and of others.

Scott and co-workers established that during the first 7 days of development in the cotton rat, third-stage larvae were the principal antigenic source and also the target of an immune response that caused retardation of the growth of a challenge infection (Scott and MacDonald, 1958; MacDonald and Scott, 1958; Scott *et al.*, 1958*a*). Interestingly, fourth-stage larvae and young adult worms were also effective, but less so in inducing this type of immunity (Scott *et al.*, 1958*b*).

Rao *et al.* (1977, 1980) found that resistance measured by a significantly reduced adult worm recovery and microfilaremia was induced in albino rats by an injection of irradiated infective larvae. The irradiation dose was critical, and effective resistance was obtained only when larvae were exposed to a dose (40 krad) that, although not lethal, prevented moulting. Similar results were obtained by Storey and Al-Mukhtar (1982). Successful filarial vaccines based on irradiated larval products have been reported against *Dirofilaria immitis* in dogs and against *Brugia* spp. in dogs, monkeys, and cats (reviewed by Denham, 1980), but it is unlikely that this approach will yield results of direct practical benefit (Ogilvie and Mackenzie, 1981). At the experimental level, however, such studies serve to establish which developmental stage may be a source of protective antigens, and whether a particular mode of antigen presentation associated with larval migra-

Table I. Rodent

Parasite	Natural host	Location of: Adult	Microfilariae	Vector
Breinlia booliati	*Rattus sabanus*	Thoracic and peritoneal cavities	Blood	Mosquito
Brugia malayi	Humans	Lymphatics	Blood	Mosquito
Brugia pahangi	Cat, dog	Lymphatics	Blood	Mosquito
Brugia patei	Cat, dog	Lymphatics	Blood	Mosquito
Brugia timori	Humans	Lymphatics	Blood	Mosquito
Dipetalonema vitae	Jird	Subcutaneous tissues	Blood	Ornithodorid ticks
Litomosoides carinii	Cotton rat (*Sigmodon hispidus*)	Pleural cavity	Blood	Dermanyssid mite, *Ornithonyssus bacoti*
Loa loa	Humans	Subcutaneous tissues	Blood	Tabanid flies
Mansonella ozzardi	Humans	Body cavities	Blood	*Culicoides* spp.
Monanema globulosa	*Lemniscomys striatus*	Pulmonary arteries	Skin	Ixodid tick, *Haemaphysalis leachii*
Onchocerca gutturosa	Cattle	Nuchal ligaments	Skin	*Simulium* spp.
Onchocerca lienalis	Cattle	Gastrosplenic tissue	Skin	*Simulium* spp.
Onchocerca volvulus	Humans	Subcutaneous tissues	Skin	*Simulium* spp.
Wuchereria bancrofti	Humans	Lymphatics	Blood	Mosquito

[a]Abbreviations: +++, full development similar to natural host; ++, less development but life cycle completed;

Filarial Systems[a]

Development in rodent hosts							
Mouse	Rat	Hamster	Jird	Cotton rat	Multimammate rat	Guinea pig	References
	+++						Ho *et al.* (1976*a*)
Mf, –	+	++	+++	++	+	–	Ahmed (1967*a,b*); Ash and Riley (1970*b*); Edeson *et al.* (1962); Laing *et al.* (1961); Petrányi *et al.* (1975)
Mf,/–	+/++	+	+++	++	+	–	Ash and Riley (1970*a*); Laing *et al.* (1961); Ramachandran and Pacheco (1965)
			+++				Ash (1973*a*)
			+				Partono *et al.* (1977)
Mf, +/–	Mf, –	++	+++		+		Chabaud (1954); Worms *et al.* (1961) and unpublished data; Weiss (1970); Haque *et al.* (1980*a,b*)
Mf,/–	++	+/–	+++	+++	+++	–	Hawking and Sewell (1948); Pringle and King (1968); Siddiqui (1979); M.J. Worms (unpublished data, 1982)
			–				Suswillo *et al.* (1977)
			–				Suswillo *et al.* (1977)
			++				Muller and Nelson (1975); Bianco (1975); A.E. Bianco, R.L. Muller, and G.S. Nelson (personal communication, 1982)
Mf/ –	Mf	Mf				Mf	Townson and Bianco (1982); Nelson *et al.* (1966)
Mf, –			–		–		Townson *et al.* (1981); A.E. Bianco (personal communication, 1982)
Mf		–	–				Suswillo *et al.* (1977); Aoki *et al.* (1980)
–		–	+,–				Ash and Schacher (1971); Suswillo *et al.* (1977); Ramachandran *et al.* (1966); Yokogawa 1939)

+, partial development; –, Refractory; /, Strain dependence; Mf, will accept microfilariae.

tion is essential to effect host protection. A recent study by Mehta *et al.* (1981) indicates that with *L. carinii* the latter is not the case, nor do protective antigens appear to be stage exclusive. In albino rats immunized with either sonicated microfilariae or infective larvae in complete Freund's adjuvant (CFA), adult worm recovery was reduced from 12% to 1%, and microfilariae were not detected. Homogenates of adult worms that did not contain microfilarial material were ineffective. These results are encouraging from the point of view of vaccine production because they imply that protective antigenic stimuli need not require the migration of larvae nor indeed infective larval material at all.

b. Developing Larvae and the Possible Induction of Immune Unresponsiveness to Adult Worms. Perhaps the most significant example of immunologic interaction between different developmental stages in infection with *L. carinii* is provided by the results obtained when adult worms are transplanted into naive and infected animals. Adult worms transplanted into the pleural cavity of cotton rats (Wharton, 1946; Fujita and Kobayashi, 1969; Storey and Al-Mukhtar, 1983), jirds (Weiner and Soulsby, 1975), or multimammate rats (Weiner and Soulsby, 1976) were killed within 10–20 days if the animals were naive, but survived if they were transplanted into animals previously infected with irradiated (Storey and Al-Mukhtar, 1983) or normal infected larvae. Splenectomized naive hosts, however, did not reject the adult worm transplant (Weiner and Soulsby, 1978). It is thus conceivable that developing larvae, or third stage larvae, induce a state of immune unresponsiveness that permits survival of the adult worms in an otherwise immunologically hostile environment. This phenomenon and the nature of the protective responses stimulated by antigens of developing larvae or microfilariae have not been further investigated.

2. *Immune Mechanisms in D. viteae Infections*

a. Importance of Antigens on the Surface of Larval Stages. Immune responses to developing larvae of *D. viteae* have been recently investigated in jirds (*M. libycus, M. unguiculatus*) (Gass *et al.*, 1979; Tanner and Weiss, 1981*b*). In this rodent, the worms develop to maturity mostly in the subcutaneous tissues. Third-stage larvae moult about 7 days after infection, and the fourth-stage larvae become young adult worms by day 21 of infection. The prepatent period lasts for about 55 days, and microfilaremia persist for as long as 17 months (Chabaud, 1954; Worms *et al.*, 1961; Weiss, 1970). In hamsters, infections with *D. viteae* have a course similar to that in jirds, but although some strains of hamsters may be even more susceptible to infection than jirds, microfilaremia is short-lived (Weiss, 1970). The hamster is therefore used as a model to investigate amicrofilaremic filariasis (Section III).

Tanner and Weiss (1981*a*) studied the effect of previous infections or immunization of jirds, with irradiated or dead infective larvae, on the survival and development of larvae implanted subcutaneously inside micropore chambers. The chambers were implanted 2 weeks after immunization with dead or irradiated (34 krad) living L3, or infected with nonirradiated living L3. Larvae recovered from infected jirds showed a significant reduction in length as compared with controls. A similar effect was observed in jirds immunized with irradiated larvae, but not in animals given dead larvae as immunogens. Interestingly, the latter had no anti-cuticular serum antibodies to L3, whereas sera from jirds given living irradiated or nonirradiated L3 showed anti-cuticular antibodies by fluorescence. Moreover, jirds given irradiated L3 were largely resistant to a challenge infection, whereas hamsters similarly immunized showed no resistance to reinfection and were without anti-cuticular serum antibodies (Tanner and Weiss, 1981*b*). A similar correlation between presence of anti-cuticular serum antibodies and impairment of larval growth and motility was found by Gass *et al.* (1979). These investigators used mice, which are insusceptible to *D. viteae* infections, but which support larval development at least until the fourth stage. Micropore chambers containing L3 were subcutaneously implanted into mice (inbred C57BL/6 strain) previously immunized with two intraperitoneal injections of L3. As before, inhibition of larval growth and impaired larval motility correlated with the appearance of anticuticular antibodies. We should mention, however, that an experiment of the same type designed to investigate the role of serum components in larval growth inhibition gave inconsistent results (Tanner and Weiss, 1981*a*). These investigators found that arrested development occurred both in naive and in immunized jirds. This effect was obtained when using micropore chambers with a pore diameter that would only permit passage of host fluids, but not cells (0.4 μm). The use of micropore chambers makes it possible to investigate otherwise inaccessible host–parasite interactions; however, the technique may introduce artifacts arising from the restrictions imposed on parasite migration, or from the inflammatory response induced by the chamber itself; also, worm recovery rates may differ markedly in similarly susceptible hosts. For example, larval recovery rates from micropore chambers implanted into naive hamsters were only a fraction of the rate observed in similarly implanted jirds (Glass *et al.*, 1979).

Further insight into the role of anti-larval antibodies in development was obtained by Tanner and Weiss (1981*b*), using an *in vitro* culture system in which infective larvae developed up to the fourth stage (Tanner, 1981). Third-stage larvae were incubated in medium BHK-21 supplemented with 10% tryptose phosphate and 10% serum of either normal or immunized jirds. Larval development was triggered by previously implanting the worms for 5 days into normal jirds, keeping the parasites enclosed in subcutaneous micropore chambers.

By day 1 of incubation, 3% of the larvae incubated with normal serum and 3% of those treated with immune serum were dead. By day 3, mortality within the latter group had risen to 31% but remained at 4% in the control group. Six days after the beginning of the incubation all larvae treated with immune serum had died, and none of them had moulted. In contrast, 50% of the control larvae had moulted, and only 16% were dead. Larvae that had not been triggered into development by implantation into jirds were not killed by immune serum in culture, thereby suggesting that relevant antigens are not present on the larvae before triggering, or that the larvicidal effect may arise as a result of an interference with development (Tanner and Weiss, 1981*b*).

Neither normal nor immune sera mediated adherence to the worms *in vitro* of peritoneal exudate or spleen cells of jirds, but immune sera reacted with the larval surface, as shown by immunofluorescence (Weiss and Tanner, 1981). Larval killing by immune sera *in vitro* was eliminated by heat treatment of the sera for 30 min at 56°C (Tanner and Weiss, 1981*b*). In view of these results, these workers suggested that larval killing in immunized jirds is mediated by antibody (probably IgM) directed at the surface of the worms and is effected by complement. However, the involvement of IgE antibodies, which would also be affected by heat, was not formally excluded.

b. Survival of the Adult Worms. The mechanism whereby adult worms can survive for ≤15 months in jirds and hamsters is less clear. Unlike *L. carinii*, adult worms of *D. viteae* survive transplantation into naive susceptible hosts, such as hamsters, for >24 weeks (Weiss, 1970; Neilson, 1979) and for about 30–35 days when transplanted into naive naturally resistant hosts, such as mice and rats (Thompson *et al.*, 1979; Haque *et al.*, 1980*b*). Thus adult worms of *D. viteae* do not seem to require a previous, possibly immune-suppressive course of infection in order to survive transfer to a new host. In fact, infection of mice with *D. viteae* L3 before a subcutaneous transplant of adult female worms had the effect of reducing both duration and intensity of microfilaremia in these mice, but did not affect the adult worm recovery rate (Haque *et al.*, 1980*a*). It is possible that adult worms themselves produce substances that induce immune unresponsiveness. For example, hamsters infected with *D. viteae* showed a strain-dependent inhibition of *in vitro* blastogenesis of spleen and lymph node cells by adult worm extracts (Weiss, 1978*b*). In hamsters of the LSH/SSLak inbred strain, cellular unresponsiveness started by week 12 of infection and by week 30 in animals of the outbred LAKZ strain. However, this antigen-specific suppression did not correlate with adult worm recoveries in either strain (Weiss, 1978*b*). Male worms released a substance *in vitro* that enhanced the microfilaremia of infected hamsters when injected by day 40 of infection (Haque *et al.*, 1978*b*). However, the effect of this material on adult worm survival was not studied.

3. *Two Potential Models of Experimental Filariasis Monanema globulosa and Breinlia booliati*

Monanema globulosa (Anderson and Bain, 1976), formerly classified within the genus *Ackertia*, is a parasite of East African rodents (Muller and Nelson, 1975), which is probably transmitted by a hard tick (*Haemaphysalis leachi*). This experimentally transmitted rodent filaria has microfilariae that are skin dwelling, and it may therefore be useful as a model for *Onchocerca volvulus* infections. Furthermore, it has been successfully transmitted to jirds (*M. unguiculatus*) in which microfilariae have been detected in the skin by day 70–80 of infection (Bianco, 1975), but it has not yet been adopted as a model system.

Breinlia booliati naturally infects the Malayan forest rat (*Rattus sabanus*) and a few other wild rat species of Southeast Asia (Singh and Cheong, 1971; Singh and Ho, 1973; Mak and Lim, 1974). Third-stage larvae, which are mosquito-borne, have been shown to infect a few strains of laboratory rats successfully, including the Charles River inbred strain (Singh *et al.*, 1972; Ho *et al.*, 1976*a*). In these rats the course of infection is comparable to that in the natural host. The third moult occurs by days 6-8 of infection and the final one by days 24–28. About 5 weeks after infection, worms are found in the peritoneal and thoracic cavities, where they become sexually mature by week 11 (Singh *et al.*, 1976). Microfilariae, which are unsheathed, appear in the peripheral blood 11–14 weeks after infection and are nocturnally subperiodic in the natural host, but not in the Charles River albino rat (Yap *et al.*, 1975).

It is surprising that this system has not been further investigated. *B. booliati* has been adapted to laboratory-reared mosquito species such as *Aedes togoi* and *Armigeres subalbatus* (Ho *et al.*, 1976*b*) and to inbred strains of rats; the body cavity-dwelling and blood-borne stages can readily by counted. The fact that no other than parasitologic studies with the *B. booliati*-rat model have as yet been reported should perhaps be attributed to its recent discovery.

B. Studies in Rodents Susceptible to Infections with Nonrodent Filariae, Including Parasites of Humans

Full development in rodents of filarial nematodes that infect humans has only been achieved with parasites of the genus *Brugia*. Jirds, rats, hamsters, and multimammate rats are all susceptible to infections with *Brugia* spp., but only the jird and to some extent the rat have been employed to investigate host responses to infection.

1. *The Jird–Brugia Model*

The jird (*M. unguiculatus*) has been shown to be susceptible to *B. pahangi* (Ash and Riley, 1970*a*), *B. malayi* (Ash and Riley, 1970*b*), *B. patei* (Ash, 1973*a*), and *B. timori* (Partono *et al.*, 1977). Of these four species, *B. timori* (David and Edeson, 1965), which together with *B. malayi* is a parasite of humans (Brug, 1927; Lichtenstein, 1927), is the least adaptable to the jird: Worm recovery rates were in the order of 16% by day 95–100 of a subcutaneous infection with L3, but microfilariae were found only in the visceral blood and not in the circulation (Partono *et al.*, 1977).

B. pahangi and *B. patei*, parasites of cats and dogs (Buckley, 1958) and *B. malayi*, have a similar course of infection in the jird, which is equally susceptible to all three species, but development as determined by the occurrence of larval moults and onset of patency is most rapid with *B. pahangi* and slowest with *B. patei* (Ash and Riley, 1970*a,b*; Ash, 1973*a*). Recovery of adult worms from subcutaneous tissue infections is similar for all three *Brugia* species (~15%) in male jirds. Female jirds are less susceptible (Ash, 1971). Recovery of adult worms from infections of *B. pahangi* induced in the peritoneal cavity of male jirds can, however, be as high as 50% (McCall *et al.*, 1973). Peak microfilarial densities approach 70 microfilariae/20 μl blood in the case of *B. pahangi* and *B. malayi* (Ash and Riley, 1970*a,b*), but can be as high as 100 microfilariae/20 μl blood for *B. patei*, and microfilaremia may persist for as long as 3 years (Ash, 1973*b*).

Three basic questions relevant to human filariasis have been addressed with the jird-*Brugia* model—the more general one of how to induce a degree of resistance to infection in a susceptible host by previous exposure to parasite material, as well as the cause and nature of the immune unresponsiveness and the pathology associated with lymphatic filariasis.

a. Attempts to Immunize Jirds against Infections with Brugia spp. Some attempts have been made to immunize jirds to infections with *B. pahangi* by their previous exposure to living parasite material, but the results of these studies are controversial. In fact, there appears to be better evidence that previous exposure of jirds to a course of infection favors susceptibility rather than resistance. Kowalski and Ash (1975) found some resistance developing in jirds against infections with *B. pahangi.* They reported that a decrease in the number of established larvae could be induced by increasing the number of subcutaneous inoculations and by raising the larval dose inoculated. Adult worm recovery rates of *B. pahangi* were found to be reduced in jirds subcutaneously immunized with irradiated infective larvae of *L. carinii*, compared with unimmunized controls (Storey and Al-Mukhtar, 1982). In contrast, Suswillo (cited in Denham and McGreevy, 1977) found no difference in the mean worm recovery rates of jirds given L3 intraperitoneally every week for 5, 10, or 15 weeks, a result that perhaps indicates that the intraperitoneal route is less immunogenic than the subcutaneous route, and is consistent with the greater susceptibility of jirds thus infected (McCall *et al.*, 1973).

Finally, Klei *et al.* (1980) found evidence for increased susceptibility rather than resistance in infected jirds given a homologous challenge infection with *B. pahangi.* Adult worm recovery rates from a 30-day-old infection were twice as high in jirds already harboring worms from an older (135 or 246 days) subcutaneous infection, compared with the recovery rates in unprimed animals. In contrast, intraperitoneal infections did not augment the susceptibility to secondary subcutaneous infections (Klei *et al.*, 1981), illustrating that tissue localization of established infections may be important in dictating host responses to incoming ones.

As discussed by the Klei *et al.* (1980), the enhanced susceptibility observed with subcutaneous primary infections could have arisen from physical alterations of the host, such as dilatation of the lymphatics caused by the established worm population. These alterations could provide an environment more suitable for larval migration and development (Klei *et al.*, 1980). Alternatively, an established infection could induce a specific immune unresponsiveness to developing larvae of subsequent infections. This situation is similar to the one described in the *L. carinii* model, in which the course of larval development may induce a state of immune unresponsiveness in favor of the establishment of adult worms and supports the notion that substances from either larval, microfilarial, or adult stages may induce cross-reactive immunosuppressive effects.

b. Immune Unresponsiveness Induced by Brugia Infections in Jirds. Cellular unresponsiveness specific to lymphatic filariae has been shown both in humans (Ottesen *et al.*, 1977; Piessens *et al.*, 1980*a,b*, 1982) and during *B. pahangi* infections in jirds (Portaro *et al.*, 1976). In this latter case, spleen cells were taken from jirds infected with *B. pahangi* at different times after infection for an overall period of 20 months, and their blastogenic response to filarial antigens was tested *in vitro*. Responses were already suppressed after the first month of infection and reached a minimum by the fifth month. By days 135 and 246, which is the time when the larval challenge was given in the experiment of Klei *et al.* (1980), filarial specific cellular responsiveness was at its lowest. The stage specificity of this unresponsiveness cannot, however, be assessed unless only male worms were employed in the antigenic preparation used, and contamination with microfilarial products explicitly avoided. Immune unresponsiveness detected by 30 days' infection was clearly induced by developing larvae and young worms only, but tested for with an undefined antigenic preparation.

An interesting example of immunosuppressive effects induced by one life cycle stage in favor of another has recently been provided by Schrater and Piessens (1982). Carrier-specific immune suppression to microfilarial antigens was induced in jirds infected with *B. malayi.* Direct anti-trinitrophenol (TNP) plaque-forming cell (PFC) responses were assayed 5 days after immunization of groups of normal and infected jirds with the following immunogens: TNP-lipopolysaccharide (LPS), TNP-keyhole limpet hemocyanin (KLH), and living TNP microfilariae. PFC/10^6 spleen cells were 775 and 513 in control and experimental animals, respectively, 8 days postinfection, but 612 and 73, respectively, by day 14 of infection (before the fourth moult), 402 and 52, respectively, 1 month postinfection (by the fourth moult), and 1360 and 25, respectively, by 5 months' infection, well into the postpatent period. Thus, a carrier-specific immunosuppression specific for microfilarial antigen was already induced by developing larvae.

Both the antibody response to TNP-microfilariae and its suppression in infected animals is T-cell dependent (A. R. Schrater and W. F. Piessens, personal communication, 1982). The first aspect was demonstrated by comparing the response to TNP-LPS, TNP-KLH, and TNP-microfilariae of thymectomized and lethally irradiated, bone marrow-repopulated jirds with normal ones. While responses to TNP–LPS were unaffected in thymectomized jirds compared with control animals, responses to TNP–KLH and to TNP–microfilariae were considerably reduced.

The thymus dependence of the suppression of the anti-TNP antibody response in infected jirds was demonstrated by passive transfer of splenic T cells from prepatent infected donors to uninfected recipients (A. F. Schrater and W. R. Piessens, personal communication, 1982). Transfer of spleen cells from prepatent donor jirds significantly reduced the direct anti-TNP–PFC response of jirds immunized with TNP–microfilariae, when the donor cells were pretreated with normal rabbit serum plus complement, but transfer of cells pretreated with rabbit anti-thymocyte antiserum and complement left the response unaltered.

Brugia infections in jirds not only induce filarial antigen-specific immune unresponsiveness, but also depress spleen cell responses to mitogens *in vitro* also (Portaro *et al.*, 1976). Spleen cell responses to both phytohemagglutinin (PHA) and concanavalin A (con A) were depressed when tested 2 months after an infection with *B. pahangi.* Mitogen-induced responses of spleen cells of uninfected animals were not affected when incubated with serum of infected jirds, and the response of cells of infected animals was not restored by preincubation of the cells with normal sera for 24 hr. Removal of nylon-wool or plastic-adherent spleen cells did, however, restore the reactivity to PHA and con A of the nonadherent population (Portaro *et al.*, 1976). It therefore seems that the depressed mitogen reactivity induced by the infection is more likely to be caused by phagocytic cells or even by B cells than by the presence of immune complexes. However, circulating immune complexes have been found in jirds throughout the course of an infection with *B. pahangi* (Karavodin and Ash, 1980, 1981, 1982), and their potential to influence both B- and T-cell responses should be borne in mind when aiming at the experimental definition of a particular immune-suppressive mechanism in this and other models. For example, sera from hamsters chronically infected with *D. viteae*, inhibited *in vitro* transformation of sensitized lymphocytes stimulated by filarial antigens (Weiss, 1978*b*), an effect that could be attributed to the presence of circulating immune complexes in infected hamsters.

c. Lymphatic Pathology in the Jird–Brugia Model. Manipulation of the host-immune response in order to induce resistance to infection is a matter complicated in filariasis by the possibility that the response itself may cause pathologic symptoms. In lymphatic filariasis, there is circumstantial evidence to support this view. Schacher and Sahyoun (1967) and Rogers *et al.* (1975) have suggested that in cats and dogs infected with *B. pahangi* the lymphatic inflammation and symptoms such as lymphadenitis and thrombolymphangitis that occur in nodes adjacent to adult worms are caused by local immune responses stimulated in the lymph nodes by antigens originating in the worms and transported by the afferent lymph flow. In Bancroftian filariasis, the hypothesis has often been put forward that material from dead worms, or the immune response that might have killed them, is the main cause of lymphatic pathology (von Lichtenberg, 1957; Warren, 1971; Denham and Nelson, 1976). In Malayan filariasis, only patients suffering from elephantiasis had elevated *in vitro* blastogenic responses of blood lymphocytes to adult worm antigens. Patients with filariasis but not elephantiasis did not respond (Piessens *et al.*, 1980*a*).

Shortly after the discovery that the jird was susceptible to infection with parasites of the genus *Brugia*, it was realized that it could serve as a model for investigating the lymphatic and other pathologies concomitant with infection. Ah and Thompson (1973), found that ≤75% of adult worms recovered from a subcutaneous infection with *B. pahangi* in the groin were in the regional lymphatic vessels. These lymphatics were severely dilated when they contained worms, and the adjacent lymph nodes were enlarged. Larvae could be found in the lymphatics as early as 1-5 days postinfection (Ah and Thompson, 1973; Vincent *et al.*, 1980*a*). Dilatation of the vessels, lymphatic and perilymphatic cellular infiltrates, intravascular granulomas, and irregular fibrosis of some valves and portions of the lymphatic walls were well established by 3 or 4 months' infection and did not change significantly thereafter (Vincent *et al.*, 1980*a*).

Interestingly, there is now evidence that sensitization of jirds with filarial antigens enhances the pathologic reactions caused by a subsequent infection with *B. pahangi* (Klei *et al.*, 1981, 1982). In the latter study, jirds were presensitized either by intravenous injections of dead microfilariae or by saline extracts of adult worms in CFA. Sensitized and control animals were infected subcutaneously with 100 L3 15 days after the last inoculation with antigens, and necropsies were performed at 90-95 days postinfection. Lymphatic pathology, as judged by the number of thrombi and the fraction of the total length of infected spermatic cord lymphatic vessels showing dilatation, was most marked in the two groups immunized with filarial antigens, compared with control groups given CFA + saline or no pretreatment. Notably, of the two presensitized groups, the group treated with adult worm antigens and CFA showed the most severe pathology. Moreover, antibody titers measured before infection and at necropsy were significantly higher in the group given adult antigen, compared with the group sensitized with microfilariae. Antibody titers were measured by indirect hemagglutination of sheep red blood cells coated with *D. immitis* antigen. Both presensitized groups also showed an increased immediate and delayed-type hypersensitivity response measured by the footpad swelling assay, in response to adult worm extracts. Although these observations do not permit a precise correlation between the nature of the immunologic status and the extent of the lymphatic pathology, they do indicate that an infection superimposed on a state of immune sensitization to filarial antigen is associated with increased lymphatic pathology, a correlation that hitherto had only been suspected.

Pathologic changes in kidneys, liver, spleen, and lungs have also been described in jirds in association with infections with *B. malayi* and *B. pahangi* (Vincent *et al.*, 1976; Vincent and Ash, 1978; Klei and Crowell, 1981). Major pathologic changes in the lungs included granulomas induced by larvae and adult worms, obstructive endarteritis, and chronic interstitial inflammation associated with degenerating microfilariae. Unfortunately, the histologic picture differed from that of human eosinophilic lung in that there was an absence of thin eosinophilic membranes around degenerating microfilariae and no parabronchial eosinophils (Vincent *et al.*, 1976).

2. *Brugia spp. in Rats*

The use of the jird as a model for studying host responses to lymphatic filariae has undoubtedly been hindered by the lack of immunologic reagents capable of distinguishing classes and subclasses of immunoglobulin or cellular subpopulations and by the scarcity of inbred strains. The 1980 edition of the *International Index of Laboratory Animals* reports the existence of one undesignated inbred strain of *M. unguiculatus* at the Bernhard-Nocht Institute in Hamburg, and the use of the MON-TUM (strain 532) inbred strain from the Tumblebrook Farms in Massachusetts, has already been reported (Karavodin and Ash, 1980). Experimental filariasis would benefit enormously if jirds were better characterized immunologically, since they are susceptible not only to infections with *Brugia* spp., but also with *D. viteae*, *M. globulosa*, and *L. carinii*, and indeed to a number of other helminth and protozoal infections of medical and veterinary importance. Meanwhile, use has been made of better characterized rodents, such as mice and rats. While mice are resistant to *Brugia* infections (Section II.C2), rats are susceptible, but unfortunately parasite recovery is erratic, which is not the case in jirds.

Early attempts to infect outbred strains of rats with *B. pahangi* and *B. malayi* yielded patency rates of <20% (Laing *et al.*, 1961; Ahmed, 1967*a*; Ash and Riley, 1970*a*; Sucharit and MacDonald, 1973; Harbut, 1973). Later, the susceptibility to infections with *B. pahangi* of six inbred strains of rats was compared (Fox and Schacher, 1976). The Lewis (LEW), Fisher (F344), and Brown-Norway (BN) strains were found to be the most susceptible, WF and ACI strains moderately so, and the Buffalo (BUF) resistant. Within 2-3 months after infection, 13 of 16 Lewis rats were microfilaremic (5-1720 microfilariae/200 μl blood), and nine of 16 had adult worms, whereas of 18 Buffalo rats, only four were microfilaremic (8-1050 microfiliariae/200 μl blood), and only one to three adult worms were found in three of these rats.

Two major studies of host responses to infection with *B. pahangi* have been reported with the Lewis rat model, that of Weller (1978) concerning the pattern of filarial specific and nonspecific cellular unresponsiveness during infections, and that of Gusmao *et al.* (1981), which is an attempt to characterize host resistance to infection immunologically. Weller (1978) used the Lewis rat model to investigate *in vitro* lymphocyte reactivity to filarial antigens and to B- and T-cell mitogens during acute and chronic infections with *B. pahangi.* Saline extracts of infective larvae, microfilariae, and adult worms were used as antigens, while PHA, LPS, and pokeweed mitogen (PWM) were used as mitogens.

Parasitologic analysis revealed a mean prepatent period of ~10 weeks in these rats, which is similar to that of jirds. By 16 weeks' infection, 90% of the rats were patent, but peak microfilaremia (week 36) was of only 534 microfilariae/ml, and the adult worm recovery rate was extremely low. Only seven adult worms were found in four rats of 30 infected animals.

Studies of immune unresponsiveness to filarial antigens of mesenteric lymph node cells, peripheral blood lymphocytes, and spleen cells yielded two main results: (1) responses to microfilarial antigens were only detectable by weeks 75–80 of infection and coincided with the decline of microfilaremia; and (2) responses to L3 antigens occurred during the first 2 weeks of infection, but not at later times, indicating that cellular reactivity was stage-specific.

A similar correlation between microfilarial-specific cellular unresponsiveness and patency has been found in humans infected with *W. bancrofti* and *B. malayi* (Ottesen *et al.*, 1977; Piessens *et al.*, 1980*a*). Furthermore, in this same study, Piessens *et al.* (1980*a*) found stage specificity in the cellular responses to microfilarial and adult antigens comparable to the one found by Weller (1978) with microfilariae and L3 antigens. Thus, peripheral blood lymphocytes from asymptomatic individuals, reactive to microfilarial antigens, were unreactive when stimulated with adult worm-derived antigens. A strict comparison of these results with those in the rat system is not possible, however, because unlike the adult worm antigen used by Piessens *et al.* (1980*a*), Weller's preparation included both male and female worms and may therefore have been contaminated with microfilarial products. Cellular responsiveness to PHA and LPS also correlated with low microfilaremia, and the best correlations were found with mesenteric lymph nodes.

Although Weller (1978) found that ⩽90% of the rats were microfilaremic by week 16 of infection, it is more common to find that only two thirds of Lewis rats infected with *B. pahangi* became patent (Fox and Schacher, 1976; Gusmao *et al.*, 1981). In fact, this differential response can be so reproducible that Gusmao *et al.* (1981) compared several immunologic parameters in rats that became microfilaremic and those that did not, in an attempt to find a correlation between the immune response and the presence or absence of parasitologic evidence of infection. Patent and nonpatent rats were evaluated for the first 50 weeks of infection with respect to blood eosinophil, neutrophil, and lymphocyte levels, antifilarial IgG and IgE serum antibody production, and peripheral blood lymphocyte responses to mitogens and filarial antigens. Infective larvae were given subcutaneously either in one single infection or by trickle inoculations over 25 weeks.

With the exception of the IgE antibody response, all other parameters were comparable in microfilariae-positive and -negative animals during the prepatent period. After patency, eosinophilia was detected and higher titers of antifilarial IgG serum antibodies were found in rats that became patent, most likely in response to the antigenic stimuli originating from microfilariae. Serum IgE antibody responses, measured by passive cutaneous anaphylaxis to *D. immitis* antigens, were undetectable during the prepatent period in rats that were to become microfilaremic, but appeared at the onset of patency, both in groups with a single or with a trickle infection, and reached a peak titer of ~1:200 shortly afterwards. In contrast, rats that remained amicrofilaremic developed an early IgE response with a maximum titer of the order of 1:8 during the time corresponding to the prepatent period.

As Gusmao *et al.* (1981) themselves pointed out, two interpretations of the results are possible, depending on whether the early IgE antibody response is regarded as a cause or effect of the parasitologic course of infection. The former view, that of a protective role for IgE antibodies in this model, is undoubtedly interesting, but is somewhat difficult to reconcile with the fact that microfilariae can coexist with titers of >1:200 (PCA) in the patent rats, whereas IgE antibody titers of only 1:8 (PCA) are supposed to suppress microfilaremia, or perhaps even control the infection. The alternative interpretation—that of an IgE response to antigens, possibly microfilarial in origin, from worms killed by other causes—is less attractive, but also less controversial. In fact, in a similar experiment using PVG/c inbred rats, Cruickshank *et al.* (1982) found no "early" IgE antibody responses in rats that did not become microfilaremic. Conversely, most patent rats developed a response at the onset of patency.

Although investigation of responses to developing larvae and adult worms in the rat model is made difficult by the poor or erractic worm recovery rates, a study of the relationship among microfilarial antigens, IgE antibody responses, and eosinophilia may yield clues to the relevance in filarial infections of these responses, which are usually concomitant to infections with helminths.

3. Brugia spp. Infections in Hamsters and Multimammate Rats

On average, 20–30% of hamsters (*Mesocrisetus auratus*) infected with *B. pahangi* or *B. malayi* become patent (Laing *et al.*, 1961; Edeson *et al.*, 1962; Ash and Riley, 1970*a*; Sucharit and MacDonald, 1972; Malone *et al.*, 1974). Although this figure is not high compared with the jird, there exists no systematic study of susceptibility to *Brugia* spp. infections of the more than 30 registered strains of inbred hamsters (Festing, 1979). A detailed study of strain-dependent variation in the response of hamsters to this parasite is likely to pay off, as so few inbred strains of rodents are susceptible to filarial infection.

Multimammate rats are as susceptible to infection with *Brugia* spp. as the jird, but sustain significantly lower levels of circulating microfilariae (Ahmed, 1966;

Petrány *et al.*, 1975; Sänger *et al.*, 1981). We know of only one immunologic study in this host concerning the antibody responses during infections with *B. pahangi* (Benjamin and Soulsby, 1976).

C. Experimental Filariasis in Resistant Rodents

Mechanisms contributing to natural resistance to filarial infection have been investigated by analyzing means whereby it can be either overcome or restored in immunodeficient susceptible strains of otherwise resistant species of rodents. Both rodent filarial species and filarial genera that infect humans have been used with this purpose.

1. The D. viteae–Rat Model

The course of infection with *D. viteae* has been investigated in both normal and immunodepressed rats of the Fisher inbred strain. Development of *D. viteae* in these rats is limited. Subcutaneous inoculation of 200 L3 yielded neither microfilariae nor adult worms, and only 12–15% of the larvae could moult and be recovered 22 days after infection (Haque *et al.*, 1980*b*, 1981*a*). However, a profound alteration of such a course of infection was observed in rats given 1000 cercariae of *S. mansoni* percutaneously 2 weeks before an infection with *D. viteae.* A patent microfilaremia of 40–50 days' duration was detected from day 60 of infection, at which time a mean of 18 adult worms was present (Haque *et al.*, 1981*a*).

Immune suppression is the most likely cause for this change in the pattern of resistance of the Fisher rat. Specific cellular unresponsiveness and nonspecific suppressor cell activity have been reported in Fisher rats infected with *S. mansoni* (Camus *et al.*, 1979). Moreover, adult worms of *S. mansoni* released a low-molecular-weight substance (500–1000 daltons) that inhibited *in vitro* proliferation of lymphocytes induced either by mitogen or by mixed lymphocyte cultures (Dessaint *et al.*, 1977). When this substance was given to Fisher rats before subcutaneous infection with 200 *D. viteae* L3, an average of 24 adult worms was recovered by day 60 of infection, and circulating microfilariae were detectable by day 30 and until day 110 of infection (Haque *et al.*, 1981*a*). The strongest implication of these results is that the natural resistance of Fisher rats to infection with *D. viteae* is based on a swift and effective immune response to developing larvae. Adult worms are probably not the target of this response, for they are never found in normal infections of Fisher rats, but survive when implanted surgically (Haque *et al.*, 1980*b*, 1981*a*). Nude rats, moreover, were fully susceptible to infection with *D. viteae*, whereas the heterozygous littermates were not (Haque *et al.*, 1981*a*). This finding is consistent with an immunologic basis for the natural resistance expressed by normal rats.

On the basis of these results, Haque and Capron (1982) reasoned that it should be possible to tolerize rats neonatally if the appropriate antigenic stimulus is provided at the right time during development. These workers have shown that offspring of microfilaremic Fisher rats (by transplanted adult worms) that were shown to harbor transplacentally derived microfilariae, were also capable of supporting full development of infective larvae inoculated subcutaneously. Moreover, a microfilaremia that lasted for about 2 months was detected by about day 60 of infection, superimposed on the neonatally derived one, and seven adult worms were recovered by day 45 of an infection with 100 L3.

Perinatally transferred microfilariae have been reported in humans (Brinkmann *et al.*, 1976) and observed in the amniotic fluid of dogs infected with *D. immitis* and in the blood of their offspring (Mantovani and Jackson, 1966). Microfilariae have also been found in litters of jirds (*M. libycus*) and hamsters (*M. auratus*) infected with *D. viteae* (Geigy *et al.*, 1967; Weiss, 1970). Therefore, if the tolerizing stimulus were from microfilarial antigens, the phenomenon described by Haque and Capron (1982) could be of widespread importance. Although the source of the tolerogenic antigen was not determined in Haque's experiment, the important point it establishes is that tolerance to filarial infection can be induced in the progeny of infected mothers. Apart from demonstrating the immunologic basis of natural resistance of the Fisher rat to *D. viteae* infection, this observation may contribute to an explanation of the wide spectrum of host responses observed in human filariasis.

2. *B. pahangi* Infections in Mice

Mice are largely resistant to infections with *B. pahangi* (Laing *et al.*, 1961; Ahmed, 1967*b*; Chong and Wong, 1967). The developing larvae are the prime target of resistance, as was shown by Suswillo *et al.* (1980). These investigators compared adult worm recoveries in mice of three different strains, 2.5-3.5 months after the mice had been given an intraperitoneal innoculation of 1) preparasitic L3, 2) 5-day old parasitic L3 from the peritoneal cavity of infected jirds, 3) 17-19-day old L4, and 4) adult worms of both sexes. In order of increasing resistance to infection the strains of mice used were Balb/c_1, CBA/Ca, and AKR. The number of adult worms recovered from the transplants increased when more mature stages were inoculated. Thus, no worms were recovered from transplants of preparasitic L3, but recovery rates of 10% were obtained in Balb/c_1 and CBA/Ca mice given fourth-stage larvae, and of 50-60% in Balb/c_1 mice given adult worms.

As in the *D. viteae*-rat model, resistance to developing larvae of *B. pahangi* in mice is immunologic. This was suggested by the increased susceptibility to infection found in nude mice compared with their heterozygous littermates (Suswillo *et al.*, 1980; Vincent *et al.*, 1980*b*, 1982*a*) and confirmed by two different yet complementary lines of evidence: the successful development of *B. pahangi* larvae in T-cell-deprived mice (Suswillo *et al.*, 1981) and the resistance to infection shown by immune reconstituted nude mice (A. C. Vickery, A. L. Vincent, W. A. Sodeman, personal communication, 1982). In the first approach, 8-week-old mice of the CBA/HT6T6 inbred strain were thymectomized and treated with anti-thymocyte antisera. Twenty-eight days after thymectomy, mice were given L3 either subcutaneously or by the intraperitoneal route. Neither microfilariae nor adult worms were found in the normal euthymic control mice. In contrast, all the intraperitoneally infected athymic mice had microfilariae in the peritoneal cavity by day 62 of infection. Adult worm recovery rates by day 165 of infection fluctuated between 5% and 45%, and all the worms were confined to the peritoneal cavity. Recoveries were lower in the group of subcutaneously infected athymic mice.

In the second approach, the model developed by Vincent *et al.* (1982*a*) was used. These workers (Vincent *et al.*, 1982*a*) carefully investigated the course of subcutaneous and intraperitoneal infections of *B. pahangi* in C3H/HeN (nu/nu) mice and in their heterozygous littermates. Subcutaneous inoculation of L3 into male mice yielded a mean worm recovery rate of 14.3%. Worms were recovered live for 160 days after infection, and mean values of blood microfilarial densities were of 150 microfilariae/20 ul blood by day 200 of infection. As in jirds, worm recovery rates in mice inoculated with larvae intraperitoneally were about twice as high as those infected subcutaneously.

Heterozygous littermates of the nude mice harbored no worms as of day 40 of infection. The effect of immune reconstitution of nude mice on their susceptibility to infection with *B. pahangi* was therefore investigated 40 days after a subcutaneous infection with L3 (A. C. Vickery, A. L. Vincent, and W. A. Sodeman, personal communication, 1982). Nude C3H/HeN mice 7-10 weeks old were reconstituted either with cells or with serum of syngeneic heterozygous euthymic donor mice. Nude mice receiving thymocytes intravenously showed a significant decrease in worm recovery rates compared with control mice. Most effective in conferring resistance to infection were grafts of normal thymus, and increasingly so the longer the time period between graft and infection. In fact, no worms were recovered when thymus grafts were performed several weeks before infection. Similarly effective were transfers of primed spleen cells taken from normal syngeneic donors 40 days postinfection. In contrast to the successful reconstruction of resistance to infection with cells, transfer of sera from primed donors to nude recipients had no effect at all as when compared with control nude mice receiving normal serum.

In summary, the results show that the natural resistance to developing larvae of *B. pahangi* exhibited by mice is immunologically determined and T cell dependent. It is hoped that research with this model will further our understanding of the cellular basis of the immune response to developing larvae in mice, as it should now be possible to dissect the reconstituting cell population. Unfortunately, this approach will not be possible in the case of *W. bancrofti*, as nude mice were shown to be as refractory to infection as are normal mice (Vincent *et al.*, 1982*b*). Similarly ineffective were attempts to infect immunodeficient mice (nudes and CBA/N with third-stage larvae of *Onchocerca lienalis* (Townson *et al.*, 1981).

3. Human Filariae Other Than Brugia spp. in Jirds

The remarkable susceptibility of *M. unguiculatus* to parasites of the genus *Brugia* contrasts with that exhibited to infections with other human parasites. No worms were recovered from jirds 18 days to 18 months after either subcutaneous or intraperitoneal inoculations of infective larvae of *Onchocerca volvulus*, *Loa loa*, and *Mansonella ozzardi* (Suswillo *et al.*, 1977). *Wuchereria bancrofti* third-stage larvae moult to the fourth stage by day 10 of a subcutaneous infection, but no adult worms or microfilariae have ever been found (Ash and Schacher, 1971; Suswillo *et al.*, 1977; Zielke, 1979; Cross *et al.*, 1981). Immunosuppressive agents such as cyclophosphamide and betamethasone (Suswillo *et al.*, 1977), or anti-lymphocyte antiserum (Cross *et al.*, 1981) did not significantly alter this pattern of resistance to *W. bancrofti*, suggesting that unlike the resistance of rats to *D. viteae* infections, or that of mice to *B. pahangi* infection, resistance of jirds to *W. bancrofti* is not immunologic.

III. RESPONSES OF THE HOST TO MICROFILARIAE

Host control of microfilaremia and of microfilarial skin burdens is a question germane not only to the transmission of filariasis, but to the morbidity of the disease in humans as well.

In onchocerciasis, low microfilarial burdens stimulate the local inflammation and systemic immune responses that may lead to skin and eye disease (Connor *et al.*, 1970; Buck, 1974; Martinez-Baez, 1978). Latent or occult lymphatic filariasis, in which microfilariae are not found in the peripheral blood but may be present in the tissues, is associated with tropical pulmonary eosinophilia (Joe, 1962; Danaraj *et al.*, 1966; Beaver, 1970; Ottesen *et al.*, 1979), with microfilarial granuloma of the breast (Chandrasoma and Mendis, 1978), and with filarial arthritis (Ismail and Nagaratnam, 1973).

Considerable insight into the nature of host responses to microfilariae has been gained with two types of models:

1. Laboratory rodents susceptible to infection with rodent filariae but that eventually become resistant to microfilariae and actively control and suppress

microfilaremia in the presence of viable adult worms. These are models for studying the phenomenon of latent filariasis.

2. Laboratory rodents resistant to filarial infection, but capable of harboring circulating or skin microfilariae introduced either by injection or by implantation of gravid worms. These models that fall into two groups, depending on whether the species of microfilariae introduced are of rodent filariae (type 1) or of filariae belonging to genera that infect humans (types 2 and 3).

A. Models of Latent Filariasis

Two host-parasite systems have been extensively investigated: *L. carinii* in rats and *D. viteae* in hamsters.

1. Immune Responses Leading to Latency in the L. carinii–Rat Model

In the albino rat (*Rattus norvegicus*), *L. carinii* infections have a similar course as in the natural host, the cotton rat. Even in the presence of living fertile adult worms, however, microfilaremia is suppressed after 3-4 months' infection, leading to the condition of latency (Ramakrishnan *et al.*, 1962). At the onset of latency, microfilariae of *L. carinii* released in the rat's pleural cavity by gravid female worms become profusely covered and immobilized by cells (Bagai and Subrahmanyam, 1970; Subrahmanyam *et al.*, 1976) and can no longer be detected in other organs, such as the spleen, liver, lungs, or lymph nodes (Bagai and Subrahmanyam, 1970). Thus it seems that the disappearance of microfilariae from the peripheral blood may result from their trapping and eventual killing by leukocytes within the pleural cavity. A similar selective trapping of microfilariae also occurs in the pleural cavity of the natural host, the cotton rat, when predisposed toward latency by repeated injections of microfilariae before infection (Haas and Wenk, 1981). Both cell attachment to, and killing of, microfilariae can be similarly induced *in vitro* using rat spleen cells in the presence of latent rat sera (LRS), but not with normal rat sera (NRS), or sera of patent rats, regardless of whether spleen cells from normal or latent rats are used (Subrahmanyam *et al.*, 1976).

a. Serum Factors Involved in the Killing of Microfilariae. Some insight into the mechanism of spleen cell cytotoxicity was gained by the observation that treatment of LRS with immune complexes, EDTA, EGTA, or heat (30 min, 56°C) virtually eliminated cell adherence and cytotoxicity to microfilariae, demonstrating that complement participates in the phenomenon (Subrahmanyam *et al.*, 1976; Mehta *et al.*, 1980). Addition of NRS to heat-treated LRS partially restored the cytotoxic effect. However, when mediation of cell adherence or cytotoxicity by LRS was eliminated by heat treatment of the serum for 3 hr (rather than 30 min), addition of NRS failed to restore the reaction (Mehta *et al.*, 1980). Thus, complement contributes to the phenomenon only when another heat-labile serum factor is present. This was shown to be IgE (Mehta *et al.*, 1980): No adherence of spleen or peritoneal exudate cells was promoted by fractions of LRS depleted of IgE by affinity chromatography with anti-IgE antibodies or by IgG fractions bound to, and eluted from, the appropriate anti-IgG immunoabsorbents. Conversely, fractions respectively bound to and eluted from, or unbound to anti-IgE and anti-IgG immunoabsorbents, were comparable to LRS in their ability to promote cell adherence and cytotoxicity to microfilariae (Mehta *et al.*, 1980).

b. Cells That Adhere to and Cells That Kill Microfilariae of L. Carinii in Vitro. The nature of the effector cells was established by testing the cytotoxic activity of 90–100% pure subpopulations from peritoneal exudate or peripheral blood cells (Mehta *et al.*, 1982). Peripheral blood lymphocytes and peritoneal exudate mast cells neither adhered to nor killed microfilariae after 16 hr in culture. Peritoneal eosinophils adhered to, but failed to kill, the larvae, whereas macrophages and neutrophils were equally effective both in adhering to and killing microfilariae, and may therefore be regarded as the principal effector cells. In agreement with this result, fractions of peripheral blood cells enriched with either polymorphonuclear cells or mononuclear phagocytes were the most cytotoxic (Mehta *et al.*, 1982). It may be, therefore, that the combination of anti-parasite IgE and macrophages is the most potent means of eliminating microfilariae in this host-parasite system.

IgE-dependent macrophage-mediated *in vitro* killing of microfilariae of *D. viteae* (See Section III. C.1.) and schistosomula of *Schistosoma mansoni* has also been described in rats (Capron *et al.*, 1975). Moreover, IgE of humans infected with *S. mansoni* also mediates *in vitro* killing of schistosomula by mononuclear phagocytes (Joseph *et al.*, 1978). The killing mechanism involves macrophage stimulation by aggregated, or at least dimeric, IgE (Joseph *et al.*, 1977; Dessaint *et al.*, 1979), associated with an increase in the release of lysosomal enzymes and neutral proteases, incorporation of glucosamine, and production of superoxide (Dessaint *et al.*, 1979; Capron *et al.*, 1980; Joseph *et al.*, 1980). A role for complement in this mechanism has not been described, however, and although aggregated IgE is known to bind complement components of the alternative pathway (Ishizaka *et al.*, 1972), the fact that killing of *L. carinii* microfilariae by macrophages and LRS can be eliminated with EGTA suggests that components of the classic pathway may intervene as well (Mehta *et al.*, 1980).

A high level of circulating IgE is the serologic hallmark of infections with helminths, yet its role has proved elusive. The appearance at the onset of latency of specific IgE serum antibodies capable of mediating *in vitro* killing of microfilariae by macrophages and neutrophils strongly suggests that antibodies of this immunoglobulin class can mediate a host-defense mechanism to microfilariae.

Host responses leading to latency have also been carefully explored in the Syrian hamster-*D. viteae* model. Results from these studies, however, indicate that IgE is not an obligate mediator of microfilarial killing.

2. *Latent Infections in the D. viteae–Hamster Model*

Most features of a *D. viteae* infection in jirds are reproduced in hamsters. Thus, larval development, adult worm size, duration of embriogenesis, and the overall course of infection are similar, but in hamsters microfilaremia may last as few as 5 weeks, whereas in jirds it persists for 6 months to 1 year or more (Worms *et al.*, 1961; Weiss, 1970; Malone and Thompson, 1975).

Infective larvae moult to the fourth stage by the end of the first week of infection, the final moult taking place 2 weeks later (Weiss, 1970). Patency begins in most strains of hamsters about 7 weeks after infection, but its duration is markedly strain dependent (Neilson, 1978; Weiss, 1978*a*). This difference in susceptibility has been exploited by Weiss (1978*a*) and by Neilson *et al.* (1981) to investigate the time correlation between the onset of patency and changes in immunologic parameters, such as humoral responses to the surface of microfilariae. Two strains of hamsters were used by Weiss (1978*a*): the more susceptible LSH (inbred), in which microfilaremia is eliminated in about one-half the infected animals after 14 weeks, and the more resistant LAKZ (outbred), with microfilaremia lasting only 9 weeks in most (96%) infected animals. In both strains, a strict correlation between the onset of latency and the appearance of serum antibodies to the cuticle of microfilariae was detected by immunofluorescence (Weiss, 1978*a*). Those animals of the LSH strain in which microfilaremia did not decline also failed to produce detectable amounts of these antibodies.

The inverse relationship between microfilaremia and antibody to the microfilarial surface is noted by many investigators of human and animal filarial infections (Wong and Guest, 1969; Ponnudurai *et al.*, 1974; Grove and Davis 1978; Wong and Suter, 1979; Dissanaike and Ismail, 1980; McGreevy *et al.*, 1980; Piessens *et al.*, 1980*a*). Moreover, both in humans and in experimental animals, patent filariasis correlates with a state of cellular immune unresponsiveness specific to microfilarial antigens (Ottesen *et al.*, 1977; Weller, 1978; Piessens *et al.*, 1980*a*,*b*, 1982).

Whether the absence of circulating antibodies to the microfilarial surface in microfilaremic hosts is caused by a state of specific immune unresponsiveness or is the consequence of passive absorption of these antibodies to excess microfilariae rapidly eliminated from the circulation is unknown. Haque *et al.* (1978*a*) showed that microfilaremia was strongly suppressed in recipient hamsters by transfer of serum taken from either latent hamsters or early patent ones.

However, sera from animals with peak microfilaremia were less effective, suggesting that microfilaricidal serum factors were already produced closely after the beginning of patency, and competed with microfilarial production until the latter was overwhelmed and latency ensued. In contrast, Neilson *et al.* (1981), showed that passive transfer of resistance to microfilariae by mixtures of spleen and lymph node cells or serum of syngeneic donors of the PD4 hamster strain could only be achieved when the donors had become latent. Contrary to Haque's experiment, sera or cells from early patent donors were ineffective. The slightly different experimental approach of measuring the appearance of spleen cells secreting anti-microfilarial specific antibody before and during patency, should perhaps be explored to find out whether patency is the consequence of an insufficient antibody response or of its temporary suppression.

a. The Antibody Class That Mediates Microfilarial Killing. Serum antibodies to the cuticle of microfilariae appearing at the onset of latency have been thought to be of the IgM class, because they eluted within the first peak of a G-200 Sephadex gel-filtration column (Weiss, 1978*a*). This has now been confirmed by Neilson *et al.* (1981), using specific anti-hamster IgM and IgG antibodies. These workers investigated serum immunoglobulin levels and anticuticular antibody responses to microfilariae in three strains of hamsters differing in ability to clear microfilariae. Strains LVG (outbred) and PD4 (inbred), which are capable of controlling and suppressing microfilaremia (Neilson, 1978), showed increased serum IgM titers during patency of 250% over those of control levels. Titers were consistently lower than in the responding strains in hamsters of the CB susceptible strain, which sustain prolonged microfilaremia similar to that in the jird (Neilson, 1978).

Specific IgM antibodies to the microfilarial cuticle, titrated by immunofluorescence, increased during the prelatent and latent period in the two resistant strains, but not in the susceptible strain. Interestingly, despite a 10- to 20-fold increase in total IgG immunoglobulin levels, there were no detectable anti-cuticular antibodies of this class at any time during infection in any of the strains investigated (Neilson *et al.*, 1981). The appearance of anti-cuticular IgM antibodies at the onset of latency and the fact that passive transfer of resistance to microfilariae can be achieved both with serum and with the excluded fraction, after gel-filtration of serum with Sephacryl S-200 (Neilson *et al.*, 1981), strongly suggests that IgM antibodies are directly involved in the clearance of microfilariae. Moreover, in support of the indication that antibodies that kill microfilariae are IgM, there is some evidence that microfilaricidal activity in hamsters could be T-cell independent. Thus neonatally thymectomized hamsters were able to control microfilaremia and produced anti-cuticular antibodies (Weiss, 1978*a*). Transfer of 10^6 spleen cells from latent syngeneic donors into patent hamsters significantly reduced both the intensity and duration of microfilaremia, but transfer of the same number of B-depleted (nylon–wool-treated) cells resulted only in a reduction in peak microfilarial density, but notably, duration of microfilaremia was unmodified compared with controls (Tanner and Weiss, 1979).

b. Mechanism of Latent Filariasis in Rodents. How does anti-cuticular IgM intervene in the control of microfilaremia? Haque *et al.* (1978) have shown that serum of latent hamsters suppresses the release of microfilariae from female worms *in vitro*. It has no direct effect, however, on either microfilarial viability or worm fertility, as microfilariae are not killed by serum *in vitro*, and their release *ex utero* is resumed when the female worms are reimplanted into naive hamsters (Weiss, 1970; Haque *et al.*, 1978*a*; Tanner and Weiss, 1978; Weiss and Tanner, 1979).

Complement affects neither microfilarial output by female worms *in vitro*, nor *in vitro* motility of the larvae incubated with latent sera (Haque *et al.*, 1978*a*; Tanner and Weiss, 1978).

Apart from its effect on fecundity, latent antibody mediates leukocyte attachment to, and killing of, microfilariae both *in vivo* and *in vitro*. This was elegantly demonstrated by Weiss and Tanner (1979), who investigated the fate of microfilariae within micropore chambers implanted into naive, patent, and latent hamsters. Microfilariae were killed within 24 hr when implanted into latent hamsters, but only if the pore diameter of the chamber permitted the entry of both fluids and cells. In chambers of smaller pore diameter, from which cells were excluded, microfilariae survived for 3 weeks. Killing of microfilariae in patent hamsters at peak microfilaremia occurred in some but not all of the animals implanted with larger pore chambers and was totally absent in naive ones, unless microfilariae were preincubated with latent sera containing anti-cuticular antibodies. This opsonizing effect was also achieved with the 19S fraction of these sera and abolished by treatment with 2-mercaptoethanol, thereby adding to the evidence that the antibody class mediating cell adherence to and killing of microfilariae of *D. viteae* is IgM.

The composition of the cell population migrating into micropore chambers was independent of both the immunologic status of the recipient animals and the presence or absence of microfilariae within the chambers. However, larger numbers of cells migrated into microfilariae-containing chambers, suggesting that a chemotactic factor with no particular cell specificity was released (Weiss and Tanner, 1979). Alternatively, the inflammatory response to the chamber itself may have obscured small differences in cellular composition (Weiss and Tanner, 1979). Interestingly, the composition of cells adhering to the microfilariae was significantly different from the nonadherent population within the chamber. Among the former, the preponderant cell types were polymorphonuclear leukocytes, mainly neutrophils and a few eosinophils (Rudin *et al.*, 1980). Thus, an IgM-dependent neutrophil-mediated cytotoxic effect could be the main mechanism of microfilarial elimination in this model. Indeed, the salient feature emerging from these studies with both rodent models is that clearance of circulating microfilariae at the onset of latency is achieved by means of cytotoxic effects mediated by antibodies specific to surface antigens of microfilariae.

3. Source of Antigens That Induce Latency

In the rat-*L. carinii* model, microfilaremia induced by the intrathoracic transplantation of adult female worms is suppressed by a previous transplantation of both adult male and female worms, by female worms alone, or by repeated intrathoracic injection of microfilariae, but not by adult male worms alone or by female worms depleted of microfilariae (Bagai and Subrahmanyam, 1970). Latency is thus induced by antigens present in microfilarial but not in adult stages. Moreover, antisera with specific IgE-dependent cytotoxic properties identical to those of latent rat sera can be raised by repeated injections of sonicated microfilariae emulsified with CFA (Mehta *et al.*, 1982).

In the hamster-*D. viteae* model, both the intensity and duration of microfilaremia can be affected by injection of crushed microfilariae or by their saline extracts, but the results depend critically on the immunization regimen. Hamsters given two injections of 25,000 crushed microfilariae on days 40 and 45 after subcutaneous injection with infective larvae showed a significantly lower peak microfilarial density as compared with unimmunized controls (Haque *et al.*, 1978*a*). In contrast, animals given six injections of saline extracts of 500–600 microfilariae every 4 days, beginning on day 25 postinfection, presented a greatly increased microfilarial density. Microfilaremia also lasted longer than in control animals. This treatment however, had no effect on adult worm burdens (Haque *et al.*, 1978*a*). Thus, microfilarial antigens can both immunize and tolerize hamsters in a stage-specific manner. However, the converse is not true, as saline extracts of male worms injected before infection suppressed both intensity and duration of microfilaremia compared with control animals, albeit not as effectively as animals injected with extracts of female worms before infection (Haque *et al.*, 1978*a*).

Implantation or injection of living worms of different stages also affects the course and intensity of microfilaremia. Injection of infective larvae into hamsters 5 days after a subcutaneous implantation of gravid female worms marginally reduced both intensity and duration microfilaremia due to the implanted worms, but had no effect on the adult worm burden (Haque *et al.*, 1978*b*). In comparison with control animals, implantation of female worms 15 days before infection predictably reduced peak microfilaremia arising from the full infection (Haque *et al.*, 1978*b*). Suprisingly, a similar implantation with male worms had the opposite effect. This enhancing effect could also be achieved with material released *in vitro* from male worms cultured overnight, injected several days before an infection with L3, or just before patency (Haque *et al.*, 1978*b*). Implantation of neither male nor female worms had any influence on the establishment of infections initiated with L3.

It is thus apparent that both duration and intensity of microfilaremia are influenced by antigens of stages other than microfilariae and that therefore latency may ensue as a result of a complex network of antigenic stimuli. Nonetheless, in both models reviewed so far, antigens derived from microfilariae alone are sufficient to stimulate an immune response leading to an amicrofilaremic state. It is this type of response that is investigated in the models described in Section III. B.

B. Responses to Microfilariae in Rodents Resistant to Normal Infections

In these models the response to microfilariae is investigated after the induction of a state of patency, either by injection of microfilariae or by implantation of gravid worms. The foremost advantage of these models is the circumvention of the larval phases of infection. This permits the use of inbred strains of mice and rats that in the main are resistant to these stages, but not to adult worms or microfilariae.

1. Microfilariae of D. viteae in Inbred Mice and Rats

Results of studies with inbred mice are in parallel with those obtained in hamsters. Thompson *et al.* (1979) compared the course of microfilaremia in normal (CBA/H) and immunologically defective (CBA/N) inbred mouse strains. Mice with the CBA/N phenotype are unable to respond to certain thymus-independent antigens (Amsbaugh *et al.*, 1972) and have preferential deficiencies of IgM and IgG_3 immunoglobulin expression, both when measured in serum and in cells secreting these isotypes (Perlmutter *et al.*, 1979).

The microfilaremia induced by subcutaneous implantation of female adult worms lasted for 70–100 days after transplant to CBA/H mice. In contrast, CBA/N mice had a long-lasting microfilaremia that showed no signs of decline even by day 180 of transplant. Moreover, its intensity was of an order of magnitude higher than in normal mice. However, both strains of mouse were equally resistant to infection by subcutaneous injections with infective larvae and showed a similar time course of adult worm survival (Thompson *et al.*, 1979).

Specific IgM antibodies to the cuticle of microfilariae were detected by immunofluorescence in CBA/H but not in CBA/N mice. Fluorescence was most intense with sera taken immediately after clearance of microfilariae. Neither strain showed anti-cuticular antibodies in the IgG_1 or IgG_2 subclasses of immunoglobulin (Thompson *et al.*, 1979).

Using F_1 hybrid mice of crosses between CBA/N females and Balb/c males

(CBA/N × Balb/c) F_1, we have been able to show similar trends of microfilarial control, but also some contrasting serologic results–F_1 hybrids of defective (CBA/N) females (X^d, X^d) and normal (Balb/c) males (XY) are phenotypically normal if they are females (X, X^d) and immunodefective if they are males (X^d, Y), owing to the recessive character of the allele associated with the immune defect.

As in the experiment of Thompson *et al.* (1979), a marked difference in level and duration of microfilaremia was found in male vs. female F_1 hybrids subcutaneously transplanted with gravid female worms. Both sets of mice became patent by day 5 of transplant, but by day 25 the microfilaremia rose to 80 microfilariae/5 μl in defective male mice, and to only about 25 microfilariae/5 μl in normal female mice. Total clearance of circulating microfilariae was achieved in female mice by day 100 of transplant, whereas male mice were microfilaremic even after 1 year, with a microfilarial level of about one-third of the maximum density. Similar results were obtained when (CBA/N × Balb/c) F_1 hybrid immunodefective males (X^d, Y) were compared with Balb/c × CBA/N) F_1 hybrid normal males (X, Y), thus excluding sex differences as a cause of differential response observed. When a microfilaremia was induced by intravenous injections of microfilariae rather than by transplant of adult worms, it was also cleared only in the phenotypically normal mice. Therefore, the presence of female adult worms does not seem to influence the response to and fate of microfilariae in these mice. Consistent with this result is the fact that there was no difference in the time course of adult worm survival in normal and defective hybrid mice.

The time course of appearance of serum antibodies to the surface of microfilariae was followed by immunoprecipitation of their ^{125}I-labeled surface antigens, with sera taken from normal and defective hybrid mice at different days after adult worm transplant. In normal mice, antibodies appeared in the serum earlier than in defective mice and reached a constant level by day 45. In defective mice, however, specific anti-microfilarial surface antibodies reached only one-third of this level by day 100 of infection. Interestingly, no difference was found in the pattern of surface antigens precipitated by sera from normal and defective mice, when these were analyzed by electrophoresis on sodium dodecylsulfate polyacrylamide gel electrophoresis, i.e., both defective and normal mice were able to recognize all the major radiolabeled surface antigens.

Comparable levels of IgG_1 antibodies to radiolabeled surface antigens were found in both defective and normal mouse sera, but IgM-specific antibodies were only detected in the latter. Furthermore, the highest levels of IgM antibodies were reached at the onset of microfilarial control in normal mice, when microfilaremia begins to decline. Antibodies of low affinity are perhaps more easily detected by radioimmunoprecipitation than by immunofluorescence, which may explain our finding of IgG_1 surface-specific antibodies, in contrast to the results obtained by Thompson *et al.* (1979). Nonetheless, what is apparent is that in mice, clearance of *D. viteae* microfilariae correlates with the appearance of serum antibodies in IgM, directed to the microfilarial surface. Furthermore, it appears that recognition by antibody of a given set of surface antigens is a necessary but insufficient condition for microfilarial control. The latter will not ensue unless strict antibody-class requirements are met.

The antigenic stimulus leading to control and suppression of microfilaremia of *D. viteae* in mice is T-cell independent. That is not only implied by the results above, but is further corroborated by the fact that microfilaremia induced in outbred homozygous nude mice (nu/nu) by a subcutaneous transplant of female adult worms is controlled and suppressed with a similar kinetic in heterozygous littermates of these mice (nu/+) (Haque *et al.*, 1980*a*). Thus, two main aspects of the immune response leading to latency in hamsters, its T-cell independence, and the role of anti-cuticular IgM antibodies seem to be operational in the mouse-*D. viteae* model. It therefore satisfies the requirements of a good experimental model, i.e., to permit simplification of a phenomenon, avoiding major qualitative distortions. Further dissection of the immune response to microfilariae should be possible in the mouse and may significantly contribute to our understanding of the mechanisms underlying latency.

Immune effector mechanisms that clear microfilariae of *D. viteae* in inbred rats depart somewhat from those operating in hamsters and in mice, but resemble the mechanisms of clearance of microfilariae of *L. carinii* in rats. Inbred Fisher/Ico rats are naturally resistant to infection by *D. viteae* but can sustain a subcutaneous transplant of adult worms for 30-40 days (Haque *et al.*, 1980*b*). Microfilaremia ensues rapidly after transplant, reaches a peak value by days 20-30, and is completely suppressed by day 135. Amicrofilaremic rats are immune to further challenge with microfilariae from newly transplanted female worms (Haque *et al.*, 1980*b*). Sera from latent, but not from patent, rats mediate adherence of peritoneal macrophages of normal Fisher rats to microfilariae *in vitro*, and about 90% of the worms are killed after 24 hr in culture (Haque *et al.*, 1980*b*; Ouassi *et al.*, 1981). As with the previous models, it is most likely that these *in vitro* results reflect the way in which microfilariae are cleared *in vivo*. It departs, however, from the other *D. viteae*-rodent models in that this antibody-dependent cytotoxicity involves IgE antibodies (Haque *et al.*, 1980*b*). Thus, specific immune absorption of sera of latent rats with anti-IgE antibodies coupled to Sepharose reduced adherence to, and killing of, microfilariae by macrophages to control levels. Heat treatment of the sera for 2 hr at 56°C had the same effect. Furthermore, cell adherence and cytotoxicity were not restored by addition of fresh normal serum, thereby excluding the participation of complement in this process. Although a specific absorption of immune sera with anti-IgM antibodies was not performed, a concomitant role in cell adherence for this immunoglobulin class is unlikely, because cells did not adhere to washed microfilariae that had been incubated with immune serum, and therefore the latter has no opsonizing activity. In fact, only simultaneous incubation of cells, sera, and parasites led to a cytotoxic effect (Haque *et al.*, 1980*b*), as previously observed in the *L. carinii*-rat model (Mehta *et al.*, 1980, 1982).

In addition to peritoneal macrophages, eosinophils also appear to be involved in the overall process leading to microfilarial killing *in vitro* (Haque *et al.*, 1981*b*). However, this was only apparent when microfilariae were incubated with immune Fisher rat sera and an eosinophil-enriched (60% eosinophils) rat peritoneal cell population, containing nonadherent macrophages as well. Eosinophils adhered to the microfilariae and degranulated by 6 hr of incubation. After 16 hr of incubation, most microfilariae (90%) were killed, but only macrophages were attached to the worms by this time. Furthermore, killing of microfilariae was only dependent on the ratio of macrophages to target, and not on the ratio of eosinophils to target (Haque *et al.*, 1981*b*). Thus, there is no direct effector role for eosinophils in the killing of microfilariae, nor is there one for mast cells. These cells did not adhere to and were not toxic to the larvae, even when an enriched population of mast cells was used (85%). However, mast cells could perform an indirect role as accessory cells in a way similar to that described in the antibody-dependent killing of schistosomula of *S. mansoni* by eosinophils *in vitro* (Capron *et al.*, 1978, 1981; Haque *et al.*, 1981*b*).

2. Brugia spp. in Mice

a. Basic Model. Grove *et al.* (1979) have developed a system whereby a microfilaremia is induced in mice by an intravenous injection of microfilariae taken from the peritoneal cavity of infected jirds. A subperiodic strain of *B. malayi* was used (Grove *et al.*, 1979). Both intensity and duration of microfilaremia depended on the number of microfilariae injected, but the number of organisms found in the peripheral blood never exceeded 1-3% of the input. For example, an injection of 2×10^5 microfilariae gave rise to a microfilaremia of 100 microfilariae/100 μl blood for the first 2 months, declining sharply thereafter to reach undetectable levels by day 120 of infection, whereas a dose of 2.5×10^4 microfilariae yielded a transient maximum of about 8 microfilariae/100 μl blood, declining steadily to negligible values 2 months after infection.

More than 50% of the microfilarial input was concentrated in the lungs and only 8.5% and 2.9% in the liver and spleen, respectively, and much less in other organisms.

The mice developed a detectable immediate-type hypersensitivity reaction to microfilarial antigens (measured by footpad swelling) as early as 14 days after infection, and a marked reaction later in the infection (day 126). Clearance from a secondary infection was much faster in mice that were already free from microfilariae of a primary infection. In these mice an eosinophilia of $\leqslant 450$ eosinophils/μl blood developed by day 12 of a secondary microfilarial infection, whereas in primary injected mice, eosinophilia never departed significantly from normal levels (10 eosinophils/μl blood by day 20).

Lung damage, hypersensitivity to microfilarial antigens, high blood eosinophilia, and amicrofilaremia are associated with tropical pulmonary eosinophilia in humans (Meyers and Kunwenaar, 1939; Buckley, 1958; Danaraj *et al.*, 1959, 1966; Beaver, 1970; Neva *et al.*, 1975; Spry, 1980; Ottesen, 1980; Spry and Kumaraswami, 1982). Although lung damage has not yet been documented in the *B. malayi*-mouse system, further studies may clarify whether this and other features of the disease can be reproduced and thus establish this model as a tool to investigate the etiology of this filarial-induced pathology.

The low levels of circulating microfilariae obtained in this model demand very skillful measurements of blood microfilarial densities. Nevertheless, a number of interesting results are already available concerning the nature of the immune response that clears microfilariae of *B. malayi*, the antigens involved, and how susceptibility to this microfilaremia is genetically associated in the mouse.

b. Clearance of Microfilariae of B. malayi. This aspect has been investigated by Thompson *et al.* (1981), who compared the response of CBA/H and CBA/N (immunodeficient) strains of mice with injected microfilariae. Peak microfilaremia caused by intravenous injections of 10^5 microfilariae was significantly higher in the CBA/N mouse strain than in the normal CBA/H. In the latter strain, clearance of microfilariae took place in 30 days, while parasites persisted in CBA/N mice beyond this time (Thompson *et al.* 1981). Serum antibody measurements performed by an enzyme-linked immunoabsorbent assay (ELISA), using microfilarial extracts as antigens, and by immunofluorescence of whole microfilariae, showed no anti-microfilarial IgM antibodies in CBA/N mice. In contrast, in CBA/H mice a correlation was found between the appearance of serum-IgM antibodies to microfilariae and clearance of parasites. IgM antibodies were detected both by ELISA and by immunofluorescence. Low titers of IgG anti-microfilarial antibodies were detectable by either method in the serum of CBA/H, but not in CBA/N mice. After secondary injection of microfilariae, however, both strains of mice showed a comparable serum titers of IgG antibodies, but whereas CBA/H mice showed no reappearance of blood microfilariae, CBA/N mice displayed the secondary microfilaremia superimposed on the primary one. These data suggest that IgM anti-microfilarial antibodies are involved in clearance of *B. malayi* microfilariae in mice. Interestingly, both strains developed similar immediate-type hypersensitivity reactions and no delayed-type hypersensitivity reactions to microfilarial extracts (Thompson *et al.*, 1981).

c. Source of Antigens That May Induce Clearance of B. malayi Microfilariae in Mice. Antigenic preparations obtained by extraction of lyophilized microfilariae of *B. malayi* with phosphate-buffered saline can protect mice against challenge infections with microfilariae (Kazura and Davis, 1982). Two subcutaneous injections of 1 μg protein administered 2 weeks apart, yielded 76% protection of Swiss-Webster outbred mice, by day 40 of an injection of microfilariae given 2 weeks after immunization. In these mice, microfilariae were cleared twice as fast than in immunized controls. Immunofluorescence assays with sera of immune mice exhibited IgG antibodies directed against the body, but not the sheath, of microfilariae. Using an ELISA with the antigenic preparations employed to immunize the mice, it was possible to show that IgG_{2a} was the predominant IgG antibody subclass in these mice. Specific IgM antibodies were detected neither by fluorescence nor by the ELISA, a result that conflicts with the report by Thompson *et al.* (1981). Nonetheless, resistance to microfilariae could be passively transferred with sera from immune mice (Kazura and Davis, 1982), thereby further substantiating the role of serum antibodies in the clearance of circulating microfilariae.

Protection was species specific. Extracts of *Trichinella spiralis* infective larvae and *B. pahangi* microfilariae obtained by the same procedure used with the *B. malayi* antigenic preparation, failed to protect mice against *B. malayi* (heterologous) microfilariae, but the *B. pahangi* antigens were not tested in a homologous challenge experiment with *B. pahangi* microfilariae.

It is interesting to analyze these results against the background of a series of experiments performed by Maizels *et al.* (1982). These experiments were designed to assess the pattern of cross-reactivity between different stages and species of filariae of the genus *Brugia*. Mice were primed with living *B. pahangi* microfilariae and then challenged with an injection of parasites of one of the following species and stages: *B. pahangi* microfilariae, *B. malayi* microfilariae or infective larvae, and *N. brasiliensis*-infective larvae. With the exception of the latter, all challenge infections stimulated a very strong anamnestic antibody response reactive against radiolabeled surface antigens of both *B. pahangi* and *B. malayi* microfilariae. These and similar experiments brought to the fore the finding that surface antigens of microfilariae are cross-reactive with parasites of other species of the same genus; moreover, they are highly immunodominant in the antibody response of a resistant host, such as the mouse (Maizels *et al.*, 1983). This finding is consistent with a number of previous observations concerning cross-reactivity between crude "extracted" antigens from whole filarial worms (Subrahmanyam *et al.*, 1974; Dissanaike and Ismail, 1980; Dasgupta *et al.*, 1980), or the ability of serum antibodies from patients infected with one filarial species to bind to the surface of microfilariae of a different one (Hedge and Ridley, 1977; Subrahmanyam *et al.*, 1978). In such a framework of immunodominant and cross-reactive anti-surface antibody responses, it is puzzling that the species-specific antimicrofilarial antibody response described by Kazura and Davis (1982) can protect mice against infection with *B. malayi* microfilariae, especially when considering that this response is not directed to the larval surface.

d. Genetics of Murine Susceptibility to Microfilariae. The mouse–*B. malayi* model lends itself to the investigation of the genetic association of susceptibility to microfilaremia. This is a matter of great interest that may help explain the wide spectrum of response to infections with lymphatic filariae in endemic areas.

Clinical symptoms and parasitologic and immunologic characteristics of this disease may vary considerably among individuals, even within a single household. One such feature is microfilaremia, as only 25–50% of persons living in endemic areas will show circulating microfilariae, although a much higher percentage are exposed to infection (Sasa, 1976).

Using a variety of inbred strains of mouse in which microfilaremia had been induced by injection of microfilariae of *B. malayi*, Fanning and Kazura (1983) studied the strain dependence of parameters of susceptibility to microfilaremia, such as its duration and peak density levels. Strains of mice with microfilaremia lasting over 50 days and with peak microfilarial densities of about 60 microfilariae/100 μl blood were considered susceptible, whereas when duration and intensity of microfilaremia were below 5 days and 20 microfilariae/100 μl, respectively, the mice were labeled resistant.

It was found that susceptibility to microfilariae was strongly strain-associated, but unrelated to the major histocompatibility gene complex (H-2). Thus, mice of congenic strains expressing the k and b H-2 haplotypes were resistant, whereas mice of a different pair of congenic strains also expressing the k and b haplotypes were susceptible by the above criteria.. Finally, the pattern of resistance expressed by F_1 hybrids of crosses between susceptible and resistant mouse strains and that of their backcrosses to the parental strains suggests that resistance to microfilaremia is a dominant trait influenced by a single or a small number of genes (Fanning and Kazura, 1983).

3. *Microfilariae of the Genus Onchocerca in Inbred Mice*

Microfilariae of the genus *Onchocerca* have been successfully transferred to a variety of laboratory rodents, but studies with these models have mainly focused on determining microfilarial life span and tissue distribution (Nelson *et al.*, 1966; Rabalais, 1974; El Bihari and Hussein, 1975; Beveridge *et al.*, 1980; Aoki *et al.*, 1980).

Models suitable to investigate immune responses to *Onchocerca* microfilariae have only recently been established in inbred strains of mice: Townson and Bianco (1982) studied the comparative natural resistance of several inbred and a few outbred strains of mice to microfilariae of *O. lienalis*, a cattle species. Subcutaneous injections of 2000 microfilariae in the neck yielded ear counts of 52–55 microfilariae in Balb/c_1, and DBA/2, and CBA/HT6T6 mice; slightly lower numbers in C57BL, BKW, and AKR mice; 22–29 microfilariae in BK/TO C3H, and Sh/Sh mice. Ear counts were performed 5 days after injection.

The CBA/HT6T6 strain was chosen among the most susceptible ones to investigate a number of parasitologic parameters. Ear recoveries of microfilariae 15 days after injection, increased expotentially with microfilarial input and thus a linear semilogarithmic relationship exists between microfilarial burden and ear counts (Townson and Bianco, 1982). The course of infection could thus be followed by counting the microfilarial ear burden. Maximum ear counts after a subcutaneous injection of 5000 microfilariae were reached by day 20-40 of injection (~500 microfilariae), followed by a sharp decline in two phases. The first phase brought the microfilarial ear count down to 150 by day 70, and the second, to 10 by day 240 of injection. This course of infection was fairly reproducible and permitted an investigation of some immunologic features of the model, such as presence or absence of resistance to reinfection with microfilariae. An injection of 5000 microfilariae given 141 days after a primary injection of 10,000 organisms resulted in a significant reduction in microfilarial burdens compared with unprimed controls. Thus, by day 6 of challenge, microfilarial burdens in immunized mice were only 26% of those in unprimed ones, and only 3% by day 35. A murine model with such a pattern of resistance to reinfection with microfilariae may prove invaluable for the investigation of the basic mechanisms controlling skin microfilarial burdens of *Onchocerca* species, of the antigens stimulating protection against microfilariae, and of the mode of action of anti-microfilarial drugs such as diethylcarbamazine (DEC).

IV. CONCLUSIONS

The past decade has seen an increase of activity in experimental filariasis in rodents. The discovery of the remarkable susceptibility of the jird to full infections with both animal and human filariae, the use of inbred strains of hamsters in the study of rodent filariae, and that of rats and mice as partial or proxy hosts for adult worms and microfilariae (Table I) have all given considerable impetus to immunoparasitologic research in filariasis. The information generated can often be correlated with results from studies of the disease in humans. For example, immune unresponsiveness specific to filarial antigens in human lymphatic filariasis (Ottesen *et al.*, 1977; Piessens *et al.*, 1980*a,b,* 1982) has also been found in jirds and in rats infected with *Brugia* spp. (Portaro *et al.*, 1976; Weller, 1978). Further studies using existing inbred strains of these rodent species and cell-transfer experiments should permit analysis of the cellular basis of this phenomenon, which may play an important role in determining natural susceptibility to filarial infection and may also explain the striking longevity of the adult worms. The implication that natural susceptibility to infection may arise from a state of immune unresponsiveness induced by the early larval stages of the parasite, stems from the evidence that in some cases its counterpart, natural resistance, is immunologic. Thus, if natural resistance is based on a swift anti-larval immune response, then natural susceptibility could arise from an abrogation of this response by parasite-derived stimuli to which some host species, but not others, are sensitive.

Two examples may serve to illustrate the occurrence of immunologically based natural resistance to infection in rodents: (1) offspring of rat species naturally resistant to infection with *D. viteae* become susceptible if perinatally exposed to adult worm or microfilarial antigens (Haque and Capron, 1982); and (2) nude mice, which unlike their heterozygous littermates, are susceptible to infections with *B. pahangi*, can be rendered resistant if reconstituted with normal syngeneic lymphocytes (A.C. Vickery, A.L.Vincent, and W.A. Sodeman (personal communication, 1982). In contrast, however, jirds remain refractory to full infections with *W. bancrofti*, even when treated with immune-suppressive agents such as cyclophosphamide and betamethasone (Suswillo *et al.*, 1977), or anti-lymphocyte antiserum (Cross *et al.*, 1981). This type of resistance is therefore probably of a different nature.

Another interesting aspect that studies with rodent models have brought to the fore is the ability of larval stages to induce immune-unresponsiveness in favor of adult and microfilarial stages. This is perhaps best exemplified by experiments showing that adult worms of *L. carinii* survive transfer into infected or splenectomized hosts, but not into naive ones (Wharton, 1946; Fujita and Kobayashi, 1969; Weiner and Soulsby, 1975, 1976, 1978), and by the carrier-specific immune suppression of the antibody response to trinitrophenol bound to microfilariae of *B. pahangi*, induced in jirds by developing larvae (Schrater and Piessens, 1982).

Immunologic cross-reactivity among different developmental stages is also apparent when analyzing the source of antigens capable of inducing resistance to infection and to microfilariae. Thus, antigens that protect rats against infections with *L. carinii* are present both in extracts of third stage larvae and microfilariae (Mehta *et al.*, 1981), and antigenic material of stages other than microfilariae affects the course and intensity of microfilaremia of *D. viteae* in hamsters (Haque *et al.*, 1978*a,b*). It therefore appears that antigens shared among different developmental stages play an important role in the filariae–host interplay.

Present knowledge of immune responses that may prevent infection is scarce. In rodents the only one that has been characterized to some extent is an IgM-antibody-dependent, complement-mediated response directed against the surface of *D. viteae* third-stage larvae. This reaction causes impairment of larval growth and development *in vitro* and probably *in vivo* (Tanner and Weiss, 1981*b*; Weiss and Tanner, 1981). In contrast, host responses to microfilariae leading to latency are now well characterized in rodents, both *in vivo* and *in vitro*. The salient feature emerging from these studies is that clearance of microfilariae is effected by antibody-dependent leukocyte-mediated killing mechanisms of the type that have been well described with other nematodes and the trematode *S. mansoni* (e.g., reviews by Capron *et al.*, 1982; Ogilvie *et al.*, 1981). Interestingly, the antibody class and effector cell type that kill microfilariae seem to be determined by the host, and not by the parasite species.

Thus, microfilariae of *D. viteae* are killed by IgM antibodies and neutrophils in hamsters (Weiss and Tanner, 1979; Rudin *et al.*, 1980; Neilson *et al.*, 1981), but by IgE antibodies and macrophages in rats (Haque *et al.*, 1980*b*). The same combination of IgE and macrophages is the principal effector in the killing of microfilariae of *L. carinii* in rats (Mehta *et al.*, 1980).

The target antigens of these anti-microfilarial immune responses, which must be placed on the worm's surface, are now beginning to be characterized. In our laboratory we have analyzed surface proteins and glycoproteins of different species of filarial nematodes (reviewed by Maizels *et al.*, 1982). The results vary, depending on the species and life cycle stage in question, but the two most common features encountered are that major proteins and glycoproteins expressed on the nematode surface are few and in most cases, antigenic. Thus, in microfilariae of *B. pahangi* and *B. malayi*, five and four components have been identified, respectively, all of which are antigenic in the infected host (Philipp *et al.*, 1980; Maizels *et al.*, 1981, 1982). In contrast, the dominant surface protein of microfilariae of *L. carinii* taken from the blood of infected cotton rats (but not from larvae released *in vitro* by gravid worms) is not antigenic and has been shown biochemically and immunologically to be serum albumin from the host (Philipp *et al.*, 1983). Microfilariae of *W. bancrofti* also bear host albumin on their surface (Maizels *et al.*, 1983*a*), and a similar result has been reported with skin microfilariae of *Onchocerca gibsoni* (Mitchell *et al.*, 1982). The role, if any, that these host components may play on the microfilarial surface is unknown, but it could be investigated by analyzing the effects of their inclusion in assays of antibody-mediated leukocyte adherence to, and killing of, microfilariae *in vitro*. Purified microfilarial surface antigen preparations could also be tested in this way, as a preliminary screening of potentially protective antigens.

In vivo vaccination trials can now be performed in a number of different rodent models. Mouse models of the type developed by Grove *et al.* (1979) for *Brugia* spp. or by Townson and Bianco (1982) for *Onchocerca* spp. are suitable for testing antigenic preparations that may protect the host against microfilariae, a task particularly desirable in onchocerciasis, in which this stage is the main cause of dermal and ocular disease.

Antigenic preparations that may protect against infection with third-stage larvae have to be tested in fully susceptible rodents. The jird-*Brugia* spp. may be suitable for this purpose, but unfortunately no rodent host has been found that is fully susceptible to *W. bancrofti* or *O. volvulus.*

A problem that concerns investigators searching for anti-filarial vaccines is the likelihood that a given antigenic preparation may at the same time protect against infection and cause immunopathologic symptoms in the vaccinated host. Indeed, experiments such as those of Klei *et al.* (1981, 1982) show that sensitization of jirds with extracts of *B. pahangi* increase the pathology induced by a subsequent infection. In addition to providing evidence of the immunologic nature of lymphatic and other pathologies associated with Brugian filariasis, these experiments also show that the jird-*Brugia* model could be used as a means to discriminate between protective and pathogenic components of a given antigenic preparation.

The usefulness of the jird in experimental filariasis contrasts with the paucity of immunologic information about this rodent. Experimental filariasis would benefit enormously if reagents were made available for the identification of classes and subclasses of immunoglobulin and the dissection of lymphocyte subpopulations of the jird.

The most desirable rodent host, the mouse, still lies beyond the grasp of a full filarial infection. Considering the large number of mouse strains now available, however, continuing efforts may yield a successful host-parasite combination. Nonetheless, a great deal has already been learned using the mouse as proxy host for microfilariae.

Inbred strains of rats are now available that will support a full infection with both rodent and human filariae of the genus *Brugia* (Fox and Schacher, 1976; Ho *et al.*, 1976*a*), and a wealth of inbred strains of hamsters remains unexplored.

In conclusion, although fully comprehensive experimental systems that will mimic the intricate immunologic and immunopathologic features of the different types of human filariasis may well be an overambitious task, there exist now a number of rodent models in which significant aspects of the immunobiology of filariasis can be investigated with the level of sophistication that such a complex host-parasite relationship demands.

ACKNOWLEDGMENTS

We thank our colleagues Dr. A. E. Bianco, Dr. A. Capron, Dr. J. K. Cruickshank, Dr. D. A. Denham, Dr. M. M. Fanning, Dr. A. Haque, Dr. J. W. Kazura, Dr. C. D. Mackenzie, Dr. R. Muller, Dr. J. K. Nayar, Dr. G. S. Nelson, Dr. W. F. Piessens, Dr. K. M. Price, Dr. D. Sauerman, Dr. A. F. Schrater, Dr. W. A. Sodeman, Jr., Dr. C. J. F. Spry, Dr. S. Townson, Dr. A. C. Vickery, Dr. A. L. Vincent, Dr. A. Winters, and Dr. B. G. Yangco for generously allowing us to quote some of their data before publication. M. P. gratefully acknowledges support of the filariasis component of the UNDP/World Bank/WHO Special Programme for Research and Training in Tropical Diseases.

V. REFERENCES

Ah, H. S., and Thompson, P. E., 1973, *Brugia pahangi* infections and their effect on the lymphatic systems of Mongolian jirds, *Exp. Parasitol.* **34**:393.

Ahmed, S. S., 1966, Location of developing and adult worms of *Brugia* ssp. in naturally and experimentally infected animals, *J. Trop. Med. Hyg.* **69**:291.
Ahmed, S. S., 1967*a*, Studies on the laboratory transmission of sub-periodic *Brugia malayi* and *B. pahangi*. II. Transmission to intact and splenectomized rats and cotton rats, *Ann. Trop. Med. Parasitol.* **61**:432.
Ahmed, S. S., 1967*b*, Studies on the laboratory transmission of sub-periodic *Brugia malayi* and *B. pahangi*. I. The resistance of guinea pigs, rabbits and white mice to infection, *Ann. Trop. Med. Parasitol.* **61**:93.
Amsbaugh, D. F., Hausen, C. T., Prescott, B., Stashak, P. W., Barthold, D. R., and Baker, P. S., 1972, Genetic control of the antibody response to Type III pneumococcal polysaccharide in mice. I. Evidence that an X-linked gene plays a decisive role in determining responsiveness, *J. Exp. Med.* **136**:931.
Anderson, R. C., and Bain, O., 1976, Keys to the genera of the order Spirurida Part 3. Filarioidea, Aproctoidea and Diplotriaenoidea in: *Keys to the Nematode Parasites of Vertebrates.* No. 3. (Anderson, R. C., Chabaud, A. G., and Willmott, S., eds.), pp. 39–116, Commonwealth Agricultural Bureau, Farnham Royal, Slough.
Aoki, Y., Recinos, M. M., and Hashiguchi, Y., 1980, Life span and distribution of *Onchocerca volvulus* microfilariae in mice, *J. Parasitol.* **66**:797.
Ash, L. R., 1971, Preferential susceptibility of male jirds (*Meriones unguiculatus*) to infection with *Brugia pahangi*, *J. Parasitol.* **57**:777.
Ash, L. R., 1973*a*, Life cycle of *Brugia patei* (Buckley, Nelson and Heisch 1958) in the jird *Meriones unguiculatus*, *J. Parasitol.* **59**:692.
Ash, L. R., 1973*b*, Chronic *Brugia pahangi* and *Brugia malayi* infections in *Meriones unguiculatus*, *J. Parasitol.* **59**:442.
Ash, L. R., and Riley, J. M., 1970*a*, Development of *Brugia pahangi* in the jird *Meriones unguiculatus*, with notes on infections in other rodents, *J. Parasitol.* **56**:962.
Ash, L. R., and Riley, J. M., 1970*b*, Development of subperiodic *Brugia malayi* in the jird *Meriones unguiculatus*, with notes on infections in other rodents, *J. Parasitol.* **56**:969.
Ash, L. R., and Schacher, J. F., 1971, Early life cycle and larval morphogenesis of *Wuchereria bancrofti* in the jird *Meriones unguiculatus*, *J. Parasitol.* **57**:1043.
Bagai, R. C., and Subrahmanyam, D., 1970, Nature of acquired resistance to filarial infection in albino rats, *Nature* (*Lond.*) **228**:682.
Beaver, P. C., 1970, Filariasis without microfilaraemia, *Am. J. Trop. Med. Hyg.* **19**:181.
Benjamin, D. B., and Soulsby, E. J. L., 1976, The homocytotrophic and hemagglutinating antibody responses to *Brugia pahangi* infection in the multimammate rat (*Mastomys natalensis*), *Am. J. Trop. Med. Hyg.* **25**:266.
Beveridge, I., Kummerow, E., and Wilkinson, P., 1980, Experimental infection of laboratory rodents and calves with microfilariae of *Onchocerca gibsoni*, *Tropenmed. Parasitol.* **31**:82.
Bianco, A. E., 1975, A new rodent filaria: A possible model for onchocerciasis, *Trans. R. Soc. Trop. Med. Hyg.* **69**:429.
Brinkmann, U. K., Krämer, P., Presthus, G. T., and Sawadogo, B., 1976, Transmission *in utero* of microfilariae of *Onchocerca volvulus*, *Bull. WHO* **54**:708.
Brug, S. L., 1927, Genuieuwe Filaria-soort (Filaria malayi parasiteerende bij den rueesch (voorloo-pige mededeeling), *Geneesk. Tijdschr. Ned. Indie* **67**:750.
Buck, A. A., 1974, *Onchocerciasis. Symptomatology, Pathology, Diagnosis*, World Health Organization, Geneva.
Buckley, J. J. C., 1958, Occult filarial infections of animal origin as a cause of tropical pulmonary eosinophilia, *East Afr. Med. J.* **35**:493.
Camus, D., Dessaint, J-P., Fischer, E., and Capron, A., 1979, Non-specific suppressor cell activity and specific cellular unresponsiveness in rat schistosomiasis, *Eur. J. Immunol.* **9**:341.

Capron, A., Dessaint, J-P., Capron, M. and Bazin, H., 1975, Specific IgE antibodies in immune adherence of normal macrophages to *Schistosoma mansoni* schistosomules, *Nature (Lond.)* **253**:474.

Capron, A., Dessaint, J-P., Capron, M., Joseph, M. and Pestel, J., 1980, Role of anaphylactic antibodies in immunity to schistosomes, *Am. J. Trop. Med. Hyg.* **29**:849.

Capron, A., Dessaint, J. P., Capron, M., Joseph, M., and Torpier, G., 1982, Effector mechanisms of immunity to schistosomes and their regulation, *Immun. Rev.* **61**:41.

Capron, M., Capron, A., Torpier, G., Bazin, H., Bout, D., and Joseph, M., 1978, Eosinophil-dependent cytotoxicity in rat schistosomiasis: Involvement of IgG_{2a} antibody and role of mast cells, *Eur. J. Immunol.* **8**:127.

Capron, M., Capron, A., Goetzl, E. J., and Austen, K. F., 1981, Tetrapeptides of the eosinophil chemotactic factor of anaphylaxis (ECF-A) enhance eosinophil Fc receptor, *Nature (Lond.)* **289**:71.

Chabaud, A. G., 1954, Sur le cycle evolutif des spirurides et de nematodes ayant une biologie comparable, *Ann. Parasitol. Hum. Comp.* **29**:238.

Chandrasoma, P. T., and Mendis, K. N., 1978, Filarial infections of the breast, *Am. J. Trop. Med. Hyg.* **26**:570.

Chong, L. K., and Wong, M. M., 1967, Experimental infection of laboratory mice with *Brugia pahangi, Med. J. Malaya* **21**:382.

Connor, D. H., Morrison, N. E., Kerdelvegas, F., Berkoff, H. A., Johnson, F., Tunnicliffe, R., Failing, C. F., Hale, L. N., Lindquist, K., Thornbloom, W., McCormick, J. B., and Anderson, S. L., 1970, Onchocerciasis, onchoceral dermatitis, lymphadenitis and elephantiasis in the Ubango territory, *Hum. Pathol.* **1**:553.

Cross, J. H., Partono, F., Hsu, M-Y. K., Ash, L. R., and Oemijati S., 1981, Further studies on the development of *Wuchereria bancrofti* in laboratory animals, *Southeast Asian J. Trop. Med. Public Health* **12**:114.

Cruickshank, J. K., Price, K. M., MacKenzie, C. D., Denham, D. A., and Spry, C. J., 1982, Eosinophil and antibody responses in an inbred rat model of Brugian filariasis: Their modulation by cyclosporin A, *Parasitology Abst.* **85**:XVI.

Danaraj, T. J., da Silva, L. S., and Schacher, J. F., 1959, The serological diagnosis of eosinophilic lung (tropical eosinophilia) and its etiological implications, *Am. J. Trop. Med. Hyg.* **8**:151.

Danaraj, T. J., Pacheco, G., Shanmugaratnam, K., and Beaver, P. C., 1966, The etiology and pathology of eosinophilic lung (tropical eosinophilia), *Am. J. Trop. Med. Hyg.* **15**:183.

Dasgupta, A., Bala, S., Banerjee, A., and Chaudhury, A., 1980, Immunodiagnosis of human filariasis by counter-immunoelectrophoresis using *Litomosoides carinii* antigens, *J. Helminthol.* **54**:83.

David, H. L., and Edeson, J. F. B., 1965, Filariasis in Portuguese Timor, with observations on a new microfilaria found in man, *Ann. Trop. Med. Paraistol.* **59**:193.

Denham, D. A., 1980, Vaccination against filarial worms using radiation-attenuated vaccines, *Int. J. Nucl. Med. Biol.* **7**:105.

Denham, D. A., and McGreevy, P. B., 1977, Brugian filariasis: Epidemiological and experimental studies, *Adv. Parasitol.* **15**:243.

Denham, D. A., and Nelson, G. S., 1976, Pathology and pathophysiology of nematode infections of the lymphatic system and blood vessels, in: *Pathophysiology of Parasitic Infections* (Soulsby, E. J. L., ed.), pp. 115–132, Academic Press, New York.

Dessaint, J. P., Camus, D., Fisher, E., and Capron, A., 1977, Inhibition of lymphocyte proliferation by factors produced by *Schistosoma mansoni, Eur. J. Immunol.* **7**:624.

Dessaint, J-P., Capron, A., Joseph, M., and Bazin, H., 1979, Cytophilic binding of IgE to the macrophage. II. Immunologic release of lysosomal enzyme from macrophages by IgE and anti-IgE in the rat: A new mechanism of macrophage activation, *Cell Immunol.* **46**:24.

Dissanaike, S., and Ismail, M. M., 1980, Antigens of *Setaria digitata:* Cross reaction with surface antigens of *Wuchereria bancrofti* microfilariae and serum antibodies of *W. bancrofti*-infected subjects, *Bull WHO* **58**:649.

Duke, B. O. L., 1962, Experimental transmission of *Onchocerca volvulus* to a chimpanzee, *Trans. R. Soc. Trop. Med. Hyg.* **56**:571.

Edeson, J. F. B., Ramachandran, C. P., Zaini, M. A., Nai, R. S., and Kershaw, W. E., 1962, The transmission of Malayan filariasis to rodents (Demonstration), *Trans. R. Soc. Trop. Med. Hyg.* **56**:269.

El Bihari, S., and Hussein, H. S., 1975, Location of the microfilariae of *Onchocerca armillata, J. Parasitol.* **61**:656.

Fanning, M. M., and Kazura, J. W., 1983, Genetic association of murine susceptibility to *Brugia malayi* microfilaraemia, *Parasite Immunol.* **5**:305.

Festing, M. F. W., 1979, *Inbred Strains in Biomedical Research*, Macmillan, London.

Fox, E. G., and Schacher, J. F., 1976, A comparison of syngeneic laboratory rat strains as hosts for *Brugia pahangi, Trans. R. Soc. Trop. Med. Hyg.*, **70**:523.

Fujita, K., and Kobayashi, J., 1969, The development of antibodies in the cotton rat transplanted with the adult cotton rat filaria, *Litomosoides carinii, Jap. J. Exp. Med.* **39**:586.

Gass, R. F., Tanner, M., and Weiss, N., 1979, Development of *Dipetalonema viteae* third stage larvae (Nematoda-Filarioidea) in micropore chambers transplanted into jirds, hamsters, normal and immunized mice, *Z. Parasitenkd.* **61**:73.

Geigy, R., Deschlimann, A., and Weiss, N., 1967, Transplacentare Ubertragung von Mikrofilarien der Art *Dipetalonema witeae* bei *Meriones libycus, Acta Trop.* **24**:266.

Grove, D. I., and Davis, R. S., 1978, Serological diagnosis of Bancroftian and Malayan filariasis, *Am. J. Trop. Med. Hyg.* **27**:508.

Grove, D. I., Davis, R. S., and Warren, K. S., 1979, *Brugia malayi* microfilaraemia in mice: A model for the study of the host response to microfilariae, *Parasitology* **79**:303.

Gusmao, D' A. R., Stanley, A. H.. and Ottesen, E. A., 1981, *Brugia pahangi* : Immunologic evaluation of the differential susceptibility to filarial infection in inbred Lewis rats, *Exp. Parasitol.* **52**:147.

Haas, B., and Wenk, P., 1981, Elimination of microfilariae (*Litomosoides carinii*, Filarioidea) in the patent and in the immunized cotton rat, *Trans. R. Soc. Trop. Med. Hyg.* **75**:143.

Haque, A., and Capron, A., 1982, Transplacental transfer of rodent microfilariae induces antigen-specific tolerance in rats, *Nature* (*Lond.*) **299**:361.

Haque, A., Lefebvre, M. N., Ogilvie, B. M., and Capron, A., 1978*a*, *Dipetalonema viteae* in hamsters: Effect of antiserum or immunization with parasite extracts on production of microfilariae, *Parasitology* **76**:61.

Haque, A., Chassoux, D., Ogilvie, B. M., and Capron, A., 1978*b*, *Dipetalonema viteae* infection in hamsters: enhancement and suppression of microfilariaemia, *Parasitology* **76**:77.

Haque, A., Worms, M. J., Ogilvie, B. M., and Capron, A., 1980*a*, *Dipetalonema viteae*: Microfilariae production in various mouse strains and in nude mice, *Exp. Parasitol.* **49**:398.

Haque, A., Joseph, M., Ouaissi, M. A., Capron, M., and Capron, A., 1980*b*, IgE antibody-mediated cytotoxicity of rat macrophages against microfilariae of *Dipetalonema viteae in vitro, Clin. Exp. Immunol.* **40**:487.

Haque, A., Camus, D., Ogilvie, B. M., Capron, M., Bazin, H., and Capron, A., 1981*a, Dipetalonema viteae* infective larvae reach reproductive maturity in rats immunodepressed by prior exposure to *Schistosoma mansoni* or its products and in congenitally athymic rats, *Clin. Exp. Immunol.* **43**:1.

Haque, A., Ouassi, M. A., Joseph, M., Capron, M., and Capron, A., 1981*b*, IgE antibody in eosinophil- and macrophage-mediated *in vitro* killing of *Dipetalonema viteae* microfilariae, *J. Immunol.* **127**:716.

Harbut, C. L., 1973, The white rat and golden hamster as experimental hosts for *Brugia pahangi* and subperiodic *Brugia malayi, Southeast Asian J. Trop. Med. Public Health* **4**:487.

Hawking, F., and Sewell, P., 1948, The maintenance of a filarial infection (*Litomosoides carinii*) for chemotherapeutic investigations, *Br. J. Pharmacol. Chemother.* **3**:285.

Hedge, E. C., and Ridley, D. S., 1977, Immunofluorescent reactions with microfilariae. I. Diagnostic evaluation, *Trans. R. Soc. Trop. Med. Hyg.* **71**:304.

Ho, B. C., Singh, M., Yap, E. H., and Lim, B. L., 1976*a*, Studies on the Malayan forest rat filaria *Breinlia booliati* (Filarioidea, Onchocercidae): Transmission to laboratory rats, *Int. J. Parasitol.* **6**:113.

Ho, B. C., Singh, M., Yap, E. H., and Lim, E. P. C., 1976*b*, Studies on the Malayan forest rat filariae *Breinlia booliati* (Filarioidea, Onchocercidae): relative susceptibility of various species of mosquitoes, *J. Med. Entomol.* **13**:531.

Ishizaka, T., Soto, C. S., and Ishizaka, K., 1972, Characteristics of complement fixation by aggregated IgE, *J. Immunol.* **19**:65.

Ismail, M. M., and Nagaratnam, N., 1973, Arthritis, possibly due to filariasis, *Trans. R. Soc. Trop. Med. Hyg.* **67**:405.

Joe, L. K., 1962, Occult filariasis: Its relationship with tropical pulmonary eosinophilia, *Am. J. Trop. Med. Hyg.* **11**:646.

Joseph, M., Dessaint, J-P., and Capron, A., 1977, Characteristics of macrophages cytoxicity induced by IgE immune complexes, *Cell. Immunol.* **34**:247.

Joseph, M., Capron, A., Butterworth, A. E., Sturrock, R. F., and Houba, V., 1978, cytotoxicity of human and baboon mononuclear phagocytes against schistosomula *in vitro*: Induction by immune complexes containing IgE and *Schistosoma mansoni* antigens, *Clin. Exp. Immunol.* **33**:48.

Joseph, M., Tonnel, A. B., Capron, A., and Voisin, C., 1980, Enzyme release and superoxide anion production by human alveolar macrophages stimulated with immunoglobulin E, *Clin. Exp. Immunol.* **40**:416.

Karavodin, L. M., and Ash, L. R., 1980, Circulating immunocomplexes in experimental filariasis, *Clin. Exp. Immunol.* **40**:312.

Karavodin, L. M., and Ash, L. R., 1981, Sequential determination of circulating immune complexes in experimental filariasis, *Infect. Immun.* **34**:105.

Karavodin, L. M., and Ash, L. R., 1982, Inhibition of adherence and cytotoxicity by circulating immune complexes formed in experimental filariasis, *Parasite Immunol.* **4**:1.

Kazura, J. W., and Davis, R. S., 1982, Soluble *Brugia malayi* microfilarial antigens protect mice against challenge by an antibody-dependent mechanism, *J. Immunol.* **128**:1792.

Klei, T. R., and Crowell, W. A., 1981, Pathological changes in kidneys, livers and spleens of *Brugia pahangi*-infected jirds (*Meriones unguiculatus*), *J. Helminthol.* **55**:123.

Klei, T. R., McCall, J. W., and Malone, J. B., 1980, Evidence for increased susceptibility of *Brugia pahangi*-infected jirds (*Meriones unguiculatus*) to subsequent homologous infections, *J. Helminth.* **54**:161.

Klei, T. R., Enright, F. M., Blanchard, D. P., and Uhl, S. A., 1981, Specific hypo-responsive granulomatous tissue reactions in *Brugia pahangi*-infected jirds, *Acta Trop.* **38**:267.

Klei, T. R., Enright, F. M., Blanchard, D. P., and Uhl, S. A., 1982, Effects on presensitization on the development of lymphatic lesions in *Brugia pahangi*-infected jirds, *Am. J. Trop. Med. Hyg.* **31**:280.

Kowalski, J. C., and Ash, L. R., 1975, Repeated infections of *Brugia pahangi* in the jird, *Meriones unguiculatus, Southeast Asian J. Trop. Med. Public Health.* **6**:195.

Laing, A. B. G., Edeson, J. F. B., and Wharton, R. F., 1961, Studies on filariasis in Malaya: Further experiments on the transmission of *Brugia malayi* and *Wuchereria bancrofti, Ann. Trop. Med. Parasitol.* **55**:86.

Lämmler, G., Saupe, E., and Herzoy, H., 1968, Infektions-versuche mit der Baumwollratten -filariae *Litomosoides carinii* bei *Mastomys natalensis, Z. Parasitenkd.* **32**:281.

Lichtenberg, F. von, 1957, The early phase of endemic bancroftian filariasis in the male. Pathological study, *J. Mount Sinai Hosp.* **26**:983.

Lichtenstein, A., 1927, Filaria-onderzoek te Birenen, *Geneesk. Tijdschr. Ned. Indie* **67**:742.

MacDonald, E. M., and Scott, J. A., 1953, Experiments on immunity in the cotton rat to the filarial worm *Litomosoides carinii, Exp. Parasitol.* **2**:174.

MacDonald, E. M., and Scott, J. A., 1958, The persistence of acquired immunity to the filarial worm of the cotton rat, *Am. J. Trop. Med. Hyg.* **7**:419.

Maizels, R. M., Philipp, M., Denham, D. A., Partono, F., Oemijati, S., and Ogilvie, B. M., 1981, Antigenic analysis in Brugian filariasis: Cross reactive and restricted antigens on the surface of *Brugia* of different stages and species, *Parasitology* **84**:xxxi.

Maizels, R. M., Philipp, M., and Ogilvie, B. M., 1982, Molecules on the surface of parasitic nematodes as probes of the immune response in infection, *Immunol. Rev.* **61**:109.

Maizels, R. M., Partono, F., Oemijati, S., Denham, D. A. and Ogilvie, B. M. 1983, Cross-reactive surface antigens on three stages of *Brugi malayi, Brugia pahangi* and *Brugia timori, Parasitology* (in press).

Maizels, R. M., Philipp, M., Daseupta, A., and Partono, F., 1983*a*, Human serum albumen is a major component on the surface of microfiliariae of *Wuchereria bancrofti, Parasite Immunol.* (in press).

Mak, J. W., and Lim, B. L., 1974, New hosts of *Breinlia booliati* in wild rats from Sarawak, with further observations on its morphology, *Southeast Asian J. Trop. Med. Public Health.* **5**:22.

Malone, J. B., and Thompson, P. E., 1975, *Brugia pahangi*: Susceptibility and macroscopic pathology in golden hamsters, *Exp. Parasitol.* **38**:279.

Malone, J. B., Leininger, J. R., and Thompson, P. E., 1974, *Brugia pahangi* in golden hamsters, *Trans. R. Soc. Trop. Med. Hyg.* **68**:170.

Mantovani, A., and Jackson, R. F., 1966, Transplacental transmission of microfilariae of *Dirofilaria immitis* in the dog, *J. Parasitol.* **52**:116.

Martinez-Baez, M., 1978, La oncocercosis en Mexico, *Symp. Gac. Med. de Mex.* **114**:525.

McCall, J. W., Malone, J. B., Ah, H., and Thompson, P. E., 1973, Mongolian jirds (*Meriones unguiculatus*) infected with *Brugia pahangi* by the intraperitoneal route: A rich source of developing larvae, adult filariae and microfilariae, *J. Parasitol.* **59**:436.

McGreevy, P. B., Ratiwayanto, S., Tuti, S., McGreevy, M. M., and Dennis, D. T., 1980, *Brugia malayi*: Relationship between anti-sheath antibodies and amicrofilaraemia in natives living in an endemic area of South Kalimantan, Borneo, *Am. J. Trop. Med. Hyg.* **29**:553.

Mehta, K., Sindhu, R. K., Subrahamanyam, D., and Nelson, D. S., 1980, IgE-dependent adherence and cytotoxicity of rat spleen and peritoneal cells to *Litomosoides carinii* microfilariae, *Clin. Exp. Immunol.* **41**:107.

Mehta, K., Subrahmanyam, D., and Sindhu, R. K., 1981, Immunogenicity of homogenates of the developmental stages of *Litomosoides carinii* in albino rats, *Acta Trop.* **38**:319.

Mehta, K., Sindhu, R. K., Subrahmanyam, D., Hopper, K., and Nelson, D. S., 1982, IgE-dependent cellular adhesion and cytotoxicity to *Litomosoides carinii* microfilariae—Nature of effector cells, *Clin. Exp. Immunol.* **48**:477.

Meyers, F. M., and Kunwenaar, W., 1939, Over hypereosinophilie en overeen werkwaardigen vorm van filariassis, *Geneesk. Tijdsch. Ned. Indie* **79**:853.

Mitchell, G. F., Anders, R. F., Brown, G. V., Handman, E., Roberts-Thomson, I. C., Chapman, C. B., Forsyth, K. P., Kahl, L. P., and Cruise, K. M., 1982, Analysis of infection characteristics and antiparasite immune responses in resistant compared with susceptible hosts, *Immunol. Rev.* **61**:137.

Muller, R. L., and Nelson, G. S., 1975, *Ackertia globulosa* sp. n. (Nematoda, Filarioidea) from rodents in Kenya, *J. Parasitol.* **61**:606.

Neilson, J. T. M., 1976, A comparison of the acquired resistance to *Dipetalonema viteae* stimulated in hamsters by trickle vs tertiary infections, *Z. Tropenmed. Parasitol.* **27**:233.

Neilson, J. T. M., 1978, Primary infections of *Dipetalonema viteae* in an outbred and five inbred strains of golden hamsters, *J. Parasitol.* **64**:378.

Neilson, J. T. M., 1979, Kinetics of *Dipetalonema viteae* infections established by surgical implantation of adult worms into hamsters, *Am. J. Trop. Med. Hyg.* **28**:216.

Neilson, J. T. M., Crandall, C. A., and Crandall, R. B., 1981, Serum immunoglobulin and antibody levels and the passive transfer of resistance in hamsters infected with *Dipetalonema viteae*, *Acta Trop.* **38**:309.

Nelson, G. S., 1979, Current concepts in parasitology. Filariasis, *N. Engl. J. Med.* **300**:1136.

Nelson, G. S., Amin, M. A., Blackie, E. J., and Robson, N., 1966, The maintenance of *Onchocerca gutturosa* microfilariae *in vivo* and *in vitro*, *Trans. R. Soc. Trop. Med. Hyg.* **60**:17.

Neva, F. A., Kaplan, A. P., Pacheco, G., Gray, L., and Danaraj, T. J., 1975, A human model of parasitic immunopathology, with observations on serum IgE levels before and after treatment, *J. Allergy Clin. Immunol.* **55**:422.

Ogilvie, B. M., and MacKenzie, C. D., 1981, Immunology and immunopathology of infections caused by filarial nematodes, in: *Parasitic Diseases. The Immunology* (J. Mansfield, ed.) pp. 227–289, Marcel Dekker, New York.

Olson, L. J., 1959, The survival of migratory and post-migratory stages of *Litomosoides carinii* in white rats, *J. Parasitol.* **45**:182.

Olson, L. J., Scott, J. A., and MacDonald, E. M., 1954, Factors in the racial immunity of the white rat to cotton rat filarial worms, *J. Parasitol.* **40**(Suppl.):14.

Olson, L. J., Scott, J. A., and MacDonald, E. M., 1955, Infection of white rats with the filarial worm of cotton rats, *J. Parasitol.* **41**(Suppl.):44.

Ottesen, E. A., 1980, Immunopathology of lymphatic filariasis in man, *Springer Sem. Immunopathol.* **2**:373.

Ottesen, E. A., Weller, P. F., and Heck, L., 1977, Specific cellular immune unresponsiveness in human filariasis, *Immunology* **33**:413.

Ottesen, E. A., Neva, F. A., Paranjape, R. S., Tripathy, S. P., Thiruvengadam, K. V., and Beaven, M. A., 1979, Specific allergic sensitisation to filarial antigens in tropical eosinophilic syndrome, *Lancet* **1**:1158.

Ouaissi, M. A., Haque, A., and Capron, A., 1981, *Dipetalonema viteae*: Ultrastructural study on the *in vitro* interaction between rat macrophages and microfilariae in the presence of IgE antibody, *Parasitology* **82**:55.

Palmieri, J. R., Purnomo, Dennis, D. T., and Marwoto, H. A., 1980, Filarid parasites of South Kalimantan (Borneo) Indonesia. *Wuchereria kalimantani* sp. n. (Nematoda, Filarioidea) from the Silver Leaf monkey *Presbytis cristatus* Eschscholtz 1921, *J. Parasitol.* **66**:645.

Partono, F., Dennis, D. T., Purnomo, and Atmosoedjono, S., 1977, *Brugia timori*: Experimental infection in some laboratory animals, *Southeast Asian J. Trop. Med. Public Health* **8**:155.

Perlmutter, R. M., Nahm M., Stein, K. E., Slack, J., Zitron, I., Paul, W. E., and Davie, J. M., 1979, Immunoglobulin subclass-specific immunodeficiency in mice with an X-linked B-lymphocyte defect, *J. Exp. Med.* **149**:993.

Petrányi, G., Mieth, H., and Leitner, I., 1975, *Mastomys natalensis* as an experimental host for *Brugia malayi* subperiodic, *Southeast Asian J. Trop. Med. Public Health* **6**:328.

Philipp, M., Denham, D. A., Maizels, R. M., Ogilvie, B. M., Parkhouse, R. M. E., Taylor, P. M., and Worms, M. J., 1980, The antigens on the surface of parasitic nematodes: Studies on *Litomosoides carinii, Brugia pahangi* and *Trichinella spiralis, Parasitology* **81**:xxx.

Philipp, M., Worms, M. J., McLaren, D. J., Oglivie, B. M., Parkhouse, R. M. E., and Taylor, P. M., 1983, Surface proteins of a filarial nematode: a major soluble antigen and a host component on the cuticle of *Litomosoides carinii, Parasite Immunol.* (in press).

Piessens, W. F., McGreevy, P. B., Piessens, P. W., McGreevy, M., Koiman, I., Saroso, J. S., and Dennis, D. T., 1980*a*, Immune responses in human infections with *Brugia malayi*: Specific cellular unresponsiveness to filarial antigen, *J. Clin. Invest.* **65**:172.

Piessens, W. F., Ratiwayanto, S., Tuti, S., Palmieri, J. H., Piessens, P. W., Koiman, I., and Dennis, D. T., 1980*b*, Antigen-specific suppressor cells and suppressor factors in human filariasis with *Brugia malayi, N. Engl. J. Med.* **302**:833.

Piessens, W. F., McGreevy, P. B., Ratiwayanto, S., McGreevy, M., Piessens, P. W., Koiman, I., Saroso, J. S., and Dennis, D. T., 1980*c*, Immune responses in human infections with *Brugia malayi*: correlation of cellular and humoral reactions to microfilarial antigens with clinical status, *Am. J. Trop. Med. Hyg.* **29**:563.

Piessens, W. F., Partono, F., Hoffman, S. L., Ratiwayanto, S., Piessen, P. W., Palmieri, J. R., Koiman, I., Dennis, D. T. and Carney, W. P., 1982, Antigen-specific suppressor T-lymphocytes in human lymphatic filariasis, *N. Engl. J. Med.* **307**:144.

Ponnudurai, T., Denham, D. A., Nelson, G. S., and Rogers, R., 1974, Studies with *Brugia pahangi*. 4. Antibodies against adult and microfilarial stages, *J. Helminthol.* **48**:107.

Portaro, J. K., Britton, S., and Ash, L. R., 1976, *Brugia pahangi*: Depressed mitogen reactivity in filarial infections in the jird, *Meriones unguiculatus, Exp. Parasitol.* **40**:438.

Pringle, G., and King, D. F., 1968, Some developments in techniques for the study of the rodent filarial parasite *Litomosoides carinii*. I. A preliminary comparison of the host efficiency of the multimammate rat *Praomys (Mastomys) natalensis* with that of the cotton rat *(Sigmodon hispidus), Ann. Trop. Med. Parasitol.* **62**:462.

Rabalais, F. C., 1974, Studies on *Onchocerca cervicalis* Railliet and Henry, 1910, microfilariae in the jird *Meriones unguiculatus, J. Helminthol.* **48**:125.

Ramachandran, C. P., and Pacheco, G., 1965, American cotton rat *(Sigmodon hispidus)* as an experimental host for *Brugia pahangi, J. Parasitol.* **51**:722.

Ramachandran, C. P., Sandosham, A. A., and Sivanandam, S., 1966, Development of *Wuchereria bancrofti* in the domestic cat (Laboratory meeting), *Med. J. Malaya* **20**:333.

Ramakrishnan, S. P., Singh, D., Bhatnagar, V. N., and Raghavan, N. G. S., 1961, Infection of the albino rat with the filarial parasite *Litomosoides carinii* of cotton rats, *Ind. J. Malariol.* **15**:255.

Ramakrishnan, S. P., Singh, D., and Krishnaswamy, A. K., 1962, Evidence of acquired immunity against microfilariae of *Litomosoides carinii* in albino rats with mite-induced infections, *Ind. J. Malariol.* **16**:263.

Rao, Y. V. B. G., Mehta, K., and Subrahamanyam, D., 1977, *Litomosoides carinii*: Effect of irradiation on the development and immunogenicity of the larval forms, *Exp. Parasitol.* **43**:39.

Rao, Y. V. B. G., Mehta, K., Subrahmanyam, D., and Venkatarao, S., 1980, Effect of irradiation on the infectivity and immunogenicity of larvae of *Litomosoides carinii, Ind. J. Med. Res.* **72**:42.

Rogers, R., Davis, R., and Denham, D. A., 1975, A new technique for the study of changes in lymphatics induced by filarial worms, *J. Helminthol.* **49**:31.

Rousseaux-Prevost, R., Chassoux, D., Bazin, H., and Capron, A., 1979, Serum IgE levels in rats infected with *D. viteae* L3 larvae, *Clin. Exp. Immunol.* **38**:389.

Rudin, W., Tanner, M., Bauer, P., and Weiss, N., 1980, Studies on *Dipetalonema viteae*

(Filarioidea). 5. Ultrastructural aspects of the antibody-dependent cell-mediated destruction of microfilariae, *Tropenmed. Parasitol.* **31**:194.

Sänger, I., Lämmler, G., and Kimmig, P., 1981, Filarial infection of *Mastomys natalensis* and their relevance for experimental chemotherapy, *Acta Trop.* **38**:277.

Sasa, M., 1976, *Human Filariasis: A Global Survey of Epidemiology and Control*, University Park Press, Baltimore.

Schacher, J. F., 1973, Laboratory models in filariasis: A review of filarial life-cycle patterns, *Southeast Asian J. Trop. Med. Public Health* **4**:336.

Schacher, J. F., and Sahyoun, P. F., 1967, A chronological study of the histopathology of filarial disease in cats and dogs caused by *Brugia pahangi* (Buckley and Edeson 1956), *Trans. R. Soc. Trop. Med. Hyg.* **61**:234.

Schrater, A. F., and Piessens, W. F., 1982, Antigens present on early larval stages induce carrier-specific suppression in *Brugia malayi* infected jirds, *Fed. Proc.* **41**:371.

Scott, J. A., and MacDonald, E. M., 1958, Immunity to challenging infections of *Litomosoides carinii* produced by transfer of developing worms, *J. Parasitol.* **44**:187.

Scott, J. A., MacDonald, E. M., and Olson, L. J., 1958*a*, The early induction in cotton rats of immunity to their filarial worms, *J. Parasitol.* **44**:507.

Scott, J. A., MacDonald, E. M., and Olson, L. J., 1958*b*, Attempts to produce immunity against the filarial worms of cotton rats by transfer of developing worms, *Am. J. Trop. Med. Hyg.* **7**:70.

Siddiqui, M. A., 1979, Host-parasite relations in cotton rat filariasis. III. The quantitative transmission of *Litomosoides carinii* to unirradiated or irradiated golden hamsters and white mice, *Ann. Trop. Med. Parasitol.* **73**:377.

Singh, M., and Cheong, C. H., 1971, On a collection of nematode parasites from Malayan rats, *Southeast Asian J. Trop. Med. Public Health* **2**:516.

Singh, M., and Ho, B. C., 1973, *Breinlia booliati* sp. n. (Filarioidea, Onchocercidae), a Filaria of the Malayan forest rat, *Rattus sabanus* (Thos), *J. Helminthol.* **47**:127.

Singh, M., Ho, B. C., and Lim, B. L., 1972, Preliminary results on experimental transmission of a new filarial parasite, *Breinlia booliati* sp. n. Singh and Ho, 1973, from Malayan wild rats to laboratory albino rats, *Southeast Asian J. Trop. Med. Public Health* **3**:622.

Singh, M., Yap, E. H., Ho, B. C., Kang, K. L., and Lim, E. P. C., 1976, Studies on the Malayan forest rat filaria *Breinlia booliati* (Filarioidea, Onchocercidae). Course of development in rat host, *J. Helminthol.* **50**:103.

Spry, C. J. F., 1980, Alterations in blood eosinophil morphology, binding capacity for complexed IgG and kinetics in patients with tropical (filarial) eosinophilia, *Parasite Immunol.* **3**:1.

Spry, C. J. F., and Kumaraswami, V., 1982, Tropical eosinophilia, *Semin. Haematol.* **19**: 107.

Storey, D. M., and Al-Mukhtar, A. S., 1982, Vaccination of jirds, *Meriones unguiculatus* against *Litomosoides carinii* and *Brugia pahangi* using irradiated larvae of *L. carinii, Tropenmed. Parasitol.* **33**:23.

Storey, D. M. and Al-Mukhtar, A. S., 1983, The survival of adult *Litomosoides carinii* in cotton rats previously injected with irradiated stage 3 larvae, *Tropenmed. Parasitol.* **34**:24.

Subrahmanyam, D., Chaudbury, S., and Jain, S., 1974, Immunological investigations on filariasis, *Ind. J. Bacteriol Pathol.* **17**:135.

Subrahmanyam, D., Rao, Y. V. B. G., Mehta, K., and Nelson, D. S., 1976, Serum dependent adhesion and cytotoxicity of cells to *Litomosoides carinii* microfilariae, *Nature (Lond.)* **260**:529.

Subrahmanyam, D., Mehta, K., Nelson, D. S., Rao, Y. B. V. G., and Rao, C. K., 1978, Immune responses in human filariasis, *J. Clin. Microbiol.* **8**:228.

Sucharit, S., and MacDonald, W. W., 1972, *Brugia pahangi* in small laboratory animals: The screening of infected rats, *Southeast Asian J. Trop. Med. Public Health* **3**:347.

Sucharit, S., and MacDonald, W. W., 1973, *Brugia pahangi* in small laboratory animals: Attempts to increase susceptibility of white rats to *Brugia pahangi* by host selection, *Southeast Asian J. Trop. Med. Public Health* **4**:71.

Suswillo, R. R., Nelson, G. S., Muller, R.., McGreevy, P. B., Duke, B. O. L., and Denham, D. A., 1977, Attempts to infect jirds (*Meriones unguiculatus*) with *Wuchereria bancrofti, Onchocerca volvulus, Loa loa* and *Mansonella ozzardi, J. Helminthol.* **51**:132.

Suswillo, R. R., Owen, D. G., and Denham, D. A., 1980, Infections of *Brugia pahangi* in conventional and nude (athymic) mice, *Acta Trop.* **37**:327.

Suswillo, R. R., Doenhoff, M. J., and Denham, D. A., 1981, Successful development of *Brugia pahangi* in T-cell-deprived CBA mice, *Acta Trop.* **38**:305.

Tanner, M., 1981, *Dipetalonema viteae* (Filarioidea): Development of the infective larvae *in vitro, Acta Trop.* **38**:241.

Tanner, M., and Weiss, N., 1978, Studies on *Dipetalonema viteae* (Filarioidea). II. Antibody-dependent adhesion of peritoneal exudate cells to microfilariae *in vitro, Acta Trop.* **35**:151.

Tanner, M., and Weiss, N., 1979, Studies on *Dipetalonema viteae* (Filarioidea). 4. Passive transfer of immunity to circulating microfilariae by spleen cells, *Tropenmed. Parasitol.* **30**:371.

Tanner, M., and Weiss, N., 1981*a*, *Dipetalonema viteae* (Filarioidea): Development of the infective larvae in micropore chambers implanted into normal, infected and immunized jirds, *Trans. R. Soc. Trop. Med. Hyg.* **75**:173.

Tanner, M., and Weiss, N., 1981*b*, *Dipetalonema viteae* (Filarioidea): Evidence for a serum dependent cytotoxicity against developing third and fourth stage larvae *in vitro, Acta Trop.* **38**:325.

Townson, S., and Bianco, A. E., 1982, Experimental infection of mice with the microfilariae of *Onchocerca lienalis, Parasitology* **85**:283.

Townson, S., Bianco, A. E., and Owen, D., 1981, Attempts to infect small laboratory animals with the infective larvae of *Onchocerca lienalis, J. Helminthol.* **55**:247.

Thompson, J. P., Crandall, R. B., Crandall, C. A., and Neilson, J. T., 1979, Clearance of microfilariae of *Dipetalonema viteae* in CBA/N and CBA/H mice, *J. Parasitol.* **65**:966.

Thompson, J. P., Crandall, R. B., and Crandall, C. A., 1981, Microfilaremia and antibody responses in CBA/H and CBA/N mice following injection of microfilariae of *Brugia malayi, J. Parasitol.* **67**:728.

Vincent, A. L., and Ash, L. R., 1978, Splenomegaly in jirds (*Meriones unguiculatus*) infected with *Brugia malayi* and related species, *Am. J. Trop. Med. Hyg.* **27**:514.

Vincent, A. L., Frommes, S. P., and Ash, L. R., 1976, *Brugia malayi*, *B. pahangi*, *B. patei*: Pulmonary pathology in jirds *Meriones unguiculatus*, *Exp. Parasitol.* **40**:330.

Vincent, A. L., Ash, L. R., Rodrick, G. E., and Sodeman, W. A., 1980*a*, The lymphatic pathology of *Brugia pahangi* in the Mongolian jird, *J. Parasitol.* **66**:613.

Vincent, A. L., Sodeman, W. A., Jr., and Winters, A., 1980*b*, Development of *Brugia pahangi* in normal and nude mice, *J. Parasitol.* **66**:448.

Vincent, A. L., Vickery, A. C., and Winters, A., 1982*a*, Life cycle of *Brugia pahangi* in nude (congenitally athymic) mice, C3H/HeN (nu/nu), *J. Parasitol.* **68**:553.

Vincent, A. L., Vickery, A. C., Nayar, J. K., Sauerman, D., and Yangco, B. G., 1982*b*, Non-development of *Wuchereria bancrofti* in nude (congenitally athymic) mice, *Am. J. Trop. Med. Hyg.* **31**:1062.

Warren, K. S., 1971, Worms, in: *Immunological Diseases* (M. Samter, ed.), pp. 668-686, Little, Brown, Boston.

Weiner, D. J., and Soulsby, E. J. L., 1975, Fate of *Litomosoides carinii* adults transferred into infected and naive Mongolian jirds (*Meriones unguiculatus*), in: *Nuclear Techniques in Helminthology Research*, pp. 85–89, International Atomic Energy Agency, Vienna.

Weiner, D. J., and Soulsby, E. J. L., 1976, Fate of *Litomosoides carinii* adults transplanted into the pleural or peritoneal cavity of infected and naive multimammate rats (*Mastomys natalensis*), *J. Parasitol.* **62**:886.

Weiner, D. J., and Soulsby, E. J. L., 1978, *Litomosoides carinii*: Effect of splenectomy on the ability of naive *Mastomys natalensis* to accept transplanted worms, *Exp. Parasitol.* **45**:241.

Weiss, N., 1970, Parasitologische und immunobiologische Untersuchungen über die durch *Dipetalonema viteae* erzeugte Nagertierfilariose, *Acta Trop.* **35**:137.

Weiss, N., 1978*a*, Studies on *Dipetalonema viteae* (Filarioidea). I. Microfilaraemia in hamsters in relation to worm burden and humoral immune response, *Acta Trop.* **35**:137.

Weiss, N., 1978*b*, *Dipetalonema viteae:In vitro* blastogenesis of hamster spleen and lymph node cells to phytohaemagglutinin and filarial antigens, *Exp. Parasitol.* **46**:283.

Weiss, N., and Tanner, M., 1979, Studies on *Dipetalonema viteae* (Filarioidea). 3. Antibody-dependent cell-mediated destruction of microfilariae *in vivo, Tropenmed. Parastiol.* **30**:73.

Weiss, N., and Tanner, M., 1981, Immunogenicity of the surface of filarial larvae (*Dipetalonema viteae*), *Trans. R. Soc. Trop. Med. Hyg.* **75**:179.

Weller, P. F., 1978, Cell-mediated immunity in experimental filariasis: Lymphocyte reactivity to filarial stage-specific antigens and to B-and T-cell mitogens during acute and chronic infection, *Cell. Immunol.* **37**:369.

Wharton, D. B. A., 1946, Transplantation of adult filarial worms: *Litomosoides carinii* in cotton rats, *Science* **104**:30.

Williams, R. W., and Brown, H. W., 1945, The development of *Litomosoides carinii* filarid parasite of the cotton rat in the tropical mite, *Science* **102**:482.

Wong, M. M., and Guest, M. F., 1969, Filarial antibodies and eosinophilia in human subjects in an endemic area, *Trans. R. Soc. Trop. Med. Hyg.* **63**:796.

Wong, M. M., and Suter, M. F., 1979, Indirect fluorescent antibody test in occult dirofilariasis, *Am. J. Vet. Res.* **40**:414.

Worms, M. J., Terry, R. J., and Terry, A., 1961, *Dipetalonema witei*, filarial parasite of the jird, *Meriones libycus*. 1. Maintenance in the laboratory, *J. Parasitol.* **47**:963.

Yap, E. H., Ho, B. C., Singh, M., Kang, K. L., and Lim, B. L., 1975, Studies on the Malayan forest rat filaria *Breinlia booliati* (Filarioidea, Onchocercidae): Periodicity and microfilaraemic patterns during the course of infection, *J. Helminthol.* **49**:263.

Yogogawa, S., 1939, Studies on the mode of transmission of *Wuchereria bancrofti, Trans. R. Soc. Trop. Med. Hyg.* **32**:653.

Zein-Eldin, E. A., 1965, Experimental infection of gerbils with *Litomosoides carinii* via intravenous injections, *Texas Rep. Biol. Med.* **23**:530.

Zielke, E., 1979, Attempts to infect *Meriones unguiculatus* and *Mastomys natalensis* with *Wuchereria bancrofti* from West Africa, *Tropenmed. Parasitol.* **30**:466.

Chapter 9

Examination of Strategies for Vaccination against Parasitic Infection or Disease Using Mouse Models

Graham F. Mitchell, Robin F. Anders, Colin B. Chapman, Ian C. Roberts-Thomson, Emanuela Handman, and Kathy M. Cruise

Laboratory of Immunoparasitology
The Walter and Eliza Hall Institute of Medical Research
Melbourne, Victoria 3050, Australia

and

Michael D. Rickard and Marshall W. Lightowlers

Department of Paraclinical Sciences
University of Melbourne Veterinary Clinical Centre
Werribee, Victoria 3030, Australia

and

Edito G. Garcia

Department of Parasitology
Institute of Public Health
University of the Philippines
Ermita, Manila 2801, Philippines

I. INTRODUCTION

There are currently no prophylactic or therapeutic vaccines against parasites of medical importance, the parasitic infections with major public health consequences being concentrated in tropical, less industrially developed countries. However, a limited number of living vaccines against economically important parasites is available in veterinary medicine, and vaccination efficacy of crude antigen preparations has been demonstrated time and again in laboratory models involving helminth, protozoan, and arthropod parasites (Clegg and Smith, 1978; Cox, 1978; Péry and Luffau, 1979; Murray *et al.*, 1979; Mitchell, 1982*a*; Mitchell and Anders, 1982; Rickard and Williams, 1982).

Anti-parasite immune responses of high titer and a degree of resistance to parasitic infection, reinfection and/or disease, are detected readily in at least some individuals of the natural host species. Because most parasitic organisms have a high biotic potential in susceptible hosts, the immune response of the parasitized mammal must be a key restraining influence on parasite populations. In evolution, two types of interrelated selection pressure must operate, *viz.* selection for parasites with fitness but that do not threaten the life of the bulk of individuals of the host population in prereproductive life, and selection for hosts that can limit parasite burdens or that can withstand the deleterious effects of parasitism. Such observations and postulates provide confidence that new parasite vaccines will become available in the future, although not necessarily the immediate future. There is no doubting the need for parasite vaccines to complement other public health measures in clinical medicine, managerial practices in veterinary medicine, and chemotherapy, in the control of parasites.

The first step in assessing the feasibility of vaccination and the development of prototype vaccines by the rational approach is to identify immune responses and their target antigens necessary, or even sufficient, for host resistance/resilience to infection/disease, i.e., functional antiparasite immune responses. The quest for parasite vaccines and identification of new vaccination strategies is dominated by attempts to pinpoint host-protective or parasite-inhibitory immune responses, the isolation of target antigens, examination of recombinant DNA methods for polypeptide production, development of new adjuvants, and exploitation of newly identified mechanisms of immunoregulation in order to increase host resistance (Mitchell, 1982*a*). Obviously, possession of a hybridoma-derived antibody (or battery of such antibodies) or a polyspecific antiserum with parasite-inhibitory effects *in vitro*, host-protective effects *in vitro*, or transmission-blocking effects, will greatly facilitate the identification of antigens of parasites by serologic methods. To date, the hybridoma probes have usually been generated by the shotgun approach, i.e., screening for functional activity in a battery of hybridoma antibodies selected by binding assays with parasites or their extracts (Mitchell and Cruise, 1981).

Defined-antigen (molecular) vaccines are an ideal sought by the immunoparasitologist (Anders *et al.*, 1982*a*). Clearly, there are many years and considerable expense separating the stated objective and the realization (and adequate testing) of a safe, stable, cheap, and quality-controlled vaccine effective at inducing long-lasting protection against the bulk of any genetically diverse parasite population in the bulk of any gentically diverse host population. Combinations of effector cells, molecules, and mechanisms operate to prejudice the establishment or persistence of parasites in their natural hosts (reviewed in Mitchell, 1979*a*). The minimal goal of the defined-antigen approach to vaccination is to induce a state of sensitivity in the vaccinee that enables other beneficial immune responses to be induced more efficiently on subsequent encounter with the parasite. These responses *in toto* then militate against parasite establishment or persistence or reduce the opportunities for induction of disease, or both.

Three major assumptions underpin the emphasis in basic immunoparasitology on defined-antigen vaccines:

1. A small subset of antigens in the vast array of immunogenic parasite molecules presented to the immune system in the parasitized mammalian host are responsible for induction of immune responses, which are beneficial in terms of host resistance expressed as concomitant, if not sterilizing, immunity.
2. New biotechnology (e.g., recombinant DNA, monoclonal antibodies, parasite cultivation, parasite cell or product immortalization through hybridization techniques) will overcome many of the difficulties of parasite antigen supply, and newer molecular fractionation techniques will improve precision in identifying and isolating parasite antigens from crude mixtures.
3. New adjuvants of selective immunopotentiating activity that are safe and suitable for wide usage will become available in the future.

Several advantages of a defined-antigen vaccine include the following:

1. Quality control of a product for administration to humans or animals is simpler than is the the case for a heterogeneous mixture of molecules.
2. Methods of synthesis can be determined and improved.
3. Opportunities for induction of untoward side effects and counterproductive (e.g., immunopathologic) responses are reduced.
4. Quenching of appropriate host-protective immune responses, by whatever mechanisms are responsible for immune deviation or antigenic competition, is less likely.
5. Monitoring of immunogenicity of the vaccine is simpler than attempting to quantitate accurately an immune response to a crude mixture of antigenic molecules (e.g., an extract, attenuated organism).

There are at least two disadvantages of a defined-antigen vaccine:

1. A limited number of immune responses are initiated or a restricted state of sensitivity is induced (e.g., a limited number of cell clones are expanded) by the vaccine.
2. The possibility exists that "antigen-negative" parasite populations will be selected in the vaccinee, i.e., parasites that do not express the restricted subset of antigens used for vaccination.

Attention is focused on two types of defined antigen, i.e., natural or novel antigens, immune responses to which either inhibit the parasite directly or increase the vulnerability of the establishing or established parasite to other extant or inducible immune responses. Natural antigens are defined as those that induce readily detected immune responses (antibodies or cell-mediated immunity detected using serologic and blast transformation assays or skin tests) in at least some individuals under normal circumstances of infection or after injection into a natural host. Novel antigens are those that are poorly if at all immunogenic in their native state in natural hosts, but that are immunogenic after appropriate antigen engineering (e.g., conjugation to an immunogenic carrier molecule and exploitation of linked antigen recognition in antibody production or administration with an entity that promotes cell-mediated immunity) (Mitchell, 1979*b*). Natural and novel antigens are both amenable to serologic analysis using gel overlay, electrophoretic transfer (Western blotting), and immunoprecipitation techniques (Anders *et al.*, 1982*a*) with sera from infected or immunized hosts of various types or monoclonal antibodies raised in homologous or heterologous species. Of course, antigens recognized by serologic techniques and subsequently isolated or produced may be required to induce immune responses other than antibody production (e.g., cell-mediated immunity) to promote efficient resistance in the vaccinee (discussed in Mitchell and Anders, 1982).

This chapter presents a progress report on attempts in this laboratory to identify host-protective (i.e., parasite-inhibitory) immune responses and their target antigens and to devise strategies of vaccination that induce predetermined desired types of immune responses against several metazoan and protozoan parasites in mouse models. Headings in succeeding sections consist of the name of the parasite (and the disease caused) and the objective sought by way of vaccination using, ultimately, defined-antigen vaccines.

II. *TAENIA TAENIAEFORMIS* (MURINE CYSTICERCOSIS): IMMUNOPROPHYLAXIS AND IMMUNOTHERAPY

Larval cestode infections in rats and mice caused by *Taenia taeniaeformis* are characterized by antibody-mediated concomitant immunity. Already-infected hosts are resistant to reinfection, and high levels of protection can be transferred to naive recipients by IgA antibodies (in the intestines) and complement-fixing IgG antibodies (systemically). Relevant studies are reviewed in Williams (1979); Lloyd (1981); Mitchell *et al.* (1982*a*); Rickard and Williams (1982); Mitchell (1982*b*). Good evidence exists that establish*ed* parasites (i.e., liver cysts), in contrast to establish*ing* parasites, are relatively resistant to immune attack in part because of anticomplementary activities in the cyst. Genetically based resistance in young mice of some strains (e.g., C57BL/6 and Balb/c) results from an accelerated appropriate antibody response that prevents establishment of the larvae before the full expression of parasite-protective mechanisms. The delayed antibody response in genetically susceptible mice (e.g., C3H/He) enables the parasite to become firmly established. Mice and rats of various genotypes can be vaccinated readily against first infection using antigens extracted from the invasive form of the parasite.

The rodent cysticercosis models serve to make several points in relation to vaccination:

1. Vaccination need not induce a high-titered host-protective immune response; such a response is induced subsequent to parasite encounter. Sensitization leads to an accelerated immune response that eliminates the parasite in the critical early phase of parasite establishment (Mitchell *et al.*, 1980*a*). Thus serum from vaccinated rats fully resistant to infection may be inefficient at transferring resistance to naive recipients (Ayuya and Williams, 1979) in contrast to the effect using serum from infected rats, wherein antigen exposure and thus titers of antibody may be greater. The duration of vaccine-based resistance has not been examined systematically in rodent models, age-related resistance confounding such a study. There is evidence in veterinary cysticercoses that protective immunity is reduced when sheep are challenged many months after vaccination (Rickard and Williams, 1982).
2. Although no information is available on the nature of host-protective antigens, identification of the broad immunologic mechanisms underlying genetically based variations in susceptibility to first infection has led to the prediction, subsequently borne out by experimentation, that vaccination of mice with the appropriate antigen, and using an appropriate schedule, will protect even the most susceptible host types (e.g., young C3H/He males) against first infection (Rajasekariah *et al.*, 1980*a*,*b*).
3. Antigens that are effective as a vaccine and that simulate infection-induced concomitant immunity will be present in the invasive form of the parasite (the oncosphere) and in the established parasite (the metacestode) (Kwa and Liew, 1977; Rajasekariah *et al.*, 1980*b*) if a high degree of sensitivity to the invasive form is actually maintained in the chronically infected host. Thus, even when the invasive form or very early stages of the parasite are the proven targets of host-protective immunity and are relatively rich in appropriate antigens, this may not necessarily be as convenient a source of antigen (or mRNA if required for recDNA approaches) as the established parasite (which is often more readily available in larger amounts) for vaccine antigen production. Quantitative aspects of antigen expression in various life-cycle stages of *Taeniid* spp. have yet to be defined and are the subject of current experiments with *T. taeniaeformis*.
4. If the parasite molecules responsible for parasite protection (e.g., anticomplementary or cytotoxic molecules in the established cyst) are identified and isolated, vaccination with such molecules (with modification to ensure immunogenicity and induction of the appropriate neutralizing antibodies)

may increase the susceptibility of established parasites to extant antiparasitic immune responses. Thus, immunotherapeutic vàccines as well as immunoprophylactic vaccines may become available for larval cestode infections. It has long been proposed that immunologically based resistance in larval cestode infections operates not only at the stage of early invasion, but to some extent at the level of the established parasite (see Gemmell and Soulsby, 1968; Gemmell and MacNamara, 1972).

5. If presensitization for an accelerated appropriate antibody response is what is required to ensure resistance against first infection, then immunization with anti-idiotype (anti-Id) antibodies may acheive this goal. A particular cross-reactive Id, shared between a sufficient number of clonotypes in the host-protective antibody population (termed IdX) and present on B cells, may be useful as a target for clonal expansion by immunogenic preparations of monoclonal anti-IdX antibodies. Of course, an appropriate IdX on anti-*T. taeniaeformis* T cells required for helper functions in IgG antibody production may also serve as a marker on cells to be sensitized (expanded?) for accelerated responsiveness. The *T. taeniaeformis*-mouse system appears ideal for testing the efficacy of vaccination with appropriate anti-IdX antibodies complexed to immunogenic carrier molecules for induction of resistance to first infection. There are obvious attractions in using anti-IdX antibodies as a parasite vaccination strategy not requiring parasite antigens that uses a reagent in unlimited supply and is a quality-controlled product (Sacks *et al.*, 1982).

Absolute protection against first infection with *T. taeniaeformis* can be induced in otherwise highly susceptible young male C3H/He mice by vaccination with soluble oncospheral antigens in complete Freund's adjuvant (CFA). High-performance liquid chromatography (HPLC) fractionation of detergent-solubilized (100,000 *g* supernatant) antigens has indicated that protective activity is largely confined to the high-molecular-weight excluded fraction (Rajasekariah *et al.*, 1982). Whether protein antigens are responsible for host protection is not yet known, although vaccination efficacy of solubilized oncosphere antigens remains after treatment with periodate whereas protease treatment abrogates the vaccinating capacity of this antigen mixture in CFA (M. W. Lightowlers, unpublished observations, 1982). A major protein antigen of high MW has been reported to be contained in a strobilocercus extract effective at immunizing rats against *T. taeniaeformis.* However, other fractions were not tested for biologic activity and whatever else is contained in the biologically active fraction isolated by column chromatography is unknown (Kwa and Liew, 1977).

Rodent cysticercoses are proving to be important model systems used as an adjunct to veterinary larval cestode systems in which considerable progress has already been made in development of vaccines (Urquhart, 1980; Lloyd, 1981; Rickard, 1982; Rickard and Williams, 1982). It is hoped that information obtained from the rodent and veterinary systems will be useful in the development of vaccines against human cysticercosis in the future (Flisser *et al.*, 1982). Besides this practical output of the cysticercosis models, these host-parasite systems will ultimately provide detailed basic information on the mode of action of antiparasite antibodies in mediating parasite-inhibitory effects *in vivo* (Fig. 1) and *in vitro.*

III. *LEISHMANIA TROPICA* (MURINE CUTANEOUS LEISHMANIASIS): VACCINATION AGAINST PARASITE ESTABLISHMENT OR CHRONIC DISEASE

Several observations point to a central role for $Ly1^{+}2^{-}$ T cells in mediating resistance of mice to the intramacrophage protozoan parasite, *Leishmania tropica* (Mitchell *et al.*, 1980*b*, 1982*a*). This group of organisms causes cutaneous leishmaniasis in humans, and the mouse model has proved to be particularly useful in identifying and dissecting host factors that influence susceptibility to cutaneous disease. Recent studies of Louis and colleagues have demonstrated an inhibitory effect *in vitro* of an anti-*L. tropica* continuous Lyl^{+} T-cell line against leishmania parasites in mouse macrophages (Louis *et al.*, 1982). Moreover, lymphokines have inhibitory effects on leishmania-infected mouse macrophages *in vitro* (Handman and Burgess, 1979; Büchmüller and Mauel, 1979; Nacy *et al.*, 1981; Murray, 1981). T cells of the delayed-type hypersensitivity (DTH) or macrophage-activating type (T_D or T_{MA}), and directed against *L. tropica* antigens at the macrophage surface (Farah *et al.*, 1975; Handman *et al.*, 1979; Berman and Dwyer, 1981), are presumably the necessary effector cells in mediating host resistance to this intramacrophage parasite. No evidence for a parasite-inhibitory effect of serum has been obtained in the *L. tropica*-mouse system, and no host-protective hybridoma antibodies against promastigotes of *L. tropica* have yet been generated (Handman and Hocking, 1982).

In Balb/c mice, a strain that is highly susceptible to cutaneous and systemic disease, there is good evidence for the operation of T-cell-dependent *parasite*-protective (i.e., suppressive or proparasitic) immune responses (Howard *et al.*, 1980*b*, 1981). The T cells involved in this type of counterproductive response are also apparently of the $Ly1^{+}2^{-}$ phenotype (Mitchell *et al.*, 1981*a*; Liew *et al.*, 1982). Thus Balb/c nude mice can be protected against disease and early death by injection of a small, but not a large number of syngeneic lymphoid cells. Cutaneous lesions in Balb/c.nu/nu mice given 10^6-10^7 normal mouse lymphoid cells (spleen plus mesenteric lymph node) resolve rapidly and completely within 1-2 months of cutaneous promastigote challenge. Balb/c nudes can be protected by 10^7 lymphoid cells from Balb/c or $B10D_2$ donors, but by cells from C57BL/6 or Balb/c.H-2^b mice; however, in experiments performed to date a deterioration in health of Balb/c nude recipients of the H-2 *incompatible* cells has complicated

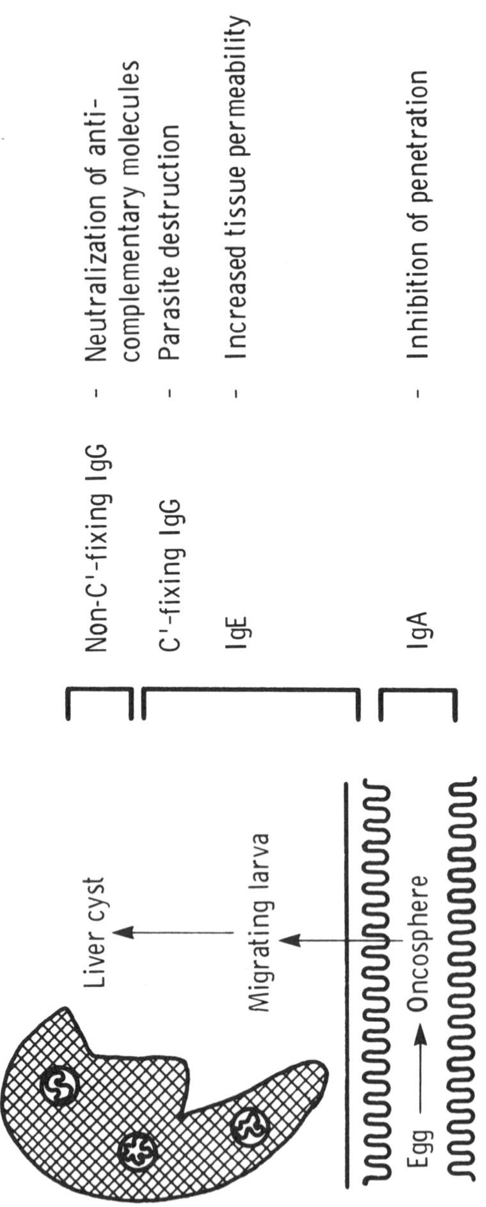

Figure 1. Postulated roles for antibodies of various isotypes in expression of resistance to *Taenia taeniaeformis* in rats and mice.

any interpretations of the findings. Nudes can be protected by allogeneic thymus grafts (Mitchell *et al.*, 1980*b*). Resistance, i.e., resolution of disease (Fig. 2) in minimally reconstituted Balb/c nude mice is abrogated readily by simultaneous injection of a small number of $Ly1^{+}2^{-}$ T cells (e.g., 10^7 cells) from chronically infected mice. Cells (6×10^6) from spleen, mesenteric lymph nodes, or cutaneous lymph nodes are effective in abrogating resistance in minimally reconstituted Balb/c nudes, whereas thymocytes are ineffective and bone marrow cells appear to delay, but do not protect against, ultimate development of lesions. On occasion, an abrogating effect of chronically infected mouse serum has also been demonstrated in minimally reconstituted nudes, but the effect with serum is by no means as reproducible or as dramatic as the effect of cells (Mitchell *et al.*, 1981*a*, 1982*a*). These data, together with those of Howard and colleagues, provide support for the notion that host-protective and parasite-protective T-cell-dependent responses operate in mice with chronic cutaneous leishmaniasis and that the outcome of this interplay, or the capacity to mount a parasite-protective response, depends largely on host genotype.

A defect in presentation of parasite antigens by infected macrophages may underlie the presumed relative insusceptibility of *L. tropica*-parasitized cells to aggressive immune attack in the Balb/c mouse (Handman *et al.*, 1979; Gorczynski and MacRae, 1982) and may lead to induction of parasite-protective responses and ready establishment of the parasite (Howard *et al.*, 1980*c*). Indirect evidence exists for a reduced expression of $H\text{-}2^d$ antigens on infected Balb/c macrophages relative to such antigens on uninfected Balb/c macrophages and $H\text{-}2^k$ antigens on either infected or uninfected CBA/H macrophages (Handman *et al.*, 1979). Technical difficulties in labeling surface molecules of infected and uninfected macrophages have hindered attempts to demonstrate quantitative differences in $H\text{-}2^d$ expression, as well as any strong association between at least $H\text{-}2^k$ and parasite molecules, using anti-H-2 and anti-leishmania hybridoma-derived antibodies and two-dimensional gel analysis of immunoprecipitates.

With the demonstration that small numbers of $Ly1^{+}$ T cells are involved in host-protective responses (e.g., in CBA/H.nu/nu mice, Mitchell *et al.*, 1980*b*), attention focuses on Ia antigens on the infected macrophage (Gorczynski *et al.*, 1981; Gorczynski and MacRae, 1982). CBA/H (and C57BL/6) mice are highly resistant to persistent cutaneous leishmaniasis, with the small swellings and lesions that do develop at the site of promastigote deposition resolving rapidly (Fig. 3). The potency of lymphoid cells in inducing resistance in C57BL/6.nu/nu mice is illustrated by the fact that 5×10^5 cells are fully protective and one-half the mice are protected by as low as 5×10^4 syngeneic cells.

Genetically based variations in susceptibility to disease caused by *L. tropica* and the related *L. mexicana* have been demonstrated in many laboratories (Preston *et al.*, 1978; Perez *et al.*, 1978; Handman *et al.*, 1979; Behin *et al.*, 1979; Bjorvatn and Neva, 1979; Modabber *et al.*, 1980; Howard *et al.*, 1980*d*; Hale

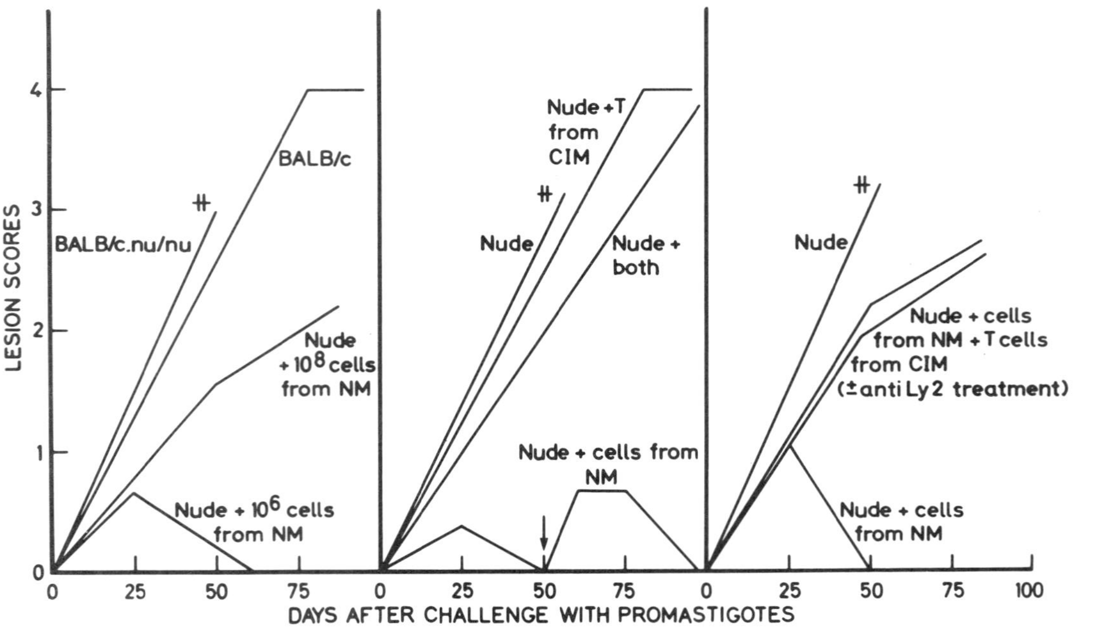

Figure 2. Reconstitution of Balb/c nude mice. Stylized representation of lesion scores (size of cutaneous lesions caused by *Leishmania tropica*) in Balb/c.nu/nu (nude) mice, minimally reconstituted nude mice injected with syngeneic lymphoid cells from normal mice (NM), and such mice injected as well with purified T cells from chronically infected Balb/c mice (CIM) with or without treatment of cells with anti Ly2 antibodies. Mice were challenged with promastigotes of *L. tropica* at time 0 and also in one group (*middle panel*) at day 50. (Data in Mitchell *et al.*, 1981*a*.)

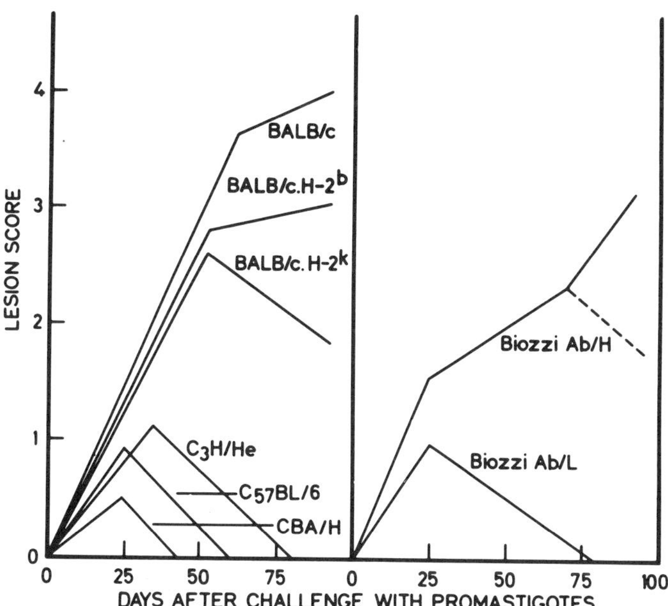

Figure 3. Mouse strain variation. Stylized representation of mouse strain variation in susceptibility to cutaneous disease caused by *Leishmania tropica.* (Data in Mitchell *et al.*, 1981*b*.)

Howard, 1981; de Tolla *et al.*, 1981). In this laboratory, and using the isolate LRC-L137 from the WHO Reference Center for Leishmaniasis, Jerusalem, Israel, Balb/c.H-2^k and Balb/c.H-2^b mice are susceptible, although resolution of cutaneous lesions does occur especially in Balb/c.H-2^k mice given relatively low numbers of promastigotes (Mitchell *et al.*, 1981*b*), and resistance is unequivocally dominant in (Balb/c × CBA/H)F_1 mice (Mitchell *et al.*, 1980*b*; Gorczynski and MacRae, 1982. However, some differences exist in the susceptibility of F_1 hybrids between resistant and susceptible parent strains, and the expression of certain Ia specificities in various F_1 hybrids (e.g., Sandrin *et al.*, 1982) may provide clues as to what specificities are involved in associative recognition by host-protective Lyl^+ T cells. Different infection characteristics in mice of various genotypes enable choices to be made when testing strategies of vaccination that are of high (e.g., in Balb/c and Balb/c.H-2 congenics) or low stringency (e.g., in CBA/H and C57BL/6 mice).

Evidence for defective major histocompatibility complex (MHC)-associative antigen recognition on *L. tropica*-infected Balb/c macrophages plus an apparent involvement of T_D cells in resolution of cutaneous leishmaniasis in genetically resistant mice and minimally reconstituted nude mice has led to a search for manipulations that might promote antigen presentation and DTH responses during experimental vaccination. *Corynebacterium parvum*, as a suspension of killed organisms, has been reported to be superior to several other adjuvants in promoting DTH responses in mice (Bomford, 1980) and in increasing the persistence (and thus availability) of antigens on mouse macrophages (Wiener and Bandieri, 1975). Although bacillus Calmette-Guerin (BCG) has been reported to increase resistance of Balb/c mice to cutaneous leishmaniasis (Weintraub and Weinbaum, 1977; cf. Grimaldi *et al.*, 1980), it was found that *C. parvum* at doses of 100–200 μg intraperitoneally had no effect on the course of cutaneous disease in genetically susceptible mice (Mitchell *et al.*, 1981*b*). This finding is consistent with results in a *L. enrietti*/guinea pig system (Bryceson *et al.*, 1972). However, when Balb/c, Balb/c.H-2^b or Balb/c.H-2^k mice were injected with crude antigen mixtures plus *C. parvum*, they were found to be relatively resistant to subsequent promastigote challenge. A proportion of mice develop no lesions and recovery is accelerated in those vaccinated mice in which swellings or lesions develop at the injection site (Mitchell *et al.*, 1981*b*, 1983*a*).

Intraperitoneal injection of crude antigen plus *C. parvum* has proved superior to subcutaneous injection in terms of subsequent resistance to infection. The crude antigen mixtures (the crudest mixture possible) were prepared by freezing and thawing *in vitro*-derived infected macrophages or spleen and lymph node cell suspensions from chronically infected Balb/c nude mice with obvious signs of visceralization of the organism.

Of the various sources of infected cells and strains of mice employed in these vaccination experiments, greatest success has been achieved with the following combination: *in vitro*-infected IC-21 cells, a transformed macrophage cell line of C57BL/6 origin (Mauel and Defendi, 1971) injected intraperitoneally with *C. parvum* into Balb/c.H-2^b mice (Fig. 4). Although not proved as yet, the possibility exists that the effectiveness of this combination reflects the use of macrophages from a genetically resistant mouse strain (C57BL/6) as the donor and, as the vaccinee, an H-2 compatible but genetically susceptible mouse strain (Balb/c. H-2^b). Frozen and thawed promastigotes plus *C. parvum* have consistently been inferior to infected cells plus *C. parvum* in this system. This result is disappointing in terms of the ready availability of large numbers of highly purified promastigotes grown in bulk cultures and as clones (E. Handman *et al.*, unpublished data, 1982) for antigen and mRNA isolation. Moreover, considerable information is available on at least the proteins of promastigotes (Handman *et al.*, 1981).

All data point to the fact that infected macrophages contain relevant antigens in sufficient quantities for vaccination. Obviously, antigen isolation from this complex starting preparation will not be simple. It is important to emphasize that neither *C. parvum* alone nor the crude antigen mixtures alone have any pro-

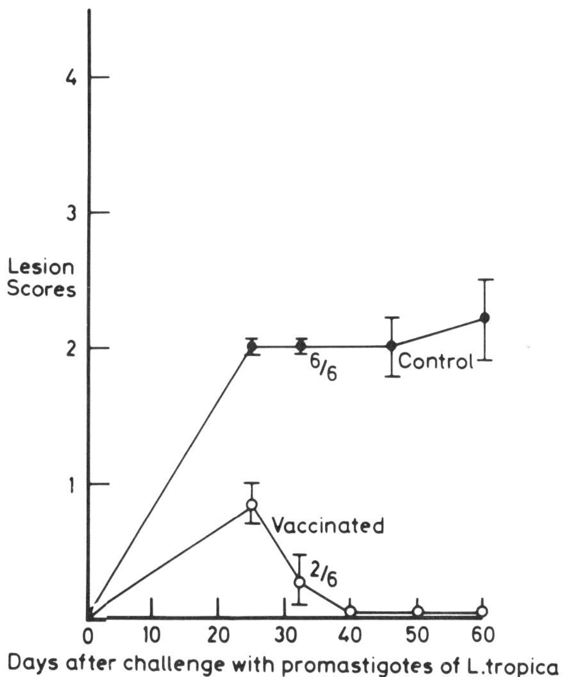

Figure 4. Protection against persistent cutaneous leishmaniasis in Balb/c.H-2^b mice preinjected intraperitoneally with frozen and thawed infected macrophages of the cell line IC-21 together with the therapeutic biologic, *Corynebacterium parvum*, at day –27. Control mice received *C. parvum* alone and showed unaltered susceptibility to persistent disease. Numbers of mice infected in the two groups are indicated at one time point. Points represent arithmetic means of lesion sizes (± SEM).

tective effects in Balb/c mice and their H-2 congenics; both are required for induction of significant and consistent protection against chronic disease initiated by cutaneous infection of vaccinated mice.

Mice of high genetically based resistance to persistent disease (C57BL/6) can be rendered even more resistant by prior injection of living promastigotes intraperitoneally. Again, killed (frozen and thawed) promastigotes have been ineffective at vaccinating for increased resistance in C57BL/6 mice (Mitchell *et al.*, 1983*a*). The results suggest that systemic infection of macrophages is required for induction of a protective response or that living promastigotes contain or release more of a particular antigen than killed promastigotes.

Although much is now known about the broad immunobiologic aspects of murine cutaneous leishmaniasis, nothing is known about what actually causes the cutaneous lesion. Moreover, no information is available on the nature of host-protective antigens and whether they differ from parasite-protective antigens (in Balb/c mice). The availability of parasite-inhibitory T-cell lines (Louis *et al.*, 1982) and suppressor T-cell lines will add much new information in the immediate future. An important question is whether Lyl$^+$ anti-Id T cells are involved in inhibition of Lyl$^+$ T cells with presumed antiparasite reactivity and host-protective effects. The system is ready for immunochemical analysis (e.g., Handman *et al.*, 1981) with the aim of identifying the spectrum of antigens of infected macrophages available to the host immune response. The data indicate that the adjuvant *C. parvum* will be useful in screening for vaccination efficacy of isolated antigens in genetically susceptible mice. Analysis of the effects of *C. parvum* on anti-leishmania immune responses will only be feasible with isolated antigens in hand, accurate quantitation of immune responses being impossible with crude antigen mixtures.

IV. *SCHISTOSOMA JAPONICUM* (MURINE SCHISTOSOMIASIS JAPONICA): VACCINATION AGAINST IMMUNOPATHOLOGIC DISEASE

Schistosomes present a formidable challenge to the immunoparasitologist interested in vaccination against establishment or persistence of systemic metazoan parasites. Nevertheless, the list of schistosomacidal effector cells and molecules in the *Schistosoma mansoni*/rat, mouse, and human systems continues to increase particularly in regard to antibody-dependent cellular mechanisms that kill young schistosomules *in vitro* (reviewed in Capron *et al.*, 1980). The suspicion exists that early schistosomule killing may, at least under some circumstances, represent a form of accelerated *in vitro* demise (because of suboptimal culture conditions for example) rather than the identified mechanisms being relevant to *in vivo* events in which parasite resilience and repair are presumably far more efficient (see discussion by von Lichtenberg, 1977). Immune evasion mechanisms appear to be highly developed in schistosomes, and the following have been proposed: blindfolding (masking) of parasite surface antigens by absorbed host molecules, innate resistance of the surface to immune attack, rapid turnover of surface molecules, and liberation of antibody-degrading enzymes (see Phillips and Colley, 1978; also Abbas *et al.*, 1981; Auriault *et al.*, 1981). The relative importance of these mechanisms presumably varies according to parasite life-cycle stage.

The vulnerability of the invasive and early schistosomule to immune attack identifies this life-cycle stage as a likely target of host-protective immunity induced by vaccination with appropriate antigens. However, care will be required in testing whether immune responses of the immediate hypersensitivity type, for example, and directed against the skin penetration stage, will lead to unacceptable side effects, such as persistent skin irritation akin to swimmer's itch or even anaphylaxis, in individuals with high parasite contact (Hsu *et al.*, 1975; von Lichtenberg, 1977; Colley *et al.*, 1977). In a *S. mansoni*-mouse model, two time periods for expression of host protection have been identified—at 1-3 days and 1-2 weeks (Smithers and Miller, 1980; Smithers and Gammage, 1980; Miller *et al.*, 1981; see also Blum and Cioli, 1981). Moreover, irradiated schistosomula (Hsu *et al.*, 1965) have proved effective in inducing resistance to subsequent infection in veterinary and laboratory situations (reviewed in Taylor, 1980). Under circumstances of induced resistance, protection against infection is rarely absolute and the possible contribution to this observation of antigenic variability within the parasite population has been raised (Smith and Clegg, 1979). In the *S. japonicum*-mouse system, resistance to reinfection in 50-60-day infected mice is ~80% (Garcia *et al.*, 1982). Recently, difficulties have arisen with respect to interpretation of resistance to *re*infection in the commonly used *S. mansoni*-mouse model. Resistance may not be attributable entirely to the operation of host-protective anti-parasite immune responses and liver (and lung) disease may contribute significantly to apparent resistance (Wilson, 1980; Dean *et al.*, 1981; Cioli, 1982).

Exploitation of mouse strain variations in susceptibility to *S. japonicum* infection (Mitchell *et al.*, 1981*c*) has not progressed far. A peculiarity exists in young 129/J mice in which 50% of mice exposed to 20 or 25 cercariae are negative for parasites 20-50 days later. The other young 129/J mice have numbers of parasites often comparable to those in the bulk of mouse strains (Garcia and Mitchell, 1982). It is unknown whether immunologic mechanisms or other factors such as hormones (Knopf and Soliman, 1980) are involved in expression of resistance in a proportion of young 129/J mice. Nude 129/J mice have yet to be developed; one observation already made is that the F_1 between 129/J and a susceptible strain, Balb/c, are uniformly susceptible. Antibodies to the target epitope of an immunodiagnostic anti-adult worm hybridoma are present in infected 129/J mice at day 50, but not in the parasite-negative mice. This result suggests that the loss of *S. japonicum* in the 50% of young 129/J mice that are not parasitized occurs early in the infection time course. Just as with another trematode, *Fasciola hepatica*, in mice (Andrews and Meister, 1978; Rajasekariah *et al.*, 1979; Chapman and Mitchell, 1982*a*), mouse strain variations in susceptibility to infection or disease are usually not pronounced or are inconsistent in the *S. japonicum*-mouse (Mitchell *et al.*, 1981*c*) and *S. mansoni*-mouse systems (Civil and Mahmoud, 1978; Smith and Clegg, 1979; Claas and Deelder, 1979; Murrell *et al.*, 1980; Bickle *et al.*, 1980; Dean *et al.*, 1981; Fanning *et al.*, 1981).

Chronic schistosomiasis mansoni and japonica are classic immunopathologic diseases in which many of the disease manifestations, such as granuloma formation and fibrosis, result from T-cell dependent immune responses to antigens emanating from eggs entrapped in the liver (reviewed in Warren, 1973; Pelley and Warren, 1978). However, such responses to focal sources of antigen are unlikely to be only pathologic, and evidence is accumulating that a failure to sequester and neutralize (with antibody) certain toxic antigens in granulomas may lead to hepatotoxic effects (von Lichtenberg, 1977; Byram and von Lichtenberg, 1977; Byram *et al.*, 1979; Dunne *et al.*, 1981). Moreover, fibrosis, although leading to severe liver damage, is likely to be an important restraint to migrating parasites in the liver, as has been postulated in *Fasciola hepatica* infections in cattle and pigs (Boray, 1967; Ross, 1967) and *Mesocestoides corti* infections in mice (Pollacco *et al.*, 1978; Mitchell, 1979*b*, 1982*c*). A useful outcome of vaccination against disease, if the more desirable vaccination against establishment or persistence of infection is difficult to attain, would be to reduce the intensity and size of granulomas formed in response to egg antigens. This may be achieved by immunologic destruction of eggs in tissues or inhibition of maturation of eggs or by induction of suppressor responses (Fig. 5).

In mice, modulation of granuloma formation is seen late in an infection time

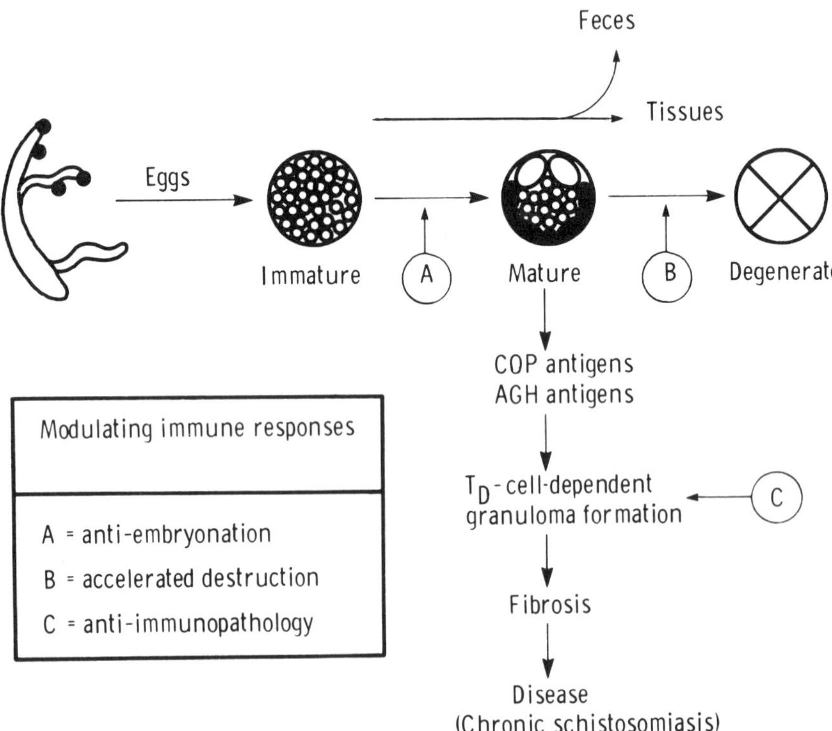

Figure 5. Scheme illustrating three postulated sites of action of granuloma-modulating immune responses in chronic schistosomiasis japonica (see text). A, antiembryonation; B, accelerated destruction; C, anti-immunopathology.

course or after multiple injections of eggs (reviewed in Warren, 1974, 1977). This modulation could result from several mechanisms probably operating in combination:

1. Accelerated destruction of eggs (James and Colley, 1976, 1978; Olds and Mahmoud, 1980) subsequent to their maturation to the maximum antigen-producing stage involving development of the miracidium (e.g., von Lichtenberg *et al.*, 1973; Hang *et al.*, 1974; Hamburger *et al.*, 1976).
2. Reduced efficiency of embryonation of eggs in tissues, i.e., antiembryonation immunity (discussed in Garcia *et al.*, 1981): Eggs vary in their sensitizing and eliciting capacity for granulomatous hypersensitivity in mice (Mitchell *et al.*, 1982*c*) as well as in their suitability for performance of the circumoval precipitation (COP) test (Oliver-Gonzalez, 1954) with sera from infected patients. Thus uterine eggs cause bleb rather than segmented precipitates with sera from chronically infected individuals, and blebs again predominate over the more usual segmented precipitates using eggs obtained from livers of rabbits infected for longer than the optimal 55–65 days (Garcia *et al.*, 1981). It is assumed that bleb reactions represent a limited number of antigen–antibody interactions in the COP test (Lewert *et al.*, 1980; Long *et al.*, 1980; Cruise *et al.*, 1981*a*), in which lyophilized eggs from tissues are simply incubated for 1–3 days with serum on slides. Accelerated destruction of mature eggs or antiembryonation effects would limit opportunities for sustained immunoresponsiveness to the diverse array of antigens liberated by mature eggs. These two postulated mechanisms should be differentiated from inhibition of oviposition (reduced egg output per worm pair) (Cheever *et al.*, 1974; Damian *et al.*, 1976; Cheever *et al.*, 1980*a*) or deviation of eggs away from the liver (Cheever *et al.*, 1980*b*). Preliminary evidence has already been obtained by E. G. Garcia and colleagues (1983) that fresh unembryonated eggs mature in the lungs of egg-sensitized mice (i.e., repeatedly injected intraperitoneally with eggs) in lower numbers than in unsensitized control mice.
3. Suppressor T-cell (T_S)-mediated inhibition of anti-egg responses presumably mediated by inhibitory effects on those T cells (e.g., T_D cells) or their products responsible for immunopathologic responses. This possibility has received greatest attention most recently because of the intense activity in cellular immunology on T-cell-dependent immunoregulation. The phenotype(s) of the effector T_S cells and the influence of other T-cell types remain confusing. Functions for $Ly2^+/I\text{-}J^+$ and $Ly1^+2^-/Ia^+$ cells have been proposed in the complex regulation of granuloma formation in modulated mice (Phillips *et al.*, 1980; Chensue *et al.*, 1981; Green and Colley, 1981; Colley, 1981; B. L. Doughty and S. M. Phillips, personal communication 1982).
4. Antibody-mediated inhibition of anti-egg responses either by diversion of

antigen away from T_D cells (e.g., opsonizing antibodies) or inhibition of antigen recognition by T_D cells (e.g., anti-idiotype antibodies): Effects of immune serum (and presumably antibodies, but of unknown specificity) on granuloma formation have been described in both *S. mansoni*-mouse (Pelley and Warren, 1978) and *S. japonicum*-mouse systems (Olds *et al.*, 1982). A role for IgM antibodies in a *S. mansoni*-baboon system has been proposed by Houba *et al.* (1979).

In most studies on granuloma formation and its modulation, the lung granuloma model introduced by von Lichtenberg (1962) has been used. A recent development has been the use of a radioisotopic assay for lung granulomatous hypersensitivity to intravenously injected *S. japonicum* eggs in immunized high-responder C57BL/6 mice. This assay measures the accumulation of radioactivity in lungs (relative to kidneys) of mice after injection of the radiolabeled DNA precursor, [^{125}I]-UdR. This radioisotopic readout has shown egg-sensitized mice to vary in their responsiveness to intravenously injected *S. japonicum* eggs, with C57BL/6 mice high responders, Balb/c and Balb/c.Ig^b intermediate responders, and CBA/H low responders (Mitchell *et al.*, 1981*c*, 1982*b*). There may well be similarities in genetic control of acute granulomatous hypersensitivity (AGH) to *S. japonicum* eggs in lungs and granulomatous hypersensitivity to BCG in mice (Allen *et al.*, 1977; Moore *et al.*, 1981; Sternick *et al.*, 1983). However, the AGH sensitization using eggs in CFA is not caused by the mycobacteria in CFA, since eggs plus incomplete Freund's adjuvant (IFA) will sensitize C57BL/6 for high AGH responsiveness (Mitchell *et al.*, 1982*b*). A feature of the system is that using a minimal sensitization regime, subcutaneous injection of eggs readily sensitizes for AGH, whereas intraperitoneal injection does not (Warren *et al.*, 1975; Mitchell *et al.*, 1982*b*). However, high-dose sensitization intraperitoneally does lead to AGH.

Various anti-*S. japonicum* hybridoma-derived monoclonal antibodies have been generated using cells from immunized Balb/c mice. Some anti-egg hybridoma antibodies are COP positive (Cruise *et al.*, 1981*a*) and one anti-adult worm antibody (designated I.134) has immunodiagnostic potential for schistosomiasis japonica in the Philippines (Mitchell *et al.*, 1981*d*; Cruise *et al.*, 1981*b*; Mitchell *et al.*, 1983*b*). Using a competitive radioimmunoassay (RIA) with crude extracted adult worm antigen (AWE) and ^{125}I-labeled I.134, more than 90% of known infected individuals (i.e., with eggs in feces) are detectable, and no false-positive reactions have yet been recorded. Sera from a limited number of *S. mekongi*-, *S. hematobium*-, or *S. mansoni*-infected individuals have also been negative in the test, although more assays with clinically defined sera and antigens from various life cycle stages are required to establish the apparent stage and species specificity of the target epitope of the I.134 antibody. This epitope is contained in a 23,000-MW surface antigen of *S. japonicum* adult worms (Cruise *et al.*, 1982). One of the COP-positive hybridoma antibodies (designated P.41) was shown previously to be incapable of detecting heavily infected individuals in a competitive RIA using crude extracted egg antigens (EA) and ^{125}I-labeled P.41. However, recent experiments have established that the ^{125}I-P.41/EA competitive RIA can detect an important subset of individuals whose sera are bleb-*only* COP reactors and who are likely to be recently infected individuals (Long *et al.*, 1980; Lewert *et al.*, 1980). Interestingly, the sera from these individuals (i.e., those with *no* segmented precipitates seen in the optimized COP test, and which are relatively rare) are negative in the ^{125}I-I.134/AWE assay (Mitchell *et al.*, 1983*b*).

Infected or egg-immunized (+CFA) mice produce antibody of the P.41 specificity as determined in the ^{125}I-P.41/EA competitive RIA. Only three independently selected COP-positive hybridoma antibodies from immunized Balb/c mice have been generated to date—two IgM proteins, one an IgG_1 protein—and all produce bleb-only COP reactions. In combination, bleb reactions are again found. Surprisingly, these three monoclonal antibodies are related in specificity (Cruise *et al.*, 1981*a*) and idiotypy (Mitchell *et al.*, 1982*b*). This remarkable coincidence suggests that the P.41 target epitope is contained on major egg antigen(s) for the mouse and that the hybridoma antibodies are representative of a major idiotypically cross-reactive antibody response on the part of the immunized mouse (Balb/c in this case). Experiments were therefore designed to test the effects of hybridoma immunization, i.e., idiotype (Id) sensitization for anti-Id responsiveness, on AGH with the expectation that anti-Id antibodies or anti-Id reactive T_S cells may inhibit T_D-cell recognition of major immunopathologic egg antigens. C57BL/6 mice were immunized with P.41 complexed to keyhole limpet hemocyanin (KLH) in CFA followed by egg sensitization and intravenous egg challenge. Control mice were injected with other IgM hybridoma antibodies complexed to KLH, P.41 being an IgM protein. Using the radioisotopic lung assay for AGH, no diminution in responsiveness to eggs was obtained in P.41 Id-immunized mice. However, and quite fortuitously, another COP-negative IgM anti-schistosome hybridoma (designated I.39) resulted in significant inhibition of AGH in three experiments when injected as a complex with KLH in CFA before egg sensitization (Mitchell *et al.*, 1982*b*). Whether anti-Id responsiveness underlies the effect of I.39:KLH immunization on lung AGH to eggs remains to be determined. In a single experiment, anti-I.39 Id antibodies, or at least sera from I.39:KLH hyperimmune C57BL/6 mice, have not inhibited AGH when given at the time of AGH *elicitation* (i.e., intravenous egg challenge). Further analysis of the effect must await testing of whether I.39:KLH immune and *infected* mice are more resistant to disease and early mortality. The approach is being explored in the expectation that vaccination against disease may be a useful adjunct to chemotherapy or eventual vaccination against infection in this serious immunopathologic disease of humans. If it evolves that egg antigen(s) isolated by P.41 affinity chromatography are efficient sensitizers for AGH in the mouse, then a battery of hybridoma antibodies raised against the antigen(s) will be examined systematically for AGH-inhibitory effects after Id sensitization.

V. *GIARDIA MURIS* (MURINE GIARDIASIS): VACCINATION FOR ACCELERATED REJECTION OF INTESTINAL PARASITES

After oral administration of cysts of the intraluminal intestinal protozoan, *Giardia muris*, trophozoites reside and multiply in the small intestines and cysts derived from them are found in the feces. In the Balb/c strain of mouse, cysts in feces are undetectable after about 4–7 weeks, whereas in C3H/He, and in particular, males, cyst excretion continues for at least 4 months (Roberts-Thomson and Mitchell, 1978). Other mice that develop chronic giardiasis are Balb/c nude mice (Roberts-Thomson and Mitchell, 1978; Stevens *et al.*, 1978) and W^f/W^f mice, which are deficient in mast cells (I. C. Roberts-Thomson, I. C. Anders, R. F. Schrader, J. W., and G. F. Mitchell, unpublished observations, 1982). Moreover, injections of the histamine and serotonin antagonist, cyproheptidine, will increase the duration of infection in Balb/c mice (Roberts-Thomson *et al.*, 1982; Mitchell *et al.*, 1982*a*).

Balb/c mice vaccinated with trophozoites plus CFA injected systemically are resistant to infection, cyst excretion being undetectable or transient after cyst challenge. However, vaccination with trophozoites in adjuvant fails to protect C3H/He mice from development of high-level persistent infection (Roberts-Thomson and Mitchell, 1979). Thus a simple vaccination protocol has been devised that is effective in mice that exhibit acute, transient giardiasis, but that is without effect in mice susceptible to development of chronic giardiasis. Naturally, the individuals most in need of protection by way of vaccination are the genetically susceptible ones. Recently, exoantigens from trophozoites maintained in short-term culture have induced a degree of resistance to subsequent infection when injected into Balb/c mice with CFA, but the protection obtained is not as impressive as that seen with trophozoites plus CFA (R. F. Anders and I. C. Roberts-Thomson, unpublished observations, 1982).

Examination of antigiardia immune responses in genetically susceptible C3H/He and genetically resistant Balb/c mice has yielded the following information: (1) titers of IgA and IgG antibodies to trophozoite extracts measured in a solid-phase RIA are comparable in the early stages of infection and are actually higher at later time points in the chronically infected C3H/He mice; (2) infected C3H/He and Balb/c mothers both transmit resistance to infection to suckling neonates; and (3) both C3H/He and Balb/c uninfected mice can be sensitized for expression of DTH to trophozoite extracts by injection of such extracts with CFA after cyclophosphamide injection (Underdown *et al.*, 1981; Anders *et al.*, 1982*b*). Importantly, C3H/He female mice are relatively resistant to reinfection after drug cure of an existing infection. Resistance is expressed when challenge cysts are obtained from either short-term infected Balb/c nude mice or chronically infected C3H/He mice (Underdown *et al.*, 1981). This observation argues against the possibility of an altered parasite population in C3H/He mice with prolonged infection. The available data indicate that C3H/He mice are in no way nonresponders to *G. muris* trophozoite antigens (Anders *et al.*, 1982*b*).

Differences have been detected between C3H/He and Balb/c mice in the *specificities* of anti-trophozoite IgG and IgA antibodies in sera. Sera from resistant or recovering Balb/c mice immunoprecipitate a radiolabeled surface antigen (~85,000 MW) and two biosynthetically labeled exoantigens more efficiently than sera from chronically infected C3H/He mice. What is cause and what is effect can only be dissected by demonstrating the vaccination efficacy of isolated antigens. Fractionation of exoantigens by HPLC and injection of Balb/c mice with fractions in CFA before oral cyst challenge has demonstrated that heightened responsiveness to the major exoantigen (MW ~30,000 as analyzed by SDS-PAGE under reducing conditions) is unlikely to account for differences in infection characteristics between Balb/c and C3H/He mice. Thus the fractions enriched for this antigen were not protective. Indeed, none of the fractions was as effective as the unfractionated exoantigen preparation, although the effects of vaccination with recombined HPLC fractions has yet to be determined (R. F. Anders and I. C. Roberts-Thomson, unpublished observations, 1982). Attention now focuses on trophozoite membrane antigens and, in particular, on the radioiodinated surface protein antigen (MW 85,000 by SDS-PAGE under reducing conditions), which is differentially immunoprecipitated by sera from Balb/c versus C3H/He mice.

Several observations point to a possible role for immediate hypersensitivity responses (and IgE antibodies) in the expression of resistance to *G. muris.* Treatment with cyproheptidine delays elimination of infection in genetically resistant Balb/c mice, and mast cell-deficient (DBA/2 × C57BL/6)F_1 · W^f/W^f mice develop prolonged infection (see also Crowle and Reed, 1981; Kojima *et al.*, 1980, cf. Uber *et al.*, 1980, re *Nippostrongylus brasiliensis*). C3H/He mice are reported to be hyporesponsive to histamine-sensitizing factors, such as pertussigen (Bergmann and Munoz, 1968). An involvement of IgE antibodies as a key component of host-protective immunity in giardiasis would be somewhat surprising, as total serum levels of IgE are usually normal in contrast to the potentiated IgE responses and IgE hypergammaglobulinemia, which are features of several helminth infections (and ectoparasitic arthropod infections in the case of IgE antibody production). In studies on the induction of resistance to infection using trophozoites given systemically, CFA was highly effective, whereas the IgE-promoting adjuvants, alum and *Bordetella pertussis* vaccine, were ineffective (Roberts-Thomson and Mitchell, 1979). Whether the effective injection protocols sensitize T_H cells for accelerated IgE responses after parasite challenge remains to be determined.

A hypothesis that accounts for several of the observations in the *G. muris*–mouse system is that genetically resistant mice are able to support prolonged immediate and delayed-type hypersensitivity responses in the intestines (see Wakelin and Donachie, 1981) through antibody-mediated neutralization of anti-inflammatory molecules produced by the parasite (see Behnke and Parish, 1979). Thus resistance will depend on T-cell-dependent and antibody-mediated inflammatory responses to certain antigens as well as neutralization of anti-inflammatory molecules. A sustained outflow of inflammatory secretions (Barth *et al.*, 1966) may further reduce access of the postulated locally active inhibitory molecules. The effects of these molecules must be short-lived in the absence of the *G. muris* parasite to account for resistance of drug-cured C3H/He mice to challenge infection (Underdown *et al.*, 1981). In *Haemonchus contortus* infections in sheep, removal of an existing intestinal worm burden was shown to be necessary before resistance could be demonstrated (Dineen and Wagland, 1966). Although this hypothesis ascribes no particular role to IgA antibodies, such antibodies may be required to mediate antiparasitic effects in the intestine of the suckling neonate (Andrews and Hewlett, 1981). In any event, the *G. muris*-mouse system, like the *Nippostrongylus brasiliensis*-rat system (Ogilvie and Love, 1974), may be one in which immune responses (and target antigens) are identified as being necessary but not sufficient for expression of efficient host resistance to persistent intestinal infection. The implication of this postulate is that a combination of antigens will be required for vaccination.

VI. *NEMATOSPIROIDES DUBIUS* (MURINE NEMATODIASIS): VACCINATION FOR ACCELERATED REJECTION OF INTESTINAL PARASITES

The intestinal nematode, *Nematospiroides dubius*, establishes a chronic infection in mice. Adult worms are found in the upper small intestine for many months after oral administration of a single dose of infective third-stage larvae (L3) although a progressive loss of worms is seen in some mice (e.g., female SJL/J and (SJL/J × C57BL/6)F_1 mice) (Table I). After oral L3 administration, there is an obligatory short period of larval encystment in the intestinal wall before emergence back into the intestines. With multiple exposures to L3, some strains of mice (e.g., SJL/J) develop striking resistance to larval development; they also reject the resident intestinal worm burden (Table I). This is in contrast to strains such as CBA/H and C57BL/6, and in particular, males, in which large worm numbers accumulate after multiple doses of L3 (reviewed in Mitchell *et al.*, 1982*a*). In resistant mice, prominent granulomata develop at sites of prior larval occupancy or larval death in intestinal wall cysts, a reaction that is T-cell dependent (Bartlett and Ball, 1974; Prowse *et al.*, 1978). Nude mice of relatively resistant genotype (i.e., Balb/c.nu/nu mice) (Prowse *et al.*, 1978), mast-cell-deficient W^f/W^f mice (Table I), and Biozzi low-line mice (Ab/L) (Jenkins and Carrington, 1981) all show defective resistance to *N. dubius* after multiple L3 challenge. Pro-

Table I. Numbers of *Nematospiroides dubius* Intestinal Worms (Arithmetic Mean ± SEM) in Various Mice Given One to Three Doses of L3

Mouse strain[a]	Sex	Injections systemically	Administration of L3 orally[b]	No. of mice	Worms in small intestines[c]	
					N	Day of assay
SJL/J	F	–	200 L3	3	122 ± 8	(21)
				3	0	(78)
C3H/He		–		3	118 ± 11	(21)
				3	47 ± 16	(78)
(SJL/J × C57BL/6)F_1	M	–	200 L3	15	57 ± 7	(30)
	F	–		18	18 ± 5	(30)
SJL/J	F	–	2 × 200 L3	8	15 ± 4	(14 + 21)
C57BL/6		–	100 + 200 L3	7	176 ± 14	(14 + 21)
(SJL/J × C57BL/6)F_1	M	–	2 × 200 L3	15	47 ± 7	(20)[d]
	F	–		15	2 ± 1	(20)
$W^f/+$	F	–	3 × 200 L3	3	<3	(18)
W^f/W^f		–		3	85 ± 38	(18)
Biozzi Ab/H	M	–	3 × 200 L3	7	85 ± 6	(12)[e]
Biozzi Ab/L		–		6	115 ± 8	(12)[e]
SJL/J	F	Eggs + CFA[f]	500 L3	3	5 ± 2	(24)
		CFA		3	129 ± 44	(24)

[a] Other than the (DBA/2 × C57BL/6)F_1.W^f/W^f or $W^f/+$ mice, which were 22 weeks of age at time of first administration of larvae, mice were 8–12 weeks old at the commencement of experiments.
[b] Two doses of L3 were given over a 14- to 20-day time period, three doses over a 30- to 50-day time period, day 0 being considered the day of last L3 administration.
[c] Day of assay designated as days after the single, or last of the multiple, injections of L3.
[d] Older F_1 male mice more closely resemble F_1 female mice in their resistance status (rejection of resident worm burden) than do the young mice used here, in which the data resemble those reported previously (Prowse and Mitchell, 1981).
[e] Although this difference is statistically significant in males, it is of doubtful biologic significance; greater differences between Biozzi high- and low-line mice have been reported by Jenkins and Carrington (1981).
[f] CFA, complete Freund's adjuvant; 400 eggs in CFA day −29 and 20,000 eggs injected into adjuvant sites (subcutaneously and intraperitoneally) on day −11 and challenged on day 0.

tective effects of immune serum have been observed in some circumstances (e.g., Dobson and Owen, 1978), and it has been proposed that one effect of antibodies may be to neutralize anti-inflammatory or immunosuppressive molecules of the parasite, which act locally in the intestines (Behnke and Parish, 1979; Dobson, 1982).

Good evidence exists that developing larvae and adult worms are both inducers and targets of host-protective immune responses in genetically resistant mouse strains although larvae *in situ* are likely to be much more efficient at inducing resistance than are adult worms *in situ* (Jacobson *et al.*, 1982). Obtaining a relatively restricted population of parasite molecules which has vaccination efficacy (i.e., something more appropriate than a macerated worm or larva as a starting antigen preparation for analysis) has proved difficult in the *N. dubius* system. Thus, although living adult worms implanted systemically sensitize for accelerated rejection of worms arising from L3 challenge, exoantigens from *in vitro*-incubated worms have proved inefficient (Day *et al.*, 1979; Hurley *et al.*, 1980). A more appropriate source of antigens may be the exoantigens of incubated larvae obtained from cysts (of nude mice given large numbers of L3) and collected using the method described by Ey *et al.* (1981). Even exoantigen preparations, at least from adults, are a very heterogeneous mixture of proteins (Day *et al.*, 1979). Unembryonated eggs of adult worms may be a less heterogeneous mixture of antigens than adults and larvae or their excretory/secretory/metabolic antigens. Female SJL/J mice were vaccinated with eggs obtained from incubated female worms and injected with CFA. Preliminary data (Table I) indicate that egg-vaccinated SJL/J mice reject intestinal worms in an accelerated manner; parasites establish in vaccinated mice and can be detected at day 15 after L3 challenge, cf. day 24. However, vaccination with sonicated egg preparations has failed to induce accelerated rejection; a possible explanation for vaccination efficacy with intact eggs (even though they had been stored frozen before injection) is that they embryonate, at least partially, in systemic sites. No success has been obtained in inducing accelerated rejection by vaccinating genetically susceptible mouse strains with eggs. Thus, to date, no real progress has been made in identifying a suitable source of host-protective antigens for fractionation (and measurement of immune response differences in genetically susceptible versus genetically resistant mice) in this important model of chronic intestinal nematode infection. Both in the *G. muris* and *N. dubius* systems, the possibility exists that genetically based mouse strain variations in susceptibility may be attributable to differences in ancillary intestinal inflammatory events rather than (or in addition to, Mitchell *et al.*, 1982*a*), antigen-recognition differences (see also Wakelin and Donachie, 1981, and Bell and McGregor, 1980, re: the *Trichinella spiralis*-mouse and rat models). However, there is now little doubt that gross parameters such as numbers of mucosal mast cells (at least in the mouse) bear little relationship to expression of resistance (e.g., Lee and Wakelin, 1982; Crowle, 1982).

VII. *FASCIOLA HEPATICA* (MURINE FASCIOLIASIS): VACCINATION FOR NEUTRALIZATION OF PARASITE-PROTECTIVE MECHANISMS

In mice, resistance to reinfection with the economically important trematode parasite, *F. hepatica* is weak if it exists at all (literature cited in Chapman and Mitchell, 1982*a*). Partial resistance to reinfection in rats has been demonstrated on numerous occasions, and a degree of protection (by no means absolute) has been obtained in rats, but not mice, using exoantigens of *F. hepatica* larvae in CFA (Rajasekariah *et al.*, 1979; see also Howell and Sandeman, 1979; Burden *et al.*, 1982). The lack of resistance to reinfection in mice cannot be ascribed readily to intraspecies variations in infectivity (antigenicity?) of *F. hepatica.* Thus, in experiments using clones of *F. hepatica* metacercariae derived from snails, each infected with one miracidium, rats developed partial resistance to reinfection using either homologous or heterologous clone challenge whereas mice developed no resistance under any circumstances (Chapman *et al.*, 1981).

The possibility exists that *F. hepatica* possesses very efficient mechanisms of immune evasion thus accounting for its ability to infect a large number of different mammalian hosts. One of these mechanisms may be the elaboration of antibody-cleaving proteases ("fabulating" enzymes) of the type described by Eisen and Tallon (1977) in the protozoan *Tetrahymena pyriformis.* More completely degradative enzymes of *Schistosoma mansoni* schistosomules have also been described recently (Auriault *et al.*, 1981) and, as might be expected of a complex metazoan, disrupted immature *F. hepatica* contain proteolytic enzymes that drastically alter SDS-PAGE protein profiles of immature parasites surface labeled using lactoperoxidase-catalyzed radioiodination. Provided protease inhibitors are present at the time of parasite disruption after the iodination of intact parasites, a dominant surface molecule with the molecular weight of mouse μ chains is demonstrable, but only when the parasites are obtained from intact rather than nude mice (R. J. Howard *et al.*, 1980*a*). [This comparative approach of surface labeling parasites from parasitized intact and nude mice for the demonstration of surface immunoglobulin (antibodies?) and masked antigens has also been used in a *Mesocestoides corti* larval cestode system (Mitchell *et al.*, 1977).]

Labeled IgG (and to a lesser extent IgM) proteins are cleaved by products elaborated by intact immature and adult *F. hepatica* maintained *in vitro*: labeled γ heavy chains being reduced to fragments of the molecular weight of light chains as analyzed by SDS-PAGE (Chapman and Mitchell, 1982*b*). The responsible enzyme(s) are thiol proteases, as determined by inhibitor studies; the full range of possible substrates has yet to be determined. Attempts are in progress to isolate the enzyme(s) from exoantigen preparations by affinity chromatography, to conjugate them to immunogenic carriers (e.g., KLH), and to vaccinate to test for immunogenicity and induction of any host resistance (initially in rats). If Ig-cleaving enzymes were a key component of parasite protection against aggressive immune attack, parasites might be rendered more susceptible to other existing or inducible antiparasite immune responses by immunologic neutralization of these enzymes. The major unknown, and as yet untested assumption, is that high-titered antibodies of certain isotypes and directed against certain epitopes of enzyme molecules may lead to their accelerated removal through opsonizing antibody effects, if not enzyme neutralization, and prejudice the establishment of parasites in the early stages of infection/reinfection. If immunologic responses are not up to the task of subverting immune evasion mechanisms of *F. hepatica*, precise definition of such mechanisms may at least lead to novel chemotherapeutic approaches (e.g., injection of selective protease inhibitors) to complement other antifasciola drugs in the control of fascioliasis, which is of economic importance in sheep and cattle.

VIII. CONCLUDING COMMENTS

Identification and isolation of the so-called functional antigens of parasites is progressing but at a slow rate, particularly in the case of parasites of medical and economical importance. One major limitation is still the availability of various parasite life-cycle stages adequately purified and/or in quantities sufficient for mRNA and antigen isolation purposes. The parasite molecular biologist and immunochemist will be relying heavily in the future on the work of parasitologists developing *in vitro* parasite culture techniques or cell hybridization and other forms of parasite cell immortalization.

Numerous examples of vaccination efficacy of crude antigen mixtures exist in the literature on immunoparasitology. Several of the systems outlined emphasize the importance of host genotype in dictating the outcome of vaccination against infection. New trends toward vaccination are being pursued, examples being anti-Id immunization, transmission-blocking and antivector immunity, as well as the use of novel antigens, and pathology-inhibiting vaccines (reviewed in Mitchell, 1982*a*). Nevertheless, most research activity still centers around devising strategies for identification, isolation, and eventual production of target antigens of proven host-protective or parasite-inhibitory immune responses. The application of the hybridoma technique to the production of antiparasite monoclonal antibodies will be the key to identification and isolation of defined antigens of immunoparasitologic importance in many systems. A major purpose of model systems of the type described will be the testing of methods of vaccination with defined-antigen (molecular) vaccines that (1) induce immune responses of predetermined desirability, (2) negate undesirable responses and consequences of infection, and (3) do not automatically select for "antigen-negative" parasite populations.

ACKNOWLEDGMENTS

Work performed in the Laboratory of Immunoparasitology, The Walter and Eliza Hall Institute, and referred to in the text is currently supported by the National Health and Medical Research Council of Australia, The Rockefeller Foundation Great Neglected Diseases Program, the UNDP/World Bank/WHO Special Programme for Research and Training in Tropical Diseases, and the National Institutes of Health, grant 17119.

IX. REFERENCES

Abbas, A. K., James, S. L., and Sher, A., 1981, Immunogenicity of haptenated schistosomula *in vitro*, *J. Immunol.* **126**:1022–1024.

Allen, E. M., Moore, V. L., and Stevens, J. O., 1977, Strain variation in BCG-induced chronic pulmonary inflammation in mice. I. Basic model and possible genetic control by non-H-2 genes, *J. Immunol.* **119**:343–347.

Anders, R. F., Howard, R. J., and Mitchell, G. F., 1982*a*, Parasite antigens and methods of analysis, in: *Immunology of Parasitic Infections*, 2nd ed. (S. Cohen and K. S. Warren, eds.) pp. 28–73, Blackwell Scientific, Oxford.

Anders, R. F., Roberts-Thomson, I. C., and Mitchell, G. F., 1982*b*, Giardiasis in mice: Analysis of humoral and cellular immune responses to *Giardia muris*, *Parasite Immunol.* **4**: 47–57.

Andrews, J. S., and Hewlett, E. L., 1981, Protection against infection with *Giardia muris* by milk containing antibody to giardia, *J. Infect. Dis.* **143**:242–246.

Andrews, P., and Meister, G., 1978, Differences in susceptibility to infection with *Fasciola hepatica* between mouse strains, *Z. Parasitenkd.* **56**:305–308.

Auriault, C., Ouaissi, M. A., Torpier, G., Eisen, H., and Capron, A., 1981, Proteolytic cleavage of IgG bound to the Fc receptor of *Schistosoma mansoni* schistosomula, *Parasite Immunol.* **3**:33–44.

Ayuya, J. M., and Williams, J. F., 1979, The immunological response to the rat to infection with *Taenia taeniaeformis.* VII. Immunization by oral and parenteral administration of antigens, *Immunology* **36**:825–834.

Barth, E. E. E., Jarrett, W. F. H., and Urquhart, G. M., 1966, Studies on the mechanism of the self-cure reaction in rats infected with *Nippostrongylus brasiliensis*, *Immunology* **10**:459–464.

Bartlett, A., and Ball, P. A. J., 1974, The immune response of the mouse to larvae and adults of *Nematospiroides dubius*, *Int. J. Parasitol.* **4**:463–470.

Behin, R., Mauel, J., and Sordat, B., 1979, *Leishmania tropica*: Pathogenicity and *in vitro* macrophage function in strains of inbred mice, *Exp. Parasitol.* **48**:81–91.

Behnke, J. M., and Parish, H., 1979, Expulsion of *Nematospiroides dubius* from the intestine of mice treated with immune serum, *Parasite Immunol.* **1**:13–26.

Bell, R. G., and McGregor, D. D., 1980, Requirement for two discrete stimuli for induction of the intestinal expulsion response against *Trichinella spiralis* in rats, *Infect. Immun.* **29**:186–193.

Bergmann, R. K., and Munoz, J., 1968, Action of the histamine sensitizing factor from *Bor-*

detella pertussis in inbred and random bred strains of mice, *Int. Arch. Allergy Appl. Immunol.* **34**:331–338.

Berman, J. D., and Dwyer, D. M., 1981, Expression of Leishmania antigen on the surface membrane of infected human macrophages *in vitro*, *Clin. Exp. Immunol.* **44**:342–348.

Bickle, Q. D., Long, E. G., James, E. R., Doenhoff, M. J., and Festing, M., 1980, *Schistosoma mansoni:* Influence of the mouse host's sex, age and strain on resistance to reinfection, *Exp. Parasitol.* **50**:222–232.

Bjorvatn, B., and Neva, F. A., 1979, A model in mice for experimental leishmaniasis with a West African strain of *Leishmania tropica*, *Am. J. Trop. Med. Hyg.* **28**:472–479.

Blum, K., and Cioli, D., 1981, *Schistosoma mansoni:* Age-dependent susceptibility to immune elimination of schistosomula artificially introduced into pre-infected mice, *Parasite Immunol.* **3**:13–24.

Bomford, R., 1980, The comparative selectivity of adjuvants for humoral and cell-mediated immunity. II. Effect on delayed-type hypersensitivity in the mouse of Freund's incomplete and complete adjuvants, alhydrogel, *Corynebacterium parvum*, *Bordetella pertussis*, muramyl dipeptide and saponin, *Clin. Exp. Immunol.* **39**:435–441.

Boray, J. C., 1967, The effect of host reaction to experimental *Fasciola hepatica* infections in sheep and cattle, in: *The Reaction of the Host to Parasitism* (E. J. L. Soulsby, ed.), pp. 84–96, Elwert, Marburg.

Bryceson, A. D. M., Preston, P. M., Bray, R. S., and Dumonde, D. C., 1972, Experimental cutaneous leishmaniasis. II. Effects of immunosuppression and antigenic competition on the course of infection with *Leishmania enriettii* in the guinea-pig, *Clin. Exp. Immunol.* **10**:305–335.

Büchmüller, Y., and Mauel, J., 1979, Studies on the mechanism of macrophage activation. II. Parasite destruction in macrophages activated by supernatants from concanavalin A-stimulated lymphocytes, *J. Exp. Med.* **150**:359–370.

Burden, D. J., Harness, E., and Hammet, N. C., 1982, *Fasciola hepatica:* Attempts to immunize rats and mice with metabolic and somatic antigens derived from juvenile flukes, *Vet. Parasitol.* **9**:261–266.

Byram, J. E., and Von Lichtenberg, F., 1977, Altered schistosoma granuloma formation in nude mice, *Am. J. Trop. Med. Hyg.* **26**:944–956.

Byram, J. E., Doenhoff, M. J., Musallam, R., Brink, L. H., and Von Lichtenberg, F., 1979, *Schistosoma mansoni* infections in T-cell deprived mice, and the ameliorating effect of administering homologous chronic infection serum. II. Pathology, *Am. J. Trop. Med. Hyg.* **28**:274–285.

Capron, A., Capron, M., and Dessaint, J-P., 1980, ADCC as primary mechanisms of defense against metazoan parasites, in: *Immunology. 80, Progress in Immunology. IV.* (M. Fougereau and J. Dausset, eds.), pp. 782–793, Academic Press, London.

Chapman, C. B., and Mitchell, G. F., 1982*a*, *Fasciola hepatica:* Comparative studies on fascioliasis in rats and mice, *Int. J. Parasitol.* **12**:81–91.

Chapman, C. B., and Mitchell, G. F., 1982*b*, *Fasciola hepatica:* Proteolytic cleavage of immunoglobulin, *Vet. Parasitol.* **11**:165–178.

Chapman, C. B., Rajasekariah, G. R., and Mitchell, G. F., 1981, Clonal parasites in the analysis of resistance to reinfection with *Fasciola hepatica*, *Am. J. Trop. Med. Hyg.* **30**:1039–1042.

Cheever, A. W., Erickson, D. G., Sadun, E. H., and Von Lichtenberg, F., 1974, *Schistosoma japonicum* infection in monkeys and baboons: Parasitological and pathological findings, *Am. J. Trop. Med. Hyg.* **23**:51–64.

Cheever, A. W., Duvall, R. H., and Minker, R. G., 1980*a*, Quantitative parasitologic findings in rabbits infected with Japanese and Phillippine strains of *Schistosoma japonicum*, *Am. J. Trop. Med. Hyg.* **29**:1307–1315.

Cheever, A. W., Duvall, R. H., Minker, R. G., and Nash, T. E., 1980*b*, Hepatic fibrosis in rabbits infected with Japanese and Philippine strains of *Schistosoma japonicum*, *Am. J. Trop. Med. Hyg.* **29**:1327–1339.

Chensue, S. W., Wellhausen, S. R., and Boros, D. L., 1981, Modulation of granulomatous hypersensitivity. II. Participation of Ly1$^+$ and Ly2$^+$ thymocytes in the suppression of granuloma formation and lymphokine production in *Schistosoma mansoni*-infected mice, *J. Immunol.* **127**:363–367.

Cioli, D., 1982, Immune protection against *Schistosoma mansoni* in permissive and nonpermissive hosts, *Scr. Varia Pont. Acad. Sci.* **47**:79–90.

Civil, R. H., and Mahmoud, A. A. F., 1978, Genetic differences in BCG-induced resistance to *Schistosoma mansoni* are not controlled by genes within the major histocompatibility complex of the mouse, *J. Immunol.* **120**:1070–1072.

Claas, F. H. J., and Deelder, A. M., 1979, H-2 linked immune response to murine experimental *Schistosoma mansoni* infections, *J. Immunogenet.* **6**:167–175.

Clegg, J. A., and Smith, M. A., 1978, Prospects for the development of dead vaccines against helminths, *Adv. Parasitol.* **16**:165–218.

Colley, D. G., Savage, A. M., and Lewis, F. A., 1977, Host responses induced and elicited by cercariae, schistosomula, and cercarial antigenic preparations, *Am. J. Trop. Med. Hyg.* **26**:88–95.

Colley, D. G., 1981, T lymphocytes that contribute to the immunoregulation of granuloma formation in chronic murine schistosomiasis, *J. Immunol.* **126**:1465–1468.

Cox, F. E. G., 1978, Specific and nonspecific immunization against parasitic infection, *Nature* (*Lond.*) **273**:623–626.

Crowle, P. K., and Reed, N. D., 1981, Rejection of the intestinal parasite *Nippostrongylus brasiliensis* by mast cell deficient W/W^V anemic mice, *Infect. Immun.* **33**:54–58.

Crowle, P., 1983, Mucosal mast cell reconstitution and *N. brasiliensis* rejection by W/W^V mice, *J. Parasitol.* (in press).

Cruise, K. M., Mitchell, G. F., Tapales, F. P., Garcia, E. G., and Huang, S-R., 1981*a*, Murine hybridoma-derived antibodies producing circumoval precipitation (COP) reactions with eggs of *Schistosoma japonicum*, *Aust. J. Exp. Biol. Med. Sci.* **59**:503–514.

Cruise, K. M., Mitchell, G. F., Garcia, E. G., and Anders, R. F., 1981*b*, Hybridoma antibody immunoassays for the detection of parasitic infection. Further studies on a monoclonal antibody with immunodiagnostic potential of schistosomiasis japonica, *Acta Trop.* **38**:437–447.

Cruise, K. M., Mitchell, G. F., Garcia, E. G., Tiu, W. U., Hocking, R. E., and Anders, R. F., 1983, Sj23, the target antigen in *Schistosoma japonicum* adult worms of an immunodiagnostic hybridoma antibody, *Parasite Immunol.* **5**:37–46.

Damian, R. T., Greene, N. D., Meyer, K. F., Cheever, A. W., Hubbard, W. J., Hawes, M. E., and Clarke, J. F., 1976, *Schistosoma mansoni* in baboons. III. The course and characteristics of infection, with additional observations on immunity, *Am. J. Trop. Med. Hyg.* **25**:299–305.

Day, K. P., Howard, R. J., Prowse, S. J., Chapman, C. B., and Mitchell, G. F., 1979, Studies on chronic versus transient intestinal nematode infections in mice. I. A comparison of responses to excretory/secretory (ES) products of *Nippostrongylus brasiliensis* and *Nematospiroides dubius* worms, *Parasite Immunol.* **1**:217–239.

Dean, D. A., Bukowski, M. A., and Cheever, A. W., 1981, Relationship between acquired resistance, partial hypertension, and lung granulomas in ten strains of mice infected with *Schistosoma mansoni*, *Am. J. Trop. Med. Hyg.* **30**:806–814.

De Tolla, L. J., Jr., Scott, P. A., and Farrell, J. P., 1981, Single gene control of resistance to cutaneous leishmaniasis in mice, *Immunogenetics* **14**:29–40.

Dineen, J. K., and Wagland, B. M., 1966, The dynamics of the host parasite relationship.

IV. The response of sheep to graded and repeated infection with *Haemonchus contortus*, *Parasitology* **56**:639–650.

Dobson, C., 1982, Passive transfer of immunity with serum in mice infected with *Nematospiroides dubius:* Influence of quality and quantity of immune serum, *Int. J. Parasitol.* **12**:207–213.

Dobson, C., and Owen, M. E., 1978, Effect of host sex on passive immunity in mice infected with *Nematospiroides dubius*, *Int. J. Parasitol.* **8**:359–364.

Dunne, D. W., Lucas, S., Bickle, Q., Pearson, S., Madgwick, L., Bain, J. and Doenhoff, M. J., 1981, Identification and partial purification of an antigen (W1) from *Schistosoma mansoni* eggs which is putatively hepatotoxic in T cell-deprived mice, *Trans. R. Soc. Trop. Med. Hyg.* **75**:54–71.

Eisen, H., and Tallon, I., 1977, *Tetrahymena pyriformis* recovers from antibody immobilization by producing univalent antibody fragments, *Nature (Lond.)* **270**:514–515.

Ey, P. L., Prowse, S. J., and Jenkin, C. R., 1981, *Heligmosomoides polygyrus:* Simple recovery of post-infective larvae from mouse intestines, *Exp. Parasitol.* **52**:69–76.

Fanning, M. M., Peters, P. A., Davis, R. S., Kazura, J. W., and Mahmoud, A. A., 1981, Immunopathology of murine infection with *Schistosoma mansoni:* Relationship of genetic background to hepatosplenic disease and modulation, *J. Infect. Dis.* **144**:148–153.

Farah, F. S., Samra, S. A., and Newayri-Salti, N., 1975, The role of the macrophage in cutaneous leishmaniasis, *Immunology* **29**:755–764.

Flisser, A., Willms, K., Lachette, J. P., Ridaura, C., Beltrán, F., and Larralde, C. (eds.), 1982, *Cysticercosis: Present State of Knowledge and Perspectives*, Academic Press, New York.

Garcia, E. G., and Mitchell, G. F., 1983, Resistance to *Schistosoma japonicum* in 129/J mice, *J. Parasitol.* (in press).

Garcia, E. G., Tapales, F. P., Valdez, C. A., Mitchell, G. F., and Tiu, W. U., 1981, Attempts to standardize the circumoval precipitin test (COPT) for schistosomiasis japonica, *Southeast Asian J. Trop. Med. Public Health* **12**:384–395.

Garcia, E. G., Tiu, W. U., Valdez, C. A., Tapales, F. P., and Mitchell, G. F., 1982, High level resistance to reinfection with *Schistosoma japonicum* in the mouse, *Southeast Asian J. Trop. Med. Pub. Health* (in press).

Gemmell, M. A., and MacNamara, F. N., 1972, Immune responses to tissue parasites. II. Cestodes, in: *Immunity to Animal Parasites* (E. J. L. Soulsby, ed.), p. 236, Academic Press, New York.

Gemmell, M. A., and Soulsby, E. J. L., 1968, The development of acquired immunity to tapeworms and progress towards active immunization, with special reference to *Echinococcus* spp., *Bull. WHO* **39**:45–55.

Gorczynski, R. M., and MacRae, S., 1982, Analysis of subpopulations of glass-adherent mouse skin cells controlling resistance/susceptibility to infection with *Leishmania tropica*, and correlation with the development of independent proliferative signals to Ly-1^+/Lyt-2^+ T lymphocytes, *Cell. Immunol.* **67**:74–90.

Gorczynski, R. M., MacRae, S., Kuba, R., and Price, G. B., 1981, Macrophage heterogeneity and Ir-gene control as factors involved in the immune response of guinea pigs to infection with *Leishmania enrietti*, *Cell. Immunol.* **60**:367–375.

Green, W. F., and Colley, D. G., 1981, Modulation of *Schistosoma mansoni* egg-induced granuloma formation: I-J restriction of T cell-mediated suppression in a chronic parasitic infection, *Proc. Natl. Acad. Sci. USA* **78**:1152–1156.

Grimaldi, G. F., Moriearty, P. L., and Hoff, R., 1980, *Leishmania mexicana* in C3H mice: BCG and levamisole treatment of established infections, *Clin. Exp. Immunol.* **41**:237–242.

Hale, C., and Howard, J. G., 1981, Immunological regulation of experimental cutaneous leishmaniasis. II. Studies with Biozzi high and low responder lines of mice, *Parasite Immunol.* **3**:45–55.

Hamburger, J., Pelley, R. P., and Warren, K. S., 1976, *Schistosoma mansoni* soluble egg antigens: Determination of the stage and species specificity of their serologic reactivity by radioimmunoassay, *J. Immunol.* **117**:1561–1566.

Handman, E., and Burgess, A. W., 1979, Stimulation by granulocyte-macrophage colony stimulating factor of *Leishmania tropica* killing by macrophages, *J. Immunol.* **122**:1134–1137.

Handman, E., and Hocking, R. E., 1982, Stage-specific, strain-specific and cross-reactive antigens of *Leishmania* identified by monoclonal antibodies, *Infect. Immun.* **37**:28–33.

Handman, E., Ceredig, R., and Mitchell, G. F., 1979, Murine cutaneous leishmaniasis: Disease patterns in intact and nude mice of various genotypes and examination of some differences between normal and infected macrophages, *Aust. J. Exp. Biol. Med. Sci.* **57**:9–29.

Handman, E., Mitchell, G. F., and Goding, J. W., 1981, Identification and characterization of protein antigens of *Leishmania tropica* isolates, *J. Immunol.* **126**:508–512.

Hang, L. M., Warren, K. S., and Boros, D. L., 1974, *Schistosoma mansoni:* Antigenic secretions and the etiology of egg granulomas in mice, *Exp. Parasitol.* **35**:288–298.

Houba, V., Sturrock, R. F., and Butterworth, A. E., 1979, Significance of local formation of immune complexes within granulomata in schistosomiasis, in: *Function and Structure of the Immune System* (W. Müller-Ruckholtz and K. H. Müller-Hermelink, eds.), pp. 683–685, Plenum Press, New York.

Howard, R. J., Chapman, C. B., and Mitchell, G. F., 1980*a*, A difference in surface proteins of *Fasciola hepatica* larvae from intact and nude mice, *Aust. J. Exp. Biol. Med. Sci.* **58**:201–205.

Howard, J. G., Hale, C., and Liew, F. Y., 1980*b*, Immunological regulation of experimental cutaneous leishmaniasis. III. The nature and significance of specific suppression of cell mediated immunity in mice highly susceptible to *Leishmania tropica*, *J. Exp. Med.* **152**:594–607.

Howard, J. G., Hale, C., and Liew, F. Y., 1980*c*, Genetically determined susceptibility to *Leishmania tropica* infection is expressed by haemopoietic donor cells in mouse radiation chimaeras, *Nature* (*Lond.*) **288**:161–162.

Howard, J. G., Hale, C., and Chan-Liew, W. L., 1980*d*, Immunological regulation of experimental cutaneous leishmaniasis. I. Immunogenetic aspects of susceptibility to *Leishmania tropica* in mice, *Parasite Immunol.* **2**:303–314.

Howard, J. G., Hale, C., and Liew, F. Y., 1981, Immunological regulation of experimental cutaneous leishmaniasis. IV. Prophylactic effect of sub-lethal irradiation due to abrogation of suppressor T cell generation in mice genetically susceptible to *Leishmania tropica*, *J. Exp. Med.* **153**:557–568.

Howell, M. J., and Sandeman, R. M., 1979, *Fasciola hepatica:* Some properties of a precipitate which forms when metacercariae are cultured in immune rat serum, *Int. J. Parasitol.* **9**:41–45.

Hsu, S. Y. L., Hsu, H. F., and Osborne, J. W., 1965, Immunization of rhesus monkeys against schistosome infection by cercariae exposed to high doses of x-irradiation, *Proc. Soc. Exp. Biol. Med.* **131**:1146–1149.

Hsu, S. Y. L., Hsu, H. F., Penick, G. D., Lust, G. L., Osborne, J. W., and Cheng, H. F., 1975, Mechanisms of immunity to schistosomiasis: histopathologic study of lesions elicited in rhesus monkeys during immunizations and challenge with cercariae of *Schistosoma mansoni*, *J. Reticuloendothel. Soc.* **18**:167–185.

Hurley, J. C., Day, K. P., and Mitchell, G. F., 1980, Accelerated rejection of *Nematospiroides dubius* intestinal worms in mice sensitized with adult worms, *Aust. J. Exp. Biol. Med. Sci.* **58**:231–240.

Jacobson, R. H., Brooks, B. O., and Cypess, R. H., 1982, Immunity to *Nematospiroides dubius*-parasite stages responsible for and subject to resistance in high responder (LAF_1/J) mice, *J. Parasitol.* **68**:1053–1059.

James, S. L., and Colley, D. G., 1976, Eosinophil-mediated destruction of *Schistosoma mansoni* eggs, *J. Reticuloendothel. Soc.* **20**:359–374.

James, S. L., and Colley, D. G., 1978, Eosinophil-mediated destruction of *Schistosoma mansoni* eggs *in vitro.* II. The role of cytophilic antibody, *Cell. Immunol.* **38**:35–47.

Jenkins, D. C., and Carrington, T. S., 1981, *Nematospiroides dubius:* the course of primary, secondary and tertiary infections in high and low responder Biozzi mice, *Parasitology* **82**:311–318.

Knopf, P. M., and Soliman, M., 1980, Effect of host endocrine gland removal on the permissive status of laboratory rodents to infection with *Schistosoma mansoni, Int. J. Parasitol.* **10**:197–204.

Kojima, S., Kitamura, Y., and Takatsu, K., 1980, Prolonged infection of *Nippostrongylus brasiliensis* in genetically mast-cell depleted W/W^v mice, *Immunol. Lett.* **2**:159–162.

Kwa, B. H., and Liew, F. Y., 1977, Immunity in taeniasis-cysticercosis. I. Vaccination against *Taenia taeniaeformis* in rats using purified antigen, *J. Exp. Med.* **146**:118–131.

Lee, T. D. G., and Wakelin, D., 1982, The use of host strain variation to assess the significance of mucosal mast cells in the spontaneous cure response of mice to the nematode *Trichuris muris, Int. Arch. Allergy Appl. Immunol.* **67**:302–305.

Lewert, R. M., Yogore, M. G., Jr., Martin, W. R., and Blas, B. L., 1980, Schistosomiasis japonica in Barrio San Antonio, Basay, Samar in the Philippines. IV. Atypical precipitates in the circumoval precipitation test—A reaction of recent infection, *Am. J. Trop. Med. Hyg.* **29**:431–434.

Liew, F. Y., Hale, C., and Howard, J. G., 1982, Immunologic regulation of experimental cutaneous leishmaniasis. V. Characterization of effector and specific suppressor T cells, *J. Immunol.* **128**:1917–1922.

Lloyd, S., 1981, Progress in immunization against parasitic helminths, *Parasitology* **83**:225–240.

Long, G. W., Pelley, R. P., Blas, B. L., Yogore, M. G., Jr., and Lewert, R. M., 1980, Analysis of the immunoglobulins responsible for the circumoval precipitation reaction, *Am. J. Trop. Med. Hyg.* **29**:1241–1245.

Louis, J. A., Zubler, R. H., Coutinho, S. G., Lima, G., Behin, R., Mauel, J., and Engers, H. D., 1982, The *in vitro* generation and functional analysis of murine T cell populations and clones specific for a protozoan parasite, *Leishmania tropica, Immunol. Rev.* **61**: 215–243.

Mauel, J., and Defendi, V., 1971, Infection and transformation of mouse peritoneal macrophages by simian virus 40, *J. Exp. Med.* **134**:335–350.

Miller, K. M., Smithers, S. R., and Sher, A., 1981, The response of mice immune to *Schistosoma mansoni* to a challenge infection which bypasses the skin: Evidence for two mechanisms of immunity, *Parasite Immunol.* **3**:25–31.

Mitchell, G. F., 1979*a*, Effector cells, molecules and mechanisms in host-protective immunity to parasites, *Immunology* **38**:209–223.

Mitchell, G. F., 1979*b*, Responses to infection with metazoan and protozoan parasites in mice, *Adv. Immunol.* **28**:451–511.

Mitchell, G. F., 1982*a*, New trends towards vaccination against parasites, in: *Clinics in Immunology and Allergy*, Vol. 2, No. 2: *Immunoparasitology* (A. Capron, ed.) pp. 721–737, W. B. Saunders, Eastbourne.

Mitchell, G. F., 1982*b*, Genetic variation in resistance of mice to *Taenia taeniaeformis:*

Analysis of host-protective immunity and immune evasion, in: *Cysticercosis: Present State of Knowledge and Perspectives.* (A. Flisser, K. Willms, J. P. Lachette, C. Ridaura, F. Beltrán, and C. Larralde, eds.) pp. 575–584, Academic Press, New York.

Mitchell, G. F., 1982*c*, The nude mouse in immunoparasitology, in: *The Nude Mouse in Experimental and Clinical Research*, Vol. II. (J. Fogh and B. C. Giovanella, eds.), pp. 262–289, Academic Press, New York.

Mitchell, G. F., and Anders, R. F., 1982, Parasite antigens and their immunogenicity in infected hosts, in: *The Antigens*, Vol. 6 (M. Sela, ed.), pp. 69–149, Academic Press, New York.

Mitchell, G. F., and Cruise, K. M., 1981, Monoclonal antiparasite antibodies: A shot-in-the-arm for immunoparasitology, in: *Monoclonal Antibodies and T Cell Hybridomas* (G. J. Hämmerling, U. Hämmerling, and J. F. Kearney, eds.), p. 303, Elsevier, Amsterdam.

Mitchell, G. F., Marchalonis, J. J., Smith, P. M., Nicholas, W. L., and Warner, N. L., 1977, Studies on immune responses to larval cestodes in mice. Immunoglobulins associated with the larvae of *Mesocestoides corti*, *Aust. J. Exp. Biol. Med. Sci.* **55**:187–211.

Mitchell, G. F., Rajasekariah, G. R., and Rickard, M. D., 1980*a*, A mechanism to account for mouse strain variation in resistance to the larval cestode, *Taenia taeniaeformis*, *Immunology* **39**:481–489.

Mitchell, G. F., Curtis, J. M., Handman, E., and McKenzie, I. F. C., 1980*b*, Cutaneous leishmaniasis in mice: Disease patterns in reconstituted nude mice of several genotypes infected with *Leishmania tropica*, *Aust. J. Exp. Biol. Med. Sci.* **58**:521–532.

Mitchell, G. F., Curtis, J. M., Scollay, R. G., and Handman, E., 1981*a*, Resistance and abrogation of resistance to cutaneous leishmaniasis in reconstituted BALB/c nude mice, *Aust. J. Exp. Biol. Med. Sci.* **59**:539–554.

Mitchell, G. F., Curtis, J. M., and Handman, E., 1981*b*, Resistance to cutaneous leishmaniasis in genetically susceptible BALB/c mice, *Aust. J. Exp. Biol. Med. Sci.* **59**:555–565.

Mitchell, G. F., Garcia, E. G., Anders, R. F., Valdez, C. A., Tapales, F. P., and Cruise, K. M., 1981*c*, *Schistosoma japonicum:* Infection characteristics in mice of various strains and a difference in the response to eggs, *Int. J. Parasitol.* **11**:267–276.

Mitchell, G. F., Cruise, K. M., Garcia, E. G., and Anders, R. F., 1981*d*, Hybridoma-derived antibody with immunodiagnostic potential for schistosomaisis japonica, *Proc. Natl. Acad. Sci. USA* 78:3165–3169.

Mitchell, G. F., Anders, R. F., Brown, G. V., Handman, E., Roberts-Thomson, I. C., Chapman, C. B., Forsyth, K. P., Kahl, L. P., and Cruise, K. M., 1982*a*, Analyis of infection characteristics and antiparasite immune responses in resistant compared with susceptible hosts, *Immunol. Rev.* **61**:137–188.

Mitchell, G. F., Garcia, E. G., Cruise, K. M., Tiu, W. U., and Hocking, R. E., 1982*b*, Lung granulomatous hypersensitivity to eggs of *Schistosoma japonicum* in mice analysed by a radioisotopic assay and effects of hybridoma (idiotype) sensitization, *Aust. J. Exp. Biol. Med. Sci.* **60**:401–416.

Mitchell, G. F., and Handman, E., 1983*a*, *Leishmania tropica major* in mice: vaccination against cutaneous leishmaniasis in mice of high genetic susceptibility, *Aust. J. Exp. Biol. Med. Sci.* **61**:11–25.

Mitchell, G. F., Garcia, E. G., and Cruise, K. M., 1983*b*, Competitive radioimmunoassays using hybridoma and anti-idiotype antibodies in the analysis of antibody responses to, and antigens of *Schistosoma japonicum*, *Aust. J. Exp. Biol. Med. Sci.* **61**:27–63.

Modabber, F. Z., Alemohammadion, A., Khamesipour, A., Pourmand, M. H., Kamali, M., and Nasseri, M., 1980, Studies on the genetic control of visceral leishmaniasis in BALB/c mice by *L. tropica*, in: *Genetic Control of Natural Resistance to Infection and Malignancy* (E. Skamene, P. A. L. Kongshavn, and M. Landy, eds.), pp. 29–37, Academic Press, New York.

Moore, V. L., Schnur, D. J., Sternick, J. L., and Allen, E. M., 1981, Genetic control of granulomatous inflammation, in: *Basic and Clinical Aspects of Granulomatous Diseases* (D. L. Boros and T. Yoshida, eds.), pp. 201–214, Elsevier/North Holland Biomedical Press, Amsterdam.

Murray, H. W., 1981, Susceptibility of *Leishmania* to oxygen intermediates and killing by normal macrophages, *J. Exp. Med.* **153**:1302–1315.

Murray, M., Robinson, P. B., Guerson, C., and Crawford, R. A., 1979, Immunization against *Nippostrongylus brasiliensis* in the rat: A study on the use of antigen extracted from adult parasites and the parameters which influence the level of protection. A review, *Acta Trop.* **36**:297–322.

Murrell, K. D., Clark, S., Dean, D. A., and Vannier, W. E., 1980, Influence of mouse strain on induction of resistance with irradiated *Schistosoma mansoni* cercariae, *J. Parasitol.* **65**:829–831.

Nacy, C. A., Meltzer, M. S., Leonard, E. J., and Wyler, D. J., 1981, Intracellular replication and lymphokine-induced destruction of *Leishmania tropica* in C3H/He mouse macrophages, *J. Immunol.* **127**:2381–2386.

Ogilvie, B. M., and Love, R. J., 1974, Cooperation between antibodies and cells in immunity to a nematode parasite, *Transplant. Rev.* **19**:147–168.

Olds, G. R., and Mahmoud, A. A. F., 1980, Role of host granulomatous response in murine schistosomiasis mansoni. Eosinophil-mediated destruction of eggs, *J. Clin. Invest.* **66**: 1191–1199.

Olds, G. R., Olveda, R., Tracy, J. W., and Mahmoud, A. A. F., 1982, Modulation of immunopathology in chronic murine schistosomiasis japonica. I. Amelioration of disease by injection of chronic serum into mice with acute infection, *J. Immunol.* **128**:1391–1393.

Oliver-Gonzalez, J., 1954, Anti-egg precipitins in serum of humans infected with *Schistosoma mansoni*, *J. Infect. Dis.* **95**:86–91.

Pelley, R. P., and Warren, K. S., 1978, Immunoregulation in chronic infectious disease: Schistosomiasis as a model, *J. Invest. Dermatol.* **71**:49–55.

Perez, H., Arredondo, B., and Gonzalez, M., 1978, Comparative study of American cutaneous leishmaniasis and diffuse cutaneous leishmaniasis in two strains of inbred mice, *Infect. Immun.* **22**:301–307.

Péry, P., and Luffau, G., 1979, Antigens of helminths, in: *The Antigens*, Vol 5. (M. Sela, ed.), pp. 83–172, Academic Press, New York.

Phillips, S. M., and Colley, D. G., 1978, Immunological aspects of host responses to schistosomiasis: Resistance, immunopathology and eosinophil involvement, *Prog. Allergy* **24**: 49–182.

Phillips, S. M., Reid, W. A., Doughty, B. L., and Beutley, A. G., 1980, The immunologic modulation of morbidity in schistosomiasis. Studies in athymic mice and *in vitro* granuloma formation, *Am. J. Trop. Med. Hyg.* **29**:820–831.

Pollacco, S., Nicholas, W. L., Mitchell, G. F., and Stewart, A. C., 1978, T cell-dependent collagenous encapsulating response in the mouse liver to *Mesocestoides corti* (Cestoda), *Int. J. Parasitol.* **8**:457–462.

Preston, P. M., Behbehani, K., and Dumonde, D. C., 1978, Experimental cutaneous leishmaniasis. VI. Anergy and allergy in the cellular response during non-healing infection in different strains of mice, *J. Clin. Lab. Immunol.* **1**:207–219.

Prowse, S. J., and Mitchell, G. F., 1981, On the choice of mice for dissection of strain variations in the development of resistance to infection with *Nematospiroides dubius*, *Aust. J. Exp. Biol. Med. Sci.* **58**:603–605.

Prowse, S. J., Mitchell, G. F., Ey, P. L., and Jenkin, C. R., 1978, *Nematospiroides dubius:* Susceptibility to infection and the development of resistance in hypothymic (nude) BALB/c mice, *Aust. J. Exp. Biol. Med. Sci.* **56**:561–570.

Rajasekariah, G. R., Mitchell, G. F., Chapman, C. B., and Montague, P. E., 1979, *Fasciola hepatica:* Attempts to induce protection against infection in rats and mice by injection of excretory/secretory products of immature worms, *Parasitology* **79**:393–400.

Rajasekariah, G. R., Mitchell, G. F., and Rickard, M. D., 1980*a*, *Taenia taeniaeformis* in mice: Protective immunization with oncospheres and their products, *Int. J. Parasitol.* **10**:155–160.

Rajasekariah, G. R., Rickard, M. D., and Mitchell, G. F., 1980*b*, Immunization of mice against infection with *Taenia taeniaeformis* using various antigens prepared from eggs, oncospheres, developing larvae and strobilocerci, *Int. J. Parasitol.* **10**:315–324.

Rajasekariah, G. R., Rickard, M. D., Anders, R. F., and Mitchell, G. F., 1982, Immunization of mice against *Taenia taeniaeformis* using solubilized oncospheral antigens, *Int. J. Parasitol.* **12**:111–116.

Rickard, M. D., 1982, Immunization against infection with larval taeniid cestodes using oncospheral antigens, in: *Cysticercosis: Present State of Knowledge and Perspectives* (A. Flisser, K. Willms, J. P. Lachette, C. Ridaura, F. Beltrán, and C. Larralde, eds.) pp. 633–646, Academic Press, New York.

Rickard, M. D., and Williams, J. F., 1982, Hydatidosis/cysticercosis: Immune mechanisms and immunization against infection, *Adv. Parasitol.* **21**:229–296.

Roberts-Thomson, I. C., and Mitchell, G. F., 1978, Giardiasis in mice. I. Prolonged infection in certain mouse strains and hypothymic (nude) mice, *Gastroenterology* **75**:42–46.

Roberts-Thomson, I. C., and Mitchell, G. F., 1979, Protection of mice against *Giardia muris* infection, *Infect. Immun.* **24**:971–973.

Ross, J. G., 1967, A comparison of the resistance status of hosts to infection with *Fasciola hepatica*, in: *The Reaction of the Host to Parasitism* (E. J. L. Soulsby, eds), pp. 96–105, Elwert, Marburg.

Sacks, D. L., Esser, K. M., and Sher, A., 1982, Immunization of mice against African trypanosomiasis using anti-idiotype antibodies, *J. Exp. Med.* **155**:1108–1119.

Sandrin, M. S., Tobias, G. H., McKenzie, I. F. C., and Hämmerling, G. J., 1982, Alterations in the expression of Ia antigens in F_1 hybrid mice, *Immunogenetics* **14**:507–516.

Smith, M. A., and Clegg, J. A., 1979, Different levels of immunity to *Schistosoma mansoni* in the mouse: The role of variant cercariae, *Parasitology* **78**:311–321.

Smithers, S. R., and Gammage, K., 1980, Recovery of *Schistosoma mansoni* from the skin, lungs and hepatic portal system of naive mice and mice previously exposed to *S. mansoni:* Evidence for two phases of parasite attrition in immune mice, *Parasitology* **80**:289–300.

Smithers, S. R., and Miller, K. M., 1980, Protective immunity in murine schistosomiasis mansoni: Evidence for two distinct mechanisms, *Am. J. Trop. Med. Hyg.* **29**:832–841.

Sternick, J. L., Allen, E. M., Schnur, D. J., and Moore, V. L., 1983, Genetic control of BCG-induced granulomatous inflammation in mice. The development of granulomatous pulmonary inflammation and splenomegaly is under polygenic control and influenced by genes linked to the Igh complex, *J. Immunol.* (in press).

Stevens, D. P., Frank, D. M., and Mahmoud, A. A. F., 1978, Thymus dependency of host resistance to *Giardia muris* infection: Studies in nude mice, *J. Immunol.* **120**:680–682.

Taylor, M. G., 1980, Vaccination against trematodes, in: *Vaccination against Parasites* (A. E. R. Taylor and R. Muller, eds.), pp. 115–140, Blackwell Scientific, Oxford.

Uber, C. L., Roth, R. L., and Levy, D. A., 1980, Expulsion of *Nippostrongylus brasiliensis* by mice deficient in mast cells, *Nature* (*Lond.*), **287**:226–228.

Underdown, B. J., Roberts-Thomson, I. C., Anders, R. F., and Mitchell, G. F., 1981, Giardiasis in mice: Studies on the characteristics of chronic infection in C3H/He mice, *J. Immunol.* **126**:669–672.

Urquhart, G., 1980, Immunity to cestodes, in: *Vaccination against Parasites* (A. E. R. Taylor and R. Muller), pp. 107–114, Blackwell Scientific, Oxford.

Von Lichtenberg, F., 1962, Host response to eggs of *S. mansoni*. I. Granuloma formation in the unsensitized laboratory mouse, *Am. J. Pathol.* **41**:711–731.

Von Lichtenberg, F., 1977, Experimental approaches to human schistosomiasis, *Am. J. Trop. Med. Hyg.* **26**:79–87.

Von Lichtenberg, F., Ericksen, D. G., and Sadun, E. H., 1973, Comparative histopathology of schistosome granulomas in the hamster, *Am. J. Pathol.* **72**:149–178.

Wakelin, D., and Donachie, A. M., 1981, Genetic control of immunity to *Trichinella spiralis*. Donor bone marrow cells determine responses to infection in mouse radiation chimaeras, *Immunology* **43**:787–792.

Warren, K. S., 1973, The pathology of schistosome infections, *Helminth. Abst.* **42**:591–633.

Warren, K. S., 1974, Modulation of immunopathology in schistosomiasis, in: *Parasites in the Immunized Host: Mechanisms of Survival* (R. Porter and J. Knight, eds.), pp. 243–251, Associated Scientific Publishers, Amsterdam.

Warren, K. S., 1977, Modulation of immunopathology and disease in schistosomiasis, *Am. J. Trop. Med. Hyg.* **26**:113–118.

Warren, K. S., Boros, D. L., Hang, L. M., and Mahmoud, A. A. F., 1975, The *Schistosoma japonicum* egg granuloma, *Am. J. Pathol.* **80**:279–293.

Weiner, E., and Bandieri, A., 1975, Modifications in the handling *in vitro* of ^{125}I-labelled keyhole limpet haemocyanin by peritoneal macrophages from mice pretreated with the adjuvant *Corynebacterium parvum*, *Immunology* **29**:265–274.

Weintraub, J., and Weinbaum, F. I., 1977, The effect of BCG on experimental cutaneous leishmaniasis of mice, *J. Immunol.* **118**:2288–2290.

Williams, J. F., 1979, Recent advances in the immunology of cestode infections, *J. Parasitol.* **65**:337–349.

Wilson, R. A., 1980, Is immunity to *S. mansoni* in the chronically-infected laboratory mouse an artefact of pathology? in: *Proceedings of the Third European Multicolloquium on Parasitology*, p. 37, Cambridge, England.

Chapter 10

Immunity in Schistosomiasis: A Holistic View

Raymond T. Damian

Department of Zoology
University of Georgia
Athens, Georgia 30602

I. INTRODUCTION

Schistosomes are parasites of ancient lineage. Mammalian and avian schistosomes, in the family Schistosomatidae, are members of a natural, monophyletic group of digenetic trematodes, the superfamily Schistosomatoidea, which also includes other families of blood flukes found in lower vertebrates (LaRue, 1957). These are the Aporocotylidae and Sanguinicolidae in fishes and the Spirorchiidae in reptiles. Byrd (1939) has suggested that the Aporocotylidae are the most primitive of the blood flukes, later giving rise to the Spirorchiidae (and Sanguinicolidae), which then gave rise to the schistosomes. Blood flukes probably evolved as such in the Permian (G. M. Davis, personal communication, 1982), which means they have had something on the order of $230\text{-}280 \times 10^6$ years in which to perfect survival tactics appropriate to their intravascular location in their vertebrate hosts and (of particular interest to this readership) their coexistence with their vertebrate host's immune system.

On the other hand, schistosomes became known to science only in 1851, when Theodor Bilharz first communicated his great discovery of a new human "distomum" in the blood of the portal vein of a cadaver in Egypt to his mentor, Theodor von Siebold. Publications relating to serology and immunity began to appear in Japan in the early 1900s, so the field of schistosome immunology has existed for a mere three-quarters of a century—small wonder then that our ignorance is still great, and prophylactic immunization of the millions of people at risk to infection with these pathogens is still only a dream.

Adult schistosomes live *in copula* (Fig. 1) within their definitive intravascular locations, which are species-specific characteristics. The major human-infecting species, *Schistosoma mansoni* and *S. japonicum*, inhabit the inferior and superior mesenteric veins and their branches, respectively, while the third major human-infecting species, *S. haematobium*, is located in the veins of the vesical plexus. The slender female inserts the anterior part of her body (Fig. 1) into small venules and deposits her eggs, which either pass through the wall of the hollow organ to break free into the lumen, thence out of the body in the excreta, or are lodged in the tissues until they are destroyed. Each entrapped egg becomes a nidus for a granuloma; much of the pathology of chronic schistosomiasis has been traced from this initial inflammatory response (Warren, 1972). Successful eggs reach fresh water, where they hatch to liberate a short-lived, nonfeeding, swimming stage called a miracidium, which is male or female in gender. Well equipped with cilia, sensory receptors, penetration apparatus, and a behavioral repertoire, the miracidium has the ability to seek, find, and invade a snail of the appropriate species (another species-specific trait). The successful miracidium metamorphoses into a primary sporocyst, a parasitic stage living in the head–foot of the snail. Through an asexual process of internal budding, primary sporocysts each produce many secondary sporocysts, which are born to continue their existence also as snail parasites. Asexual reproduction occurs in secondary sporocysts as well, but the progeny are different, and are called cercariae (Fig. 1). This process of clonal expansion may yield more than 100,000 cercariae per miracidium. These are of a single sex, but since multiple miracidial infections of snails ofter occur, both sexes may be represented in the cercarial population developing within a single snail.

Cercariae are the second free-living, swimming, infective dispersal stage. They leave the snail and, through a combination of taxes and tropisms, are often well placed in a stream or lake to encounter potential vertebrate hosts. These too are penetrated and, if compatible, the life cycle continues. The forked tail (Fig. 1) or swimming organ of the cercaria is shed, penetration gland contents emptied, and profound physiologic and biochemical changes undergone during or immediately after the penetration process to yield the next stage, the schistosomulum.

The schistosomulum is basically an undeveloped, sexually immature male or female fluke. After what is currently believed to be a route entirely confined to the vascular system, the schistosomula reach their preferred sites where, probably in response to specific stimuli, they shorten and thicken, cease their migration, grow, and develop. Through some process involving chemoattraction, male and female juvenile flukes pair up, the female being held in the gynecophoric canal of the male (Fig. 1). Sexual maturity is reached; fertilization then takes place in some unknown way—the male lacks an intromittent organ—and egg production begins. The rate of egg production is another species-specific characteristic and varies from several hundred to several thousand per day per fe-

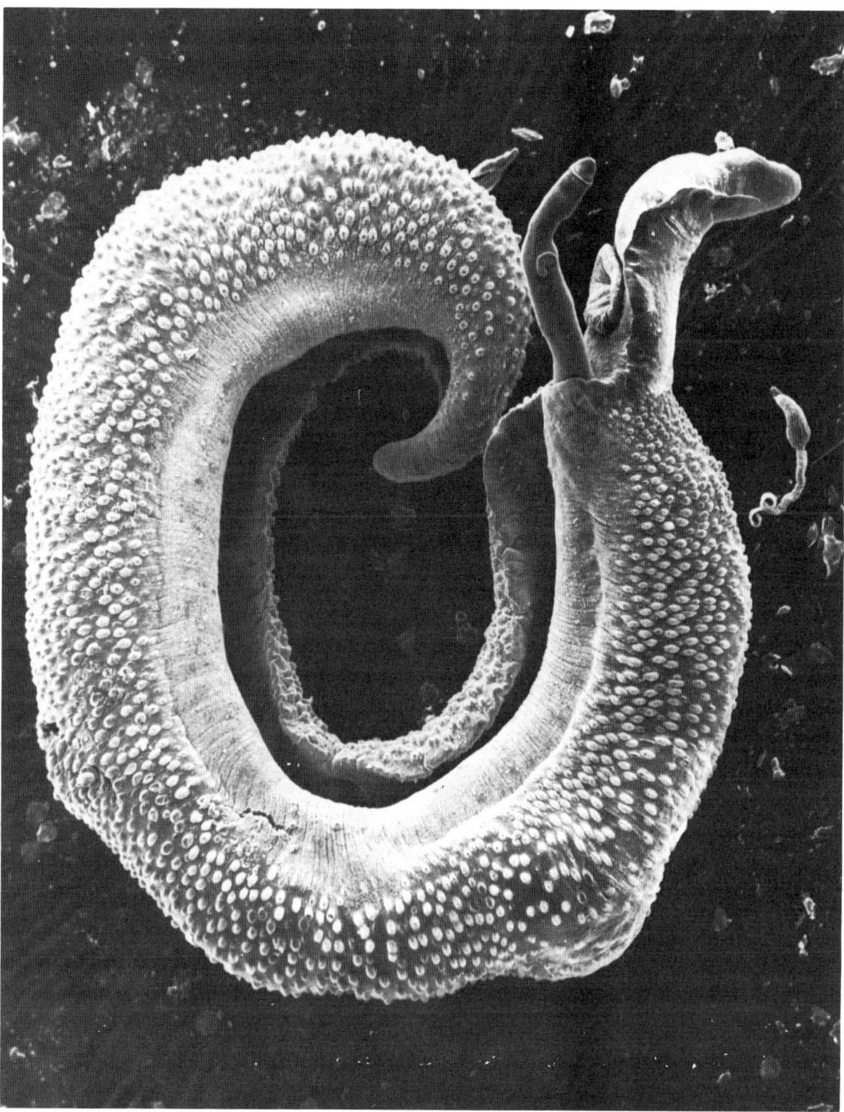

Figure 1. *Schistosoma mansoni* adult worm pair and cercaria; (×100). Upon penetration of the mammalian host, the cercaria sheds its tail. Full growth from cercarial body to adult fluke is accomplished in ~ 5 weeks. (Photograph courtesy of Dr. Robert E. Miller.)

Figure 2. Thin section of the dorsal tegument of adult male *S. mansoni*. The outer membrane (OM) is invaginated, forming surface pits (P), and encloses the tegumental spines (S). The basal membrane (BM) separates the muscle fibers (M) from the tegument, ×24,000.

male. Thus the cycle is completed, all within a time frame of approximately 2 or 3 months, including both vertebrate and molluscan phases.

Positioning of the adult male flukes in their rheoactive environment is facilitated by their muscular structure, suckers, and warty surface (Fig. 1), as well as their possession of tegumental spines (Fig. 2). Nutrition is by ingestion of blood through the mouth and absorption of nutrients through the tegumental surface, an area greatly expanded by pits and channels (Fig. 2).

Great advances have been made during the modern era relating to the general biology of schistosomes, including the discoveries (1) that adult worms are homolactic fermenters with incredibly high glucose-utilization rates (Bueding, 1950), (2) that adult schistosomes have a nearly unique* surface covering—a dou-

*It is nearly unique in that it occurs also in the related families of blood flukes, Aporocotylidae and Spirorchiidae (McLaren and Hockley, 1977), hence probably represents a specific adaptation to intravascular existence (McLaren and Hockley, 1977) as well as providing powerful additional evidence of a monophyletic origin for the Schistosomatoidea.

ble outer membrane (Hockley and McLaren, 1973) (Fig. 3), (3) of a powerful surface inhibitor of the contact activation pathway of blood coagulation (Tsang *et al.*, 1977), (4) of endogenous ecdysteroid production in *S. mansoni* and its possible connection with tegumental shedding (Torpier *et al.*, 1982), and new insights into male-female interactions (Imperia *et al.*, 1980; Atkinson and

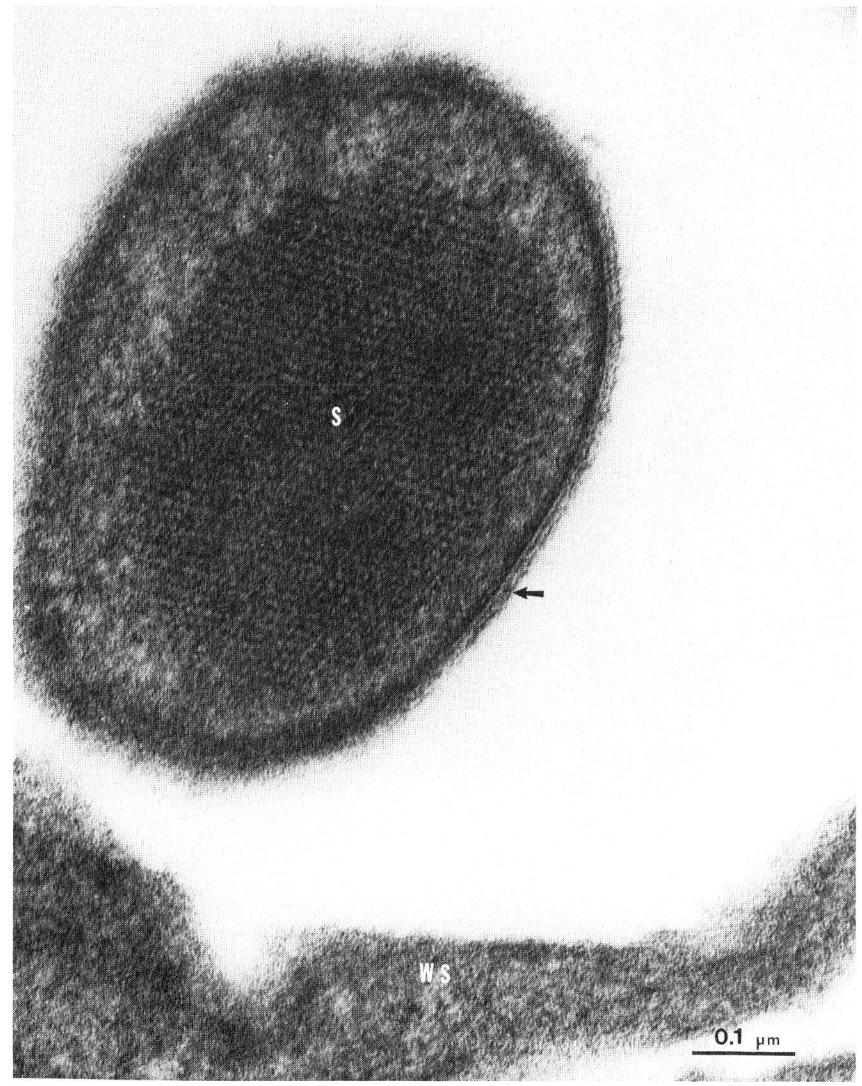

Figure 3. Transverse section through a tegumental spine (S) of adult male *S. mansoni*, demonstrating the double outer membrane (*arrow*) of the tegumental surface. The crystalline lattice structure of the spine core can be seen. The worm surface (WS) appears below, ×224,000.

Atkinson, 1980; Cornford and Huot, 1981; Shirazian and Schiller, 1982).

But equally great gaps in knowledge of some of the most basic aspects of the biology of these organisms still exist. However, the belated general recognition of schistosomiasis as one of the great plagues of humankind, with its attendant influx of research funding, and new, diverse scientific talent almost guarantee the continued accelerated pace of knowledge accretion

This Chapter presents a selective, unabashedly biased, and personal attempt to construct a coherent view of the incredibly diverse, fast-moving, and complicated literature dealing with immunity in schistosomiasis. I begin by amalgamating the models, then I apply Occam's razor to schistosome antigens, and end by looking at the slashed and bleeding corpus through the wrong end of a telescope. It looks whole to me, but whether the view is idiosyncratic remains to be determined.

II. HOST QUALITY AND IMMUNITY

Fundamental to any discussion of immunity in schistosomiasis is the concept of host specificity. The various species of schistosomes are optimally adapted to various species of hosts, differentially tolerating a range of host species. Investigators have long been aware of this and have distinguished "good" from "bad," or "normal" from " abnormal" hosts. Good hosts were considered those that permit parasite development and prolonged egg-laying and bad hosts those that give some degree of suboptimal development, survival, and reproduction (Kuntz and Malakatis, 1955; Newsome, 1956; von Lichtenberg *et al.*, 1962; Stirewalt 1963). Cioli *et al.* (1977) were apparently the first to apply the terms "permissive" and "nonpermissive" to schistosomiasis, terms that have gained popular usage in this field. Their definition of a nonpermissive host was one in which "development of worms is incomplete and/or oviposition is interrupted after a short period." Later, Knopf (1982) used more stringent definitions, a permissive host being one "from which viable parasite ova can be isolated from the feces." I will adhere to the earlier definition. In fact, I consider the rhesus monkey a nonpermissive host for *S. mansoni*, since fecal egg passage after a moderate single infection is usually terminated after only 1 year (Naimark *et al.*, 1960; P. E. Thompson *et al.*, 1965; McMullen *et al.*, 1967). Compare this with our experience with baboons: Six animals exposed to 1000 *S. mansoni* cercariae each are still passing moderate to large numbers of viable eggs more than 9 years (at the time of this writing) since they were infected.

Two separate but interacting phenomena underlie nonpermissiveness: (1) the inadequacy of a host as an environment for providing all the parasite's life needs (Read, 1958), and (2) the host's immune response to the parasite. The connections between the two are beginning to be understood. This understanding in turn will continue to provide valuable insight into the host–parasite relationship, including immunity, of permissive combinations, foremost of which are the human schistosomes in man.

The most intensively studied nonpermissive system is *S. mansoni* in the rat. Early work established the basic features of this relationship: In primary infections the worms are atypically located, stunted, short-lived, and infertile. More recent work suggests immaturity of the tegumental surface of rat worms, both morphologic (Senft *et al.*, 1978) and functional, in that they poorly acquire host antigens (Cioli, 1976). Challenge infections arc rapidly eliminated, despite the persistence of a few residual primary infection worms (Knopf *et al.*, 1977; Phillips *et al.*, 1977). It is clear that the immune response of the rat plays a major role in reinfection immunity (Phillips *et al.*, 1977; Knopf *et al.*, 1977). Nonimmune factors appear to predominate in the rat's handling of primary infections, however. Transplantation experiments have been particularly revealing (Michaels, 1970; Cioli *et al.*, 1977) in this regard. Stunted rat worms, upon transfer into permissive hamsters, found their definitive location in the mesenteric veins, resumed their growth, and became fertile (Cioli *et al.*, 1977). Similar regain of health and normal activity was recorded by Michaels (1970) after transplantation of guinea pig worms into hamsters. Reversal of the direction of transplantation, from permissive to nonpermissive hosts, yielded variable results in the two investigations, but worms transplanted from permissive to nonpermissive hosts did not generally do well. As Cioli *et al.* (1977) have pointed out, these experiments indicate that even prolonged residence in a nonpermissive host fails to block the maturation process irreversibly. Furthermore, the permissive environment permits not only full maturation, but maintenance of that state as well.

The critical time at which adversity to a primary infection is manifested in the rat is at about 1 month, with spontaneous elimination beginning shortly thereafter (Ritchie *et al.*, 1963; Smithers and Terry, 1965*b*; Maddison *et al.*, 1970*a*). Cioli *et al.* (1978), again using transplantation experiments, elegantly demonstrated this critical time to be an inherent property of the worm when *in the rat*. That is, a 4-week maturation period for the rat's immune response was not the basis for the critical time, nor was age per se of the worms. Three explanations were considered by these investigators: (1) stage-specific immunity, which they thought unlikely in view of Cioli and Dennert's (1976) results showing that arrested development also occurs in thymectomized rats; (2) stage-specific sensitivity to the presence or absence of some critical factor in the nonpermissive host; or (3) inappropriate location (the liver) in the rat. The latter may be more consequence than cause if the ability of worm pairs to migrate to the mesenteric veins is maturation-dependent. Evidence for the second alternative was provided by Knopf and Soliman (1980), who found hypophysectomy and thyroidectomy of rats to result in improved worm maturation and survival. Conversely, these procedures failed to alter the permissiveness of mice. Further evidence for host hormonal influence on schistosome metabolism and growth was provided by various *in vivo* and *in vitro* approaches (Abdel Wahab *et al.*, 1971; Cornford, 1974*a,b*; Levi-Schaffer and Smolarsky, 1981). Knopf (1982) recently developed a stimulating hypothesis in which interactions between thyroid hormones and the host immune system are advanced as explanations for the various blocks to *S. mansoni* development and survival in the rat. Knopf (1982) also asked whether schistosomes might directly influence their host's hormonal levels by producing host hormone-like substances.

Superimposed on this innate immunity to primary *S. mansoni* infections in the rat is an immune response. This was first demonstrated by adoptive cell-transfer experiments (Phillips *et al.*, 1975) and by thymectomy (Cioli and Dennert, 1976). Both approaches agreed in implicating T cells as the main factors in protective immunity in primary infections. Thus, it appears that components of innate immunity, perhaps related to improper parasite nutrition in the rat, result in an unhealthy, immature, and vulnerable parasite. This damaged worm is susceptible to elimination by the T-cell-dependent immune response of the rat, which is simultaneously being induced. The strength of this elimination response is dose-dependent (Stirewalt *et al.*, 1951; Ritchie *et al.*, 1963; Phillips *et al.*, 1975) with respect to the infecting cercariae and has been suggested to relate to total antigenic load by Phillips *et al.*, (1975), a point discussed in more detail in Section V.

The acquired immunity component becomes dominant in secondary infections. On the basis of the results of a clever series of experiments, Knopf *et al.* (1977) argued against the possibility that infected rats are deficient in factors such as nutrients that might be required for the development and survival of reinfecting worms.* Their work identified the appearance in once-infected rats of factors, probably immune, that are toxic to secondary infection schistosomes. That these toxic factors are indeed immune effectors, more precisely T-cell-mediated effectors, was amply demonstrated by Phillips *et al.* (1977) and Cioli *et al.* (1980). Just what these effectors might be, in the rat, has been recently and thoroughly reviewed by A. Capron *et al.* (1982).

Is it merely coincidental that the rat is at once nonpermissive and one of the most easily immunized, by primary infection and living or dead vaccines (Phillips

*There has been an unfortunate confusion of terms in the literature of schistosomiasis relating to this. Knopf *et al.* (1977) resurrected Paul Ehrlich's term "athrepsia", but used it to denote a condition in which some nutrient or other deficiency limits an infection. The immune system is not involved, in their view. This new definition has since often been repeated. Actually, Ehrlich himself, in his Nobel Lecture (Ehrlich, 1910), used the name *atrepsins* for *antibodies* "whose action is purely antinutritive"—the exact opposite of the current derived meaning.

et al., 1978*a*), of hosts? Do these characteristics go hand in hand in other host-schistosome combinations, or are they separated in some? Although incomplete, the data do suggest a correlation between host quality and ease of immunity induction. For example, it is known that mice develop a much less effective immunity than rats, both to primary and to secondary infections (Phillips and Colley, 1978).* Examination of another closely related pair of hosts, the rhesus monkey (*Macaca mulatta*) and the savanna baboon (*Papio cynocephalus* superspecies), shows the same correlation. *Schistosoma mansoni* in the baboon is very long-lived and fertile, but the infection is self-limiting in the rhesus monkey (Naimark *et al.*, 1960). Direct comparison of a single parasite strain's ability to induce reinfection immunity in the two hosts was performed in our laboratory, confirming the ability of the rhesus monkey to develop immunity more rapidly than the baboon (Damian *et al.*, 1976*a,b*). Adult worms transplanted into the mesenteric veins of baboons (Damian *et al.*, 1972) were unable to confer resistance to a subsequent percutaneous challenge infection under conditions very similar to those in which rhesus monkeys were successfully immunized (Smithers and Terry, 1967).

It seems that host quality is inversely related to the subsequent development of immunity in schistosomiasis. And it is both our misfortune and our challenge that human beings are the best, or among the best, of hosts for *Schistosoma haematobium*, *S. mansoni*, and *S. japonicum*.

III. IMMUNOGENIC STAGES

A. Concomitant Immunity and the Immunogenic Role of Adult Worms

Since 1969, one idea has dominated the field of schistosome immunity. That year, Smithers and Terry published their seminal paper, in which the term "concomitant immunity," borrowed from the literature of tumor immunity, was introduced and applied to schistosomiasis. As originally stated, concomitant immunity refers to the situation in which "invading schistosomula are destroyed by the immune response of the host, while the adult worms that engendered this response are unaffected." These investigators considered whether another term, "premunition," first used in 1924 by Sergent and his collaborators (Sergent, 1963), might not suffice for schistosomiasis. Premunition was coined to describe resistance in latent protozoan infections to superinfections by the same species

*This generalization has recently become complicated by the deliberate search among many inbred strains of mice for nonpermissive and high-responder strains. Such strains have been identified, which approach, and in some cases surpass, rats in these qualities (Murrell *et al.*, 1979*a*; Smith and Clegg, 1979; Mitchell *et al.*, 1981*b*). However, a parallel effort has not been made with rats.

(Sergent, 1963). Smithers and Terry (1969) held that "until it has been definitely established that the living worm is essential in order to maintain resistance, it is more profitable to look upon the existence of immunity in the presence of the living adult worm not as premunition but as concomitant immunity." As a matter of fact, later research has definitely established that immunity wanes and even disappears when the resident population of schistosomes is chemotherapeutically eliminated. This was shown for *Schistosomatium douthitti* in mice (Kagan and Lee, 1953) and for *S. mansoni* in mice (Doenhoff *et al.*, 1980; Andrade and Azevedo de Brito, 1982) and in rats (Phillips *et al.*, 1978*a*); it may also occur in baboons (R. T. Damian, unpublished data 1982). Despite these findings, the choice of a special term for schistosomiasis was a wise one, since primary infection schistosomes are hardly latent, in the sense that a hypnozoite, for example, is latent.

The evidence upon which the concomitant immunity hypothesis rested was the finding, in rhesus monkeys infected with *S. mansoni*, that egg production by the primary infection adult worms persisted even while a challenge infection was apparently being destroyed (Smithers and Terry, 1965*a*; Naimark *et al.*, 1960).

Observations consistent with concomitant immunity have now been made in many schistosome and host species combinations. These include *S. mansoni* in mice (Doenhoff *et al.*, 1978*a*), baboons (M. G. Taylor *et al.*, 1973*b*; Damian *et al.*, 1974), and rats (Knopf *et al.*, 1977; Phillips *et al.*, 1977); *S. haematobium* in baboons (Webbe and James, 1973; Webbe *et al.*, 1976); and *S. japonicum* in the rhesus monkey (Vogel and Minning, 1953; Hsü and Hsü, 1963). Evidence for the occurrence of concomitant immunity in human schistosomiasis haematobia was provided by McCullough and Bradley (1973), whereas Yogore *et al.* (1971) made observations in Filipinos infected with *S. japonicum* that were not inconsistent with concomitant immunity in that population.

In a later review, Smithers and Terry (1976) modified their original definition of concomitant immunity in schistosomiasis to read, "It describes a situation in which the host is resistant to reinfection but cannot at the same time rid itself of an established parasite population." Thus, Smithers and Terry's modified definition of concomitant immunity deemphasized a role for persisting adult worms in its maintenance. This was consistent with evidence that (1) immature female worms in unisexual *S. japonicum* or *S. mansoni* infections could stimulate resistance in rhesus monkeys (Vogel and Minning, 1953; Smithers, 1962; Hsü, 1969), and (2) primary infection with poorly or nonmaturing species of schistosomes in "abnormal" hosts could induce partial protection against challenge with pathogenic (egg-laying) species in nonhuman primates (Amin *et al.*, 1968; Eveland *et al.*, 1969; M. G. Taylor *et al.*, 1973*b*). Similar findings have been reported in other hosts (reviewed in Dean *et al.*, 1978*b*).

The experiments using nonhuman primates were also important in that they indicated that eggs were not necessary for the development of immunity in these hosts. This was partially confirmed by the unsuccessful attempts to immunize rhesus monkeys directly with eggs (Smithers, 1962; Smithers and Terry, 1967) and also by conferral of protection against challenge by surgical introduction of "half-worms" into naive rhesus monkeys (Smithers and Terry, 1967). This conclusion stands in sharp contrast to the situation in murine *S. mansoni* infections, in which bisexual egg-laying infections are necessary for the induction of immunity (Olivier and Schneidermann, 1953; J. H. Thompson, 1954; Dean *et al.*, 1978*a,b*; Bickle *et al.*, 1979*a*).

It is apparent from the work reviewed that juvenile or adult worms can induce concomitant immunity and that, at least in nonhuman primates, this is independent of egg-lying activity. The importance of earlier stages as immunogens has been minimized in these models by the results of surgical transfer experiments, in which the young, migrating stages were bypassed by direct introduction of adult worms into the portal system. Induction of a considerable degree of immunity by transplanted adult *S. mansoni* worms has been achieved in rhesus monkeys (Smithers and Terry, 1967), rats (Knopf and Cioli, 1980), and mice (Peresan and Cioli, 1980). A measure of protection against a homologous percutaneous cercarial challenge was also observed when *S. haematobium* worms (Webbe *et al.*, 1976), but not *S. mansoni* worms (Damian *et al.*, 1972), were transplanted into baboons. The latter failure may have been simply the result of inadequate size and/or residence time of the transplanted "dose," since these factors have been shown to be crucial for immunity induction in the murine surgical transfer system (cf. Boyer and Kalfayen, 1978; Peresan and Cioli, 1980). Thus, it is clear from transplantation experiments that juvenile and adult worms are immunogenic in both nonpermissive and permissive hosts. However, an immunogenic role for larval stages is not ruled out by these experiments.

B. Immunogenic Role of Larval Stages

That cercariae and schistosomula are also immunogenic was abundantly demonstrated by means of the use of irradiated cercarial and schistosomular vaccines, in a variety of hosts, for the induction of resistance (reviewed by Minard *et al.*, 1978*a*). Other approaches led to the same conclusion. For example, Phillips *et al.* (1977) elegantly demonstrated acquisition of functional immunity to *S. mansoni* by rats as early as 1 week after primary infection by means of adoptive transfer experiments.

Dean *et al.* (1981*b*) and Sher and Benno (1982) compared the abilities of various irradiated *S. mansoni* stages to stimulate immunity in mice. The former study found that cercariae and lung-stage schistosomula were about comparable (at least in one strain of mouse), and both were more immunogenic than 3- or

4-week-old worms. The results in the second report were somewhat at variance in that lung-stage schistosomula were less immunogenic than either cercariae or skin-stage schistosomula. Nevertheless, they stimulated resistance in three of six experiments; thus the difference in lung-worm immunogenicity noted by Sher and Benno (1982) was quantitative, not qualitative.

Unquestionably, living larval stages, either normal or irradiated, can stimulate resistance in both nonpermissive and permissive hosts.

C. Immunogenic Role of Eggs

1. Specific Acquired Immunity

Abortive infections with both "heterologous" and artificially attenuated larvae have shown that, for many hosts, oviposition and the presence of eggs in the tissues are not required for the development of acquired immunity.

Immunization with isolated eggs, egg incubates, or egg fractions also failed to give any large or consistent degrees of protection in a number of systems: in mice injected with eggs by various routes (Hunter and Crandall, 1962; von Lichtenberg *et al.*, 1963; Moore *et al.*, 1963; Bickle *et al.*, 1979*a*; Long *et al.*, 1980; Harrison *et al.*, 1982), in monkeys after intravenous egg injection (Smithers, 1962; Smithers and Terry, 1967), and in rats after injection of soluble egg antigen (SEA) (Phillips *et al.*, 1978*a*,*b*).

The relative inefficiency of eggs as sources of immunogens was also found to be true in carrier-priming experiments (Ramalho-Pinto *et al.*, 1976*a*). Of 11 antigens from various life-cycle stages that were used to prime mice, whole eggs and SEA were the poorest primers, as determined by the subsequent stimulation of an anti-TNP plaque-forming cell response by TNP-haptenated schistosomula.

2. Nonspecific Acquired Immunity

The finding that unisexual infections conferred no protection to subsequent challenge infections in mice (Dean *et al.*, 1978*a*; Bickle *et al.*, 1979*a*) prompted these two groups to reexamine the role of eggs in stimulating acquired resistance in murine schistosomiasis mansoni. Both groups reported strong correlations between the extent of tissue egg load and the degree of challenge infection reduction (Dean *et al.*, 1978*b*; Harrison *et al.*, 1982). Furthermore, Dean *et al.* (1978*b*) found that *intravenous* injection of eggs did provoke resistance, in contrast to other reports in which eggs were introduced by that route (von Lichtenberg *et al.*, 1963; Bickle *et al.*, 1979*a*; Harrison *et al.*, 1982). In further investigation of the relationship of tissue egg burdens and immunity, Dean *et al.* (1981*a*) made the significant discovery that the correlation of resistance was not to worm burdens or numbers of tissue eggs per se, but rather to the degree of portal pressure elevation with concomitant shunting of eggs to the lungs, and to the presence of egg granulomas in the lungs. These results indicate that, at least in some murine infections, eggs play a profound, although presumably nonspecific, role in acquired immunity. How they do this is still conjectural, although their presence in organs (i.e., the lungs) traversed by migrating schistosomula has suggested to several investigators that they might stimulate inflammatory reactions in which schistosomula are damaged or destroyed as "innocent bystanders" (M. A. Smith *et al.*, 1975; Dean *et al.*, 1978*b*; K. L. Miller *et al.*, 1981). M. A. Smith *et al.* (1975) reported experimental evidence that dead *Escherichia coli*-induced inflammation in hamster and mouse lungs was followed by decreased adult worm recovery from an infection with *S. mansoni* given 3 days before the bacteria. In addition, von Lichtenberg and Byram (1980) showed an inverse relationship between pulmonary granulocytic and monocytic reactions to schistosomula and innate or acquired resistance to *S. mansoni* in a range of different hosts.

3. Induction of Granulomatous Response

Despite a considerable degree of antigen sharing between the eggs of *S. mansoni* and other life-cycle stages (Kagan and Norman, 1963; A. Capron *et al.*, 1965; Sadun *et al.*, 1965), the *induction* of granulomatous hypersensitivity to *S. mansoni* eggs is a stage-specific phenomenon, since exposure of mice to irradiated cercariae, unisexual infections, or dead 4-week-old worms did not sensitize them to subsequent intravenous challenge with eggs (Warren and Domingo, 1970*a*). These investigators did find that egg sensitization was effective, for both *S. mansoni* and *S. haematobium* (Warren and Domingo, 1970*a*,*b*) but, oddly, not for *S. japonicum* (Warren and Domingo, 1970*b*). Later studies indicated a requirement for subcutaneous priming with eggs or SEA in order to obtain a granulomatous response around injected *S. japonicum* eggs (Warren *et al.*, 1975; Olds and Mahmoud, 1981), but the stage specificity of induction has apparently yet to be investigated for either *S. japonicum* or *S. haematobium*.

4. Conclusions

It is apparent from the foregoing that eggs and egg antigens are generally inefficient immunogens with respect to the induction of acquired resistance. It is also apparent that their overall effects on acquired resistance may nonetheless be great. Their immunologic roles appear to be twofold: (1) they play a role in the nonspecific expression of concomitant immunity, at least in some murine strain infections with *S. mansoni*; and (2) they specifically induce granulomatous hypersensitivity, a system of immune reactivity largely separate, except for nonspecific crossover phenomena, from anti-worm immunity (Section VIII).

D. Stage-Specific or Stage-Common Antigens?

In the immunologic sense, stage-restricted phenomena such as concomitant immunity must be traceable to either (1) the presentation to the host's immune system of stage-specific immunogens that could make their producers and bearers uniquely vulnerable to the response they set off, or (2) immune stimulation by immunogens common to two or more stages, but the stages expressing them are differentially sensitive to the resulting immune response.

Both stage-common and stage-specific antigens were disclosed by the inquiries into the antigenic composition of schistosomes—those using predominantly agar gel-diffusion techniques (Kent, 1963; Kagan and Norman, 1963; A. Capron *et al.*, 1965; Sadun *et al.*, 1965). These workers all agreed in showing that most antigens were shared among the stages, particularly between cercariae and adult worms. The extent of antigenic communion was generally less between eggs and the other stages, except in the work of Sadun *et al.* (1965), who found most of the cross-reactivity to be between the eggs and cercariae (Section IX.B).

More recent work using modern techniques of protein and antigen identification has confirmed and extended the early observations, particularly with respect to the striking parity between larval and adult stages (Murrell *et al.*, 1977; Ruppel and Cioli, 1977; Brink *et al.*, 1980; Snary *et al.*, 1980; Dissous *et al.*, 1981; Atkinson and Atkinson, 1982; M. A. Smith *et al.*, 1982; Butterworth *et al.*, 1982). The latter group, however, believe that, although present in older schistosomula, the stage-common antigens are not well expressed on the older parasite's surface (Butterworth *et al.*, 1982; D. W. Taylor and Butterworth, 1982). Common antigens between schistosomula and adult worms have also been revealed by absorption of antibodies cytotoxic for schistosomula by adult worm fractions (Clegg and Smithers, 1972; Sher *et al.*, 1974*a*; A. Capron *et al.*, 1975; MacKenzie *et al.*, 1977; M. Capron *et al.*, 1977). Antigens shared by cercariae and adult worms have also been shown by cross-absorption of *Cercarienhüllen-reaktion* (cercarial envelope reaction) antibodies (Kemp *et al.*, 1974). Several hybridoma antibodies have now been produced that react with epitopes borne on more than one life-cycle stage (M. A. Smith *et al.*, 1982; Zodda and Phillips, 1982, Aronstein *et al.*, 1983). Finally, Ramalho-Pinto *et al.* (1976*a*) showed the existence of cross-reacting carrier components in eggs, miracidia, cercariae, schistosomula, and adults of *S. mansoni.*

The relative antigenic disjunction of eggs is reconcilable with the known stage specificity of granuloma induction (Warren and Domingo, 1970*a*); modern immunochemistry has confirmed the stage specificity of some of the major soluble *S. mansoni* egg antigens (Hamburger *et al.*, 1976). Although the major proteinases of *S. mansoni* eggs (Asch and Dresden, 1979) and adults (Dresden and Deelder, 1979) are both thiol-dependent, they are biochemically different. A monoclonal antibody produced against the egg proteinase indicates some antigenic differences with the adult worm enzyme (Dresden *et al.*, 1983).

Unique immunogenic roles for eggs are also indicated by the stage specificity of induction of antibodies that can (1) prevent putative egg-induced hepatotoxicity in *S. mansoni*-infected mice, and (2) enhance the rate of egg excretion in heavily infected T-cell-deprived mice (Doenhoff *et al.*, 1981; Dunne *et al.*, 1981).

But stage-specific antigens are also found in cercariae and adult worms. Some of these are associated with unique larval or adult features. The cercaria loses its tail, its glycocalyx, and its penetration gland contents when it is transformed into a schistosomulum. As it progresses to adulthood, the schistosomulum gains its outermost membrane and tubercle spines, its rudimentary gut fully differentiates, and its reproductive organs mature. Example of structurally associated stage-specific schistosome antigens follow: The major proteinases of *S. mansoni* cercariae and adult worms are biochemically distinct, the cercarial enzyme being a serine proteinase (Landsperger *et al.*, 1982) while the adult enzyme is thiol-dependent (Dresden and Deelder, 1979). Preliminary results indicate that they are also antigenically distinct (M. H. Dresden, personal communication 1982). Cercariae have unique proteins, perhaps glycocalyx associated (Kusel, 1972; Ruppel and Cioli, 1977; Cordeiro and Gazzinelli, 1979; Snary *et al.*, 1980), which may be their unique antigens (Kemp, 1970; Bogitsh and Katz, 1976) or their alternate pathway complement activators (Machado *et al.*, 1975). According to preliminary data, a major 58,000-dalton con A-binding surface glycoprotein of adult *S. mansoni* (Hayunga *et al.*, 1983) may be a stage-specific antigen (E. G. Hayunga, personal communication). Mitchell *et al.* (1981*a*) have reported on a monoclonal antibody that reacts with an adult worm extract, but not an egg extract, of *S. japonicum.** Two polysaccharide antigens that circulate in host blood have been characterized as originating from the ceca of adult worms. These are the gut-associated proteoglycan (GASP) of Nash *et al.* (1981) and the M antigen of Carlier *et al.* (1980). Although stage distribution has apparently not yet been thoroughly delineated, one might expect their presence to be restricted to stages with at least a rudimentary gut. In fact, von Lichtenberg *et al.* (1974) found GASP to be absent from eggs by immunofluorescence, and Andrade and Sadigursky (1978) found epitopes cross-reactive with circulating schistosomal antigens on cercarial and schistosomular gut primordia, but not within any developmental stages in eggs, also by immunofluorescence.

Thus, stage-specific antigens do occur, and their identification and characterization will continue to be an important endeavor.

Perhaps greater emphasis should be placed instead on the extensive degree of antigenic similarity among cercariae, schistosomula, juvenile, and adult worms. Indeed, the evidence argues that all postmolluscan stages of the schistosome,

*Norden *et al.* (1982) produced a monclonal antibody specific for a spine glycoprotein of schistosomula and adults of *S. mansoni*. Interesting, cercarial, but not egg, extract contains a cross-reactive molecular of lower molecule weight.

except the eggs, can stimulate some measure of specific acquired resistance. Therefore, the shared antigens may indeed be the most important ones for stimulating immunity to schistosomes (Peresan and Cioli, 1980). Indeed, a monoclonal IgM antibody that detects a surface epitope common to *S. mansoni* skin schistosomula, adult worms, and miracidia was able to confer partial protection against a challenge infection, after its passive transfer into mice (M. A. Smith *et al.*, 1982). Zodda and Phillips (1982) have produced a protective IgG_1 monoclonal antibody that precipitates 96,000- and 160,000-dalton antigenic moieties from cercarial membranes and a 33,000-dalton moiety from egg antigen. But on a cautionary note, stage cross-reactivity of monoclonal antibodies says nothing of itself about the relative immunogenicity of the shared antigens during an infection.

IV. VULNERABLE STAGES

A. Concomitant Immunity and the Vulnerability of Larval Stages

The concomitant immunity hypothesis (Smithers and Terry, 1969) suggested that the stages subject to immune attack were the invading schistosomula of secondary infections. Subsequent work was highly consistent with this concept (Smithers *et al.*, 1977). More recently, two experimental approaches have yielded data directly suporting the concomitant immunity hypothesis, at least in rodents. In the first case, early perfusion of the hosts, 2–4 weeks after challenge infection, is accomplished. Any new invaders breaking through the immunity barriers would theoretically be distinguishable from primary infection worms by their size and immaturity. This approach was introduced by Knopf *et al.* (1977), who found that the number of mature worms was similar in both once- and twice-infected rats. The same procedure yielded an identical conclusion in murine *S. mansoni* infections (Doenhoff *et al.*, 1978*a*; Colley and Freeman, 1980). In the second approach, Phillips *et al.* (1977) used endogenously radiolabeled cercariae to challenge once-infected rats. These retained enough label through their subsequent development to permit their positive identification by autoradiography. These workers found that the challenge infection was reduced by about 90% and that survival of the primary infection was virtually unaffected. These studies broadly confirmed the concomitant immunity hypothesis and indicated that the invading parasites of a secondary infection are indeed highly vulnerable. The questions of just when, where, and how these organisms are destroyed have been among the most intensively investigated in this field during the last decade.

1. *When are Schistosomula Most Vulnerable?*

a. In Vitro Studies. A large body of literature has accumulated on the *in vitro* vulnerability of schistosomula to specific and nonspecific immune effectors. The subject has been reviewed extensively (Butterworth, 1977*a,b*; A. Capron *et al.*, 1977, 1982; Phillips and Colley, 1978; McLaren, 1980; Incani and McLaren, 1981; Colley, 1981*b*; Butterworth *et al.*, 1982). Evidence is accumulating that antibody-dependent cell-mediated cytotoxicity mechanisms involving anaphylactic antibodies and phagocytic cells may be the main mediators of *in vivo* schistosomular death (A. Capron *et al.*, 1982). Recently Ellner *et al.* (1982) showed *in vitro* killing of schistosomula by T lymphocytes after their activation by mitogens, schistosome antigens, or alloantigens. Other nonspecific mechanisms involving complement, phagocytes, and natural killer cells appear to have minor roles. More germane to the present discourse are studies concerned with the timing of the development of susceptibility or insusceptibility of schistosomula to the cytotoxic mechanisms.

With respect to the present question, studies showed that 4- or 5-day-old lung-stage schistosomula or their similarly aged *in vitro*-maintained counterparts had become highly refractory to *in vitro* cytotoxicity mediated by the alternate complement pathway (Santoro *et al.*, 1979*a*), by hyperimmune rhesus monkey antibodies and complement (Clegg and Smithers, 1972), by rat eosinophils and complement (Ramalho-Pinto *et al.*, 1978), or by rat neutrophils, antibody, and complement (Incani and McLaren, 1981). Refractoriness to cytotoxic mechanisms was coincident with the presence of host antigens (McLaren *et al.*, 1975, 1978; Goldring *et al.*, 1977) and the absence of antibody and complement binding ability (McLaren *et al.*, 1975, 1978; Samuelson *et al.*, 1980). Later studies showed this refractoriness to develop in as little as 24–48 hr in culture after transformation from the cercarial stage (Dessein *et al.*, 1981; McLaren and Incani, 1982). It is now known that schistosomula begin to develop their insusceptibility to cytotoxic mechanisms almost as soon as they penetrate the skin, since 3-hr-old schistosomula, *recovered from skin*, had already manifested considerable resistance to eosinophil- and complement-mediated killing *in vitro* (Bickle and Ford, 1982).

In vitro experiments therefore show that it is the newly transformed schistosomulum that is maximally vulnerable to cytotoxic mechanisms. Thereafter, it rapidly becomes refractory to attack.

b. In Vivo Studies. Dean *et al.* (1978*b*) examined resistance in mice induced by bisexual *S. mansoni* infections by comparing lung schistosomula and adult worm recoveries after challenge. These workers found that in 6- or 8-week-old primary infections, the vulnerable stages were older than 6 days. More information relative to a prolonged period of vulnerability was gained by transplanting various stages into their "correct" locations in either chronically immune or vaccinated mice. Blum and Cioli (1981) and Dean *et al.* (1981*b*) generally agreed in showing that the critical age at which the worms became refractory to immune elimination in the host was at or after 12 days in either model; invulnerability was well established by 12–20 days of age. These results are at variance with the conclusion drawn from the *in vitro* experiments, which showed refractoriness to be virtually complete by 5 days. But one *in vivo* study was in close agreement with the *in vitro* result: K. L. Miller *et al.* (1981) found 5-day lung schistosomula to be protected after intravenous injection into vaccinated mice, in contrast to the results from similar experiments by Dean *et al.* (1981*b*). The reasons for this discrepancy are unknown, but could be related to the different vaccinating doses used (as discussed in Section V).

2. *Where Are Schistosomula Killed?*

Knowledge of the exact sites of preadult worm death within the host has been elusive. Part of the difficulty has stemmed from uncertainties in the migration route taken by the worms from the skin to their definitive intravascular location (Wilks, 1967; Kruger *et al.*, 1969; Pereira *et al.*, 1972; Bruce *et al.*, 1974; Wilson *et al.*, 1978; Wheater and Wilson, 1979; P. Miller and Wilson, 1980). The most recent work has favored the hypothesis of an entirely intravascular migration route (Wheater and Wilson, 1979; P. Miller and Wilson, 1980; Dean *et al.*, 1982). Another difficulty has been that sites of parasite death have had to be inferred from viable larva recovery patterns from various organs, permitting only minimum estimates of the total population (Georgi, 1982). Also, direct histologic evidence of larval death has not permitted determination of the relative importance of the various implicated sites. Finally, one recovery method, in which viable schistosomula are allowed to crawl out of the chopped lungs (Clegg, 1965), was adapted for use as an assay for immunity (Perez *et al.*, 1974; Sher *et al.*, 1974*b*; Lewis and Colley, 1977). Interpretations using this method are questionable, however, since various types of artifacts are now known to exist in this system (Dean *et al.*, 1978*b*; Minard *et al.*, 1978*b*; Blum and Cioli, 1981).

Therefore, the hypothesis of dual immunity in murine models (Sher *et al.*, 1974*b*), in which early-phase attrition and late-phase attrition were distinguished (Smithers and Gammage, 1980) and described as differentially expressed in chronic and vaccinated murine models (K. L. Miller and Smithers, 1980; K. L. Miller *et al.*, 1981) may require reassessment. Successful labeling of cercariae with radioselenium (Knight *et al.*, 1968; Christensen, 1977; Georgi, 1982) and application of new methods to locate and quantify schistosomulum-bound label within the host (Georgi, 1982) are potentially highly useful advances in this area. They are currently being applied to determine the fate of a challenge infection in immune mice (Dean *et al.*, 1982).

3. *Conclusions*

Worm age and location appear to be critical factors in larval vulnerability. *In vitro* studies suggest almost immediate transformation to refractoriness after skin penetration, with lung stages being particularly hardy. *In vivo* studies tend to support the idea of a longer period of larval vulnerability. Further work using methods for precisely locating in space and time the death of radiolabeled schistosomula *in vivo* should clarify this question.

B. Immune Evasion

The other side of the coin of concomitant immunity is the protected status of primary infection worms. The catch phrase that has come best to signify this phenomenon is evasion of the immune response (Terry and Smithers, 1975; Ogilvie and Wilson, 1976).

Two distinct sets of theories of immune evasion in schistosomiasis have predominated: (1) the host antigen theories, and (2) the schistosome-induced suppression theories. These theories are currently being supplemented (or supplanted) by a third idea—the intrinsic insusceptibility theory (Simpson and Cioli, 1982).

The host antigen theories have taken various forms: antigenic mimicry (Sprent, 1962, 1963; Dineen, 1963*a,b*; Damian, 1962, 1964), the shared antigen induction hypothesis (A. Capron *et al.*, 1968), and the antigen acquisition or antigen masquerade hypothesis (Smithers *et al.*, 1968, 1969). The differences and similarities among these ideas have been discussed (Smyth, 1973; Damian, 1979). In schistosomiasis, the acquired host antigen hypothesis has received the greatest experimental support, although evidence for antigen mimicry has also been gained (Damian *et al.*, 1973; Bout *et al.*, 1974; Kemp *et al.*, 1976). They are not mutually exclusive.

As touched upon in Section IV.A, the correlations between the presence of host antigens and the *in vitro* inability to bind antibody and complement and the insusceptibility to cytotoxicity mechanisms have lent much circumstantial support to the concept that host antigen acquisition is responsible for the increasingly privileged status of maturing schistosomula (reviewed by Clegg, 1974; Smithers *et al.*, 1977; Phillips and Colley, 1978; Damian, 1979; McLaren, 1980; Colley, 1981*b*). More recently, a direct approach has been taken in attempts to settle the question of whether host macromolecular uptake is responsible for insusceptibility to cytotoxic mechanisms, with very interesting results. In fact, as is common in scientific inquiry, such "definitive" experiments raised more questions than they answered. The approach was a simple one and consisted of maintaining the organisms in media free of host components (Dean, 1977; Tavares *et al.*, 1978, 1980; Samuelson *et al.*, 1980; Dessein *et al.*, 1981; McLaren and Incani, 1982). The answer obtained was that schistosomula intrinsically became refractory during serum-free medium incubation. Therefore the antigen masquerade hypothesis of invulnerability was not upheld. However, the intriguing additional result was that added serum factors enhance the ability of schistosomula to protect themselves (Tavares *et al.*, 1978, 1980; Dessein *et al.*, 1981; McLaren and Incani, 1982). Tavares *et al.* (1978) investigated the nature of the stimulatory factor in rabbit serum and concluded that it is a non-immunoglobulin 7-19S macromolecule.

Using an entirely different approach, Moser *et al.* (1980) have also obtained evidence for intrinsic changes in the schistosomular outer membrane, independent of host antigen acquisition, being responsible for the developing refractoriness of the schistosomulum. These investigators attempted to override any potential protective function of acquired host antigens by haptenating larvae with trinitrophenyl, then comparing the survival of haptenated skin- and lung-stage larvae in both *in vivo* and *in vitro* anti-TNP situations. They found that lung-stage schistosomula remained resistant to the cytotoxic mechanisms, arguing again for some intrinsic structural change as the basis for the effect.

The structural changes themselves are currently under investigation in several laboratories. Many studies are pointing to membranal lipid modifications and modulation by serum components as factors in the development of intrinsic insusceptibility (Simpson and Cioli, 1982).

There is recent evidence, however, from an *in vitro* anti-host system, that 3-week and older mouse worms may derive their protection from the cytotoxic effects of rat anti-mouse erythrocyte antibodies and rat eosinophils by acquiring host antigens (McLaren and Terry, 1982). These investigators believe that host antigen masquerade is the major evasive tactic of the older worm.

One of the most intriguing observations possibly related to host antigen acquisition was provided by Caulfield *et al.* (1980*a*,*b*, 1982). These workers found that, under certain conditions, schistosomula could actually fuse their surface membranes with the surface membranes of human neutrophils and eosinophils. These fusions were promoted by lectins or anti-parasite antibodies and complement, which were thought to act by bringing the heterologous membranes into close enough apposition for parasite fusigens to induce the membrane coalescence (Caulfield *et al.*, 1982). If this process also occurs *in vivo*, then a mechanism for host antigen acquisition is provided.

The mimicry and masquerade hypotheses of immune evasion originally assumed a passive role for the parasite-associated host antigens—either an overall reduction in immunogenicity or else protection from effector binding. Reconsideration of these assumptions was necessitated by the discovery on schistosomes of host or hostlike molecules having known immunoregulatory roles in their normal environments (Damian, 1979). These include α_2-macroglobulin (Damian *et al.*, 1973; Kemp *et al.*, 1976; Hubbard and Damian, 1976), murine MHC antigens (Sher *et al.*, 1978; Sher and Moser, 1981; Gitter and Damian, 1982; Gitter *et al.*, 1982; Simpson *et al.*, 1982), and immunoglobulins bound to the tegumental surface with their idiotypes exposed (Kemp *et al.*, 1977; Torpier *et al.*, 1979*a*; Tarleton and Kemp, 1981).

Expression or display of immunoregulatory molecules on the parasite surface could possibly lead to active intervention in the host's immune response. For this to occur, however, one might imagine that rather stringent topochemical requirements must be met. There is as yet no experimental evidence to suggest that MHC gene products or other immunoregulatory-type molecules are arranged on the schistosome's tegument in such a way as to promote any host immune function. They fail to elicit cytotoxicity either *in vivo* (Damian *et al.*, 1973; Boyer *et al.*, 1976; Cohen and Eveland, 1980) or *in vitro* (Butterworth *et al.*, 1979). Gitter *et al.* (1982) attempted to stimulate naive, allogeneic murine lymphocytes with schistosomula bearing alloreactive H-2 antigens in a system analogous to a mixed lymphocyte reaction. The acquired H-2 antigens were incapable of stimulating the lymphocytes (Fig. 4), indicating that their display on the schistosomular surface was in some way different from their normal display on lymphocytes. This would seem to make any immune interventionary

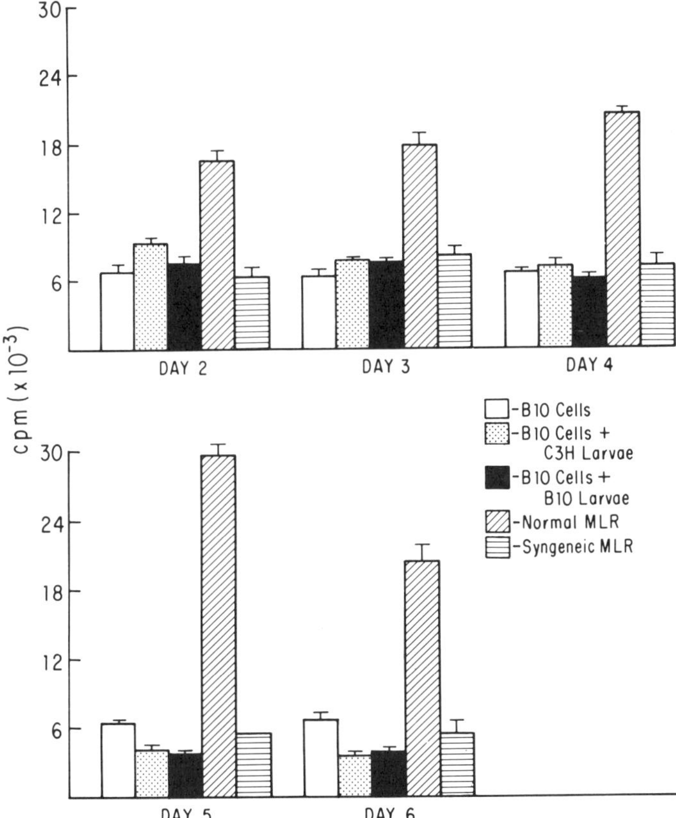

Figure 4. Inability of *S. mansoni* lung-stage schistosomula from C3H(H-2^k) mice to stimulate normal C57BL/10 (H-2^b) lymphoid cells in a mixed lymphocyte-type reaction. Graph illustrates [^{3}H] thymidine uptake by 2×10^5 C57BL/10 lymphoid cells in the presence of 2×10^5 mitomycin-C treated C57BL/10 or C3H spleen cells, or with 50 schistosomula recovered from either strain of mouse. (With permission from Gitter *et al.*, 1982.)

role highly unlikely, at least while these molecules are *in situ* on or in the outer membrane.

However, the fact that the tegument is a dynamic structure, constantly being sloughed off and renewed (Kusel *et al.*, 1975; Wilson and Barnes, 1977; Torpier and Capron, 1980), suggests that *release* of immunoregulatory-type molecules may be the way in which they could function. This may seem to be an odd process for acquired host molecules. Why should they be bound only to be released? However, it is possible that they may be altered in some way before or as they are released, as is known to occur in the binding and release of IgG by *S. mansoni* schistosomula, through a process of proteolytic cleavage of the IgG molecules (Auriault *et al.*, 1981). Peptides resulting from proteinase action often have immunomodulatory roles, as is well illustrated by the complement system (Lachmann, 1979). Indeed; the IgG peptides released from schistosomula have already been shown to inhibit macrophage activity (Auriault *et al.*, 1980). Another way in which acquisition and subsequent release of host molecules could have survival value to the schistosome would be in locally concentrating certain molecules, as could well occur in local hemostasis resulting from vessel occlusion during oviposition. Finally, bound host molecules could be released as complexes, much like *in vitro* antigen modulation (Kemp *et al.*, 1980). Circulating immune complexes are known to be associated with schistosomiasis, and evidence for the participation of schistosome antigens in these complexes exists (Bout *et al.*, 1977; Santoro *et al.*, 1981). It has been suggested that complexes between regurgitated host α_2-macroglobulin and worm proteinases may subserve an evasive function in schistosomiasis (Hubbard, 1978). Evidence that murine α_2-macroglobulin-like molecules are both synthesized and acquired in mice was presented by Damian *et al.* (1973). Hubbard *et al.* (1981) later provided evidence of an immunoregulatory role for proteinase-complexed α_2-macroglobulin, but this finding has not yet been related to schistosomiasis.

Immunoregulatory effects of immune complexes are manifold (Theofilopoulis, 1980). In schistosomiasis, there is evidence that they may block complement-dependent cytotoxic antibodies (M. Capron *et al.*, 1980). Another way in which immune complexes could exert an immunoregulatory role is through the stimulation of anti-immunoglobulin antibodies (Goidl *et al.*, 1979). In this regard, the report by Wistar *et al.* (1975) on the development of anti-immunoglobulin antibodies in rhesus monkeys repeatedly exposed to *S. japonicum* is of considerable interest. This may have relevance for human infections, since Wilkins and Brown (1977) found increased rheumatoid factor incidence among Gambians living in an area endemic for schistosomiasis haematobia.

Substances of apparent *parasite* origin are known to perturb host immune responses (reviewed by A. Capron and Camus, 1979). Several of these substances, derived from schistosomes, have been described, including schistosome-derived low-molecular-weight inhibitors of rat and human lymphocyte proliferation

(Dessaint *et al.*, 1977; Camus *et al.*, 1981) and *in vivo* rat delayed-type hypersensitivity (Camus *et al.*, 1981), adult worm membrane fraction inhibitors of *in vivo* anti-SRBC responses in mice (Mota-Santos *et al.*, 1977), and human suppressor cell generation *in vitro* by *S. mansoni* soluble antigen preparations (Colley *et al.*, 1978). These substances are extremely interesting in their own right, but in the present context, it is not known whether any of them bears any biochemical or antigenic relationship to host molecules.

Of the various hypotheses advanced to give host antigens some role in evasion, none has met with overwhelming success. Unresponsiveness in the sense of nonrecognition in immune induction (Damian, 1964) is difficult to prove. Unresponsiveness in the sense of effector blockade (Smithers and Terry, 1969) has met with limited success, only in the case of the highly artificial *in vivo* (Smithers *et al.*, 1969; Erickson *et al.*, 1973; Cioli, 1976; Boyer *et al.*, 1977; Lewert *et al.*, 1977) or *in vitro* (Perez and Terry, 1973; Torpier *et al.*, 1979*b*; McLaren and Terry, 1982) anti-host systems. It may be that host antigens have no biologic role, but this too seems unlikely in view of the recent demonstrations of specific host antigen receptors on schistosomes (Torpier *et al.*, 1979*a*; Santoro *et al.*, 1979*b*, 1980; Ouaissi *et al.*, 1980; McGuinness and Kemp, 1981; Tarleton and Kemp, 1981). The possibility must be admitted that their true functions are yet to be discovered. This answer too must await the results of the intense inquiry being mounted on the biochemistry, modulation, and maturation of the surface membrane.

C. Adult Worm Vulnerability

1. Adult Worm Death and Self-Cure

The concomitant immunity hypothesis has been broadly confirmed by many investigations showing both relative adult worm refractoriness and young-stage susceptibility to immune attack. The purpose of this section is to reexamine the question of adult worm vulnerability. Is it as absolute as much of the current literature seems to imply? Or are adult worms also subject to immune attack in their hosts?

As reviewed in Section III.A, the original evidence upon which the concomitant immunity hypothesis was based was the persistence, at constant levels, of fecal egg excretion even after a large challenge infection that obviously failed to establish in rhesus monkeys (Smithers and Terry, 1969). We suggested that this same pattern of egg excretion could also occur if a steady state or threshold of adult worm burden were permitted in an immune host (Damian *et al.*, 1974). Thus, some death in the primary infection worm population would be compensated for by a comparable degree of recruitment from the new, invading population. The studies in rodents using early perfusion after a challenge to separate primary from secondary infection worms on the basis of their size and maturity (Knopf *et al.*, 1977; Doenhoff *et al.*, 1978*a*; Colley and Freeman, 1980) broadly supported concomitant immunity. Yet they failed to negate the steady-state hypothesis, since there is a possibility that all worms may not reach their definitive location synchronously if recirculation in the vascular system truly occurs (P. Miller and Wilson, 1980). This would complicate any simple interpretation. Similarly, the report by Phillips *et al.* (1977), in which challenging cercariae had been radiolabeled, also confirmed concomitant immunity, but without proving the absolute refractoriness of established primary infection worms. The fecal egg excretion pattern of previously infected baboons reexposed to cercariae is similar to that of rhesus monkeys and is consistent with concomitant immunity (Damian *et al.*, 1974, 1976*a*). Although adult worm death is not a typical or characteristic feature in immune baboons (Damian *et al.*, 1976*a*; Sturrock *et al.*, 1976), we nevertheless did find portal endophlebitis in one baboon (Damian *et al.*, 1976*a*), a condition associated with recent adult worm destruction in rhesus monkeys (Cheever and Powers, 1972). Cheever (1965) found a few dead worms in the livers of mice with *S. mansoni* infections of long duration. Living worms were simultaneously present in these mice. In three mice, examined 60–85 weeks after infection, no worm pairs were evident in the tissues. Kloetzel (1967) found numerous dead worms in mouse infections of lengthy duration. Mice of the A/J strain lose primary infection adult worms with time (D. A. Dean, personal communication, 1982). The question of the invulnerability of adult worms remains open, even in highly permissive hosts such as baboons and mice.

In nonpermissive species, the early death of juvenile or adult worms of the primary infection (often called "self-cure") is well known. *Schistosoma mansoni* infections in rats have been most intensively studied in this respect. In this system most of the worms die between the fourth and sixth or eighth week of infection (Smithers and Terry, 1965*b*; Maddison *et al.*, 1970*a*; Phillips *et al.*, 1975; Cioli and Dennert, 1976; Cioli *et al.*, 1978). Although earlier experiments involving immunosuppression and transfer of cells and serum failed to support an immune elimination hypothesis (Maddison *et al.*, 1970*b*), later work using adoptive transfer (Phillips *et al.*, 1975), and T-cell deprivation (Cioli and Dennert, 1976), did prove immune system involvement in worm elimination. The report by Phillips *et al.* (1975) was particularly instructive on this point, since a synergism of innate and immune mechanisms in self-cure was indicated.

Adult worms gradually die in *S. japonicum* infections in rabbits, with about two-thirds of the worms surviving the first year of infection (Cheever *et al.*, 1980). Rhesus monkey infections with *S. japonicum* and *S. mansoni* may also be categorized as nonpermissive systems, in which self-cure occurs. This interpretation was usually made on the basis of decreased fecal egg excretion, which does not, however, distinguish by itself, between the effects of worm death and suppressed oviposition. But several reports do show decreasing worm burdens with time (P. E. Thompson *et al.*, 1965; Cheever and Powers, 1969, 1972).

Thompson's report indicated that during the first 35 weeks after heavy primary infection, the decrease in egg excretion had a strong positive correlation with the death of worms. But the later report by Cheever and Powers (1972) provided strong evidence that a decrease in oviposition rate among the surviving adult worms occurred at some time after the 15th week of infection. The data of P. E. Thompson *et al.* (1965) are not incompatible with this interpretation.

2. *Condition of Surviving Worms after Self-Cure*

Meleney and Moore (1954) found residual *S. mansoni* worms in an immune rhesus monkey to be "abnormally short." Moreover, according to these workers, "some of the females did not contain an egg and showed atrophy of the ovary and vitelline glands, similar to the atrophy seen in worms from mice following treatment with an antimony compound." These observations were particularly interesting in view of the fact that this monkey's stools had been negative for *S. mansoni* eggs for at least 6 months before its death, and that examination of the tissues revealed only "minimal pathology and [that] most of the eggs were mere shells."

Ritchie *et al.* (1967) made the prescient suggestion that worms may regress in size in their hosts. Unfortunately, their measurements of rhesus monkey worms did not support their hypothesis, since the monkeys had been twice reinfected. Such a phenomenon has more recently been qualitatively described for *S. japonicum* in rabbits (Cheever *et al.*, 1980).

Cioli *et al.* (1977) stated that 8-week-old mouse worms, surgically transferred into rats, regressed in size, without giving data. Size regression must be distinguished from "stunting," an often-used term in the literature of schistosomiasis. Stunting refers to the growth lag of worms in nonpermissive (Cioli and Dennert, 1976) or immune (Vogel, 1962; Sadun, 1963; Knopf *et al.*, 1977; Doenhoff *et al.*, 1978*b*) hosts, or to some effect associated with overcrowding (Peña de Grimaldo and Kershaw, 1961; Lennox and Schiller, 1972; Bruckner and Schiller, 1974; Coelho *et al.*, 1976), or to the failure of female worms to grow and mature in the absence of males (Vogel, 1942). Size regression may be absent over long periods of time in permissive hosts, i.e., *S. mansoni* in *Cercopithecus* monkeys (Ritchie *et al.*, 1967). Newer, more sophisticated measurements indicate slight but significant size reductions in female *S. mansoni* and *S. japonicum* worms recovered from infections of $>$ 90 days' duration in mice (Cornford *et al.*, 1982).

3. *Antifecundity Effects*

Reduced fecundity, dissociable from adult worm death, is now recognized as occurring in various experimental and natural schistosome infections, including *S. mansoni* in rhesus monkeys (Cheever and Powers, 1969), *Cercopithecus* monkeys (Meisenhelder and Thompson, 1963; Cheever and Duvall, 1974), and baboons (Damian *et al.*, 1976*a*); *S. japonicum* in rhesus monkeys (Vogel and Minning, 1953; Murrell *et al.*, 1973; Cheever *et al.*, 1974) and rabbits (Cheever *et al.*, 1980); *S. haematobium* in baboons (Webbe *et al.*, 1976); and *S. bovis* (Bushara *et al.*, 1980, 1983) and *S. mattheei* (Lawrence, 1973) in cattle. Studies on humans infected with *S. mansoni* (Cheever, 1968) or *S. haematobium* (Cheever *et al.*, 1977) have yielded no evidence that reduced worm fecundity occurs in humans. However, as pointed out by Cheever *et al.* (1980), most of the people necropsied in these studies had chronic infections, so that a possible earlier phase of higher fecundity could have been missed. Therefore, the results from human infections are inconclusive in this regard.

At least three hypotheses can explain reduced fecundity: (1) crowding effects, (2) worm senescence, and (3) host effects (Damian *et al.*, 1976*a*). Although little hard experimental evidence exists to support any of these, there is evidence that tends to militate against the first two. With respect to parasite crowding, Cheever and Duvall (1974) have remarked on the common occurrence of decreased oviposition in monkeys with heavy or light infections, it does occur sooner in heavy infections, however. Worm senescence also appears unlikely, for two reasons. First, since reduced oviposition occurs more rapidly in rhesus monkeys than in baboons, it is difficult to imagine why *S. mansoni* should age more rapidly in the former host than in the latter (Damian *et al.*, 1976*a*). Second, it is also difficult to imagine why worms should age more rapidly in heavy infections (Cheever and Powers, 1972). Direct support for the conclusion that senescence is relatively unimportant comes from the demonstration that *S. haematobium* worms showing decreased fecundity in baboons resumed greater egg-laying activity after transplantation into naive baboons (Webbe *et al.*, 1976). A similar result has also been recently obtained with *S. bovis* in cattle (Bushara *et al.*, 1983). Therefore, the third hypothesis, that of host effects, emerges as the one most likely to be true. The host immune response has been singled out by several investigators as the probable agent (Foster and Broomfield, 1971; Damian *et al.*, 1976*a*; Webbe *et al.*, 1976; Cheever *et al.*, 1980; Bushara *et al.*, 1980, 1983).

Evidence favoring immunity as the agent of fecundity suppression is both circumstantial and direct. Circumstantial evidence includes the apparent dose effect in the rate of appearance of fecundity reduction (Cheever and Duvall, 1974) and the coincident phenomena of fecundity reduction, anterior shift, and immunity found by Damian *et al.* (1976*a*) in baboon schistosomiasis mansoni. We suggested that the reduced oviposition we found to occur in chronically infected baboons was a consequence of host immunity, perhaps the result of slow starvation of the worms in a partially immune host. In this scenario, the adult worms, "confronted with immune interference to absorption of glucose or other essential nutrients, might respond by migrating to the nutritionally richer environment of the small intestinal mesenteric veins from the colonic vessels." The so-called "anterior shift" is an interesting phenomenon. It is a change in location of oviposition, known to occur in *S. mansoni* infections in presumably immune rhesus and grivet monkeys and in baboons (Cheever and Powers, 1969; Foster and Broomfield, 1971; Cheever and Duvall, 1974; Damian *et al.*, 1976*a*) and perhaps in human schistosomiasis mansoni as well (Cheever, 1968). The anterior shift is an actual change in location of the ovipositing adult worms (Cheever and Powers, 1969) and may be an example of an immunity-induced behavioral change in schistosomes (Damian, 1982).

Additional circumstantial evidence for immune system involvement in oviposition perturbations was provided by Bruce *et al.* (1966), who noted that splenectomized rhesus monkeys passed more *S. mansoni* eggs in the feces than did intact monkeys, even though their worm burdens were similar. However, Cheever and Powers (1969) also found higher fecal egg excretion in splenectomized rhesus monkeys, but attributed this to their worm burdens, which were higher than those of intact animals. Moreover, Maddison *et al.* (1971) could not alter the fecal egg excretion rate in rhesus monkeys infected with *S. mansoni* by treatment with anti-lymphocyte globulin and/or cyclophosphamide. Still, the possibility that more thorough immunosuppression could increase egg production is suggested by an interesting observation made by Vogel (1962) on *S. japonicum* in rhesus monkeys. A female monkey, made immune by repeated unisexual (male) cercarial exposures followed by a bisexual challenge infection, was shown to be completely negative for fecal egg passage after the bisexual challenge. Thirty-three days after a second, large bisexual challenge, this monkey began to pass eggs in its feces. Apparently unknown to Vogel, this female, caged with males, had become pregnant at some time before the second challenge infection, since she aborted a still-born *Junges* 2 weeks after eggs had first appeared in the stools. Eggs continued to be passed for 2.5 months, then ceased for an ensuing 11 months. Two explanations for this observation seem reasonable, both based upon a pregnancy-induced state of immunosuppression. Either a strong degree of concomitant immunity was abrogated, enabling the second bisexual challenge infection to mature, or else suppressed worms from the first bisexual infection resumed oviposition during the host's pregnancy. In this regard, Damian *et al.* (1976*a*) did find some evidence for a dissociation between concomitant immunity and reduced oviposition in baboon *S. mansoni* infections.

Vaccination trials in nonhuman primates with irradiated larvae (Murrell *et al.*, 1979*b*; Stek *et al.*, 1981) and in mice with cross-reacting *Fasciola hepatica* antigens (Hillyer and Serrano, 1982) also provide circumstantial evidence of this nature, in that vaccinated-challenged animals had either reduced fecal or tissue egg counts when compared with controls.

More direct evidence for the immunity hypothesis was recently provided by Bushara *et al.* (1983). Using *S. bovis* in cattle, these investigators showed passive transfer of immune serum into recipient calves to result in a reduction in tissue egg counts after percutaneous challenge, although worm burdens and fecal egg counts were unaffected.

Thus, both circumstantial and direct evidence favors the hypothesis of immune-mediated antifecundity in schistosomiasis, at least in some hosts. It is not clear whether this phenomenon occurs in the much-investigated murine-*S. mansoni* model. Two investigators failed to find any evidence for worm fecundity suppression in mice (Kloetzel, 1967; Cheever, 1969). Doenhoff *et al.* (1978*b*) found that *S. mansoni* infections in T-cell-deprived mice were characterized by markedly lower egg counts in both feces and tissues, even though worm burdens were similar to those in intact mice. Egg production was partially restored by injection of chronic infection serum into the deprived mice. Although initially favoring the hypothesis that antibodies in chronic infection serum acted on the adult worms to *increase* their fecundity, Byram *et al.* (1979) later ascribed this result to a developmental lag of the worms in the deprived mice. This could have led to the reduced egg production observed within the time span of their experiments. More recently, Harrison *et al.* (1982) obtained the intriguing result that an excess of either male or female worms in murine *S. mansoni* infections, in a dose-dependent way, inhibited the rate of egg laying by the paired worms. It is not known whether this phenomenon is in any way related to the immune response.

4. Conclusions

Undoubtedly, juvenile and adult worms of primary infections, even in permissive hosts, are subject to a certain level of harassment. The protected status of established worms is not absolute. The adverse effects vary from the subtle to the severe, even including death of the worms. The most interesting of these effects is reduction in fecundity, which could well be related to starvation, manifested as size reduction and sexual organ atrophy. Reduced fecundity most likely is an effect attributable to the host immune response, but this remains to be unequivocally proved. If true, a new avenue of disease control is suggested, since reduced fecundity should result in reduced pathogenesis of an infection with schistosomes.

V. DOSE AND TIME EFFECTS IN INDUCTION OF IMMUNITY

A. Effect of Cercarial Dose Size on Development of Primary Infections in Nonpermissive Hosts

The clearest demonstration of a dose effect in this situation was provided by Phillips *et al.* (1975, 1978*a*), who found that the greater the number of *S. mansoni* cercariae used for exposure, the fewer the percentage of worms that could be recovered 6 weeks later. However, the response was not linear (Fig. 5). Deviations from the theoretical expectation based on a linear dose–response relationship were even more pronounced at 16 weeks postexposure (Phillips

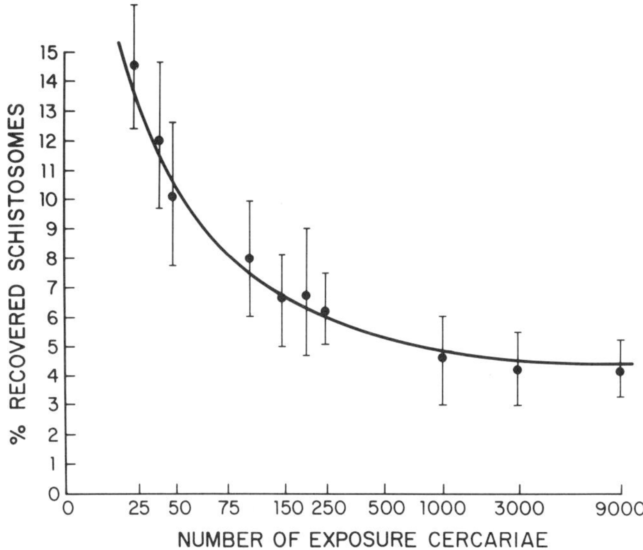

Figure 5. Dose effect on primary infection in nonpermissive host. Rats were infected with various numbers of *S. mansoni* cercariae and perfused 6 weeks later. (With permission from Phillips *et al.*, 1975.)

et al., 1978*a*), suggesting to these workers the play of "internal modulatory" phenomena.

Cheever and Powers (1972) also presented some evidence for an effect of exposure size on subsequent adult worm burdens in another nonpermissive system, *S. mansoni* in the rhesus monkey.

B. Effect of Cercarial Dose Size on Reinfection Immunity

In rats infected with *S. mansoni*, Knopf *et al.* (1977) and Phillips *et al.* (1977, 1978*a*) have demonstrated a relationship between the size of the initial cercarial infection dose and the strength of resistance to reinfection. Both studies agreed in showing a plateau or optimum dose size (Fig. 6).

The picture is not as clear in mice, apparently complicated by the development of egg-induced nonspecific resistance (Section III.C.2), but Dean (1983) has made a careful comparative study of the numerous published mouse experiments and has reached the general conclusion that resistance in mice increases with increasing average worm burdens. Recently, Long *et al.* (1980) have found adult worm burdens to correlate approximately with the size of the infecting cercarial dose. They went on to show that, on both group and individual bases, the degree of resistance to reinfection increased approximately linearly with the size of the primary worm burden. In assessing immune performance of individual

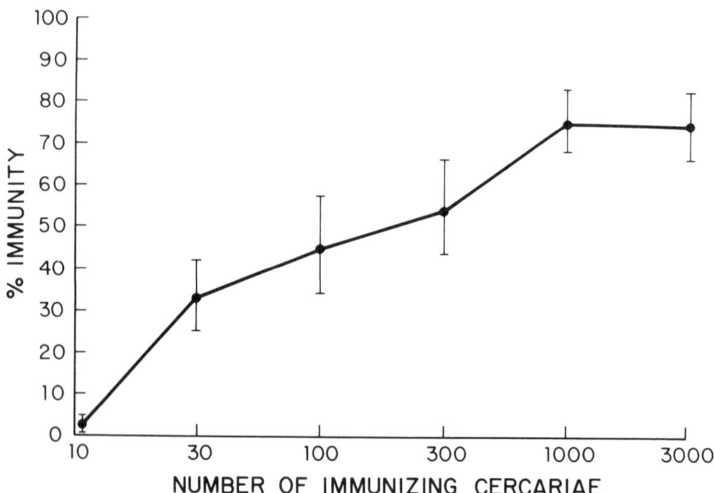

Figure 6. Dose effect on secondary infection. Rats were infected with various numbers of *S. mansoni* cercariae, then challenged after 6 weeks with radiolabeled cercariae. Percentage reduction in the challenge worm population was determined. (With permission from Phillips *et al.*, 1977.)

mice, Colley and Freeman (1980) also found evidence of dose dependency (measured by primary adult worm burden) in two strains of mice, when reinfection occurred 8 weeks after primary infection (Fig. 7).

In experiments involving rhesus monkeys and *S. mansoni*, Smithers and Terry (1965*a*) found no cercarial dose size influence on subsequent reinfection immunity strength, within the range tested of 100–1600 immunizing cercariae, but some evidence for a dose effect at a lower infection level exists (McMullen *et al.*, 1967; Smithers, 1967). The minimum dose of ~100 cercariae (and 16 weeks' residence time) needed to stimulate immunity reliably to reinfection in the British work led Smithers and Terry (1965*a*) to suggest a threshold value for immunity induction in this system.

In experiments in which a non-human-infecting strain *S. japonicum* from Taiwan was used to immunize rhesus monkeys against a human-infecting Japanese strain of *S. japonicum*, Hsü and Hsü (1963) found increasing levels of protection with increasing cercarial doses, but a plateau was reached as well.

C. Effect of Adult Worm Dose Size in Surgical Transfer Experiments on Reinfection Immunity

Peresan and Cioli (1980) found better stimulation of immunity to challenge after introducing eight worm pairs into the mesenteric veins of mice than with four worm pairs.

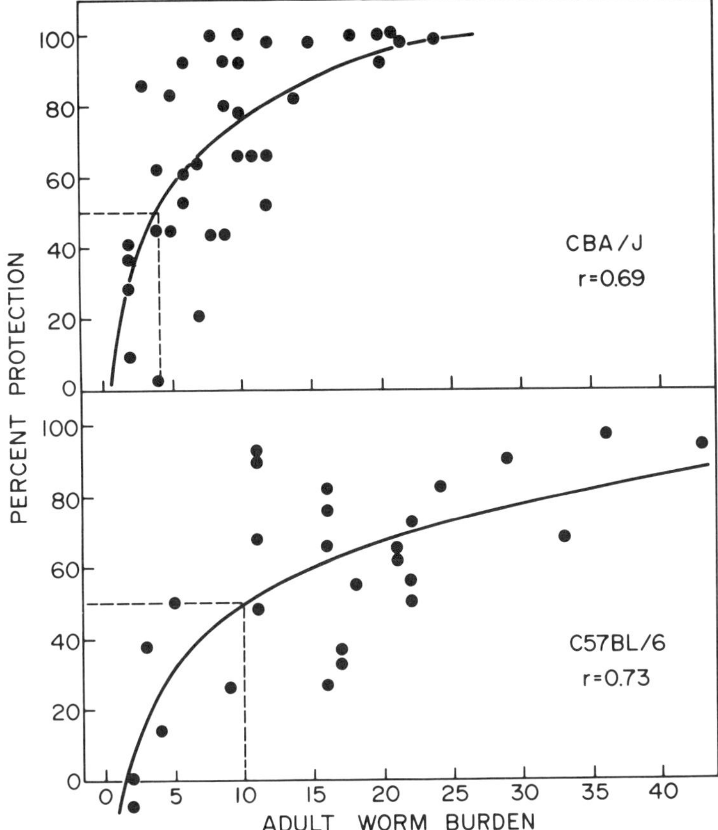

Figure 7. Dose effect of primary adult worm burden in two strains of mice on subsequent immunity to challenge infection given 8 weeks afterward. The relationship breaks down for the C57BL/6 strain when assessed 20 weeks afterward (data not shown). (With permission from Colley and Freeman, 1980.)

D. Dose Effects Using Irradiated Larvae and Dead Vaccines

Dose effects are not easy to ascertain with irradiated vaccines. One difficulty has been the propensity of investigators to vary many conditions simultaneously along with dose size, e.g., irradiation intensity, number of doses, time intervals, and routes of injection. Using γ-irradiated *S. mansoni* cercariae in mice, Minard *et al.* (1978*a*) found a definite dose–response relationship, with a threshold and a plateau, when irradiation was at 16 krad, but not at 56 krad. In contrast, Bickle *et al.* (1979*b*), using γ-irradiated schistosomula, found no evidence for dose dependency in mice. Another complication in obtaining dose–response curves in these models is the relationship between irradiation dose size and survival of the attenuated larvae after penetration or injection (Bickle *et al.*, 1979*c*). Bickle (1982) has described a relationship between survival time of the immunizing irradiated larvae and the development of optimum protection. He found that irradiated cercariae had to persist at least 1–2 weeks to be highly immunogenic. Such results remind one of the residence time effect in the induction of immunity with normal (unirradiated) stages.

Further evidence for a dose effect in immunity induction in mice after vaccination with a dead vaccine was recently provided by Hillyer and Serrano (1982). They used a preparation of *Fasciola hepatica* tegumental (Fht) antigens that cross-reacts with *S. mansoni* and found three injections per mouse of 30 μg Fht more efficacious than a single 30-μg injection.

In mice primed with formalin-fixed *S. mansoni* schistosomula, subsequently challenged, no protection was noted after injection of 30 schistosomula, but injection of 1000 schistosomula resulted in marginal protection (Ramalho-Pinto *et al.*, 1976*b*).

E. Interaction of Worm Burden and Residence Time in the Development of Reinfection Immunity

Several reports, using different systems, point to a synergism between worm burden and duration of infection in the induction of immunity. This was noted in simian infections (Smithers and Terry, 1965*a*; Ritchie *et al.*, 1966), in the surgical transfer model (Peresan and Cioli, 1980), and in murine *S. mansoni* infections (Long *et al.*, 1978; Colley and Freeman, 1980; Harrison *et al.*, 1982). Interpretation of the murine system is complicated by the egg-associated influences on immunity (Colley and Freeman, 1980; Dean *et al.*, 1981*a*; Harrison *et al.*, 1982).

F. Conclusions

Three broad generalizations emerge out of the work reviewed in this section: (1) a dose dependency exists in the induction of both primary and secondary immunity, and with respect to a variety of inducing stages; (2) smaller doses, given enough time in which to act, may achieve the same effect as larger doses; and (3) plateaus in dose dependency, at which immunity no longer increases, or may even decrease, are characteristic and imply immunoregulation. The combined factors of dose × time will hereafter be referred to as effective antigenic load, when a protective specific immune response is stimulated by their synergistic action.

VI. INFLUENCES OF HOST AND PARASITE QUALITY ON EFFECTIVE ANTIGENIC LOAD

A. Host Quality

The premises on which this argument is constructed are that: (1) schistosome eggs, because of less antigenic cross-reactivity with other stages (Section III.D) and also because of other factors (Section III.C) play a unique role in immunostimulation of the host. The role is unique in that eggs do not readily enter into the effective antigenic load equation for induction of host immunity; rather, they preferentially stimulate a largely separate set of immune responses, as indicated by the stage specificity of granuloma induction (Warren and Domingo, 1970*a*). (2) Depending on their own state of health, schistosomes will produce preponderantly one of two basic products: eggs or antigens. This switch is determined by the interaction of host permissiveness and host immunity.

Therefore, effective antigenic load is influenced by host quality in at least two ways. First, nonpermissive hosts increase effective antigenic load by killing invading larvae and adult worms (self-cure) and by stressing the survivors. Stresses of various types–nutritional, immune, chemotherapeutic–probably have similar end results with respect to immune stimulation. This can be illustrated by considering the effects of ligand binding to the surfaces of larval and adult schistosomes. In several *in vitro* systems, the result of ligand binding has been tegumental hyperactivity, deterioration, and/or sloughing of both ligand and ligand receptors. This has been shown to occur with antibodies (Perez and Terry, 1973; Kemp *et al.*, 1977) and with lectins (Simpson and McLaren, 1982; Samuelson *et al.*, 1982). Tegumental disruptions also occur after *in vitro* drug treatment (Wilson and Barnes, 1974) and after surgical transfer of worms into anti-host (Smithers and Terry, 1969) or hyperimmune (Hockley and Smithers, 1970) recipients. Finally, maintenance of schistosomes *in vitro* also leads to tegumental sloughing (Samuelson and Caulfield, 1982; Simpson and McLaren, 1982). These latter observations are particularly relevant because they suggest that similar tegumental disruptions, perhaps less pronounced but more sustained, may be taking place in the physiologically inadequate environment provided by a nonpermissive host.

Second, permissive hosts promote large, healthy worm biomass, with minimal antigenic stimulation. The worms' energy is channeled into egg production, hence removal of worm products from contribution to an effective antigenic load occurs, by virtue of the overall disparate antigenicity of the eggs (Section III.D). As adjunct mechanisms worth considering, high egg production and resultant tissue lodgement could result in general immunodepression through the nonspecific mechanisms considered below (Section IX.C). Also, healthy worms may be better able to produce the direct inhibitors of the immune response

(Section IV.B), if evolution did indeed favor the appearance of such inhibitors as evasion mediators. Thus, permissiveness could enhance parasite survival by permitting the diversion of energy from the production of effective antigenic stimuli into "nonantigenic" and suppressive pathways.

B. Parasite Quality

The stimulating paper by Smith and Clegg (1979) elegantly illustrated the great variability among *S. mansoni* cercariae, both in their ability to stimulate immunity and in their susceptibility to the immune response. These investigators could find no evidence for antigenic variation or polymorphism as the basis for the observed variations in immunogenicity and susceptibility. Despite this failure to find antigenic differences among the cercarial clones, it is possible that genetic differences underlie the variations. However, the study by Fletcher *et al.* (1981) on genetic variation in adult *S. mansoni* worms indicates a very low level of genetic heterozygosity in isoenzymes among most laboratory strains of *S. mansoni* tested, particularly among those maintained in mice for many years. If intraclonal variations could be shown to exist, some somatic mechanism for diversity generation would be suggested. Otherwise, the cercarial variations would appear more likely to be a function of other factors, such as differences in cercarial vitality. Such variation probably stems from factors within the snail (Evans and Stirewalt, 1951) e.g., host nutrition, sporocyst infection intensity, and variations in length of egress pathway, plus elapsed time since emergence from the snail. Whatever the reasons, variations within the infecting cercariae must influence their subsequent survival in their new host, and consequently the total antigenic load to which they contribute. This in turn could affect the subsequent development of immunity in the host (Phillips *et al.*, 1975; Colley and Freeman, 1980).

VII. INTERACTION OF EFFECTIVE ANTIGENIC LOAD AND HOST GENOTYPE IN INDUCTION OF IMMUNITY

A. Influence of Host Genotype

The knowledge of immunogenetic systems and their control over various aspects of the immune response may provide one rational explanation for individual host variability in immune responses and immunity to schistosomes. Immunogenetic systems are not the only components in immune variability, however. Other genetic or phenotypic variations in the host population, as well as variations in the infecting schistosomes (Smith and Clegg, 1979) must also contribute. But the immunogenetic systems, particularly as they are fixed among inbred strains of animals, offer a rich possibility for gaining an understanding at this level. Some studies already point in this direction. Murrell *et al.* (1979*a*) and James *et al.* (1981) found differences in the degrees of immunity expressed after vaccination with irradiated *S. mansoni* cercariae in various inbred strains of mice, and Smith and Clegg (1979) noted similar strain variability after primary infections. Colley and Freeman (1980) found that twice the number of adult worms were required to stimulate comparable levels of reinfection resistance in C57BL/6 as in CBA/J mice. Dean *et al.* (1981*b*), in comparing quality of various larval immunogens, found that in C57BL/6J mice irradiated cercariae were much better inducers of resistance than were irradiated lung-stage schistosomula, whereas in NIH/Nmri mice, they were about equal. That observed differences in mouse strain behavior are not necessarily simple consequences of different immunogenetic constitutions, however, is well illustrated by the results of Dean *et al.* (1981*a*), who examined immunopathologic factors in relationship to acquired resistance in 10 strains of mice. They suggested that varying strain predilections for nonspecific responses to inflammatory agents in general could play a role in the varying levels of resistance which they had observed, since a similar ranking of strains with respect to BCG-induced resistance occurred (Civil and Mahmoud, 1978). Evidence for MHC-linked immune response variations in murine *S. mansoni* infections also exists (Claas and Deelder, 1979).

The problems are complex and will require much more intensive investigation to sort out. However, as an approximation, we can postulate that host genetic constitution operates on at least two levels in the host-parasite relationship with *Schistosoma.* First, species characteristics determine the degree of permissiveness (Section II), and thus influence the effective antigenic load (Section VI). Second, individual characteristics influence the response to any given antigenic load, which can be simplified, for heuristic purposes, as a switch between effector and suppressor mechanisms.

This is not to say that any sharp demarcation must exist between the effects of species and individual host characteristics in the host-parasite relationship. This is well illustrated by the experiences with experimental infections in nonhuman primates. The baboon (*Papio cynocephalus* superspecies) is characteristically a highly permissive host for *S. mansoni* (Newsome, 1956; Damian *et al.*, 1976*a*; Sturrock *et al.*, 1976). Yet, the rare baboon "self-cures" its infection (M. G. Taylor *et al.*, 1973*a*), acting for all the world like a rhesus monkey. Similarly, the rhesus monkey, which usually develops resistance to reinfection rather rapidly, is a relatively nonpermissive host for *S. japonicum* (Vogel and Minning, 1953). Yet we have Vogel and Minning's record of their monkey "Otto," which maintained unabated fecal egg output for 9 months after an exposure to 200 cercariae, and then showed no immunity to a challenge infection. It is of course hazardous to draw conclusions regarding the influence of host genotype from a few atypical animals. But studies using inbred strains of hosts add weight to the overall argument. Series of inbred strains compared by means of some criterion often show ranking within a range of response (Civil and Mahmoud, 1978; Murrell *et al.*, 1979*a*; Dean *et al.*, 1981*a*). In a sense, genetic diversity among inbred strains reflects that of individuals within an outbred population, but should be less because of their homozygosity and the recent common ancestry of many inbred lines (Klein, 1975).

Individual variability in immune responses is almost a cliché, yet it deserves an occasional reminder. Powell and Damian (1982) studied the primary *in vitro* antibody response to a heterologous antigen, sheep erythrocytes (SRBCs), by peripheral blood mononuclear cells from normal baboons and a group of six baboons chronically infected with *S. mansoni*. Although their infections were comparable, we found great individual variability in response suppression in cells from the infected animals (Fig. 8). Analysis of individual responses is important in studying immunity in outbred populations, including humans (Todd *et al.*, 1980). Individual responses are increasingly being considered even in work using inbred hosts, with important results (Dean *et al.*, 1978*b*; Long *et al.*, 1980; Colley and Freeman, 1980; Garcia *et al.*, 1983).

Individual variability in the host must have its influence on effective antigenic load. This can be inferred from reports that individual variation in the strength of reinfection immunity is much more pronounced at lower immunizing doses (primary infections) for *S. mansoni* in rhesus monkeys (Smithers and Terry, 1965*a*; McMullen *et al.*, 1967; Smithers, 1967).

B. Effector-Suppressor Switches

Immunoregulation has been abundantly shown to occur in human and experimental schistosomiasis (reviewed by Colley, 1981*b*). It generally remains poorly understood, except for those responses involved in granuloma modulation, a vast area largely beyond the scope of this discussion (reviewed by Boros, 1978; Phillips and Colley, 1978; Colley, 1981*b*; Warren, 1982).

Suppression of the cellular response to various schistosome antigen preparations was found in human (Colley *et al.*, 1977*a*,*b*; Todd *et al.*, 1979; Ottesen *et al.*, 1978; Ottesen, 1979; Ottesen and Poindexter, 1980), simian (Maddison *et al.*, 1979), and rodent (Boros *et al.*, 1975; Pelley *et al.*, 1976; Coulis *et al.*, 1978; Camus *et al.*, 1979; Attallah *et al.*, 1979) infections.

Immunoregulation of the antibody response also occurs. Mice in acute (Mota-Santos *et al.*, 1976; Attallah *et al.*, 1979) and chronic (Boros *et al.*, 1975; Mota-Santos *et al.*, 1977) stages of *S. mansoni* infection were also depressed in their ability to form antibodies to heterologous or schistosome antigens. More recently, Powell and Damian (1982) showed that the peripheral blood mononuclear

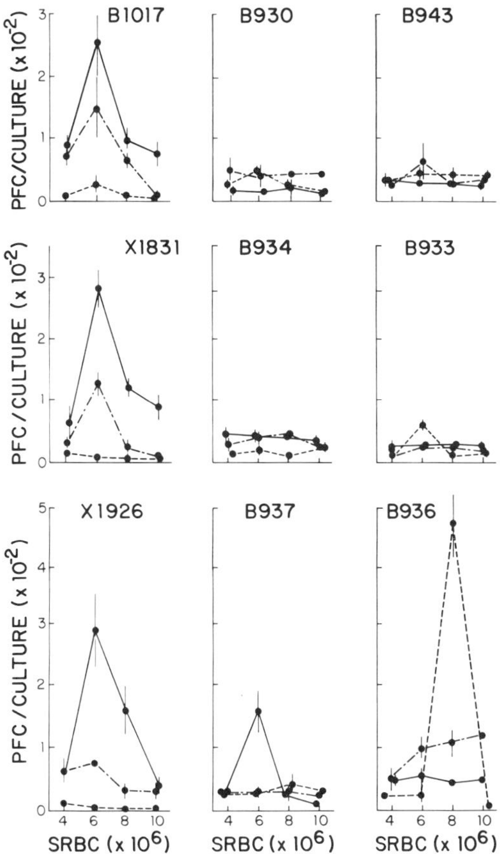

Figure 8. Individual variability in baboon peripheral blood mononuclear cell ability to respond *in vitro* to SRBC. B1017, X1831, and X1926 are normal animals. The other six baboons have had long-standing *S. mansoni* infections. One baboon (B936) was completely unsuppressed, another (B937) was partially suppressed, and the remaining four were completely suppressed. (With permission from Powell and Damian, 1982.)

cells of baboons chronically infected with *S. mansoni* were depressed in their ability to generate anti-SRBC plaque-forming cells after primary *in vitro* immunization.

It is evident that immunoregulatory mechanisms are stimulated during acute and chronic schistosome infections. As stressed by Colley (1981*b*), the net immunologic result of infection must rest on a balance of effector and regulatory mechanisms. It is likely that the switches between effector and suppressor mechanisms in schistosomiasis are strongly influenced by antigenic load. Antigen dose effects in immune stimulation are extremely complex, as is well illustrated by the phenomenon of high–low zone tolerance induction (reviewed by Weigle,

1973 and Cinader, 1982), or by the various forms taken by even simple models for antigen-cell interactions (Bell and Perelson, 1978).

That antigenic load plays a large role in the effector-suppressor balance in schistosomiasis is suggested by several of the phenomena reviewed: (1) the existence of optima or plateaus of immunity development to cercariae, (2) the "internal modulatory phenomena" of Phillips *et al.* (1978*a*), (3) the threshold value for immune stimulation of Smithers and Terry (1965*a*), and (4) the dose × residence time relationship. That this may also be true in human schistosomiasis is suggested by the report of Ellner *et al.* (1981) who found that the peripheral blood mononuclear cells of heavily, but not lightly, infected Egyptians showed depressed blastogenic responses to an *S. mansoni* adult worm antigen preparation.

VIII. DUAL-IMMUNITY THEORIES IN SCHISTOSOMIASIS

The complexity of host-parasite relationships in schistosomiasis, as exemplified by immune responses, has recently been likened to a rabbit warren (Colley, 1981*b*). Such complexity stimulates attempts at its understanding, of which this contribution is but one of many. Prominent in this collective effort has been the emergence of various dual-immunity theories. One of the first of these was enunciated by Hsü and Hsü (1963), who thought that "breakthrough" *S. japonicum* worms of a challenge infection given to immunized rhesus monkeys "had escaped the immune effect produced by the schistosomulic stage . . . and might not be further influenced by the anti-schistosomulic immune bodies in these monkeys. Because the worms which escaped were maturing into adult worms, the host needs a different kind of immune body to combat the adult worms." The concomitant immunity hypotheses (Smithers and Terry, 1967) is the conceptual successor to Hsü and Hsü's hypothesis. More recently (reviewed in Section IV.A.2), a new form of dual-immunity hypothesis has emerged, on the basis of varying patterns of parasite attrition in different immune murine models. These ideas may require modification, if the results to be gained from the application of newer *in vivo* methods for following schistosomular migration and for pinpointing sites of the destruction warrant it. This form of dual-immunity hypothesis has recently received experimental support from another quarter, with the discovery of Bickle and Ford (1982) of apparent qualitative differences in antigen recognition between "vaccinated serum" and chronic infection serum.

Another form of dual-immunity hypothesis was put forth by Dean *et al.* (1978*b*), who separated egg-influenced acquired resistance from that stimulated by abortive infections. As reviewed in Section III.C.2, the first probably relates to nonspecific mechanisms, and the second to specific mechanisms.

My position in the development of these topics is to favor a more unitary point of view, through the emphasis I have placed on (1) cross-reacting over stage-specific schistosome antigens, and (2) the differential sensitivity of the various life-cycle stages to the effects of the immune response stimulated by the predominant stage-common antigens.

However, one aspect of dual immunity stubbornly fails to give ground to any such attempt at full integration. This is the immunologic response to the egg, culminating in the modulated and fibrosing granuloma.

The historic development of schistosome immunoparasitology has been along two separate paths, and this, in my opinion, reflects more the fact that they are in fact separate paths than any artificial development of schools of research.

With the enabling hindsight provided by the collective research effort in schistosomiasis, one can infer the evolutionary logic in the dichotomous development of these responses. The selective pressures on the common ancestral host and on the common ancestral blood fluke must have been enormous to separate these responses. The development of specific egg antigens involved in granulomatous hypersensitivity induction and elicitation is thought to have permitted the host's survival as a home for the parasite by sequestering the toxic substances in the egg (Colley, 1981*b*). This viewpoint is supported by several studies showing increased morbidity in T-cell-deficient mice (Buchanan *et al.*, 1973; Byram and von Lichtenberg, 1977; Doenhoff *et al.*, 1979; Epstein *et al.*, 1979) as well as recent evidence for hepatotoxic substances in eggs (Dunne *et al.*, 1981).

This evolutionary process could have promoted the schistosome's survival in a more direct way, in that not only was much of the host's response diverted into granulomatous hypersensitivity, but also the worm found a haven from that very response—side-stepping it as it were—by the neat trick of being, in important respects, antigenically distinct. One does not see a granulomatous response around healthy adult worms *in situ* in the mesenteric veins (Smith and von Lichtenberg, 1974).

Of course, the two major response pathways are not absolutely distinct. Some cross-walks exist, and a few that might prove important are considered in the next section.

IX. CONNECTIONS BETWEEN ACQUIRED RESISTANCE AND EGG-INDUCED HYPERSENSITIVITY

A. Immunopathologic Host Changes

An important principle emerges, with renewed vigor, from the research reviewed in Section III.C. The disease process itself includes modifications in the host, such that it may become very different from its uninfected, naive, and healthy counterpart. Besides having an immune system in some new equilibrium state (Jerne, 1974), the infected diseased host presents a vastly different anatomic and physiologic entity to its guests. This is perhaps especially true in schistosomiasis, in which various anatomic vascular changes, related to portal hypertension (itself a consequence of immunopathogenic responses to the entrapped tissue eggs) have been recorded during the course of the disease, both in animal models and in humans. These include the development of arteriovenous anastomoses in the liver, spleen, and intestine (Cheever and Warren, 1964; Nagy *et al.*, 1981). The hepatic anastomoses are interesting in view of the opportunity they afford for shunting of eggs to the lungs, but Dean *et al.* (1981*a*) also reiterated the additional possibility that schistosomula may, by these same new avenues, be diverted from the liver to the lungs for multiple circuits through the vascular system, as was also thought to occur by Miller and Wilson (1980). The latter investigators also suggested that the innate resistance of some host species might be related to their vascular organization. The report of Nagy *et al.* (1981) is extremely provocative, since the changes in vessel tortuosity and the arteriovenous anastomoses which they found in the enteric vasculature of mice were anatomically confined to areas of oviposition. If similar changes could be shown to occur also in infected primates, they may in some way relate to the migration of adult worms from colonic to more anterior mesenteric venules–the anterior shift (Section IV.C.3).

B. Cercarial–Egg Antigenic Cross-Reactions and Their Possible Significance

The report by Sadun *et al.* (1965) was exceptional in its genre in that these workers observed more similarity between egg and cercarial extracts than either with adult antigens. However, the adult antigen used was an excretion–secretion preparation rather than a somatic extract, so that the results were not strictly comparable. Other cross-reactions between cercariae and eggs have been reported, however. For example, Suzuki and Damian (1981), in seeking an explanation for the preoviposition appearance of IgM antibodies reacting with SEA in baboons infected with *S. mansoni*, demonstrated shared epitopes between the cercarial and egg extracts used. They suggested the possibility that, based on immunologic cross-reactivity, the exposure pattern to cercariae could influence subsequent or concomitant anti-egg responses, including granuloma formation. Similarly, Khoury and Phillips (1981) found an early SEA-specific IgM response in murine splenic lymphocytes, which they attributed to larval immunogens shared with eggs. They speculated that the splenic T-RFC (T-antigen-specific rosette-forming cell) response could have "resulted in the generation of the populations of modulating cells which ultimately controlled the perpetuation of the anti-SEA response and host morbidity."

In a recently completed experiment (Damian *et al.*, 1983), we have found immunization of baboons with highly irradiated, cryopreserved *S. mansoni* schistosomula to result in markedly smaller granulomas after challenge infection as compared to these in unvaccinated baboons. Interestingly, this phenomenon occurred in the absence of protection against the invading schistosomula of the challenge, reaffirming the separation of acquired immunity and granuloma responses, at least at the effector level.

Reciprocity of specific responses from eggs to cercariae seems unlikely, however, in view of the poor performance of eggs as inducers of acquired resistance (Section III.C.1).

Another similarity between eggs and cercariae was uncovered by the work of Colley *et al.* (1979). These investigators found that *S. mansoni* SEA and soluble cercarial antigenic preparation (CAP) were both able to inhibit the *in vitro* responsiveness of normal or *S. mansoni*-infected persons' peripheral blood mononuclear cells to the T-cell mitogen, phytohemagglutinin (PHA) when the cells were simultaneously exposed to the schistosome antigens and the mitogen. A soluble adult worm antigenic preparation (SWAP) did not suppress PHA responsiveness, however. This observation is discussed further in Section IX.C.

C. Granuloma-Associated Nonspecific Modulatory Phenomena

Spontaneous modulation was the name given (Warren, 1974) to the observed phenomenon of active suppression of the *in vivo* granulomatous response to eggs. Some evidence exists for spontaneous modulation in human schistosomiasis mansoni (Rocklin *et al.*, 1980). It has been most intensively studied in experimental, chronic murine *S. mansoni* infections (see reviews by Phillips and Colley, 1978; Boros, 1978; Colley, 1981*b*; and Warren, 1982). The field has been progressing through use of adoptive suppression models (Colley, 1981*a*; Chensue *et al.*, 1981) and a new *in vitro* model (Doughty and Phillips, 1982). Recently, the *S. japonicum* egg granuloma in mice has also come under scrutiny, with the very different result that chronic infection serum, not T cells, transfers granuloma suppression (Olds *et al.*, 1982).

Colley (1981*b*) has thoroughly reviewed immunoregulatory events in spontaneous modulation. I wish here only to dwell on a single aspect of this phenomenon: the nonspecific suppressive responses associated with granuloma modulation. In mice, generalized immunodepression at this time has been reflected by reduced lymphocyte responses to mitogens (Pelley *et al.*, 1976) and reduced ability to generate MLC-stimulated alloantigen-specific cytotoxic T lymphocytes *in vitro* (Coulis *et al.*, 1978). Since SEA can suppress the cellular response to PHA of normal or schistosomiasis mansoni patients (Colley *et al.*, 1979) non-antigen-specific *induction* of suppressor cells by egg products may occur.

D. Conclusions

The coexisting immune responses of acquired resistance and granulomatous hypersensitivity are largely, although not entirely, separated. At least three connecting links are known:

1. Egg-induced immunopathogenesis has a nonspecific influence on acquired resistance. This may be so great as to largely replace specific acquired resistance in at least the system of murine schistosomiasis mansoni (Dean *et al.*, 1981*a*). The nonspecific effects may be fairly direct, as when migrating schistosomula are damaged while encountering pulmonary inflammation (M.A. Smith *et al.*, 1975) or indirect, involving egg-related hemodynamic changes.
2. Special antigenic relationships between cercariae and eggs suggest a possible specific, interactive pathway. Cercarial exposure patterns could modify ongoing and subsequent granulomatous hypersensitivity, but the poor immunogenicity of eggs for stimulating acquired immunity suggests that the relationship is a nonreciprocal one.
3. Egg-induced generalized, nonspecific immunodepression could promote the survival of adult worms by muting specific anti-worm responses. This by itself may not be sufficient for adult worm evasion, but could summate with other mechanisms, for example, antigen-specific suppressor generation, direct suppressor substances, host antigen disguise, reduced antigenic disparity with the host, and diversion of the immune response to the granulomatous hypersensitivity channel to account for the phenomenon of immune evasion.

X. CONCLUDING REMARKS

A holistic view of immunity in schistosomiasis was developed through use of the following devices:

1. The various experimental host–schistosome combinations and human schistosomiasis were considered together as a continuum on the nonpermission to permission scale.
2. Antigens shared among the cercariae, schistosomula, and juvenile and adult worms were considered relatively more important than stage-specific antigens in immunity induction.
3. Host quality itself was considered a large determiner of effective antigenic load for immunity induction, by channeling worm products into either (1) eggs and suppressors or (2) antigens.
4. Host quality was thought also to interact with antigenic load by channeling responses into either effector or suppressor pathways.

Evidences supporting the following conclusions were brought together and reviewed:

1. The disparate antigenicity of eggs and worms has great significance in that

two largely distinct pathways of specific immune induction result. This is the most fundamental aspect of dual immunity in schistosomiasis.
2. Differential stage sensitivity to immune effectors underlies observed stage-specific phenomena such as concomitant immunity.
3. No stage, including the adult worm, is entirely refractory to immune attack. The potentially most important aspect of adult worm vulnerability is an immune-mediated anti-fecundity effect.

Concomitant immunity may be considered the end result of a long history of coevolution between blood flukes and their vertebrate hosts. The main element in concomitant immunity is the antigenic uniqueness of eggs and the immune responses they engender. These immune responses favor host survival. They may even facilitate egg passage through the tissues (Doenhoff *et al.*, 1981). Moreover, they do these things without harming the worms. In fact, since eggs induce generalized nonspecific immunodepression, their presence in the tissues probably even enhances worm survival. The worms themselves gain additional privileges by various evasive strategies, among the most interesting of which are parasite-produced, or parasite-altered-host immunomodulatory factors.

Although the introduction to this chapter stated the goal of vaccination against schistosomiasis as being still but a dream, I can close on a more optimistic note. It is apparent that the 250×10^6 year head start enjoyed by the schistosome is rapidly being cut down by an evolutionary brand new form of host nonpermissiveness.

ACKNOWLEDGMENTS

Preparation of this chapter was supported by the United States–Japan cooperative Medical Sciences Program administered by the National Institutes of Allergy and Infectious Diseases, National Institutes of Health, Public Health Service, Department of Health and Human Services, grant AI 18906. I am grateful to Dr. Robert E. Miller and Dr. Walter M. Kemp for the scanning electron micrograph, to University of Georgia graduate students Susan L. McCormick and Clifford S. Shuman for the transmission electron micrographs, and to Dr. S. Michael Phillips and Dr. Daniel G. Colley and to George L. Freeman, Jr., for permission to reproduce their published material. Dr. David A. Dean and Dr. Philip T. LoVerde critically read the manuscript; however, flaws in concepts, coverage, and interpretations remain as my own. I am also indebted to Dr. John P. Caulfield, Dr. David A. Dean, Dr. Marc H. Dresden, Dr. Eugene G. Hayunga, Dr. Graham H. Mitchell, and Dr. S. Michael Phillips for access to as yet unpublished material. My graduate students, Stephen C. Bosshardt, Doreen A. Dalesandro, Hernando del Portillo, Susan L. McCormick, A. Kwame Nyame, and Malcolm R. Powell, helped in many ways. Finally, to the typists, Ann B. Clark, Kathy Glisson, and Marty Ledford, a special word of thanks for your forbearance.

XI. REFERENCES

Abdel Wahab, M. F., Warren, K. S., and Levy, R. P., 1971, Function of the thyroid and the host–parasite relation in murine schistosomiasis mansoni, *J. Infect. Dis.* **124**:161.

Amin, M. A., Nelson, G. S., and Saoud, M. F. A., 1968, Studies on heterologous immunity in schistosomiasis. 2. Heterologous immunity in rhesus monkeys, *Bull. WHO* **38**:19.

Andrade, Z. A., and Azevedo de Brito, P., 1982, Curative chemotherapy and resistance to reinfection in murine schistosomiasis, *Am. J. Trop. Med. Hyg.* **31**:116.

Andrade, Z. A., and Sadigursky, M., 1978, Immunofluorescence studies of schistosome structures which share determinants with circulating schistosome antigens, *Trans. R. Soc. Trop. Med. Hyg.* **72**:316.

Aronstein, W. S., Norden, A. P., and Strand, M., 1983, Tegumental expression in larval and adult stages of a major schistosome structural glycoprotein, *Am. J. Trop. Med. Hyg.* **32**:334.

Asch, H. L., and Dresden, M. H., 1979, Acidic thiol proteinase activity of *Schistosoma mansoni* egg extracts, *J. Parasitol.* **65**:543.

Atkinson, K. H., and Atkinson, B. G., 1980, Biochemical basis for the continuous copulation of female *Schistosoma mansoni*, *Nature (Lond.)* **282**:478.

Atkinson, B. G., and Atkinson, K. H., 1982, *Schistosoma mansoni:* One- and two-dimensional electrophoresis of proteins synthesized *in vitro* by males, females, and juveniles, *Exp. Parasitol.* **53**:26.

Attallah, A. M., Smith, A. H., Murrell, K. D., Fleischer, T., Woody, J., Vannier, W. E., Scher, I., Ahmed, A., and Sell, K. W., 1979, Characterization of the immunosuppressive state during *Schistosoma mansoni* infection, *J. Immunol.* **122**:1413.

Auriault, C., Joseph, M., Dessaint, J. P., and Capron, A., 1980, Inactivation of rat macrophages by peptides resulting from cleavage of IgG by schistosoma larvae proteases, *Immunol. Lett.* **2**:135.

Auriault, C., Ouaissi, M. A., Torpier, G., Eisen, H., and Capron, A., 1981. Proteolytic cleavage of IgG bound to the Fc receptor of *Schistosoma mansoni* schistosomula, *Parasite Immunol.* **3**:33.

Bell, G. I., and Perelson, A. S., 1978, An historical introduction to theoretical immunology, in: *Theoretical Immunology* (G. I. Bell, A. S. Perelson, and G. H. Pimbley, Jr., eds.), pp. 3–41, Marcel Dekker, New York.

Bickle, Q. D., 1982, Studies on the relationship between the survival of *Schistosoma mansoni* larvae in mice and the degree of resistance produced, *Parasitology* **84**:111.

Bickle, Q. D., and Ford, M. J., 1982, Studies on the surface antigenicity and susceptibility to antibody-dependent killing of developing schistosomula using sera from chronically infected mice and mice vaccinated with irradiated cercariae, *J. Immunol.* **128**:2101.

Bickle, Q., Bain, J., McGregor, A., and Doenhoff, M., 1979*a*, Factors affecting the acquisition of resistance against *Schistosoma mansoni* in the mouse. III. The failure of primary infections with cercariae of one sex to induce resistance to reinfection, *Trans. R. Soc. Trop. Med. Hyg.* **73**:37.

Bickle, Q. D., Taylor, M. G., Doenhoff, M. J., and Nelson, G. S., 1979*b*, Immunization of mice with gamma-irradiated intramuscularly injected schistosomula of *Schistosoma mansoni*, *Parasitology* **79**:209.

Bickle, Q. D., Dobinson, T., and James, E. R., 1979*c*, The effects of gamma-irradiation on migration and survival of *Schistosoma mansoni* schistosomula in mice, *Parasitology* **79**:223.

Blum, K. and Cioli, D., 1981, *Schistosoma mansoni:* Age-dependent susceptibility to immune elimination of schistosomula artificially introduced into preinfected mice, *Parasite Immunol.* **3**:13.

Bogitsh, B. J., and Katz, S. P., 1976, Immunocytochemical studies on *Schistosoma mansoni*. II. Soluble cercarial antigens in cercariae and schistosomules, *J. Parasitol.* **62**:709.

Boros, D. L., 1978, Granulomatous inflammation, *Prog. Allergy* **24**:183.

Boros, D. L., Pelley, R. P., and Warren, K. S., 1975, Spontaneous modulation of granulomatous hypersensitivity in schistosomiasis mansoni, *J. Immunol.* **114**:1437.

Bout, D., Capron, A., Dupas, H., and Capron, M., 1974, Characterization of *Schistosoma mansoni* antigens, in: *Proceedings of the Third International Congress on Parasitology, Munich*, Vol. 2, p. 1146, Facta, Vienna.

Bout, D., Santoro, F., Carlier, Y., Bina, J. C., and Capron, A., 1977, Circulating immune complexes in schistosomiases, *Immunology* **33**:17.

Boyer, M. H., and Kalfayan, L. J., 1978, Effect of transferred adult *Schistosoma mansoni* on resistance of mice to cercarial challenge, *Am. J. Trop. Med. Hyg.* **27**:542.

Boyer, M. H., Ketchum, D. G., and Palmer, P. D., 1976, The host antigen phenomenon in experimental murine schistosomiasis: The transfer of 3-week old *Schistosoma mansoni* between two inbred strains of mice, *Int. J. Parasitol.* **6**:235.

Boyer, M. H., Kalfayen, L. J., and Ketchum, D. G., 1977, The host antigen phenomenon in experimental murine schistosomiasis. III. Destruction of parasites transferred from mice to hamsters, *Am. J. Trop. Med. Hyg.* **26**:254.

Brink, L. H., Krueger, K. L., and Harris, C., 1980, Stage-specific antigens of Schistosoma mansoni, in: *The Host–Invader Interplay* (H. Van den Bossche, ed.) pp. 393–404, Elsevier/North Holland, Amsterdam.

Bruce, J. I., von Lichtenberg, F., Schoenbechler, M. J., and Hickman, R. L., 1966, The role of splenectomy in the natural and acquired resistance of rhesus monkeys to infection with *Schistosoma mansoni*, *J. Parasitol.* **52**:831.

Bruce, J. I., Pezzlo, F., Yagima, Y., and McCarthy, J. E., 1974, *Schistosoma mansoni:* Pulmonary phase of schistosomule migration studied by electron microscopy, *Exp. Parasitol.* **35**:150.

Bruckner, D. A., and Schiller, E. L., 1974, Some biological characteristics of Liberian and Puerto Rican strains of *Schistosoma mansoni*, *J. Parasitol.* **60**:551.

Buchanan, R. D., Fine, D. P., and Colley, D. G., 1973, *Schistosoma mansoni* infection in mice depleted of thymus-dependent lymphocytes. II. Pathology and altered pathogenesis, *Am. J. Pathol.* **71**:207.

Bueding, E., 1950, Carbohydrate metabolism of *Schistosoma mansoni*, *J. Gen. Physiol.* **33**: 475.

Bushara, H. O., Majid, A. A., Saad, A. M., Hussein, M. F., Taylor, M. G., Dargie, J. D., Marshall, T. F. de C., and Nelson, G. S., 1980, Observations on cattle schistosomiasis in the Sudan, a study in comparative medicine. II. Experimental demonstration of naturally acquired resistance to *Schistosoma bovis*, *Am. J. Trop. Med. Hyg.* **29**:442.

Bushara, H. O., Hussein, M. F., Majid, M. A., Musa, B. E. H., and Taylor, M. G. 1983, Observations on cattle schistosomiasis in the Sudan, a study in comparative medicine,

IV. Preliminary observations on the mechanism of naturally-acquired resistance, *Am. J. Trop. Med. Hyg.* **32**:(in press).

Butterworth, A. E., 1977*a*, Effector mechanisms against schistosomes in vitro, *Am. J. Trop. Med. Hyg.* **26** (6, part 2) :29.

Butterworth, A. E., 1977*b*, The eosinophil and its role in immunity to helminth infection, *Curr. Topics Microbiol. Immunol.* **77**:127.

Butterworth, A. E., Vadas, M. A., Martz, E., and Sher, A., 1979, Cytolytic T lymphocytes recognize alloantigens on schistosomula of *Schistosoma mansoni*, but fail to induce damage, *J. Immunol.* **122**:1314.

Butterworth, A. E., Taylor, D. W., Veith, M. C., Vadas, M. A., Dessein, A., Sturrock, R. F., and Wells, E., 1982, Studies on the mechanisms of immunity in human schistosomiasis, *Immunol. Rev.* **61**:5.

Byram, J. E., and von Lichtenberg, F., 1981, Altered schistosome granuloma formation in nude mice, *Am. J. Trop Med. Hyg.* **26**:944.

Byram, J. E., Doenhoff, M. J., Musallam, R., Brink, L. H., and von Lichtenberg, F., 1979, *Schistosoma mansoni* infections in T-cell deprived mice, and the ameliorating effect of administering homologous chronic infection serum. II. Pathology, *Am. J. Trop. Med. Hyg.* **28**:274.

Byrd, E. E., 1939, Studies on the blood flukes of the family Spirorchidae. Part II. Revision of the family and description of new species, *J. Tenn. Acad. Sci.* **14**:116.

Camus, D., Dessaint, J.-P., Fischer, E., and Capron, A., 1979, Nonspecific suppressor cell activity and specific cellular unresponsiveness in rat schistosomiasis, *Eur. J. Immunol.* **9**:341.

Camus, D., Nosseir, A., Mazingue, C., and Capron, A., 1981, Immunoregulation by *Schistosoma mansoni*, *Immunopharmacology* **3**:193.

Capron, A., and Camus, D., 1979, Immunoregulation by parasite extracts, *Springer Semin. Immunopathol.* **2**:69.

Capron, A., Biguet, J., Rose, F., and Vernes, A., 1965, Les antigenes de *Schistosoma mansoni.* II. Ètude immunoélectrophorétique comparée de divers stades larvaires et des adultes des deux sexes. Aspects immunologiques des relations hote-parasite de la cercaire et de l'adulte de *S. mansoni*, *Ann. Inst. Pasteur* **109**:798.

Capron, A., Biguet, J., Vernes, A., and Afchain, D., 1968, Structure antigénique des helminthes. Aspects immunologiques des relations hote-parasite, *Pathol. Biol.* **16**:121.

Capron, A., Dessaint, J.-P, Capron, M., and Bazin, H., 1975, Specific IgE antibodies in immune adherence of normal macrophages to *Schistosoma mansoni* schistosomules, *Nature (Lond.)* **253**:474.

Capron, A., Dessaint, J.-P., Joseph, M., Torpier, G., Capron, M., Rousseaux, R., Santoro, F., and Bazin, H., 1977, IgE and cells in schistosomiasis, *Am. J. Trop. Med. Hyg.* **26**(6, Part 2):39.

Capron, A., Dessaint, J.-P., Capron, M., Joseph, M., and Torpier, G., 1982, Effector mechanisms of immunity to schistosomes and their regulations, *Immunol. Rev.* **61**:41.

Capron, M., Camus, D., Carlier, Y., Figueiredo, J. F. M., and Capron, A., 1977, Immunological studies in human schistosomiasis. II. Antibodies cytotoxic for *Schistosoma mansoni* schistosomules, *Am. J. Trop. Med. Hyg.* **26**:248.

Capron, M., Carlier, Y., Nzeyimana, H., Minoprio, P., Santoro, F., Sellin, B., and Capron, A., 1980, *In vitro* study of immunological events in human and experimental schistosomiasis: Relationships between cytotoxic antibodies and circulating Schistosoma antigens, *Parasite Immunol.* **2**:223.

Carlier, Y., Bout, D., Strecker, G., Debray, H., and Capron, A., 1980, Purification, immunochemical, and biologic characterization of the *Schistosoma* circulating M antigen, *J. Immunol.* **124**:2442.

Caulfield, J. P., Korman, G., Butterworth, A. E., Hogan, M., and David, J. R., 1980*a*, The adherence of human neutrophils and eosinophils to schistosomula: Evidence for membrane fusion between cells and parasites, *J. Cell Biol.* **86**:46.

Caulfield, J. P., Korman, G., Butterworth, A. E., Hogan, M., and David, J. R., 1980*b*, Partial and complete detachment of neutrophils and eosinophils from schistosomula: Evidence for the establishment of continuity between a fused and normal parasite membrane, *J. Cell Biol.* **86**:64.

Caulfield, J. P., Korman, G., and Samuelson, J. C., 1982, Human neutrophils endocytose multivalent ligands from the surface of schistosomula of *Schistosoma mansoni* before membrane fusion, *J. Cell Biol.* **94**:370.

Cheever, A. W., 1965, A comparative study of *Schistosoma mansoni* infections in mice, gerbils, multimammate rats and hamsters. II. Qualitative pathological differences, *Am. J. Trop. Med. Hyg.* **14**:227.

Cheever, A. W., 1968, A quantitative post-mortem study of schistosomiasis mansoni in man, *Am. J. Trop. Med. Hyg.* **17**:38.

Cheever, A. W., 1969, Quantitative comparison of the intensity of *Schistosoma mansoni* infections in man and experimental animals, *Trans. R. Soc. Trop. Med. Hyg.* **63**:781.

Cheever, A. W., and Duvall, R. H., 1974, Single and repeated infections of grivet monkeys with *Schistosoma mansoni*: Parasitological and pathological observations over a 31-month period, *Am. J. Trop. Med. Hyg.* **23**:884.

Cheever, A. W., and Powers, K. G., 1969, *Schistosoma mansoni* infection in rhesus monkeys: Changes in egg production and egg distribution in prolonged infections in intact and splenectomized monkeys, *Ann. Trop. Med. Parasitol.* **63**:83.

Cheever, A. W., and Powers, K. G., 1972, *Schistosoma mansoni* infection in rhesus monkeys: Comparison of the course of heavy and light infections, *Bull. WHO* **46**:301.

Cheever, A. W., and Warren, K. S., 1964, Hepatic blood flow in mice with acute hepatosplenic schistosomiasis mansoni, *Trans. R. Soc. Trop. Med. Hyg.* **58**:406.

Cheever, A. W., Erickson, D. G., Sadun, E. H., and von Lichtenberg, F., 1974, *Schistosoma japonicum* infection in monkeys and baboons: Parasitological and pathological findings, *Am. J. Trop. Med. Hyg.* **23**:51.

Cheever, A. W., Kamel, I. A., Elwi, A. M., Mosimann, J. E., and Danner, R., 1977, *Schistosoma mansoni* and *S. haematobium* infections in Egypt. II. Quantitative parasitological findings at necropsy, *Am. J. Trop. Med. Hyg.* **26**:702.

Cheever, A. W., Duvall, R. H., and Minker, R. G., 1980, Quantitative parasitologic findings in rabbits infected with Japanese and Philippine strains of *Schistosoma japonicum*, *Am. J. Trop. Med. Hyg.* **29**:1307.

Chensue, S. W., Wellhausen, S. R., and Boros, D. L., 1981, Modulation of granulomatous hypersensitivity. II. Participation of Ly 1^+ and 2^+ lymphocytes in the suppression of granuloma formation and lymphokine production in *Schistosoma mansoni* infected mice, *J. Immunol.* **127**:363.

Christensen, N. O., 1977, A method for the in vivo labeling of *Schistosoma mansoni* and *S. intercalatum* cercariae with radioselenium, *Z. Parasitenkd.* **54**:275.

Cinader, B., 1982, Immunological tolerance, in: *The Antigens* (M. Sela, ed.), Vol. VI, pp. 152–406, Academic Press, New York.

Cioli, D., 1976, *Schistosoma mansoni:* A comparison of mouse and rat worms with respect to host antigens detected by the technique of transfer into hamsters, *Int. J. Parasitol.* **6**:355.

Cioli, D., and Dennert, G., 1976, The course of Schistosoma mansoni infection in thymectomized rats, *J. Immunol.* **177**:59.

Cioli, D., Knopf, P. M., and Senft, A. W., 1977, A study of *Schistosoma mansoni* transferred into permissive and nonpermissive hosts, *Int. J. Parasitol.* **7**:293.

Cioli, D., Blum, K., and Ruppel, A., 1978, *Schistosoma mansoni*: Relationship between parasite age and time of spontaneous elimination from the rat, *Exp. Parasitol.* **45**:74.

Cioli, D., Malorni, W., DeMartino, C., and Dennert, G., 1980, A study of *Schistosoma mansoni* reinfection in thymectomized rats, *Cell. Immunol.* **53**:246.

Civil, R. H., and Mahmoud, A. A. F., 1978, Genetic differences in BCG-induced resistance to *Schistosoma mansoni* are not controlled by genes within the major histocompatibility complex of the mouse, *J. Immunol.* **120**:1070.

Claas, F. H. J., and Deelder, A. M., 1979, H-2 linked immune response to murine experimental *Schistosoma mansoni* infections, *J. Immunogen.* **6**:167.

Clegg, J. A., 1965, *In vitro* cultivation of *Schistosoma mansoni*, *Exp. Parasitol.* **16**:133.

Clegg, J. A., 1974, Host antigens and the immune response in schistosomiasis, in: *Parasites in the Immunized Host: Mechanisms of Survival, Ciba Foundation Symposium*, Vol. 25, p. 161, Elsevier, Amsterdam.

Clegg, J. A., and Smithers, S. R., 1972, The effects of immune rhesus monkey serum on schistosomula of *Schistosoma mansoni* during cultivation *in vitro*, *Int. J. Parasitol.* **2**:79.

Coelho, P. M. Z., Souza, R. C. A., Bredt, A., and Souza-Neto, J. A., 1976, The crowding effect in *Schistosoma mansoni* infection of hamsters: Influence on worm size, *Rev. Inst. Med. Trop. São Paulo* **18**:440.

Cohen, L. M., and Eveland, L. K., 1980, Effect of graft-versus-host reaction on *Schistosoma mansoni* infection, American Society of Parasitologists, 55th Annual Meeting, Berkeley, Abstracts, p. 57 (Abst.).

Colley, D. G., 1981*a*, T lymphocytes that contribute to the immunoregulation of granuloma formation in chronic murine schistosomiasis, *J. Immunol.* **126**:1465.

Colley, D. G., 1981*b*, Immune responses and immunoregulation in experimental and clinical schistosomiasis, in: *Parasitic Diseases, The Immunology*, Vol. 1 (J. M. Mansfield, ed.), pp. 1–83, Marcel Dekker, New York.

Colley, D. G., and Freeman, F. L., Jr., 1980, Differences in adult *Schistosoma mansoni* worm burden requirements for the establishment of resistance to reinfection in inbred mice I. CBA/J and C57BL/6 mice, *Am. J. Trop. Med. Hyg.* **29**:1279.

Colley, D. G., Cook, J. A., Freeman, G. L., Bartholomew, R. K., and Jordan, P., 1977*a*, Immune responses during human schistosomiasis mansoni. I. *In vitro* lymphocyte blastogenic responses to heterogeneous antigenic preparations from schistosome eggs, worms and cercariae, *Int. Arch. Allergy Appl. Immunol.* **53**:420.

Colley, D. G., Hieny, S. E., Bartholomew, R. K., and Cook, J. A., 1977*b*, Immune responses during schistosomiasis mansoni. III. Regulatory aspects of patient sera on human lymphocyte blastogenic responses to schistosome antigen preparations, *Am. J. Trop. Med. Hyg.* **26**:917.

Colley, D. G., Lewis, F. A., and Goodgame, R. W., 1978, Immune responses during human schistosomiasis mansoni. IV. Induction of suppressor cell activity by schistosome antigen preparations and concanavalin A, *J. Immunol.* **120**:1225.

Colley, D. G., Todd, C. W., Lewis, F. A., and Goodgame, R. W., 1979, Immune responses during human schistosomiasis mansoni. VI. *In vitro* nonspecific suppression of phytohemagglutinin responsiveness induced by exposure to certain schistosomal preparations, *J. Immunol.* **122**:1447.

Cordeiro, M. N., and Gazzinelli, G., 1979, *Schistosoma mansoni*: Resolution and molecular weight estimates of tegument glycoproteins by polyacrylamide gel electrophoresis, *Exp. Parasitol.* **48**:337.

Cornford, E. M., 1974*a*, *Schistosomatium douthitti*: Effects of thyroxine, *Exp. Parasitol.* **36**:210.

Cornford, E. M., 1974*b*, Effects of insulin on *Schistosomatium douthitti*, *Gen. Comp. Endocrinol.* **23**:286.

Cornford, E. M., and Huot, M. E., 1981, Glucose transfer from male to female schistosomes, *Science* **213**:1269.

Cornford, E. M., Huot, M. E., Diep, C. P., and Rowley, G. A., 1982, Protein, glycogen, and water content in schistosomes, *J. Parasitol.* **68**:1010.

Coulis, P. A., Lewert, R. M., and Fitch, F. W., 1978, Splenic suppressor cells and cell-mediated cytotoxicity in murine schistosomiasis, *J. Immunol.* **120**:1074.

Damian, R. T., 1962, A theory of immunoselection for eclipsed antigens of parasites and its implications for the problem of antigenic polymorphism in man, *J. Parasitol.* **48**:16 (Abst.).

Damian, R. T., 1964, Molecular mimicry: Antigen sharing by parasite and host and its consequences, *Am. Nat.* 98:129.

Damian, R. T., 1979, Molecular mimicry in biological adaptation, in: *Host-Parasite Interfaces: At Population, Individual, and Molecular Levels* (R. B. Nichol, ed.), pp. 103–126, Academic Press, New York.

Damian, R. T., 1982, The influence of host immune responses on parasite behavior, in: *Cues That Influence Behavior of Internal Parasites* (W. S. Bailey, ed.), pp. 149–160, U.S. Department of Agriculture, Research Service (Southern Region), New Orleans.

Damian, R. T., Greene, N. D., and Fitzgerald, K., 1972, Schistosomiasis mansoni in baboons. The effect of surgical transfer of adult *Schistosoma mansoni* upon subsequent challenge infection, *Am. J. Trop. Med. Hyg.* **21**:951.

Damian, R. T., Greene, N. D., and Hubbard, W. J., 1973, Occurrence of mouse α_2- macroglobulin antigenic determinants on *Schistosoma mansoni* adults, with evidence on their nature, *J. Parasitol.* 59:64.

Damian, R. T., Greene, N. D., and Fitzgerald, K., 1974, Schistosomiasis mansoni in baboons. II. Acquisition of immunity to challenge infection after repeated small exposures to cercariae of *Schistosoma mansoni, Am. J. Trop. Med. Hyg.* **23**:78.

Damian, R. T., Greene, N. D., Meyer, K. F., Cheever, A. W., Hubbard, W. J., Hawes, M. E., and Clark, J. D., 1976*a*, *Schistosoma mansoni* in baboons. III. The course and characteristics of infection, with additional observations on immunity, *Am. J. Trop. Med. Hyg.* **25**:299.

Damian, R. T., Greene, N. D., Meyer, K. F., 1976*b*, Immunizing efficacy of a Kenyan strain of *Schistosoma mansoni* in rhesus monkeys (*Macaca mulatta*), *Am. J. Trop. Med. Hyg.* **25**:355.

Damian, R. T., Roberts, M. L., Powell, M. R., Clark, J. D., Lewis, F. A., and Stirewalt, M. A., 1983, Small *Schistosoma mansoni* egg granulomas in challenged baboons after vaccination with irradiated, cryopreserved schistosomula (submitted).

Dean, D. A., 1977, Decreased binding of cytotoxic antibody by developing *Schistosoma mansoni*. Evidence for a surface change independent of host antigen adsorption and membrane turnover, *J. Parasitol.* **63**:418.

Dean, D. A., 1983, A review, *Schistosoma* and related genera: Acquired resistance to schistosome infection in mice, *Exp.Parasitol.* **55**:1.

Dean, D. A., Minard, P., Stirewalt, M. A., Vannier, W. E., and Murrell, K. D., 1978*a*, Resistance of mice to secondary infection with *Schistosoma mansoni*. I. Comparison of bisexual and unisexual initial infections, *Am. J. Trop. Med. Hyg.* **27**:951.

Dean, D. A., Minard, P., Murrell, K. D., and Vannier, W. E., 1978*b*, Resistance of mice to secondary infection with *Schistosoma mansoni*. II. Evidence for a correlation between egg deposition and worm elimination, *Am. J. Trop. Med. Hyg.* **27**:957.

Dean, D. A., Bukowski, M. A., and Cheever, A. W., 1981*a*, Relationship between acquired resistance, portal hypertension, and lung granulomas in ten strains of mice infected with *Schistosoma mansoni, Am. J. Trop. Med. Hyg.* **30**:806.

Dean, D. A., Cioli, D., and Bukowski, M. A., 1981*b*, Resistance induced by normal and irradiated *Schistosoma mansoni*: Ability of various worm stages to serve as inducers and targets in mice, *Am. J. Trop. Med. Hyg.* **30**:1026.

Dean, D. A., Mangold, B. L., Georgi, J. R., and Jacobson, R. J., 1982, Tracking of *Schisto-*

soma mansoni in irradiated cercaria-immunized, previously infected and normal mice by means of autoradiography, *Mol. Biochem. Parasitol. Suppl. (Abstracts of the 5th Int. Congr. Parasitol. Toronto)* p. 101.

Dessaint, J.-P., Camus, D., Fischer, E., and Capron, A., 1977, Inhibiton of lymphocyte proliferation by factor(s) produced by *Schistosoma mansoni, Eur. J. Immunol.* 7:624.

Dessein, A., Samuelson, J. C., Butterworth, A. E., Hogan, M., Sherry, B. A., Vadas, M. A., and David, J. R., 1981, Immune evasion by *Schistosoma mansoni*: Loss of susceptibility to antibody or complement-dependent eosinophil attack by schistosomula cultured in medium free of macromolecules, *Parasitology* **82**:357.

Dineen, J. K., 1963*a*, Immunological aspects of parasitism, *Nature (Lond.)* **197**:268.

Dineen, J. K., 1963*b*, Antigenic relationship between host and parasite, *Nature (Lond.)* **197**:471.

Dissous, C., Dissous, C., and Capron, A., 1981, Isolation and characterization of surface antigens from *Schistosoma mansoni* schistosomula, *Mol. Biochem. Parasitol.* **3**:215.

Doenhoff, M., Bickle, Q., Long, E., Bain, J., and McGregor, A., 1978*a*, Factors affecting the acquisition of resistance against *Schistosoma mansoni* in the mouse. I. Demonstration of resistance to reinfection using a model system that involves perfusion of mice within three weeks of challenge, *J. Helminthol.* **52**:173.

Doenhoff, M., Musallam, R., Bain, J., and McGregor, A., 1978*b*, Studies on the host-parasite relationship in *Schistosoma mansoni*-infected mice: The immunological dependence of parasite egg excretion, *Immunology* **35**:771.

Doenhoff, M., Musallam, R., Bain, J., and McGregor, A., 1979, *Schistosoma mansoni* infections in T-cell deprived mice, and the ameliorating effect of administering homologous chronic infection serum. I. Pathogenesis, *Am. J. Trop. Med. Hyg.* **28**:260.

Doenhoff, M., Bickle, Q., Bain, J., Webbe, G., and Nelson, G., 1980, Factors affecting the acquisition of resistance against *Schistosoma mansoni* in the mouse. V. Reduction in the degree of resistance to reinfection after chemotherapeutic elimination of recently patent primary infections, *J. Helminthol.* **54**:7.

Doenhoff, M. J., Pearson, S., Dunne, D. W., Bickle, Q., Lucas, S., Bain, J., Musallam, R., and Hassounah, O., 1981, Immunological control of hepatotoxicity and parasite egg excretion in *Schistosoma mansoni* infections: stage specificity of the reactivity of immune serum in T-cell deprived mice, *Trans. R. Soc. Trop. Med. Hyg.* **75**:41.

Doughty, B. L., and Phillips, S. M., 1982, Delayed hypersensitivity granuloma formation and modulation around *Schistosoma mansoni* eggs *in vitro*. II. Regulatory T cell subsets, *J. Immunol.* **128**:37.

Dresden, M. H., and Deelder, A. M., 1979, *Schistosoma mansoni*: Thiol proteinase properties of adult worm "hemoglobinase," *Exp. Parasitol.* **48**:190.

Dresden, M. H., Sung, C. K., and Deelder, A. M., 1983, A monoclonal antibody from infected mice to a Schistosoma monsoni egg proteinase, *J. Immunol.* **130**:1.

Dunne, D. W., Lucas, S., Bickle, Q., Pearson, S., Madgwick, L., Bain, J., and Doenhoff, M. J., 1981, Identification and partial purification of an antigen (ω_1) from *Schistosoma mansoni* eggs which is putatively hepatotoxic in T-cell deprived mice, *Trans. R. Soc. Trop. Med. Hyg.* **75**:54.

Ehrlich, P., 1910, The partial-functions of cells, The Nobel Lecture, Stockholm, Dec. 11, 1908. Reprinted from Münchener Wochenschrift, No. 5, 1909, in: *Studies in Immunity* (P. Ehrlich, ed.), 2nd ed., pp. 676–693, Wiley, New York.

Ellner, J. J., Olds, G. R., Osman, G. S., Elkholy, A., and Mahmoud, A. A. F., 1981, Dichotomies in the reactivity to worm antigen in human schistosomiasis mansoni, *J. Immunol.* **126**:309.

Ellner, J. J., Olds, G. R., Lee, C. W., Kleinhenz, M. E., and Edmonds, K. L., 1982, Destruction of the multicellular parasite *Schistosoma mansoni* by T lymphocytes, *J. Clin. Invest.* **70**:369.

Epstein, W. L., Fukuyama, K., Danno, K., and Kevan-Wong, E., 1979, Granulomatous inflammation in normal and athymic mice infected with *Schistosoma mansoni*: An ultrastructural study, *J. Pathol.* **127**:207.

Erickson, D. G., Beattie, R. J., Yamaguchi, S., Miyasaka, E., and Williams, J. E., 1973, Host–parasite relationships in schistosomiasis mansoni and japonica in rhesus monkeys: Interhost worm transfers, *Jap. J. Parasitol.* **22**:307.

Evans, A. S., and Stirewalt, M. A., 1951, Variations in infectivity of cercariae of *Schistosoma mansoni, Exp. Parasitol.* **1**:19.

Eveland, L. K., Hsü, S. Y. L., and Hsü, H. F., 1969, Cross-immunity of *Schistosoma japonicum, S. mansoni*, and *S. bovis* in rhesus monkeys, *J. Parasitol.* **55**:279.

Fletcher, M., LoVerde, P. T., and Woodruff, D. S., 1981, Genetic variation in *Schistosoma mansoni*: enzyme polymorphisms in populations from Africa, Southwest Asia, South America, and the West Indies, *Am. J. Trop. Med. Hyg.* **30**:406.

Foster, R., and Broomfield, K. E., 1971, Preliminary studies on the development of *Schistosoma mansoni* in rhesus monkeys following different regimens of infection, *Ann. Trop. Med. Parasitol.* **65**:367.

Garcia, E. G., Tiu, W. U., and Mitchell, G. F., 1983, Innate resistance to *Schistosoma japonicum* in a proportion of 129/J mice, *J. Parasitol.* (in press).

Georgi, J. R., 1982, *Schistosoma mansoni*: Quantification of skin penetration and early migration by differential external radioassay and autoradiography, *Parasitology* **84**:263.

Gitter, B. D., and Damian, R. T., 1982, Murine alloantigen acquisition by schistosomula of *Schistosoma mansoni*: Further evidence for the presence of K, D, and I region gene products on the tegumental surface, *Parasite Immunol.* **4**:383.

Gitter, B. D., McCormick, S. L., and Damian, R. T., 1982, Murine alloantigen acquisition by *Schistosoma mansoni*: Presence of H-2K determinants on adult worms and failure of allogeneic lymphocytes to recognize acquired MHC gene products on schistosomula, *J. Parasitol.* **68**:513.

Goidl, E. A., Schrater, A. F., Siskind, G. W., and Thorbecke, G. J., 1979, Production of auto-anti-idiotypic antibody during the normal immune response to TNP-ficoll. II. Hapten-reversible inhibition of anti-TNP plaque-forming cells by immune serum as an assay for auto-anti-idiotypic antibody, *J. Exp. Med.* **150**:154.

Goldring, O. L., Sher, A., Smithers, S. R., and McLaren, D. J., 1977, Host antigens and parasite antigens of murine *Schistosoma mansoni, Trans. R. Soc. Trop. Med. Hyg.* **71**:144.

Hamburger, J., Pelley, R. P., and Warren, K. S., 1976, *Schistosoma mansoni* soluble egg antigens: Determination of the stage and species specificity of their serologic reactivity by radioimmunoassay, *J. Immunol.* **117**:1561.

Harrison, R. A., Bickle, Q., and Doenhoff, M. J., 1982, Factors affecting the acquisition of resistance against *Schistosoma mansoni* in the mouse. Evidence that the mechanisms which mediate resistance during early patent infections may lack immunological specificity, *Parasitology* **84**:93.

Hayunga, E. G., Sumner, M. P., Stek, M. Jr., Vannier, W. E., and Chesnut, R. Y., 1983, Purification of the major concanavalin A binding surface glycoprotein from adult *Schistosoma mansoni, Proc. Helminthol. Soc. Wash.* (in press).

Hillyer, G. V., and Serrano, A. E., 1982, Cross protection in infection due to *Schistosoma mansoni* using tegument antigens of *Fasciola hepatica, J. Infect. Dis.* **145**:728.

Hockley, D. J., and McLaren, D. J., 1973, *Schistosoma mansoni*: Changes in the outer membrane of the tegument during development from cercaria to adult worm, *Int. J. Parasitol.* **3**:13.

Hockley, D. J., and Smithers, S. R., 1970, Damage to adult *Schistosoma mansoni* after transfer to a hyperimmune host, *Parasitology* **61**:95.

Hsü, S. Y. L., 1969, Sex of schistosome cercariae as a factor in the immunization of rhesus monkeys, *Exp. Parasitol.* **25**:202.

Hsü, S. Y. L., and Hsü, H. F., 1963, Further studies on rhesus monkeys immunized against *Schistosoma japonicum* by administration of cercariae of the Formosan strain, *Z. Tropenmed. Parasitol.* **14**:506.

Hubbard, W. J., 1978, Hypothesis: Alpha-2 macroglobulin-enzyme complexes as suppressors of cellular activity, *Cell. Immunol.* **39**:388.

Hubbard, W. J., and Damian, R. T., 1976, Isolation of mouse alpha$_2$-macroglobulinlike antigen from *Schistosoma mansoni*, American Society of Parasitologists, 52nd Annual Meeting, San Antonio, Abstracts, p. 44 (Abst.).

Hubbard, W. J., Hess, A. D., Hsia, S., and Amos, D. B., 1981, The effects of electrophoretically "slow" and "fast" α-2 macroglobulin on mixed lymphocyte cultures, *J. Immunol.* **126**:292.

Hunter, G. W. III, and Crandall, R. B., 1962, Studies on schistosomiasis. XIX. Results of preliminary experiments on the antigenic significance of the schistosome egg, *Mil. Med.* **127**:101.

Imperia, P. S., Fried, B., and Eveland, L. K., 1980, Pheromonal attraction of *Schistosoma mansoni* females toward males in the absence of worm-tactile behavior, *J. Parasitol.* **66**:682.

Incani, R. N., and McLaren, D. J., 1981, Neutrophil-mediated cytotoxicity to schistosomula of *Schistosoma mansoni in vitro*; Studies on the kinetics of complement and/or antibody-dependent adherence and killing, *Parasite Immunol.* **3**:107.

James, S. L., Labine, M., and Sher, A., 1981, Mechanisms of protective immunity against *Schistosoma mansoni* infection in mice vaccinated with irradiated cercariae. 1. Analysis of antibody and T-lymphocyte responses in mouse strains developing differing levels of immunity, *Cell. Immunol.* **65**:75.

Jerne, N. K., 1974, Towards a network theory of the immune system, *Ann. Immunol. Inst. Pasteur* **125c**:373.

Kagan, I. G., and Lee, C. L., 1953, Duration of acquired immunity of *Schistosomatium douthitti* infections in mice following treatment, *J. Infect. Dis.* **92**:52.

Kagan, I. G., and Norman, L., 1963, Analysis of helminth antigens (*Echinococcus granulosus* and *Schistosoma mansoni*) by agar gel methods, *Ann. N.Y. Acad. Sci.* **113**(art. 1): 130.

Kemp, W. M., 1970, Ultrastructure of the Cercarienhüllen Reaktion of *Schistosoma mansoni, J. Parasitol.* **56**:713.

Kemp, W. M., Greene, N. D., and Damian, R. T., 1974, Sharing of Cercarienhüllen Reaktion antigens between *Schistosoma mansoni* cercariae and adults and uninfected *Biomphalaria pfeifferi, Am. J. Trop. Med. Hyg.* **23**:197.

Kemp, W. M., Damian, R. T., Greene, N. D., and Lushbaugh, W. B., 1976, Immunocytochemical localization of mouse α_2-macroglobulinlike antigenic determinants on *Schistosoma mansoni* adults, *J. Parasitol.* **62**:413.

Kemp, W. M., Merritt, S. C., Bogucki, M. S., Rosier, J. G., and Seed, J. R., 1977, Evidence for adsorption of heterospecific host immunoglobulin on the tegument of *Schistosoma mansoni, J. Immunol.* **119**:1849.

Kemp, W. M., Brown, P. R., Merritt, S. C., and Miller, R. E., 1980, Tegument-associated antigen modulation by adult male *Schistosoma mansoni, J. Immunol.* **124**:806.

Kent, N. H., 1963, Comparative immunochemistry of larval and adult forms of *Schistosoma mansoni, Ann. N.Y. Acad. Sci.* **113**(art. 1):100.

Khoury, P. B., and Phillips, S. M., 1981, Kinetics and characterization of antigen-binding and antibody-producing cells in the regional draining lymph nodes and spleen during initial murine schistosomiasis. II. Cellular responses against egg immunogens, *Cell. Immunol.* **59**:246.

Klein, J., 1975, *Biology of the Mouse Histocompatibility-2 Complex. Principles of Immunogenetics Applied to a Single System*, Springer-Verlag, New York.

Kloetzel, K., 1967, Egg and pigment production in *Schistosoma mansoni* infections of the white mouse, *Am. J. Trop. Med. Hyg.* **16**:293.

Knight, W. B., Liard, F., Ritchie, L. S., Pellegrino, J., and Chiriboga, J., 1968, Labeling of *Biomphalaria glabrata* and cercariae of *Schistosoma mansoni* with radioselenium, *Exp. Parasitol.* **22**:309.

Knopf, P. M., 1982, The role of host hormones in controlling survival and development of *Schistosoma mansoni*, *Parasitology* **65**:35.

Knopf, P. M., and Cioli, D., 1980, *Schistosoma mansoni*: Resistance to an infection with cercariae induced by the transfer of live adult worms to the rat, *Int. J. Parasitol.* **10**:13.

Knopf, P. M., and Soliman, M., 1980, Effects of host endocrine gland removal on the permissive status of laboratory rodents to infection by *Schistosoma mansoni, Int. J. Parasitol.* **10**:197.

Knopf, P. M., Nutman, T. B., and Reasoner, J. A., 1977, *Schistosoma mansoni*: Resistance to reinfection in the rat, *Exp. Parasitol.* **41**:74.

Kruger, S. P., Heitman, L. P., Van Wyk, J. A., and McCully, R. M., 1969, The route of migration of *Schistosoma mattheei* from lungs to liver in sheep, *J. S. Afr. Vet. Med. Assoc.* **40**:39.

Kuntz, R. E., and Malakatis, G., 1955, Susceptibility studies in schistosomiasis. II. Susceptibility of wild mammals to infection by *Schistosoma mansoni* in Egypt, with emphasis on rodents, *Am. J. Trop. Med. Hyg.* **4**:75.

Kusel, J. R., 1972, Protein composition and protein synthesis in the surface membranes of *Schistosoma mansoni*, *Parasitology* **65**:35.

Kusel, J. R., MacKenzie, P. E., and McLaren, D. J., 1975, The release of membrane antigens into culture by adult *Schistosoma mansoni, Parasitology* **71**:247.

Lachmann, P. J., 1979, Complement, in: *The Antigens* (M. Sela, ed.), Vol. V, pp. 284–353, Academic Press, New York.

Landsperger, W. J., Stirewalt, M. A., and Dresden, M. H., 1982, Purification and properties of a proteolytic enzyme from the cercariae of the human parasite, *Schistosoma mansoni, Biochem. J.* **201**:137.

LaRue, G. R., 1957, The classification of digenetic Trematoda: A review and a new system, *Exp. Parasitol.* **52**:378.

Lawrence, J. A., 1973, *Schistosoma mattheei* in cattle: The host–parasite relationship, *Res. Vet. Sci.* **14**:400.

Lennox, R. W., and Schiller, E. L., 1972, Changes in dry weight and glycogen content as criteria for measuring the postcercarial growth and development of *Schistosoma mansoni, J. Parasitol.* **58**:489.

Levi-Schaffer, F., and Smolarsky, M., 1981, *Schistosoma mansoni*: Effect of insulin and a low-molecular-weight fraction of serum on schistosomula in chemically defined media, *Exp. Parasitol.* **52**:378.

Lewert, R. M., Yogore, M. G. Jr., Para, J. M., and Özcel, M. A., 1977, Rejection of mouse derived *Schistosoma japonicum* and serum lethality of hosts immunized with mouse globulin or mouse erythrocytes, *J. Parasitol.* **63**:825.

Lewis, F. A., and Colley, D. G., 1977, Modification of the lung recovery assay for schistosomula and correlations with worm burdens in mice infected with *Schistosoma mansoni, J. Parasitol.* **63**:413.

Long, E., Doenhoff, M., and Bain, J., 1978, Factors affecting the acquisition of resistance against *Schistosoma mansoni* in the mouse. 2. The time at which resistance to reinfection develops, *J. Helminthol.* **52**:187.

Long, E., Harrison, R., Bickle, Q., Bain, J., Nelson, G., and Doenhoff, M., 1980, Factors affecting the acquisition of resistance against *Schistosoma mansoni* in the mouse. The effects of varying the route and the number of primary infections, and the correlation between the size of the primary infection and the degree of resistance that is acquired, *Parasitology* **81**:355.

Machado, A. J., Gazzinelli, G., Pellegrino, J., and Dias da Silva, W., 1975, The role of C3-activating system in the cercaricidal action of normal serum, *Exp. Parasitol.* **38**:20.

MacKenzie, C. D., Ramalho-Pinto, F. J., McLaren, D. J., and Smithers, S. R., 1977, Antibody-mediated adherence of rat eosinophils to schistosumula of *Schistosoma mansoni in vitro, Clin. Exp. Immunol.* **30**:97.

Maddison, S. E., Norman, L., Geiger, S. J., and Kagan, I. G., 1970*a, Schistosoma mansoni* infection in the rat. I. Worm burden and serological response in infected, reexposed, and antigen-sensitized animals, *J. Parasitol.* **56**:1058.

Maddison, S.E., Geiger, S. J., Botero, B. and Kagan, I. G., 1970*b*, *Schistosoma mansoni* infection in the rat. II. Effects of immunosuppression and serum and cell transfer on innate and acquired immunity, *J. Parasitol.* **56**:1066.

Maddison, S. E., Geiger, S. J., and Kagan, I. G., 1971, *Schistosoma mansoni*: Immunity in *Macaca mulatta, Exp. Parasitol.* **29**:463.

Maddison, S. E., Slemenda, S. B., Hillyer, G. V., Chandler, F. W., and Kagan, I. G., 1979, Immune response to *Schistosoma mansoni* in rhesus monkeys with multiple chronic and early primary infections, *Infect. Immun.* **25**:249.

McCullough, F. S., and Bradley, D. J., 1973, Egg output stability and the epidemiology of *Schistosoma haematobium.* I. Variation and stability in *Schistosoma haematobium* egg counts, *Trans. R. Soc. Trop. Med. Hyg.* **67**:475.

McGuiness, T. B., and Kemp, W. M., 1981, *Schistosoma mansoni*: A complement-dependent receptor on adult male parasites, *Exp. Parasitol.* **51**:236.

McLaren, D. J., 1980, *Schistosoma Mansoni: The Parasite Surface in Relation to Host Immunity,* Research Studies Press, Chichester.

McLaren, D. J., and Hockley, D. J., 1977, Blood flukes have a double outer membrane, *Nature (Lond.)* **269**:147.

McLaren, D. J., and Incani, R. N., 1982, *Schistosoma mansoni*: Acquired resistance of developing schistosomula to immune attack *in vitro, Exp. Parasitol.* **53**:285.

McLaren, D. J., and Terry, R. J., 1982, The protective role of acquired host antigens during schistosome maturation, *Parasite Immunol.* **4**:29.

McLaren, D. J., Clegg, J. A., and Smithers, S. R., 1975, Acquisition of host antigens by young *Schistosoma mansoni* in mice: Correlation with failure to bind antibody *in vitro*, *Parasitology* **70**:67.

McLaren, D. J., Hockley, D. J., Goldring, O. L., and Hammond, B. J., 1978, A freeze fracture study of the developing tegumental outer membrane of *Schistosoma mansoni, Parasitology* **76**:327.

McMullen, D. B., Ritchie, L. S., Oliver-González, J., and Knight, W. B., 1967, *Schistosoma mansoni* in *Macaca mulatta*. Long-term studies on the course of primary and challenge infections, *Am. J. Trop. Med. Hyg.* **16**:620.

Meisenhelder, J. E., and Thompson, P. E., 1963, Comparative observations on experimental *Schistosoma mansoni* infections in African green and rhesus monkeys, *J. Parasitol.* **49**:567.

Meleney, H. E., and Moore, D. V., 1954, Observations on immunity to superinfection with *Schistosoma mansoni* and *S. haematobium* in monkeys, *Exp. Parasitol.* **3**:128.

Michaels, R. M., 1970, *Schistosoma mansoni*: Alteration in ovipositing capacity by transplanting between heterologous hosts, *Exp. Parasitol.* **27**:217.

Miller, K. L., and Smithers, S. R., 1980, *Schistosoma mansoni*: The attrition of a challenge infection in mice immunized with highly irradiated live cercariae, *Exp. Parasitol.* **50**: 212.

Miller, K. L., Smithers, S. R., and Sher, A., 1981, The response of mice immune to *Schistosoma mansoni* to a challenge infection which bypasses the skin: Evidence for two mechanisms of immunity, *Parasite Immunol.* **3**:25.

Miller, P., and Wilson, R. A., 1980, Migration of the schistosomula of *Schistosoma mansoni* from the lungs to the hepatic portal system, *Parasitology* **80**:267.

Minard, P., Dean, D. A., Jacobson, R. H., Vannier, W. E., and Murrell, K. D., 1978*a*, Immunization of mice with cobalt-60 irradiated *Schistosoma mansoni* cercariae, *Am. J. Trop. Med. Hyg.* **27**:76.

Minard, P., Dean, D. A., Vannier, W. E., and Murrell, K. D., 1978*b*, Effect of immunization on migration of *Schistosoma mansoni* through lungs, *Am. J. Trop. Med. Hyg.* **27**:87.

Mitchell, G. F., Cruise, K. M., Garcia, E. G., and Anders, R. F., 1981*a*, Hybridoma-derived antibody with immunodiagnostic potential for schistosomiasis japonica, *Proc. Natl. Acad. Sci. USA* **78**:3165.

Mitchell, G. F., Garcia, E. G., Anders, R. F., Valdez, C. A., Tapales, F. P., and Cruise, K. M., 1981*b*, *Schistosoma japonicum:* Infection characteristics in mice of various strains and a difference in the response to eggs, *Int. J. Parasitol.* **11**:267.

Moore, D. V., Crandall, R. B., and Hunter, G. W., 1963, Studies on schistosomiasis. XX. Further studies on the immunogenic significance of *Schistosoma mansoni* eggs in albino mice when subjected to homologous challenge, *J. Parasitol.* **49**:117.

Moser, G., Wassom, D. L., and Sher, A., 1980, Studies of the antibody-dependent killing of schistosomula of *Schistosoma mansoni* employing haptenic target antigens. I. Evidence that the loss of susceptibility to immune damage undergone by developing schistosomula involves a change unrelated to the masking of parasite antigens by host molecules, *J. Exp. Med.* **152**:41.

Mota-Santos, T. A., Gazzinelli, G., Ramalho-Pinto, F. J., Pellegrino, J., and Dias da Silva, W., 1976, Immunodepression in mice following *Schistosoma mansoni* infection, *Rev. Inst. Med. Trop. São Paulo* **18**:246.

Mota-Santos, T. A., Tavares, C. A. P., Gazzinelli, G., and Pellegrino, J., 1977, Immunosuppression mediated by adult worms in chronic schistosomiasis mansoni, *Am. J. Trop. Med. Hyg.* **26**:727.

Murrell, K. D., Yogore, M. G., Jr., Lewert, R. M., Clutter, W. G. and Vannier, W. E., 1973, Immunization with a zoophilic strain of *Schistosoma japonicum:* A re-evaluation of the Formosan strain of *S. japonicum* in rhesus monkeys, *Am. J. Trop. Med. Hyg.* **22**:723.

Murrell, K. D., Vannier, W. E., Minard, P., and Schinski, V. D., 1977, *Schistosoma mansoni:* Extraction and partial characterization of membrane antigens using an assay based on competitive inhibition of human antibodies binding to schistosomules, *Exp. Parasitol.* **41**:446.

Murrell, K. D., Clark, S., Dean, D. A., and Vannier, W. E., 1979*a*, Influence of mouse strain on induction of resistance with irradiated *Schistosoma mansoni* cercariae, *J. Parasitol.* **65**:829.

Murrell, K. D., Clark, S. S., Dean, D. A., and Vannier, W. E., 1979*b*, *Schistosoma mansoni:* Immunization of cynomolgus monkeys by injection of irradiated schistosomula, *Exp. Parasitol.* **48**:415.

Nagy, B. A., File, S. K., and Smith, J. H., 1981, Changes in the enteric vasculature of mice infected with *Schistosoma mansoni*, *Am. J. Trop. Med. Hyg.* **30**:999.

Naimark, D. H. Benenson, A. S., Oliver-González, J., McMullen D. B., and Ritchie, L. S., 1960, Studies of schistosomiasis in primates: Observations of acquired resistance (progress report), *Am. J. Trop. Med. Hyg.* **9**:430.

Nash, T. E., Lunde, M. N., and Cheever, A. W., 1981, Analysis and antigenic activity of a carbohydrate fraction derived from adult *Schistosoma mansoni*, *J. Immunol.* **126**:805.

Newsome, J., 1956, Problems of fluke immunity: with special reference to schistosomiasis, *Trans. R. Soc. Trop. Med. Hyg.* **50**:258.

Norden, A. P., Aronstein, W. S., and Strand, M., 1982, *Schistosoma mansoni:* Identification, characterization, and purification of the spine glycoprotein by monoclonal antibody, *Exp. Parasitol.* **54**:432.

Ogilvie, B. M., and Wilson, R. J. M., 1976, Evasion of the immune response by parasites, *Br. Med. Bull.* **32**:177.

Olds, G. R., and Mahmoud, A. A. F., 1981, Kinetics and mechanisms of pulmonary granuloma formation around *Schistosoma japonicum* eggs injected into mice, *Cell. Immunol.* **60**:251.

Olds, G. R., Olveda, R., Tracy, J. W., and Mahmoud, A. A. F., 1982, Adoptive transfer of modulation of granuloma formation and hepatosplenic disease in murine schistosomiasis japonica by serum from chronically infected animals, *J. Immunol.* **128**:1391.

Olivier, L., and Schneidermann, M., 1953, Acquired resistance to *Schistosoma mansoni* infection in laboratory animals, *Am. J. Trop. Med. Hyg.* **2**:298.

Ottesen, E. A., 1979, Modulation of the host response in human schistosomiasis. I. Adherent suppressor cells which inhibit lymphocyte proliferative responses to parasite antigens, *J. Immunol.* **123**:1639.

Ottesen, E. A., and Poindexter, R. W., 1980, Modulation of the host response in human schistosomiasis. II. Humoral factors which inhibit lymphocyte responses to parasite antigens, *Am. J. Trop. Med. Hyg.* **29**:592.

Ottesen, E. A., Hiatt, R. A., Cheever, A. W., Sotomayor, Z. R., and Neva, F. A., 1978, The acquisition and loss of antigen-specific cellular immune responsiveness in acute and chronic schistosomiasis in man, *Clin. Exp. Immunol.* **33**:38.

Ouaissi, M. A., Santoro, F., and Capron, A., 1980, Interaction between *Schistosoma mansoni* and the complement system. Receptors for C3b on cercariae and schistosomula, *Immunol. Lett.* **1**:197.

Pelley, R. P., Ruffier, J. J. and Warren, K. S., 1976, Suppressive effect of a chronic helminth infection, schistosomiasis mansoni, on the *in vitro* responses of spleen and lymph node cells to the T cell mitogens phytohemagglutinin and concanavalin A, *Infect. Immunol.* **13**:1176.

Peña de Grimaldo, E., and Kershaw, W. E., 1961, Results obtained by intensive exposure of white mice to *Schistosoma mansoni* infection. I. Recovery and distribution of adult *S. mansoni* from white mice seven weeks after percutaneous infection, the relation between the size of individual worms and the load of infection, and the longevity of heavily infected mice, *Ann. Trop. Med. Parasitol.* **55**:107.

Pereira, L. H., Coelho, P. M. Z., Fonesca, J. J. A., Bredl, A., and Pellegrino, J., 1972, Migration of *Schistosoma mansoni* larvae in the albino mouse, *Rev. Inst. Med. Trop. São Paulo* **14**:306.

Peresan, G., and Cioli, D., 1980, Resistance to cercarial challenge upon transfer of *Schistosoma mansoni* into mice, *Am. J. Trop. Med. Hyg.* **29**:1258.

Perez, H., and Terry, R. J., 1973, The killing of adult *Schistosoma mansoni in vitro* in the presence of antisera to host antigenic determinants and peritoneal cells, *Int. J. Parasitol.* **3**:499.

Perez, H., Clegg, J. A., and Smithers, S. R., 1974, Acquired immunity to *Schistosoma mansoni* in the rat: Measurement of immunity by the lung recovery technique, *Parasitology* **69**:349.

Phillips, S. M., and Colley, D. G., 1978, Immunologic aspects of host responses to schistosomiasis: Resistance, immunopathology, and eosinophil involvement, *Prog. Allergy* **24**:49.

Phillips, S. M., Reid, W. A., Bruce, J. I., Hedlund, K., Colvin, R. C., Campbell, R., Diggs, C. L., and Sadun, E. H., 1975, The cellular and humoral immune response to *Schistosoma mansoni* infections in inbred rats. I. Mechanisms during initial exposure, *Cell. Immunol.* **19**:99.

Phillips, S. M., Reid, W. A., and Sadun, E. H., 1977, The cellular and humoral immune response to *Schistosoma mansoni* infections in inbred rats. II. Mechanisms during reexposure, *Cell Immunol.* **28**:75.

Phillips, S. M., Reid, W. A., Doughty, B., and Khoury, P. B., 1978*a*, The cellular and humoral immune response to *Schistosoma mansoni* infections in inbred rats III. Development of optimal protective immunity following natural infections and artificial immunizations, *Cell. Immunol.* **38**:225.

Phillips, S. M., Reid, W. A., Doughty, B., and Khoury, P. B., 1978*b*, The cellular and humoral immune response to *Schistosoma mansoni* infections in inbred rats V. Prerequisite mechanisms for the development of optimal protective immunity, *Cell. Immunol.* **38**:239.

Powell, M. R., and Damian, R. T., 1982, Schistosomiasis mansoni in baboons. VI. Plastic-adherent-cell suppression of the primary *in vitro* antibody response of peripheral blood mononuclear cells to a heterologous antigen (SRBC) during chronic infection, *Am. J. Trop. Med. Hyg.* **31**:790.

Ramalho-Pinto, F. J., Goldring, O. L., Smithers, S. R., and Playfair, J. H. L., 1976*a*, T-cell helper response to antigens of *Schistosoma mansoni* in CBA mice, *Clin. Exp. Immunol.* **26**:327.

Ramalho-Pinto, F. J., Goldring, O. L., Playfair, J. H. L., and Smithers, S. R., 1976*b*, T-cell response to surface components of *Schistosoma mansoni*, in: *Biochemistry of Parasites and Host-Parasite Relationships* (H. Van den Bossche, ed.), pp. 291–298, Elsevier North-Holland Amsterdam.

Ramalho-Pinto, F. J., McLaren, D. J., and Smithers, S. R., 1978, Complement-mediated killing of schistosomula of *Schistosoma mansoni* by rat eosinophils *in vitro*, *J. Exp. Med.* **147**:147.

Read, C. P., 1958, Status of behavioral and physiological "resistance," *Rice Institute Pamphlet* **45**:36.

Ritchie, L. S., Garson, S., and Knight, W. B., 1963, The biology of *Schistosoma mansoni* in laboratory rats, *J. Parasitol.* **49**:571.

Ritchie, L. S., Knight, W. B., McMullen, D. B., and von Lichtenberg, F., 1966, The influence of infection intensity of *Schistosoma mansoni* on resistance against existing and subsequent infections in *Macaca mulatta*, *Am. J. Trop. Med. Hyg.* **15**:43.

Ritchie, L. S., Knight, W. B., Oliver-González, J., Frick, L. P., Morris, J. M., and Croker, W. L., 1967, *Schistosoma mansoni* infections in *Cercopithecus sabaeus* monkeys, *J. Parasitol.* **53**:1217.

Rocklin, R. E., Brown, A. P., Warren, K. S., Pelley, R. P., Houba, V., Siongok, T. K. A., Ouma, J., Sturrock, R. F., and Butterworth, A. E., 1980, Factors that modify the cellular-immune response in patients infected by *Schistosoma mansoni*, *J. Immunol.* **125**:1916.

Ruppel, A., and Cioli, D., 1977, A comparative analysis of various developmental stages of *Schistosoma mansoni* with respect to their protein composition, *Parasitology* **75**:339.

Sadun, E. H., 1963, Immunization in schistosomiasis by previous exposure to homologous and heterologous cercariae by inoculation of preparations from schistosomes and by exposure to irradiated cercariae, *Ann. N.Y. Acad. Sci.* **113**(art. 1):418.

Sadun, E. H., Schoenbechler, M. J., and Bentz, M., 1965, Multiple antibody response in *Schistosoma mansoni* infections: Antigenic constituents in eggs, cercariae, and adults (excretions and secretions) determined by flocculation reactions, cross absorption and double diffusion studies, *Am. J. Trop. Med. Hyg.* **14**:977.

Samuelson, J. C., and Caulfield, J. P., 1982, Loss of covalently labeled glycoproteins and glycolipids from the surface of newly transformed schistosomula of *S. mansoni*, *J. Cell. Biol.* **94**:363.

Samuelson, J. C., Sher, A., and Caulfield, J. P., 1980, Newly transformed schistosomula spontaneously lose surface antigens and C3 acceptor sites during culture, *J. Immunol.* **124**:2055.

Samuelson, J. C., Caulfield, J. P., and David, J. R., 1982, Schistosomula of *S. mansoni* clear concanavalin A from their surface by sloughing, *J. Cell. Biol.* **94**:355.

Santoro, F., Lachmann, P. J., Capron, A., and Capron, M., 1979*a*, Activation of complement by *Schistosoma mansoni* schistosomula: Killing of parasites by the alternative pathway and requirement of IgG for classical pathway activation, *J. Immunol.* **123**:1551.

Santoro, F., Ouaissi, M. A., and Capron, A., 1979*b*, Receptors for complement (Clq and C3b) on the immature forms of *Schistosoma mansoni*, *IRCS Med. Sci.* **7**:576.

Santoro, F., Ouaissi, M. A., Pestel, J., and Capron, A., 1980, Interaction between *Schistosoma mansoni* and the complement system. Binding of Clq to schistosomula, *J. Immunol.* **124**:2886.

Santoro, F., Prata, A., Silva, A. E., and Capron, A., 1981, Correlation between circulating antigens detected by the radioimmunoprecipitation-polyethylene glycol assay (RIPEGA) and Clq-binding immune complexes in human schistosomiasis mansoni, *Am. J. Trop. Med. Hyg.* **30**:1020.

Senft, A. W., Gibler, W. B., and Knopf, P. M., 1978, Scanning electron microscope observations on tegument maturation in *Schistosoma mansoni* grown in permissive and nonpermissive hosts, *Am. J. Trop. Med. Hyg.* **27**:258.

Sergent, E., 1963, Latent infection and premunition. Some definitions of microbiology and immunology, in: *Immunity to Protozoa. A Symposium of the British Society for Immunology*, (P. C. C. Garnham, A. E. Pierce, and I. Roitt, eds.), pp. 39–47, F. A. Davis, Philadelphia.

Sher, A., and Benno, D., 1982, Decreasing immunogenicity of developing schistosome larvae, *Parasite Immunol.* **4**:101.

Sher, A., and Moser, G., 1981, Immunologic properties of developing schistosomula, *Am. J. Pathol.* **120**:121.

Sher, A., Kusel, J. R., Perez, H., and Clegg, J. A., 1974*a*, Partial isolation of a membrane antigen which induces the formation of antibodies lethal to schistosomes cultured *in vitro*, *Clin. Exp. Immunol.* **18**:357.

Sher, A., MacKenzie, P., and Smithers, S. R., 1974*b*, Decreased recovery of invading parasites from the lungs as a parameter of acquired immunity to schistosomiasis in the mouse, *J. Infect. Dis.* **130**:626.

Sher, A., Hall, B. F., and Vadas, M. A., 1978, Acquisition of murine major histocompatibility complex gene products by schistosomula of *Schistosoma mansoni*, *J. Exp. Med.* **148**:46.

Shirazian, D., and Schiller, E. L., 1982, Mating recognition by *Schistosoma mansoni in vitro*, *J. Parasitol.* **68**:650.

Simpson, A. J. G., and Cioli, D., 1982, The key to the schistosome's success, *Nature* (*Lond.*) **296**:809.

Simpson, A. J. G., and McLaren, D. J., 1982, *Schistosoma mansoni:* Tegumental damage as a consequence of lectin binding, *Exp. Parasitol.* **53**:105.

Simpson, A. J. G., Singer, D., Sher, A., and McCutchan, T. M., 1982, Investigation of possible homologies between the genome of *Schistosoma mansoni* and that of its hosts, *Fed. Proc.* **41**:729 (Abst.).

Smith, J. H., and von Lichtenberg, F., 1974, Observations on the ultrastructure of the tegument of *Schistosoma mansoni* in mesenteric veins, *Am. J. Trop. Med. Hyg.* **23**:71.

Smith, M. A., and Clegg, J. A., 1979, Different levels of immunity to *Schistosoma mansoni* in the mouse: the role of variant cercariae, *Parasitology* **78**:311.

Smith, M. A., Clegg, J. A., Kusel, J. R., and Webbe, G., 1975, Lung inflammation in immunity to *Schistosoma mansoni*, *Experientia* **31**:595.

Smith, M. A., Clegg, J. A., Snary, D. and Trejdosiewicz, A. J., 1982, Passive immunization of mice against *Schistosoma mansoni* with an IgM monoclonal antibody, *Parasitology* **84**:83.

Smithers, S. R., 1962, Stimulation of acquired resistance to *Schistosoma mansoni* in monkeys: Role of eggs and worms, *Exp. Parasitol.* **12**:263.

Smithers, S. R., 1967, Acquired resistance to *Schistosoma mansoni* in the rhesus monkey (*Macaca mulatta*), *Ann. Soc. Belg. Med. Trop.* **47**:87.

Smithers, S. R., and Gammage, K., 1980, Recovery of *Schistosoma mansoni* from the skin, lungs and hepatic portal system of naive mice and mice previously exposed to *S. mansoni:* Evidence for two phases of parasite attrition in immune mice, *Parasitology* **80**:289.

Smithers, S. R., and Terry, R. J., 1965*a*, Naturally acquired resistance to experimental infections of *Schistosoma mansoni* in the rhesus monkey (*Macaca mulatta*), *Parasitology* **55**:701.

Smithers, S. R., and Terry, R. J., 1965*b*, Acquired resistance to experimental infections of *Schistosoma mansoni* in the albino rat, *Parasitology* **55**:711.

Smithers, S. R., and Terry, R. J., 1967, Resistance to experimental infection with *Schistosoma mansoni* in rhesus monkeys induced by the transfer of adult worms, *Trans. R. Soc. Trop. Med. Hyg.* **61**:517.

Smithers, S. R., and Terry, R. J., 1969, Immunity in schistosomiasis, *Ann. N.Y. Acad. Sci.* **160**:826.

Smithers, S. R., and Terry, R. J., 1976, The immunology of schistosomiasis, in: *Advances in Parasitology* (B. Dawes, ed.), Vol. 14, pp. 399–411, Academic Press, New York.

Smithers, S. R., Terry, R. J., and Hockley, D. J., 1968, Do adult schistosomes masquerade as their hosts? *Trans. R. Soc. Trop. Med. Hyg.* **62**:466.

Smithers, S. R., Terry, R. J., and Hockley, D. J., 1969, Host antigens in schistosomiasis, *Proc. R. Soc. B.* **171**:483.

Smithers, S. R., McLaren, D. J., and Ramalho-Pinto, F. J., 1977, Immunity to schistosomes: The target, *Am. J. Trop. Med. Hyg.* **26**(6, part 2):11.

Smyth, J. D., 1973, Some interface phenomena in parasitic protozoa and platyhelminths, *Can. J. Zool.* **51**:367.

Snary, D., Smith, M. A., and Clegg, J. A., 1980, Surface proteins of *Schistosoma mansoni* and their expression during morphogenesis, *Eur. J. Immunol.* **10**:573.

Sprent, J. F. A., 1962, Parasitism, immunity and evolution, in: *The Evolution of Living Organisms* (G. W. Leeper, ed.), Symposium of the Royal Society, Victoria, Melbourne, pp. 149–165, Melbourne University Press, Melbourne, Australia.

Sprent, J. F. A., 1963, *Parasitism. An Introduction to Parasitology and Immunology for Students of Biology, Veterinary Science and Medicine*, University of Queensland Press, St. Lucia.

Stek, M., Jr., Minard, P., Dean, D. A., and Hall, J. E., 1981, Immunization of baboons with *Schistosoma mansoni* cercariae attenuated by gamma irradiation, *Science* **212**:1518.

Stirewalt, M. A., 1963, Seminar on immunity to parasitic helminths. IV. Schistosome infections, *Exp. Parasitol.* **13**:18.

Stirewalt, M. A., Kuntz, R. E., and Evans, A. S., 1951, The relative susceptibilities of the commonly-used laboratory mammals to infection by *Schistosoma mansoni*, *Am. J. Trop. Med. Hyg.* **31**:57.

Sturrock, R. F., Butterworth, A. E., and Houba, V., 1976, *Schistosoma mansoni* in the baboon (*Papio anubis*): parasitological responses of Kenyan baboons to different exposures of a local parasite strain, *Parasitology* **73**:239.

Suzuki, T., and Damian, R. T., 1981, Schistosomiasis mansoni in baboons. IV. The development of antibodies to *Schistosoma mansoni* adult worm, egg, and cercarial antigens during acute and chronic infections, *Am. J. Trop. Med. Hyg.* **30**:825.

Tarleton, R. L., and Kemp, W. M., 1981, Demonstration of IgG-Fc and C3 receptors on adult *Schistosoma mansoni*, *J. Immunol.* **126**:379.

Tavares, C. A. P., Soares, R. C., Coelho, P. M. Z., and Gazzinelli, G., 1978, *Schistosoma mansoni*: Evidence for a role of serum factors in protecting artificially transformed schistosomula against antibody-mediated killing *in vitro*, *Parasitology* **77**:225.

Tavares, C. A. P., Cordeiro, M. N., Mota-Santos, T. A., and Gazzinelli, G., 1980, Artificially transformed schistosomula of *Schistosoma mansoni:* Mechanism of acquisition of protection against antibody-mediated killing, *Parasitology* **80**:95.

Taylor, D. W., and Butterworth, A. E., 1982, Monoclonal antibodies against surface antigens of schistosomula of *Schistosoma mansoni*, *Parasitology* **84**:65.

Taylor, M. G., Nelson, G. S., Smith, M., and Andrews, B. J., 1973*a*, Comparison of the infectivity and pathogenicity of six species of African schistosomes and their hybrids. 2. Baboons, *J. Helminthol.* **47**:455.

Taylor, M. G., Nelson, G. S., Smith, M., and Andrews, B. J., 1973*b*, Studies on heterologous immunity in schistosomiasis. 7. Observations on the development of acquired homologous and heterologous immunity to *Schistosoma mansoni* in baboons, *Bull. WHO* **49**:57.

Terry, R. J., and Smithers, S. R., 1975, Evasion of the immune response by parasites, *Symp. Soc. Exp. Biol.* **29**:453.

Theofilopoulos, A. N., 1980, Immune complexes in humoral immune responses: Suppressive and enhancing effects, *Immunol. Today* **1**:1.

Thompson, J. H., 1954, Host-parasite relationships of *Schistosoma mansoni*, *Exp. Parasitol.* **3**:140.

Thompson, P. E., Meisenhelder, J. E., Moore, A. K., and Waitz, J. A., 1965, Laboratory studies on the joint effects of certain Tris (*p*-aminophenyl) carbonium salts and antimonials as antischistosomal drugs, *Bull. WHO* **33**:517.

Todd, C. W., Goodgame, R. W., and Colley, D. G., 1979, Immune responses during human schistosomiasis mansoni. V. Suppression of schistosome antigen-specific lymphocyte blastogenesis by adherent/phagocytic cells, *J. Immunol.* **122**:1440.

Todd, C. W., Goodgame, R. W., and Colley, D. G., 1980, Immune responses during human schistosomiasis mansoni. VII. Further analysis of the interactions between patient sera and lymphocytes during in vitro blastogenesis to schistosome antigen preparations, *Am. J. Trop. Med. Hyg.* **29**:875.

Torpier, G., and Capron, A., 1980, Intramembrane particle movements associated with binding of lectins on *Schistosomia mansoni* surface, *J. Ultrastruct. Res.* **72**:325.

Torpier, G., Capron, A., and Ouaissi, M. A., 1979*a*, Receptor of IgG (Fc) and Human β_2-microglobulin on *S. mansoni* schistosomula, *Nature* (*Lond.*) **278**:447.

Torpier, G., Ouaissi, M. A., and Capron, A., 1979*b*, Freeze-fracture study of immune-induced *Schistosoma mansoni* membrane alterations. I. Complement-dependent damage in the presence of antisera to host antigenic determinants, *J. Ultrastruct. Res.* **67**:276.

Torpier, G., Hirn, M., Nirde, P., DeReggi, M., and Capron, A., 1982, Detection of ecdysteroids in the human trematode, *Schistosoma mansoni*, *Parasitology* **84**:123.

Tsang, V. C. W., Hubbard, W. J., and Damian, R. T., 1977, Coagulation Factor XIIa (activated Hageman Factor) inhibitor from adult *Schistosoma mansoni*, *Am. J. Trop. Med. Hyg.* **26**:243.

Vogel, H., 1942, Ueber den Einfluss des Geschletspartners auf Wachstum und Entwicklung bei *Bilharzia mansoni* und bei Kreuzpaarungen zwischen vierschiedenen Bilharzia-Arten, *Zentrabl. Bakteriol.* **148**:78.

Vogel, H., 1962, Beobachtungen über die etworbene immunität von Rhesusaffen gegen Schistosoma-Infektionen, *Z. Tropenmed. Parasitol.* **13**:397.

Vogel, H., and Minning, W., 1953, Über die erworbene Resistenz von Macacus rhesus gegenüber *Schistosoma japonicum*, *Z. Tropenmed. Parasitol.* **4**:418.

von Lichtenberg, F., and Byram, J. E., 1980, Pulmonary cell reactions in resistance to *Schistosoma mansoni*, in: *The Host-Invader Interplay* (H. Van den Bossche, ed.), pp. 405–416, Elsevier/North-Holland, Amsterdam.

von Lichtenberg, F., Sadun, E. H., and Bruce, J. I., 1962, Tissue responses and mechanisms of resistance in schistosomiasis mansoni in abnormal hosts, *Am. J. Trop. Med. Hyg.* **11**:347.

von Lichtenberg, F., Sadun, E. H., and Bruce, J. I., 1963, Host response to eggs of *Schistosoma mansoni*. III. The role of eggs in resistance, *J. Infect. Dis.* **113**:113.

von Lichtenberg, F., Bawden, M. P., and Shealey, S. H., 1974, Origin of circulating antigen from the schistosome gut. An immunofluorescent study, *Am. J. Trop. Med. Hyg.* **23**:1088.

Warren, K. S., 1972, The immunopathogenesis of schistosomiasis: A multidisciplinary approach, *Trans. R. Soc. Trop. Med. Hyg.* **66**:417.

Warren, K. S., 1974, Modulation of immunopathology in schistosomiasis, in: *Parasites in the Immunized Host: Mechanisms of Survival*, Ciba Found. Symp. **25**(new series):243.

Warren, K. S., 1982, The secret of the immunopathogenesis of schistosomiasis: *in vivo* models, *Immunol. Rev.* **61**:189.

Warren, K. S., and Domingo, E., 1970*a*, *Schistosoma mansoni*: Stage specificity of granuloma formation around eggs after exposure to irradiated cercariae, unisexual infections, or dead worms, *Exp. Parasitol.* **27**:60.

Warren, K. S., and Domingo, E. O., 1970*b*, Granuloma formation around *Schistosoma mansoni*, *S. haematobium*, and *S. japonicum* eggs. Size and rate of development, cellular composition, cross-sensitivity, and rate of egg destruction, *Am. J. Trop. Med. Hyg.* **19**: 292.

Warren, K. S., Boros, D. L., Hang, L. M., and Mahmoud, A. A. F., 1975, The *Schistosoma japonicum* egg granuloma, *Am. J. Pathol.* **80**:279.

Webbe, G., and James, C., 1973, Acquired resistance to *Schistosoma haematobium* in the baboon (*Papio anubis*), *Trans. R. Soc. Trop. Med. Hyg.* **67**:151.

Webbe, G., James, C., Nelson, G. S., Smithers, S. R., and Terry, R. J., 1976, Acquired resistance to *Schistosoma haematobium* in the baboon (*Papio anubis*) after cercarial exposure and adult worm transplantation, *Ann. Trop. Med. Parasitol.* **70**:411.

Weigle, W. O., 1973, Immunological unresponsiveness, *Adv. Immunol.* **16**:61.

Wheater, P. R., and Wilson, R. A., 1979, *Schistosoma mansoni*: A histological study of migration in the laboratory mouse, *Parasitology* **79**:46.

Wilkins, H. A., and Brown, J., 1977, *Schistosoma haematobium* in a Gambian community. II. Impaired cell-mediated immunity and other immunological abnormalities, *Ann. Trop. Med. Parasitol.* **71**:59.

Wilks, N. E., 1967, Lung-to-liver migration of schistosomes in the laboratory mouse, *Am. J. Trop. Med. Hyg.* **16**:599.

Wilson, R. A., and Barnes, P. E., 1974, An *in vitro* investigation of dynamic processes occurring in the schistosome tegument, using compounds known to disrupt secretory processes, *Parasitology* **68**:259.

Wilson, R. A., and Barnes, P. E., 1977, The formation and turnover of the membrano-calyx on the tegument of *Schistosoma mansoni*, *Parasitology* **74**:61.

Wilson, R. A., Draskau, T., Miller, P., and Lawson, J. R., 1978, *Schistosoma mansoni:* The activity and development of the schistosomulum during migration from the skin to the hepatic portal system, *Parasitology* **77**:57.

Wistar, R., Murrell, K. D., Lewert, R. M., Yogore, M. C., Cole, C., and Clutter, W. G., 1975, The development of anti-immunoglobulin antibodies in rhesus monkeys repeatedly exposed to *Schistosoma japonicum*, *Am. J. Trop. Med. Hyg.* **24**:632.

Yogore, M. G., Jr., Lewert, R. M., and Santos, A. T., 1971, Precipitins in serum from men with presumptive reexposure to cercariae of *Schistosoma japonicum*, *Am. J. Trop. Med. Hyg.* **20**:54.

Zodda, D. M., and Phillips, S. M., 1982, Monoclonal antibody-mediated protection against *Schistosoma mansoni* infection in mice, *J. Immunol.* **129**:2326.

Chapter 11

Immunopathology of Parasitic Diseases: A Conceptual Approach

S. Michael Phillips and Edward G. Fox

Allergy and Immunology Section
University of Pennsylvania School of Medicine
Philadelphia, Pennsylvania 19104

I. INTRODUCTION

The pathology caused by parasites, the most prevalant infectious diseases of humankind, can be divided into two major aspects: the first relates to the effects of the parasite per se, e.g., iron deficiency anemia, secondary to blood loss caused by intestinal nematodes; hemolytic anemia, secondary to intracellular protozoans; or hepatobiliary fibrosis, secondary to toxic products produced by intraductal trematodes. However, in many parasitic conditions there is limited pathology directly attributable to the parasite. Most host morbidity is related to the immunoinflammatory response of the host to the parasite. This reaction is the principal subject of this chapter.

The immunopathology of parasitic disease represents an extremely complex and diverse series of phenomena. This complexity is dependent on a wide variety of factors including both characteristics of the infecting agent and the nature of the host's response to these agents.

To illustrate the immunopathologic spectrum that results as a consequence of parasitic infection or infestation, a number of approaches may be taken. Clearly, it is beyond the logistics of this review to study the clinical manifestations, pathology, pathophysiology, and immunopathogenesis associated with each parasitic disease. Rather, we attempt to view immunopathology in a conceptual manner, attempting to use immunologic phenomena to produce a mechanistic model for the clinical spectrum of parasitic disease (Phillips and Fox, 1982) (Fig.1). Viewed teleologically, this hypothesis suggests that the body seeks an optimal immune response that will maximize resistance and minimize morbidity. To achieve such an optimal homeostatic mechanism, certain postulates are suggested:

1. Total immune response is composed of two major aspects. The first is related to an immunoprotective phenomenon and principally designed to control parasite growth and development. As a consequence of these responses, host pathology caused by the parasite is minimized. The second aspect of the immune response is immunopathogenic. This latter category produces pathology per se. The two aspects of the immune response need not be consonant, either kinetically or mechanistically. In this context, clinical disease would be most salient when immunopathogenic mechanisms are dominant (A) or the effective control of the parasite is not attained (C).
2. The spectrum of the immune response ranges from a preponderance of supraoptimal immunologic reactivies, associated with a hyperergic response, to suboptimal immune reactivity, associated with hypoergic dominance states. A variety of helper or enhancing influences dominate hyperergia whereas suppressor or modulatory influences, stemming either from the parasite per se or from various host components, dominate hypoergia.
3. The hyperergic state is associated with the immunopathologic consequences of a variety of host-dependent immunologic mechanisms, while the parasite is well controlled. The hypoergic state is associated with relatively uncontrolled parasite growth and pathologic consequences are more closely related to this latter phenomenon.

To illustrate this conceptual model of immune pathology, two apparently dissimilar parasitic diseases–filariasis and schistosomiasis–are described. It is intended in this paper to suggest that the majority of immunopathology in these diseases and, by analogy, most other parasitic diseases, are related to a broad spectrum of immunologic reactions. The relative efficiency with which these reactions are capable of controlling the parasitic disease, while being optimally self-modulated, determines the degree of host pathology (B).

II. FILARIASIS

Filariasis is a disease complex etiologically resulting from a variety of parasitic nematodes belonging to the order Filarioidea. Most morbidity is associated with various developmental and adult stages of the endolymphatic-dwelling

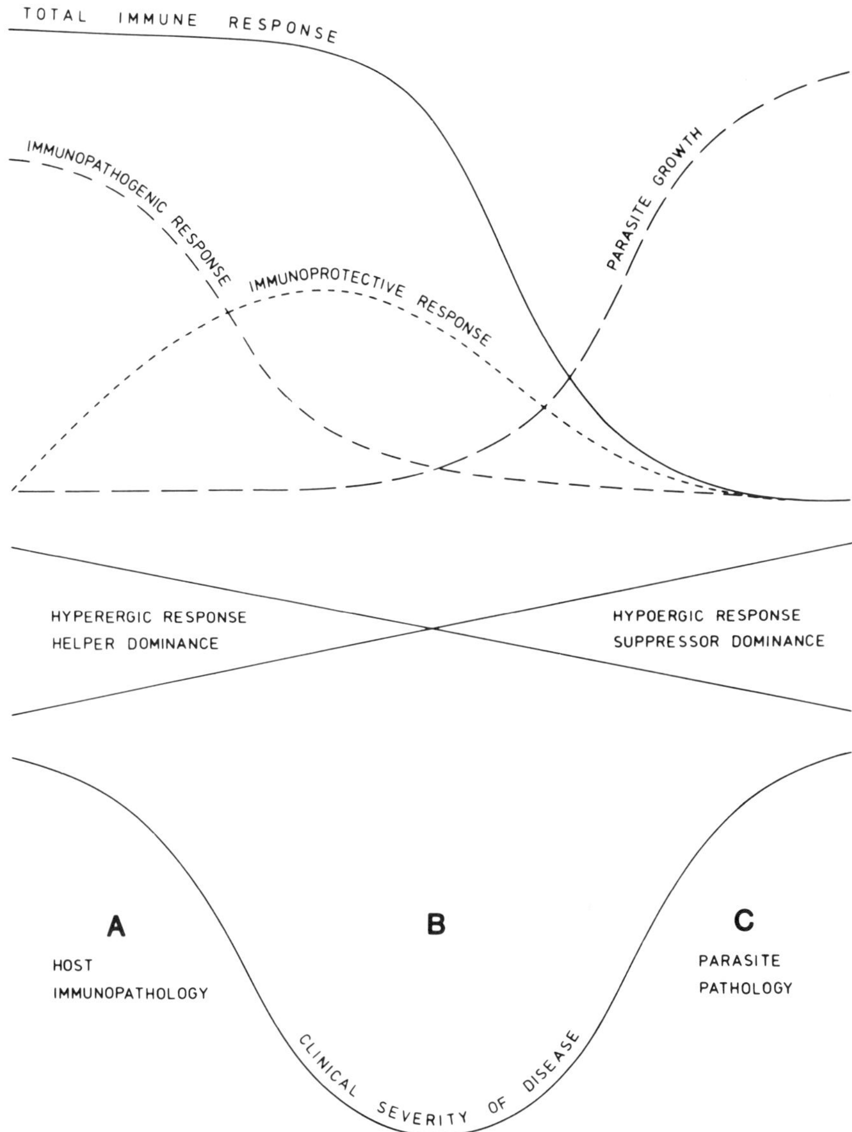

Figure 1. Mechanistic mode for the clinical spectrum of parasitic disease. (Phillips and Fox, 1982.)

Wuchereria bancrofti, W. lewisi, Brugia malayi, and *B. timori,* and with the subcutaneous-dwelling *Onchocerca volvulus*. Transmission of bancroftian and malayan filariasis is accomplished by mosquito vectors including many species of the genera *Culex, Aedes, Mansonia,* and *Anopheles,* while onchocerciasis is transmitted by black flies, *Simulium* species.

The pathology associated with other filarial species is less severe. *Loa loa,* the West African eyeworm, migrates in the skin and subcutaneous tissues and may incite Calabar swellings, apparent hypersensitivity reactions to the parasite. *Dipetalonema perstans* adults inhabit pleural, peritoneal, and pericardial areas and produce minimal symptoms. *D. streptocerca* adults inhabit dermal areas and local pathology is correlated with the degree of skin microfilariae (Mff). *Mansonella ozzardi* adults may inhabit pleural or peritoneal and perhaps lymphatics and apparently induce limited pathology.

The filarial life cycle is of the cyclic nonpropagative form in that multiplication only occurs in the final host and not in the intermediate host, which is clearly differentiable from the life cycle of *Schistosoma* species. In the final host, infection begins with the deposition of the infective stage larvae (L_3) on the skin surface. In the case of these lymphatic dwelling worms, the L_3 penetrate and migrate through lymphatic channels to the draining lymph node. Here, in or near the subcapsular sinus, the larvae molt twice, cast off both cuticle and highly allergenic molting fluids, and become juvenile adults. Pairing of male and female worms may occur, and they move distally in the lympatic channels. The mature females are ovoviviparous, giving birth to active sheathed embryos, (Mff), that circulate in the blood. Patency, observation of Mff in blood, may be observed as early as two months postinfection for the brugian infections. Epidemiologic investigations suggest that patency of bancroftian infections may occur as early as 6 months postinfection and may last up to 40 years (Carme and Laigret, 1979).

In onchocerciasis, the localization of adult worms occurs in subcutaneous tissues, where the parasites are securely encapsulated by the host. The migratory pathways from infective bites to final anatomic sites are unknown. Microfilariae are released by the female and migrate by unknown mechanisms through the host capsule and circulate in layers of host skin. Patency is suggested to take 18-20 months (World Health Organization, 1981), but this time estimate and the precise nature of early development remain to be delineated definitively owing to the present lack of a suitable experimental host.

The life cycles of all filariae are completed when the Mff are ingested by the appropriate arthropod vector. The larvae molt twice in the vector to become infective third-stage larvae.

In the infected human population at risk, a wide spectrum of disease, both in terms of severity and specific manifestations, is observed. This spectrum varies from apparent entirely refractory individuals, to asymptomatic individuals with circulating Mff, to microfilaremic patients with episodic fevers, lymphangitis, lymphadenitis, funiculitis, and hydrocoels, to chronically debilitated patients demonstrating chyluria, and/or elephantiasis. Tropical pulmonary eosinophilia (TPE), an occult filarial condition, incorporates elements of several of these stages. Endemic onchocercine populations will provide similar spectral patterns as well as evidence of severe urticaria and chronic ophthalmic pathology.

This observed melange of human manifestations may be attributable to several factors. One is that the host encounters a multitude of potentially soluble and particulate immunogenic materials from four distinct developmental stages (L_3, L_4, adults, Mff) and the products of two molts. Second, epidemiologic evidence suggests that a given endemic population may receive from none to several hundred infective bites per year; thus the exposure patterns to the parasite are asynchronous and quantitatively highly variable. Third, genetically determined differences, both as regards the nature of the challenge provided by the parasite and the response of the human host to that challenge, play as yet largely undetermined roles. Socioeconomic and nutritional differences and concomitant infections may also affect disease expression.

A. Clinical and Histologic Pathology

The earliest events associated with filariid infections cannot be practically studied in exposed human populations. This has necessitated reliance on experimental animal models which, parenthetically, may not be highly analogous to events in man. Three to 4 days postinfection (p.i.), the lymphatic channels, which contain migrating L_3, demonstrate flow reductions. Hyperplasia of Bilroth's cords and paracortical areas, anatomic sites of thymic dependent lymphocyte responses, are observed in draining lymph nodes (Schacher and Sahyoun, 1967; Rogers *et al.*, 1975). With maturation to L_4 and adults, molted cuticles and fluids cause additional localized inflammatory reactions through both direct effects and secondary to antigenic stimuli. Newly ecdysed females extrude uterine products, and initial microfilarial deposition also incites additional acute inflammatory reactions. Soon after the appearance of patency, splenic and lymphatic nodules demonstrate their greatest activity regarding proliferation of T and B cells. Specific T- and B-cell responses are also detected at this time (Portaro *et al.*, 1976; Benjamin and Soulsby, 1976). In time, chronic inflammatory mononuclear cell types dominate and T-dependent areas are less prominent. In spleens and lymph nodes and other reticuloendiothelial system (RES) tissues, increased numbers of fibroblasts and the deposition of collagen and amyloid (Crowell *et al.*, 1973) are observed.

In the lymphatic channels, early reactions to molting fluids are later augmented by host responses to dead and moribund larvae and adults. Radiography demonstrates hypertrophic lymph nodes, hypertrophic, dilated lymphatic channels, and resultant lymphangiectasia. In animal models, the production of clinical lymphedema has been observed, but only rarely. Schacher *et al.* (1973) innoculated *B. pahangi* L_3 into the dorsal metatarsal hind limbs of dogs and were able to demonstrate transient lymphedema of the rear legs as early as 3 months p.i. Some of the animals developed elephantiasis-like lesions. Other studies have suggested the presence of regional immunodecompensation with the propensity toward secondary infections. Bosworth *et al.* (1973) described a reaction wherein the infected and noninfected limbs of cats were compared after challenge with group G *B*-hemolytic streptococcus. While the host easily controlled the bacterial infection in the noninfected limb, the *B. pahangi* infected limb with lymphangitis permitted bacterial proliferation, eventually associated with cellulitis and exfoliation of the skin. These confounding bacterial infections and other nonfilarial insults are of unknown significance *vis-à-vis* lymphedema and more chronic problems (Ottesen, 1980). As a result of these insults, the lymphatic system demonstrates increased hydrostatic pressure and a number of secondary changes. First, collateral lymphatic circulation develops to reestablish patent lymph circulation. Second, recanalization of thrombolymphatic lesions permits additional lymph flow. Parenthetically, *W. bancrofti* and the newly described species, *W. kalimantani*, have recently been studied in the silver-leafed monkey, *Presbytis cristatus* (Palmieri *et al.*, 1980, 1982). When infected for at least 1 year, these animals develop early lymphadenopathic changes similar to those in the human condition (Palmieri *et al.*, 1982). Thus, it is hoped that in the future this model may provide much useful information relative to pathogenesis.

Reactions in endemic populations to early infections are difficult to detect and evaluate. Manifestations include headache, backache, myalgia, anorexia, urticaria, and especially episodic fevers. These transient signs and symptoms, lasting a few days to a week, reoccur several times per year, ostensibly after reexposure or death of adult parasites. Knowledge of these early manifestations comes from newly introduced individuals with no previous exposure to the parasites (Webster, 1946). They develop clinical signs of acute filariasis within a few months, but only rarely develop patent infections. If these individuals remain for extended periods, the usual long-term chronic obstructive changes are manifested in much-reduced time spans in comparison to disease time frames in endemic populations. Acute attacks include retrograde lymphangitis with swelling, lymphadenitis and possible abscesses. Other findings may include funiculitis, orchitis, epididymitis, and hydrocoels.

With continued insult of the lymphatic system, the effects of lymphangitis and obstructive fibrosis may excede the host's compensatory abilities to maintain patent lymphatic flows. Several sequelae result from these episodes of lymphedema. The commonest sign of chronic infection, particularly with *W.*

bancrofti, is hydrocoels (Nelson, 1979). If lymphangiectasis and lymphedema lead to obstruction of drainage into the thoracic duct, particularly of the abdominal and renal lymphatics, ascites and chyluria may result. *W. bancrofti* has often been associated with these conditions (Nelson, 1966). Ultrastructural analyses have shown that normal glomeruli may be located adjacent to areas evidencing glomerular pathologic changes including mesangial proliferation and intraposition (Date *et al.*, 1979). These changes are thought to be a response to immune complex deposition and episodes of glomerulonephritis.

A number of concomitant laboratory hematologic changes are noted, including hypochlyomicronemia, indicative of severe chyluria. Immune components such as lymphocytes are often lost, in some cases leading to a mild lymphocytopenia (Date *et al.*, 1980). Granulocyte numbers are maintained. Complement components are found in the chylous urine, but hypocomplementemia is rare (Date *et al.*, 1982). This decrease in total lymphocytes, and particularly T lymphocytes, is associated with a concomitant depression of DTH to a variety of antigens. While patients with untreated filariasis demonstrate immunosuppression (Grove and Forbes, 1979), their responses will return toward normal when the patients are treated with diethylcarbamazine (DEC) or surinam. Several months after DEC treatment, however, chyluria as well as immunodepression may still be evident (Date *et al.*, 1981).

The most dramatic representation of chronic filariasis, elephantiasis, afflicts a very small percentage of the endemic population. This condition is believed to be a consequence of recurrent edematous insults to the hypodermis. With replacement of elastic fibers by fibrous tissue, thickening and inelasticity of the skin occurs. With decreased lymphatic drainage, secondary infections may become increasingly manifest. Circulating Mff are rare in this condition, indicating that there is no obvious relationship between the number of Mff and severity of pathology.

The occult filarial syndrome known as tropical pulmonary eosinophilia (TPE) is typified by cough and wheezing, particularly at night, hypereosinophilia, and transient interstitial pulmonary infiltrates, particularly at the base of the lungs. Systemic signs and symptoms in young children include low-grade fever, malaise, lymphadenopathy, and hepatomegaly. Laboratory findings include eosinophilia, high total IgE levels, and high titers of specific antibodies against filarial antigens. Complete amelioration of symptoms are produced by the use of antimicrofilarial or antimacrofilarial drugs (Neva and Ottesen, 1978). Evidence provided by epidemiologic inference and specific immunologic profiles suggested that filarial worms, particulary *W. bancrofti* and *B. malayi*, are the etiologic agents of this condition (Ottesen *et al.*, 1979). Circulating Mff are usually not observed in confirmed TPE cases.

In bancroftian and malayan filariasis, most disease is attributable to host reactions to the lymphatic dwelling adults. Reaction to Mff are more innocuous.

The major exceptions to this generalization are cases of TPE. In onchocerciasis, the situation is reversed in that most pathology is referable to the reaction to Mff circulating in the skin. If high densities of Mff are present, parasites will be observed in blood and urine, lymph nodes, visceral organs, and the cerebrospinal fluid. Signs and symptoms are related to localized reactions to the Mff and include urticaria, pruritis, and scratching. Complicating secondary infections, hyper- or hypopigmentated plaquelike lesions, *mal morado* (purplish skin discolorations), and *erisipela de la costa* (a pale, red edema of the skin) (World Health Organization, 1976) are also observed. With advanced disease, fibrous replacement of elastic tissues and obstruction of lymphatics may lead to onchocercal elephantiasis, particularly of the groin. In extremely heavy infections, severe cachexia and dwarfism may be observed. Ocular lesions produced consequent to Mff deposition include "fluffy" punctate keratitis; sclerosing keratitis; anterior uveitis, secondary glaucoma, and cataracts; choroidoretinitis; optic atrophy; and blindness. Problems associated with the encapsulated adult worms, the onchocercomas, are usually minor; however, regional release of Mff from nodules in the head and neck is often associated with subsequent ocular complications.

B. Immunopathology of Filariasis

The broad clinical spectrum and associated immune responses suggest that immunopathogenic mechanisms are complex and as yet poorly understood. However, certain mechanisms appear to be emerging.

Immediate hypersensitivity (IH) responses to the parasite, Arthus and serum sickness reactions, are considered part of the acute pathologic syndrome. Lymphadenitis, retrograde lymphangitis, and episodic fever are observed (Piessens, 1981). In addition, these reactions are reported to occur subsequent to specific antifilarial drug therapy. With the destruction of parasites and the release of large amounts of antigen, particulary in onchocerciasis, skin IH reactions to dying Mff may debilitate populations. Profound pruritus, exfoliation, and possible death may result. In Malayan filariasis, this reaction may also be a consequence of damage to larvae and adult stages.

TPE patients demonstrate IgE antibody to both adult and Mff antigens. However, evidence for the involvement of immunoglobulin G (IgG) and cell-mediated immunity (CMI) has also been reported. Pleural pathology suggests a reaction involving both IgG and IgE. Associated with TPE infection is the M-K lesion, a microabscess originally described by Meyers and Kouwenaar (1939). These well-differentiated bodies consist of a degenerating microfilarial nidus intimately surrounded by an eosinophilic material (similar to Splendore-Hoeppli substance, as also seen in schistosomiasis) encircled by eosinophils and epithelioid cells. The structures have been identified in various lymphoid, lung, and hepatic tissues from diagnosed TPE patients and from various experimental animals (Webb *et al.*, 1960; Fox, 1979; Crandall *et al.*, 1982).

Whereas the importance of eosinophils in chronic filariasis is uncertain, interactions as regards TPE have stimulated interesting speculations. Circulating eosinophils were morphologically abnormal, containing many vacuoles. These cells apparently released eosinophil cationic proteins that might locally or systemically produce pathology. Eosinophils demonstrated increased ability to bind complexed IgG, all findings that were associated with intensity of clinical signs (Spry, 1981). Eosinophils and their related components have also been examined before and after DEC treatment of non-TPE patients. Pretreatment levels of eosinophils and serum levels of Charcot-Leyden crystal protein were unrelated to Mff intensity; after treatment, however, the levels of eosinophils, serum Charcot-Leyden crystal protein and major basic protein increased and correlated with the pretreatment microfilaremic densities. Nonpatent filarial patients and endemic controls demonstrated no such fluctuations after DEC treatment. Ottesen and Weller, (1979) and Ackerman *et al.*(1981) suggested that after pretreatment, eosinophil-regulating factors were operative. Interestingly, niridazole treatment of patients with *S. mansoni* demonstrated that pre- and post-therapeutic eosinophil levels were also related to the degree of infection (Ottesen and Weller, 1979).

The important contributions to immunopathologic reactions must include immune complexes. Circulating parasite antigens and immune complexes have been identified in both experimental (Karavodin and Ash, 1980) and human filarial infections (Lambert, 1978; des Moutis *et al.*, 1982). These antigens may be derived from Mff, in the case of onchocercal infections, and from larval and adult forms in lymphatic infections. Alone or in combination with antibody, their antigens may induce lymphatic inflammation and vasculitis (Henson *et al.*, 1979) as a consequence of their deposition. Other possible consequences include glomerulonephritis and arthritis (Waltzing and Bioch-Michel, 1971).

Several possible pathogenic mechanisms may be envisioned. *Dirofilaria immitis* will produce severe vascular lesions in the pulmonary arteries of dogs. These lesions may lead to life-threatening cardiopulmonary complications (Knight, 1979). Immune complexes produced efferent to the parasite may attach to specific receptors on the endothelium of the pulmonary artery and other vessels. For example, IgG, Fc, and C3b receptors have been identified on the endothelium of the aorta and aortic valves (Kasukawa *et al.*, 1980).

Henson *et al.* (1979) suggested alternate mechanisms for the pathologic effects of dying Mff. They hypothesized that alterations in Mff metabolism resulted in changes in the release of cyclic nucleotides. These in turn might lead to the production and release of prostaglandins or thromboxanes or the release of proteases which may activate complement components, all acting as pro-inflammatory agents.

One of the main consequences of lymphatic filarial infections is the development of highly characteristic granulomatous lesions. These have been identified not only in human tissues (Nelson, 1966), but also in almost all experimental animals infected with lymphatic-dwelling filariae (Schacher and Sahyoun, 1967; Vincent *et al.*, 1980). While evidence suggests that specific hypersensitivity reactions to filarial products are operative and CMI reactions appear to be very important, the precise mechanism remains unproved. A semiquantitative method to evaluate these granuloma-like lesions was recently developed by Klei and colleagues (1981). Similar techniques have been used, both *in vitro* (Doughty and Phillips, 1982*a,b*) and *in vivo* (Phillips and Colley, 1978) to investigate immunogenic and pathogenic mechanisms associated with schistosomiasis.

Klei *et al.* (1982) examined the size and composition of granulomas in lymphatic lesions in jirds resulting from infection by *B. pahangi*. The intravascular antigen bead granulomas contained fewer eosinophils than similar structures observed in natural intravascular granulomas from other host species, but they were basically of the same form—granular leukocytes with surrounding mononuclear and epitheliod cells. The nidus of these reactions has often been reported to be parasite components. Therefore, it may be suggested that any of a number of parasite products—excretory or secretory antigens, ova, Mff, adults, or immune complexes—might produce thrombi or vasculitis (Schacher and Sahyoun, 1967). These investigators discovered that preexisting infections increased the susceptibility of jirds to reinfection as judged by worm recoveries and unaltered growth of homologous challenge infections (Klei *et al.*, 1980). This hyposensitization by previous infection was substantiated by their granuloma lesion protocol (Klei *et al.*, 1981). The mean lesion size and intralymphatic thrombi number—both naturally formed and those induced by innoculated antigen-coated Sepharose 4B beads—were significantly reduced in groups with preexisting intraperitoneal infections. Their results indicated a reduced response to secondary infection—a parasite immunomodulatory effect—as witnessed by the reduced pathology and lack of protection to challenge infections.

In a somewhat different infection, i.e., *D. viteae* in hamsters, granulomatous responses were enhanced by multiple infections (Simpson and Neilson, 1976). Whereas lesions were noted in liver, lungs, and kidneys in singly infected hamsters, unlike the studies of Klei *et al.* (1982), the lesions were even more pronounced in multiply infected animals. Glomerular basement membranes and alveolar septal walls were thickened. Small inflammatory foci in liver triads and perivascular cuffing were observed. Chronic inflammatory cell types, consisting of plasma cells and lymphocytes, were noted. Subcutaneous nodules with necrotic cores containing degenerating and calcified worms, plasma cells, and lymphocytes were present in hyperinfected animals, but not in singly infected animals. Failure to observe significant immunomodulation may be explained by the observation that in this parasite host model, microfilaremia is terminated several months postinfection (p.i.) and occult infections are produced. Reinfections do not produce recrudescence of microfilaremia, which may be critical for the expression of immunodepression. However, in other conditions associated with immunodepression, caused by other agents such as tumors and parasites (*S. mansoni*), augmented growth of *D. viteae* has been observed (Haque *et al.*, 1981).

Circulating Mff are suggested to have the ability to modulate host responses by several mechanisms. In microfilaremic individuals, anti-Mff antibody is noticeably reduced or absent. Proteases have been identified that selectively cleave Fc portions. This effectively may diminish ADCC because effector cells in the reaction require the Fc portion for effective binding to the parasite. Additional complete antibodies may not be found in sufficient quantities to activate ADCC. Alternatively, they may be blocked by the Fáb fragments or by other classes and subclasses of antibody that cannot functionally participate in ADCC. Immune complexes and antibody aggregates may also directly inhibit ADCC (Walker *et al.*, 1979). The presence of Mff or their products may act in additional direct ways. For example, parasites inhibit C3 components, possibly by cuticular-associated polyanionic components (Staniunas and Hammerberg, 1982). *In vitro*, the addition of living Mff will significantly reduce *in vitro* blastogenesis.

Correlation of the presence and density of Mff with acute inactivation and chronic modulation of host responses is also observed in onchocerciasis. Pruritus, acute maculopapular eruptions, and lymphadenitis are maximum early in infection as are cellular reactions around Mff in the cornea. Thereafter, these reactions will be modified. Whereas the incidence of skin atrophy is correlated with the intensity of the infection, no correlation is observed with regard to incidence of hanging groin, lymphodenopathy, and skin depigmentation (Fuglsang *et al.*, 1979).

Specific cell-mediated immune responses, particularly to Mff, are prominent in chronic obstructive filariasis–usually nonpatent infections (Ottesen, 1980; Piessens *et al.*, 1980). This process may result in the more extensive regional lymphatic pathology evident in patients with elephantiasis, hydrocoels, and chyluria. Interestingly, CMI responses have also been reported to occur subsequent to DEC treatment. Two weeks post-treatment, responses to both specific and nonspecific antigens are diminished (Ottesen *et al.*, 1977). Two months after DEC treatment, specific antigenic responses may be augmented in previously amicrofilaremic patients, but will remain low in previous microfilaremic and endemic controls. Six and 12 months after DEC treatment, lymphocytes derived from individuals previously patent and now negative, are significantly more responsive to Mff antigens. Specific antibody to microfilarial sheaths is unchanged or only slightly increased. The initial depression in response noted post-treatment may be attributable to the known, but apparently transient, panimmunodepressive effects of DEC (Orange *et al.*, 1968). Animal studies have demonstrated the profound specific antibody depression subsequent to DEC treatment (Desowitz *et al.*, 1978). A second reason may be the formation of circulating immune complexes as a consequence of microfilarial destruction. Circulating immune complexes are widely recognized modulators of immune reactivity. For example, they have been shown to suppress mononuclear activity, interfere with T cell–antigen interactions, and induce the formation of regulatory anti-idiotypic antibody.

Thus, immunomodulation is evident in chronic patent infections. Reactivity is also evident once circulating Mff and possibly adults are eliminated, either naturally or therapeutically. The mechansism of the specific immunosuppression noted in Mff-positive patients has been explored by Piessens and co-workers (1980, 1982). These investigators initially identified an adherent cell population that suppressed lymphoid blastogenic responses to Mff antigens. When the suppressor cells were removed from culture, blastogenesis occurred. Continuous stimulation of the cells by antigen was necessary to maintain the suppressive ability. In addition, serum from donors with patent infections significantly suppressed known positive-reacting cells.

This effect was primarily, but not exclusively, parasite specific. Serum from patients with acute filariasis, without circulating Mff, also suppressed specific blastogenesis, but to a less marked degree. These workers further dissected T lymphocytes using OKT monospecific antibodies and found an increased ratio of OKT8-sensitive cells (T-suppressor cells) to OKT4-sensitive cells (T-helper cells) in patent filariasis patients. These results also correlated with *in vitro* findings. By augmenting or reducing the ratios of T_s: T_h cells, specific blastogenisis could either be increased or decreased, respectively; the reaction to nonfilarial antigens (PPD-SKSD) remained unaffected. Thus, at least three different mechanisms causing antigen-specific suppression in human filariasis have been described: adherent monocytes, T-suppressor cells and undetermined serum factors.

Other possible mechanisms that may affect host immunity include genetics and nutrition. Danaraj (1951) suggested that TPE prevalence was explained only by referring to genetic influences. Fimilial predisposition to filarial infections could not be linked to HLA by Ottesen *et al.* (1981*b*). In animal studies, inherited susceptibility was found to be polygenetically based (Fox, 1979). The nutritional aspects also remain unclear; however, with deficiencies in serum components, monocytes, and granulocytes, we must suspect that malnutrition may play a significant deterimental role in resistance to infection and resulting pathology (Greenwood and Whittle, 1981).

In summary, there is strong evidence to suggest that the immunologic consequences of antibody, immune complexes, and cellular interactions, and the relative intensity of these reactions must be interpreted in the context of their effects on both host and parasite. When appropriate and balanced, resistance is maximized and pathology is minimized. When inappropriate, resistance is suboptimal and pathology more manifest.

III. SCHISTOSOMIASIS

In contrast to filariasis, schistosomiasis has been far more extensively studied *vis-à-vis* immunopathologic manifestations. A number of excellent reviews have been previously published (Smithers and Terry, 1969; Smithers and Terry, 1976; Phillips and Colley, 1978; Colley, 1981; Nash *et al.*, 1982; Warren, 1982; Damian, 1983). In the light of the comprehensiveness of these reviews, only certain more salient aspects of schistosome immunopathology are addressed in this chapter. An understanding of the immunologic mechanisms within the context of the aforementioned mechanistic model is emphasized as an explanation for a clinical spectrum of schistosomiasis.

A. Life Cycle

Schistosomiasis is a group of diseases caused by digenetic trematodes inhabiting the circulation system. Several closely related species belonging to the family Schistosomatidae, *Schistosoma mansoni, S. japonicum,* and *S. hematobium,* will principally incite human pathology. *Schistosoma intercalatum* and *S. mekongi* have recently been described as distinct species of parasites, also capable of inducing disease in humans. These and other heterologous species of blood flukes infect a wide variety of domestic and wild animals and humans in tropical and subtropical regions. Disease transmission is highly contingent on a number of environmental factors that provide a suitable habitat for the growth of a specific species of snail, which serves as an intermediate host. The three species of schistosomes that commonly infect humans have similar life cycles characterized by a succession of stages: egg, miracidium, sporocyst, cercaria, schistosomulum, and adult. The life cycle is one of cyclic propagation with reproduction occurring in both the intermediate snail and definitive vertebrate host. Generation of adult schistosomes and sexual reproduction occurs in the definitive vertebrate host, and asexual multiplication occurs in the molluscan host. The egg hatches when deposited in water and releases a relatively short-lived, free-swimming stage, the miracidium. This stage infects the snail intermediate host and becomes a first-stage mother sporocyst, which gives rise by asexual division to multiple second-stage daughter sporocysts. These in turn produce cercariae, which escape from the snail host and, by responding to a variety of physical and chemical stimuli, locate and penetrate the definitive host, using specific digestive enzymes. This stage enters the bloodstream, becoming the schistosomulum. After intravascular migration, the schistosomulum is trapped in the lung ~ 4-7 days after initial exposure. The parasite then migrates to and develops in the liver. After 3-5 weeks, the dioecious juvenile adults mate for life and migrate to anatomic locations within peripheral vessels. *S. mansoni* localizes in the inferior mesenteric veins; *S. hematobium* localizes in submucosal venules of the ascending colon and the hemorrhoidal anastomosis of the bladder and other pelvic organs; *S. japonicum* tends to localize less specifically. Adult worms reach a length of 1-2 cm and have a life expectancy of 4-30 years. The female worm is held firmly in the gynecophoric canal of the male during copulation and ova deposition. The female produces large number of eggs, varying from approximately 300 per day for *S. mansoni* to > 3000 a day for *S. japonicum*. The eggs cross the appropriate endothelium and mucosal barriers to be passed in feces or urine, where they hatch to liberate ciliated miracidia, which penetrate the appropriate snail to complete the life cycle (Manson-Bahr and Apted, 1982).

B. Disease Spectrum

The clinical manifestations of schistosomiasis can be broadly divided into four categories, associated with invasion, maturation, establishment of infection, and the late consequences of infection (Table I).

1. Schistosome Dermatitis

Schistosomiasis begins with the initial penetration of skin by cercariae. There are generally very minimal subjective or objective manifestations of initial

Table I. Spectrum of Schistosomiasis[a]

Stage of infection	Clinical manifestations	Pathologic manifestations	Immunologic mechanisms
Invasion			
Penetration	Swimmer's itch	Cutaneous inflammation	IgE, ADCC, CMI
Migration	PIE, cough, fever	Pulmonary/hepatic inflammation	IgE, ADCC
Maturation			
Initial oviposition	Katayama fever, serum sickness	Intense local and generalized vasculitic reaction	CIC, ADCC
Acute			
Intense oviposition maximum egg production and excretion	GI, Sx, hematuria	Large granulomata	CMI (*S.m.*) B cell (*S.j.*)
Chronic			
Prolonged infection decreased egg production and excretion	Chronic disease, portal hypertension nephropathy CNS cor pulmonale	Modulated granulomata, Symmer's fibrosis	CMI (*S.m.*) B cell (*S.j.*)

[a]Abbreviations: ADCC, antibody dependent cell-mediated cytoxicity; CIC, circulating immune complexes; CMI, cell-mediated immunity; CNS, central nervous system; GI, gastrointestinal; PIE, pulmonary infiltrates and eosinophia, S.j., *Schistosoma japonica*; S.m. *Schistosoma mansoni*.

exposure. Indeed the host is usually totally unaware of the penetration phenomenon. However, after multiple exposures, especially to zoonotic schistosomes, i.e., those that normally complete their life cycles in other animals, far more intense cutaneous reactions occur. Clinically, sensations of pruritis or local erythema may occur within 15 min to 2-4 hr after penetration. A high percentage of the heterologous cercariae are quickly destroyed in the skin; however, the dermal reactions are very significant. The local dermatitis becomes progressively more severe after multiple exposures to parasites and may progress to a wide variety of clinical manifestations. These include severe macular papular eruptions, Arthus-like reactions with exfoliation, and erythema nodosum-like lesions. Histologically, there is initial tissue edema with mononuclear, neutrophilic, and eosinophilic infiltrates involving epidermis and dermis (MacFarlane, 1949; Olivier, 1949).

2. *Immunopathogenesis*

It is tacitly assumed that there is a relationship between immunologic defense mechanisms and immunopathology. The latter results from manifestations of the former. There is a great deal of evidence suggesting that the invading schistosomula are damaged by a combination of humoral and cellular responses (Smithers and Terry, 1969, 1976; Phillips and Colley, 1978; Colley, 1981; Nash *et al.*, 1982; Warren, 1982; Chapter 10, this volume). The relative importance of the various host responses, mainly established by *in vitro* studies to challenge, are not yet delineated *in vivo*. The consitutents and localization processes of the histopathologic lesions vary greatly in various definitive hosts, and it is not known whether most cellular reactions precede or follow parasite killing. Significant focal dermal leukocyte reactions, juxtaposed to penetrating schistosomula of *S. mansoni*, have been observed in a large number of naturally resistant as well as artificially immunized and challenged hosts. The nature of the infiltrates appears to be subject to genetically determined immune-response characteristics (Sher and Scott, 1982) as well as to a wide variety of characteristics of the parasite, per se (Hsü *et al.*, 1963, 1965). Initial responses begin with edema, polymorphononuclear infiltration, and the deposition of IgM, IgG, and complement. Subsequently, the reactions evolve to more complex reactions involving the deposition of several immunoglobulin subclasses, including IgE and an increasingly rich eosinophilic component.

3. *Homocytophilic Reactions*

Considerable interest has focused on the contribution of IgE and IgG2a antibodies *vis-à-vis* the acute dermatologic reactions (Hsü *et al.*, 1963; Zvaifler *et al.*, 1967; Colley 1981; Dessaint, 1982). Significant elevation of total and specific anti-schistosomular IgE have been noted in a number of parasitic conditions, although these changes are most pronounced with helminths (Sadun, 1972; Dessaint, 1982). Immediate hypersensitivity reactions have been implicated by histologic analysis as being salient in schistosome dermatitis (Olivier, 1949; Olivier and Weinstein, 1953; Hsü *et al.*, 1963; Sadun, 1963; Colley *et al.*, 1972; Oda, 1973; Askenase *et al.*, 1976). *S. mansoni* infection results in a massive IgE response. Seventy-eight percent of one reported series of patients had IgE levels in excess of 700 I.U./ml (Dessaint *et al.*, 1975). Significant specific anti-schistosome IgE antibodies have been detected in most infected humans (Weiss *et al.*, 1978; Ottesen *et al.*, 1981*a*) or primates (Suzuki and Damian, 1981). After experimental exposure of animals to schistosomiasis, anaphylactic antibodies have also been demonstrated to be composed of both IgE and IgG2a (A. Capron *et al.*, 1977, 1980). The mechanism whereby the parasites stimulate the production of immunologically specific and nonspecific reagenic antibody is not clear. Parasites per se are potent allergens (Dessaint, 1982); indeed parasitic extracts have been used as adjuvants for the production of IgE antibodies against a number of ostensibly unrelated antigens. Potentiation may be attributed to polyclonal activation of B cells by parasites (Ogilvie and Jones, 1973; Ishizaka *et al.*, 1976; Jarrett, 1978). Alternative explanations include the presence of IgE-binding factors that stimulate helper T lymphocytes (Suemura *et al.*, 1980) and recruitment of specific E (ϵ) cells (Urban *et al.*, 1977). A number of directly acting schistosome-derived factors may also be significant (A. Capron and Dessaint, 1982). It has been suggested that IgG2a and IgE antibodies contribute to the development of immunity in the infected host. Indeed, a close relationship develops between the degree of immunity in experimental rats and their anaphylactic antibody titers. In addition, the passive transfer of immunity to schistosomes can be reduced when serum is depleted of IgG2a (A. Capron *et al.*, 1980, 1982) or IgE (S.M. Phillips, unpublished observations, 1983). In addition, both monoclonal Ig2a and polyclonal IgG2a (Verwaerde *et al.*, 1979; Phillips and Zodda, 1983) have been directly implicated in the adoptive transfer of resistance.

Both IgE and IgG2a have also been implicated as participating in anti-schistosomular cytotoxic events by *in vitro* studies. The killing of schistosomula may involve antibody-dependent, eosinophil-mediated cytotoxicity (Butterworth, 1974, 1975). Mast cells and complement augment this cytotoxicity (M. Capron *et al.*, 1978). Eosinophil major basic protein, and arginase are released by activation by specific surface antigens (Butterworth *et al.*, 1979*a*; Olds *et al.*, 1980; Kazura *et al.*, 1981) and may result in cytotoxicity as well. Indeed, tetrapeptide, eosinophil chemotactic factors, have been used to replace mast cells in enhancing eosinophil-mediated killing of opsonized larvae and the expression of eosinophil Fc receptors. There is a correlation between the effect of ECF-A tetrapeptides on the expression of rat eosinophil Fc receptors and eosinophil-mediated cytotoxicity (M. Capron *et al.*, 1981*a,b*). In addition, immune complexes composed of IgE–IgG-schistosome antigens have also been shown to trigger macrophage cytotoxicity (A. Capron *et al.*, 1975, M. Capron *et al.*, 1978), and these macrophages bear specific receptors for IgE on their surface (Dessaint *et al.*, 1979). The role of ADCC-mediated cytotoxic events *vis-à-vis* parasites has been carefully reviewed in detail by Dessaint (1982). However, the relationship of anaphylactic antibodies, hyperergia, and resistance is by no means clear, as the IgE titers tend to increase with time, while other acute allergic manifestations may diminish. In addition, the parasite per se produces factors capable of modulating the reactivities of IgE antibodies and various other effector cells (A. Capron and Dessaint, 1982). It is of interest that nonspecific manifestations of allergy appear to be negatively associated with the presence of parasites; however, clearly the presence of IgE antibodies and mast cell and eosinophil degradation products would tend to implicate strongly the importance of these classes of antibody. Since the antibodies may be highly important in controlling disease, it is postulated that the relative ratios of early hyperegic responses related to disease containment processes may determine the relative balance between effective resistance and secondary acute immunopathology.

4. *Cell-Mediated Reactions*

Histologic analysis would also suggest that cell-mediated phenomena are important in the early dermatologic manifestations of schistosomiasis (Olivier, 1949; MacFarlane, 1949; Olivier and Weinstein, 1953). There currently exists some controversy as to the relative importance of T cells as immune effectors compared to other cell types. *In vitro*-mediated cytotoxicity by T cells has been found relatively inefficient (Butterworth *et al.*, 1979*b*), suggesting that antibody-dependent killing by Fc-bearing cells (A. Capron *et al.*, 1980) is a more dominant mechanism. However, recent studies by Ellner *et al.* (1982) have suggested that schistosomula can be destroyed by T lymphocytes that are either nonspecifically activated by phytomitogens, specifically activated by other unrelated antigens, or specifically activated by schistosome antigens. This killing was apparently most effectively mediated by cells of the OKT4 phenotype, but was not restricted to this population. A mechanism involving augmented NK-like activity, not obviously related to hydrogen peroxide generation, was postulated.

Other possible components of these dermal reactions have been recognized. These include cutaneous basophil hypersensitivity (Askenase *et al.*, 1976) as well as antibody-mediated Arthus reactions (Colley *et al.*, 1972). Skin biopsies have demonstrated deposition of immune complexes and basophils along capillary loops in skin. In addition, there is evidence that cercarial extracts may affect complement activation (Gazzinelli *et al.*, 1969; Machado *et al.*, 1975), possess eosinophil chemotactic activity (Colley *et al.*, 1977) or stimulate T cells indirectly to produce a wide variety of lymphokines (Colley, 1981). The ability to reduce cellular infiltrates with corticosteroids and the reduction of skin reactivity in athymic animals similarly suggest the importance of cellular components.

C. Acute Pulmonary Schistosomiasis

1. Clinical and Pathologic Manifestations

Seven to 8 days after initial exposure, concomitant with the period of parasite migration through the lung, transient symptoms of fever, malaise, cough, and wheeze associated with transient pulmonary infiltrates may occur. Generally, these events follow extremely heavy exposure to schistosomes in previously exposed hosts. At this time, the patients manifest peripheral eosinophilia and the syndrome thus qualifies as a cause of the pulmonary infiltrate with eosinophila (PIE) syndrome, in many ways similar to that observed in tropical eosinophilia secondary to filariasis. Extensive studies in humans are not available; however, studies in primates and mice have indicated that the schistosomules migrating through the lung are associated with accumulations of significant numbers of eosinophilic granulocytes and small numbers of mononuclear cells.

Although the absolute importance of parasite killing within the lung must be considered conjectural, it is undoubtedly related to pulmonary pathology. Pulmonary lesions, rich in eosinophils, have been described as surrounding schistosomula in the lungs of challenged primates (Hsü *et al.*, 1965) and mice (Magalhaes-Filho, 1959; von Lichtenberg *et al.*, 1977). Parasites may be trapped in the lung where they are destroyed (Hsü *et al.*, 1965; von Lichtenberg *et al.*, 1977). In addition, there is evidence to suggest that the intensity of the cellular inflammatory reaction reflects the immunologic status of host and is directly proportional to the resistance manifest in a wide variety of experimental models (von Lichtenberg and Byram, 1980). The mechanisms of these reactions are largely unknown. However, the propensity of this syndrome to develop in highly immunized hosts, especially upon repeated heavy exposure, the failure to occur in athymic mice (Byram and von Lichtenberg, 1981), and the tendency of signs and symptoms to resolve completely in response to corticosteroids would suggest that acute pulmonary schistosomiasis represents a form of hypersensitivity pneumonitis.

This hypersensitivity reaction appears to be different from that which occurs in the skin. For example, the parasite changes dramatically in morphologic, morphologic, functional, and antigenic characteristics as it migrates from skin to lungs. A number of host antigens are acquired. Complement receptors susceptible to a variety of putative killing reactions diminish. These findings suggest that the immunologic mechanisms responsible for the hypersensitivity reactions in the lung may not be totally analogous to those within the skin. In addition, there is strong evidence to suggest that inflammatory reactions around the schistosomules migrating through the lung can be markedly affected by route of exposure. For example, the intensity of host response in the lung against parasites that infect transcutaneously is considerably greater than the reaction against parasites injected intravenously.

D. Early Systemic Schistosomiasis

1. Clinical and Pathologic Manifestations

This most dramatic stage of schistosomiasis was best defined as acute Katayama fever in the nineteenth century (Billings *et al.*, 1946; Wei-Hsin, 1958). The syndrome was observed in individuals migrating into the Katayama Valley of Japan and was caused by the heavy initial exposure of nonimmune hosts. Eventually, the connection between Katayama fever and *S. japonicum* was firmly established. The condition is also occasionally observed with *S. mansoni* (Diaz Rivera *et al.*, 1956; Oliveria *et al.*, 1969; MMWR, 1982).

Patients develop chills, fever, sweats, anorexia, headache, cough, and diarrhea. A wide variety of dermatologic manifestations, including urticaria and vasculitis, hepatosplenomegaly, and generalized lymphadenopathy, are often observed. Mild leukocytosis with eosinophilia is common. The syndrome usually lasts for several weeks, at which point it subsides spontaneously. In severe cases, however, death may result.

2. Immunopathogenesis

Although a number of factors have been implicated in the pathogenesis of acute schistosomiasis mansoni (Hiatt *et al.*, 1979; Smithers and Terry, 1969, 1976; Phillips and Colley, 1978; Colley, 1981; Nash *et al.*, 1982; Chapter 10, this volume), the overwhelming evidence would suggest that this disease syndrome is related to the sudden production and release of large quantities of parasite antigen at the onset of egg deposition.

Ostensibly in combination with a hyperergic humoral immune response, these conditions result in the production of high titers of circulating immune complexes (Phillips and Colley, 1978; Colley, 1981). Recent studies have also clearly documented the relationship between the level of circulating immune complexes and the severity of clinical disease in humans (Lawley *et al.*, 1979) and experimental animals (Bout *et al.*, 1977). In addition, the successful treatment of the parasitoses leads to a more rapid waning of circulating antigenemia than might be expected in the presence of ongoing disease (Abdel-Hafez *et al.*, 1983). A number of other studies have also suggested that circulating antigens with clearly defined specificities occur early in human infection (Hillyer, 1969; Reis *et al.*, 1970; von Lichtenberg *et al.*, 1974; Hillyer *et al.*, 1976; Nash *et al.*, 1982) and that the level of circulating immune complexes diminishes with chronicity.

These immune complexes are composed of a number of constituents, including IgM, IgG, IgA, IgE, a variety of complement components, and parasite antigens, and can be found both in the serum and glomeruli of infected hosts (F. Houba *et al.*, 1977; A. Capron *et al.*, 1982). Schistosomiasis-induced immune complex glomerulonephritis has been used as an excellent experimental renal disease model for the dissection of pathologic mechanisms (V. Houba, 1979). Schistosome-specific antigens can be eluted from the kidney of infected animals, and unilaterally nephrectomized infected animals more rapidly developed severe disease than did their nonnephrectomized cohorts. Evidence for complement activation and consumption both systemically and within the kidney has been shown (Colley, 1981; Galvao-Castro *et al.*, 1981; Phillips and Fox, 1982). Immunoinflammatory mechanisms may also be influenced by the release of inactivating parasitic derived factors (A. Capron and Dessaint, 1982). However, it must also be noted that these pathologic manifestations occur more commonly in chronic diseases.

Finally, local tissue destruction may lead to secondary antitissue reactions, such as the production of anti-DNA antibodies (Hillyer, 1973). For example, the suboptimal confinement of eggs in an inflammatory granuloma might allow for the release of cytotoxic egg related enzymes (Gutekunst *et al.*, 1965) with direct cytotoxic effects. The intensity of local cytotoxic events may increase in the tissues of congenitally athymic mice and immunodepressed mice in which the anti-egg inflammatory response is minimal (Byram and von Lichtenberg, 1981).

E. Chronic Schistosomias

Clinically, patients with chronic schistosomiasis experience fatigue, abdominal pain, intermittant diarrhea, and ultimately the progressive signs of hepatic failure. Histologically, the major lesions observed are focal granulomatous reactions surrounding embolic eggs within the vessels. The distribution of the eggs tends to determine the principal pathology. For example, *S. mansoni* lives in the mesenteric veins; hence, most eggs will embolize to the liver leading to hepatic disease. *S. hematobium* lives in the pundendal vessels; hence, eggs tend to embolize to the organs of the lower pelvis, where pathology is observed. *S. japonicum*, because of its wider distribution within the vascular compartment and production of larger numbers of eggs, tends to cause more severe, disseminated disease.

1. Hepatic Schistosomiasis

Hepatic schistosomiasis is associated with granulomata and precapillary pipe stem fibrosis (Symmers' fibrosis), which are the principal determinants of pathology (Symmers, 1903; Warren, 1972*a,b*, 1973; Phillips and Colley, 1978; Colley, 1981; Nash *et al.*, 1982). Symmers' fibrosis is a relatively unique histologic lesion, occurring only in humans and higher primates (Sadun *et al.*, 1970) or perhaps in very chronically infected mice (S. M. Phillips and B. L. Doughty, unpublished communication, 1983). The granulomatous reaction is more ubiquitous and has been far more extensively studied, although its significance *vis-à-vis* morbidity is less clear (Cheever, 1968, 1972; Warren, 1972*b*). Ostensibly, the chronic fibrotic reaction is a result of obstruction to normal hepatic blood flow with hyperkinetic shunting of blood and subsequent perihepatic shunting (Canto *et al.*, 1977; Bloch *et al.*, 1972). This results in portal hypertension and the sequelae associated with obstructed flow.

It is of interest that the intrahepatic pressures are nearly normal in the face of elevated portal pressure. This finding suggests the importance of extrahepatic shunting and that the liver blood flow is partially maintained by a compensatory increase of arterial blood supply (Cheever and Warren, 1964; Bloch *et al.*, 1972). The earliest clinical signs of significant liver involvement are hepatomegaly and secondary splenomegaly. The most common serious presentation is that of esophageal bleeding, secondary to ruptured varices. Patients may suffer from multiple episodes of hemoptysis, with the relative maintenance of good liver function. Disease is highly variable and may terminate in liver failure with associated massive ascites. However, since the hepatic parenchyma is relatively spared, liver failure usually does not occur until late in the disease or unless complicated by surgical shunting procedures (Warren *et al.*, 1965). Endocrinologic liver disease associated with reduced estrogen catabolism is not uncommon; however, clinical jaundice and markedly abnormal liver function tests (LFTs) are relatively rare. Although terminal fibrosis has been described, evolution into true cirrhosis is somewhat controversial. It must be recognized that individuals with schistosomiasis are also at risk for complications of cirrhosis contingent on malnutrition, hepatitis, hepatotoxins, and other parasitic diseases, which may directly or indirectly affect the liver and spleen.

2. Pulmonary Schistosomiasis

Pulmonary schistosomiasis is a chronic complication that usually occurs only after significant portal hypertension and shunting has occurred. Airway obstruction, sometimes reversible with bronchodilators, is the most common clinical finding. Decreased lung volumes, impaired diffusion, and abnormal V/Q ratios are also observed. Because the pulmonary alveoli are relatively well maintained, oxygen saturation is usually normal and cyanosis is rare; however, terminal alveolar–capillary block syndromes and respiratory failure may occur. In chronic schistosomiasis, embolization of ova to small arteries of the lung result in multiple pseudotubercles. The endovascular localization of the eggs and the granulomatous reaction about them may result in a terminal arteriolitis and obliterate the lumens of small arterioles (Andrade and Andrade, 1970; McCully *et al.*, 1976). Terminally, significant pulmonary fibrosis and hypertension may occur. This latter complication is relatively rare and produces significant cor pulmonalé in fewer than 5% of patients with clinically significant schistosomiasis (Barrett-Connor, 1982).

3. Urinary Tract Schistosomiasis

Glomerulopathology may occur acutely or progressively during chronic schistosomiasis. However, urinary tract schistosomiasis usually occurs only in the latter stages of *S. haematobium* infection (Lehman, 1973). Disease is more severe with *S. haematobium*, not only because of the adult parasite localization pattern, but also because of the deposition of large numbers of eggs in masses that induce focal areas of acute inflammation and subsequent fibrosis known as sandy patches. Initial symptoms are urinary frequency, dysuria, and hematuria. The bladder often undergoes fibrosis and calcification (Lehman, 1973), and there is secondary development of both motor and sensory uropathies. Subsequently a large, flaccid, nonfunctioning bladder is often observed. Secondary complications include progressive hydronephrosis and pyelonephritis. In addition, metaplasia and cancer of the bladder are observed.

4. Gastrointestinal Schistosomiasis

Most intestinal symptoms occur with acute *S. mansoni* infections (Hiatt *et al.*, 1979). Although clinical symptoms are reported, they do not necessarily correlate with signs of disease severity. Several lesions, however, are characteristic. These include colonic polyposis (Lehman *et al.*, 1970) and focal areas of intense fibrosis known as bilharziomas (McCulley *et al.*, 1976). The occurrence of carcinoma is controversial.

5. Central Nervous System Schistosomiasis

Schistosomiasis of the brain is relatively rare and is almost always associated with *S. japonicum* (Kane and Most, 1948). It carries particularly ominous implications (Marcial-Rojas and Fial, 1963). Clinical manifestations are extremely variable. Diffuse meningitic or encephalopathic syndromes ranging from meningismus to diffuse encephalopathy secondary to microembolism of eggs may occur. Immunopathologic reactions to eggs deposited within the brain parenchyma may also result in localized space-occupying lesions, often presenting as focal epilepsy. Ectopic worms may migrate into the central nervous system, or more often into the spinal cord, more commonly observed with *S. mansoni* or *S. haematobium* (McCulley *et al.*, 1976), where cord transection lesions have been described.

F. *Schistosoma mansoni:* Immunopathogenesis–Granuloma Formation and Modulation

A number of recent reviews (Phillips and Colley, 1978; Colley, 1981; Phillips and Fox, 1982; Nash *et al.*, 1982; Chapter 10, this volume) have summarized the overwhelming evidence which suggests that granuloma formation is a T-cell-mediated immune response against the products of the egg. Histologically, the lesion resembles a delayed hypersensitivity reaction. Although minor differences in the resulting pathology induced around eggs of various schistosome species do exist (von Lichtenberg *et al.*, 1973; Nash *et al.*, 1982), general analogies are possible. Ultrastructural analysis reveals that all the constituents of a delayed hypersensitivity lesion are present in schistosome granulomas, including lymphocytes, neutrophils, eosinophils, plasma cells, and multinucleated giant cells (Stenger *et al.*, 1967). During the later stages of infection, granulomas tend to be dominated by immunoblasts and plasma cells (Smith, 1977). Experimentally, granulomatous hypersensitivity can be adoptively transferred with the use of T lymphocytes (Warren *et al.*, 1967). Granuloma formation is reduced in congenitally athymic mice (Hsü *et al.*, 1976; Phillips *et al.*, 1977), T-cell-immunosuppressed animals (Domingo and Warren, 1967, 1968; Warren, 1971, 1974; Buchanan *et al.*, 1973; Doenhoff *et al.*, 1979), and thymectomized chickens (Davis *et al.*, 1974). Bursectomized birds and B-depleted mice (Davis *et al.*, 1974; Phillips and Colley, 1978; Colley, 1981) produce granulomas.

Although the precise etiology and mechanisms of granuloma formation are not entirely clear (Smith and Terry, 1969, 1976; Phillips and Colley, 1978; Colley, 1981; Nash *et al.*, 1982; Warren, 1982), the relationship between this phenomenon and morbidity has been studied extensively (Andrade and Warren, 1964).

The egg-induced granulomatous response is clearly an immunologic phenomenon, apparently resulting from a response to materials elaborated by the maturing miracidium (Hang *et al.*, 1974). Granulomatous hypersensitivity can be induced through the use of egg-derived proteins, but not by immunization with other stages. In addition, as demonstrated both *in vivo* and *in vitro*, a certain degree of egg maturation is required to elicit strong granulomatous reactions.

The resultant T-cell-dependent reaction becomes a two-edged sword. As illustrated in athymic mice, morbidity may be reduced if the granulomas are controlled (Phillips *et al.*, 1980). When the environment was stringently controlled, infected congenitally athymic mice actually did better than the infected heterozygote controls. The athymic animals demonstrated less morbidity and mortality as a result of the granulomatous process. When exposed to ambient xenic conditions, however, the athymic mice did poorly, especially in the face of other concomitant infections. Conversely, pathology may increase if granuloma formation is impaired (Magalhaes-Filho *et al.*, 1965; von Lichtenberg, 1964). Indeed, most studies suggest that clinical morbidity increases with T-cell deprivation (Epstein *et al.*, 1979; Colley, 1981; Byram and von Lichtenberg, 1981). Therefore, it would appear that morbidity is a result of at least two phenomena. The first mode is related to adverse affects of the egg/miracidial products, which may be minimized by granulomatous localization. The second is related to adverse affects of the hyperergic host-immune response directed against these products. These findings suggest that minimum morbidity would be achieved when these two competing processes are optimally balanced.

Much less information is available *vis-à-vis* fibrosis per se. Studies suggest that this process is also a result of a T-cell/monocyte/fibroblast reaction, stimulated by complex circulating, subendothelially deposited, egg products. It is reasonable to assume that similar considerations of balanced pathologic processes are operative.

The phenomenon of granuloma modulation provides potential teleologic evidence for the body's attempt to achieve the optimal relationship between granulomatous hypersensitivity and parasite-associated morbidity. Andrade and Warren, (1964) noted a progressive increase, followed by a decrease in granulomatous reactivity, as assessed by the size of granulomas. Clinical morbidity was also noted to slowly wax and wane in parallel to granuloma formation in highly controlled experimental murine studies. These authors originally called the phenomenon "endogenous desensitization" (Domingo and Warren, 1968). Subsequently, the phenomenon has been studied in actively infected animals and via pulmonary egg embolism experimental models (Colley, 1981) and renamed "spontaneous modulation" (Warren, 1974). Spontaneous modulation has been reported in man (Rocklin *et al.*, 1977) and also inferred *vis-à-vis* the criteria of *in vitro* granuloma formation around the egg nidus by peripheral blood obtained from infected human patients (B. L. Doughty and S. M. Phillips, unpublished observations, 1983).

The mechanism of the granuloma formation and its suppression has been extensively discussed in great detail. Space does not permit but a most cursory summary. Through an elegant series of adoptive transfer studies using T-cell subpopulations and the analysis of *in vitro* proliferative responses, mediator release, and granuloma formation, the lymphocytes participating in these reactions have been characterized in terms of antigenic markers and function. To summarize briefly the contributions of a large number of studies (Boros and Warren, 1970;

Phillips and Colley, 1978; Colley, 1976, 1981; Chensue and Boros, 1979*a,b*; Chensue *et al.*, 1980, 1981, 1983; Wellhausen *et al.*, 1980; Phillips and Fox, 1982; Nash *et al.*, 1982), it is now clear that granuloma formation and its modulation represent a delicate balance between subpopulations of T lymphocytes; one population bears the phenotype of helper cells and is responsible for the formation of granulomas; the other population bears various suppressor phenotypes and is responsible for granuloma modulation. Studies using cyclophosphamide depletion (Colley *et al.*, 1979), thymectomy (Colley, 1981), and variety of other manipulations have further implicated T-cell subsets, as have elegant analyses of I/J region gene products (Green and Colley, 1981). Most recently, Chensue *et al.* (1983), using the criteria of MIF release as measurement of granulomatous hypersensitivity, have performed an elegant analysis of the subpopulations of T lymphocytes that cooperate in granuloma modulation. They have identified a T-suppressor cell-derived IC subregion-encoded suppressor factor.

Stoichiometric titration experiments (Phillips *et al.*, 1980; Doughty and Phillips, 1982*a,b*; Bentley *et al.*, 1982; and unpublished observations) using *in vitro* technology are clearly compatible with the above interpretations. These studies would suggest that the granuloma is a result of the initial activation of IA-bearing mononuclear phagocytes with antigens released from the schistosome membrane. Subsequently, there are interactions with a variety of subpopulations of lymphocytes and numerous other cells, including macrophages, eosinophils, and fibroblasts. Soluble factors, capable of inducing directed migration of these cell subpopulations and stimulating collagen synthesis have been detected in culture supernatants. Admixture experiments have provided evidence for both direct and indirect feedback suppressive influences. These subpopulations have been shown to be active *in vivo*, using adoptive transfer experiments (B. L. Doughty, S. M., Phillips, F. von Lichtenberg, and J. E. Byram, unpublished observations, 1983). Subsequently, a number of other cell types also participate in this reaction. These include plasma cells, which actively synthesize parasite specific antibody, and eosinophils, which migrate into the granulomas, mature therein and degranulate in juxtaposition to the surface of the egg. Finally, fibroblasts migrate to the lesions, divide, and synthesize both fibronectin and collagen in response to materials liberated into the culture supernatants. These studies suggest that the formation of *in vitro* granulomas closely recapitulates the *in vivo* processes and confirm the usefulness of this model for analyzing the physiologic and immunologic mechanisms involved in granulomatous hypersensitivity and fibrosis. Figure 2 illustrates some aspects of the phenomenon of *in vitro* granuloma formation.

As yet no definitive evidence for participation of antibody/B-cell-dependent mechanisms has been provided. However, it has been observed that the ratio of B cells to T cells within granulomas increases progressively within time (Chensue and Boros, 1979*a*). These changes in T- and B-cell density are not necessarily

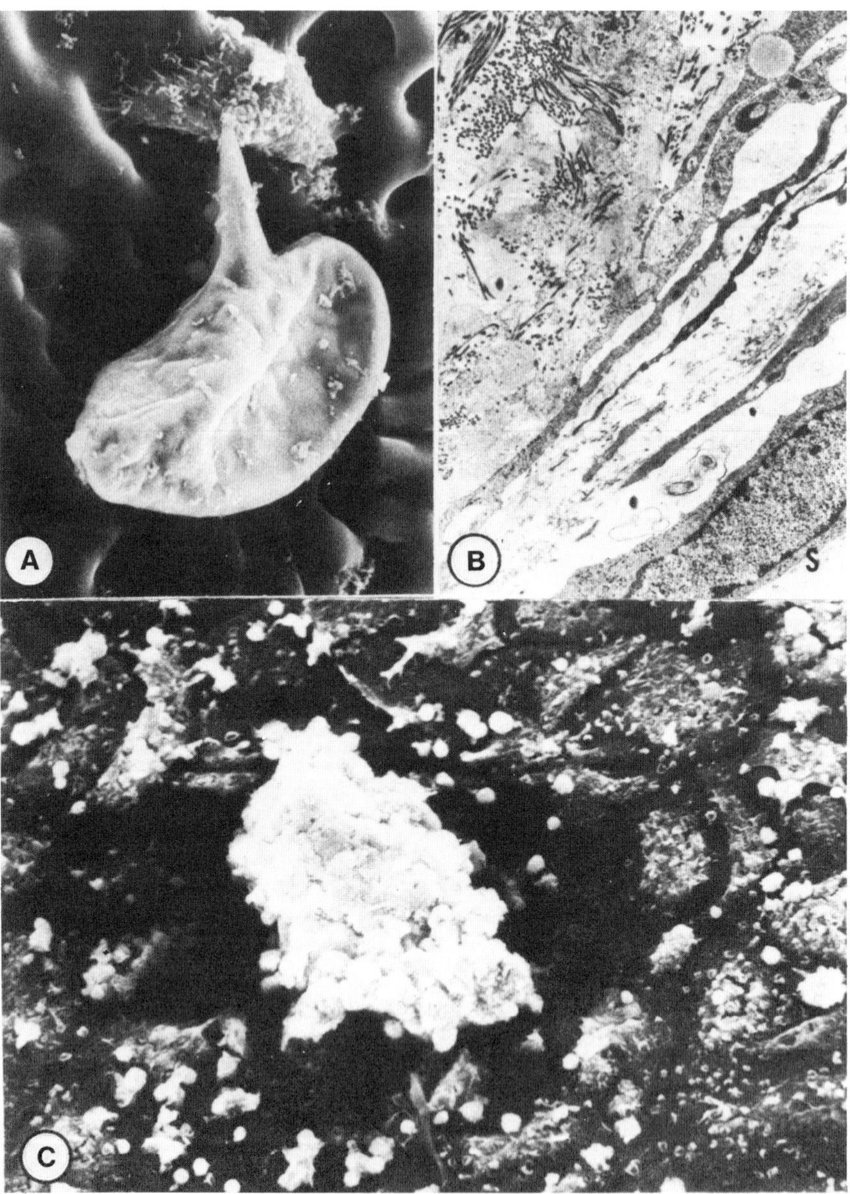

Figure 2. *In vitro* granuloma formation: (A) Initial interaction of cells with egg *in vitro*. Macrophages often adhere to the spine of the *S. mansoni* egg. Smaller cells, probably lymphocytes, bind to the surface of the macrophage and subsequently to the egg, ×840. (B) *In vitro* granuloma. Macrophages and fibroblasts layer at the surface of the forming granuloma. Note the relatively amorphous extracellular material sequestered by cells beneath the outer surface(s) and the well-organized collagen deeper in the interior, ×9800. (C) Fully formed *in vitro* granuloma. Numerous cells completely surround the egg, forming a well-organized, three-dimensional structure. These cells tend to adhere closely, forming concentric layers and effectively isolating the egg from the surrounding environment, ×520. (From Phillips *et al.*, 1980).

equally reflected in all lymphoid compartments (Weinstock and Boros, 1981). However, attempts to suppress granuloma formation using B-depletion (Davis *et al.*, 1974), antibody or B cells (Colley, 1981) have been largely unsuccessful. Recently, however, monoclonal antibody, directed against soluble egg antigen (SEA) have been shown to affect granuloma formation in actively infected animals (B. L. Doughty and S. M. Phillips, unpublished observations, 1983).

The majority of the above studies have suggested that modulation is a specific event. Specific immunosuppression of cellular reactivity has been directly demonstrated *in vitro* in both schistosomiasis (Ottesen *et al.*, 1978; Ottesen and Poindexter, 1980; Ellner *et al.*, 1980; Colley, 1981; Rocklin *et al.*, 1981; Nash *et al.*, 1982) and filariasis (Ottesen *et al.*,1977; Piessens *et al.*, 1980). However, evidence for nonspecific suppression of immunity has also been observed (Nash *et al.*, 1982). DTH may be suppressed and skin grafts prolonged in severe chronic schistosomiasis (Colley, 1981). SEA-induced stimulation of cells, capable of suppressing the responses of other T cells to nonspecific lectins, has been observed (Colley, 1981). The ability of murine spleen cells to generate cytotoxic lymphocytes in MLC reactions is impaired during infection (Coulis *et al.*, 1978) as is the response to sheep red blood cells (Powell and Damian, 1982). These modulators of the host-immune response have been shown to have characteristics of adherent cells (Todd *et al.*, 1979; Ottesen, 1979). Indeed, species-specific innate resistance may be directly related to the efficacy of *in vitro* killing of schistsosomules by mononuclear phagocytes (Peck *et al.*, 1983). The importance of these mononuclear phagocytes in the modulation of granuloma formation, although not as clearly and definitively defined as the role of lymphocyte populations, would appear certain. Immunologic specificity may occur through the specific stimulation of suppressor macrophages, by T cells, or by informational bridges provided by antibody and contingent on Fc receptor interactions. Macrophages are capable of proliferating in the granulomas *in vivo* and synthesizing angiotensin-converting enzyme, the inhibition of which may be important in the modulation of the granulomas (Weinstock *et al.*, 1981). These phagocytic cells may also be important in the generation of factors capable of affecting fibroblast reactivity (D. H. Wyler, 1983; unpublished communications, 1982) and perhaps may participate directly in cytotoxic reactions directed against the eggs (Mahmoud *et al.*, 1979).

The importance of these specific and nonspecific mechanisms of immunomodulation ("hyperergia") relative to granulomatous hypersensitivity are not clear. However, they may be important in the context of the observed depression of immunologic reactivity that occurs in a number of parasitic diseases including not only schistosomiasis (Smithers and Terry, 1969, 1976; Phillips and Colley, 1978; Colley, 1981; Warren, 1982; Nash *et al.*, 1982; Chapter 10, this volume) but also malaria (Greenwood *et al.*, 1972), trypanosomiasis (Goodwin *et al.*, 1972) leishmaniasis (Wyler *et al.*, 1979) and filariasis (Ottesen *et al.*, 1977; Piessens *et al.*, 1982). However, there are critical questions regarding the relationship of immunoreactivity, immunopathology, and parasite growth. For example, what is the timing of these events? Does the specific and/or general immunosuppression precede changes in immunopathologic events or occur as a consequence of some aspect of the infection? What is the importance of other effects, such as genetics (Abdel-Salam *et al.*, 1979; Pereira and Da, 1979; Sasazuki *et al.*, 1980), nutritional influences, or other concomitant diseases (Greenwood and Whittle, 1981)? In addition, the parasite may modulate reactions by elaborating a number of anti-inflammatory, anti-immunologic factors (Tsang *et al.*, 1977; A. Capron and Dessaint, 1982). What is the significance of the immunosuppression *vis-à-vis* general or specific hyporeactivity of the host and resultant increased parasite growth and associated parasite pathology.

However, in summary it is safe to say that as the disease progresses, suppressive immunologic reactions at a number of levels, both unrelated to and specifically directed against schistosome-specific antigens, have been increasingly observed. These mechanisms are extremely complex, based on a number of observed subpopulations of cells, and the true significance of any given interaction at the present time is largely enigmatic. However, teleologically it is contended that if a delicate balance of both specific and nonspecific internally self-modulated interactions is achieved, then immunopathology will be minimized.

G. *Schistosoma japonicum*: Immunopathogenesis

There is much less information *vis-à-vis* the immunopathogenesis of disease associated with *S. japonicum* infection. Those specific studies that have been performed are primarily in the area of granulomatous hypersensitivity and demonstrate that hepatic granulomas are produced around eggs deposited in the liver. These granulomas are far more extensive and are often more necrotic, perhaps because of the deposition of greater number of eggs, usually in aggregates (von Lichtenberg *et al.*, 1973; Nash *et al.*, 1982). There are some histologic differences between *S. japonicum* and *S. mansoni* granulomas. For example, unlike the granuloma associated with *S. mansoni*, larger numbers of plasma cells and larger deposits, compatible with antigen–antibody complexes, have been observed in the *S. japonicum* granulomata (Tsutsumi and Nakashima, 1972; von Lichtenberg *et al.*, 1973; Warren *et al.*, 1975).

There are certain experimental differences between *S. mansoni*- and *S. japonicum*-associated granuloma formation as well (Warren *et al.*, 1978). For example, the injection of *S. mansoni* eggs into the vasculature of infected mice is associated with significant granuloma formation in the lungs about the embolized eggs; conversely, *S. japonicum* eggs do not incite significant granulomatous inflammation (Warren *et al.*, 1975). This event occurs even in the face of active infection and, paradoxically, concurrent with significant inflammation around eggs deposited within the liver of the same animals. However, immunization with egg antigens does result in significant granuloma formation. Sensitization to SEA by subcutaneous injection results in little footpad swelling in response to SEA as the measurement of DTH (Warren *et al.*, 1978). Most granulomatous reactivity appears to be of B-cell origin.

Considerable recent work has suggested that the immune response of mice to *S. japonicum* is immunologically specific (Warren *et al.*, 1978; Garb *et al.*, 1981). It appears to be modulated both specifically and nonspecifically. Serum, but not lymphoid cells, from chronically infected animals can adoptively transfer accelerated granuloma formation (Olds *et al.*, 1982; Warren, 1982). Recently, the immune regulation of murine schistosomiasis, as measured by inhibition of *in vitro* antigen and mitogen-induced cellular responses, has been reported (Garb *et al.*, 1982). Activity of splenocytes obtained from infected mice was studied. Both nonspecific, primarily IgM-mediated, suppressive reactivity, which apparently could be adsorbed by T lymphocytes, and a specific immunologic suppressive activity, primarily IgG_1-mediated, were demonstrated. These results were interpreted as suggesting the importance of both nonspecific and specific immunoregulatory activity in modulating granulomatous responses to *S. japonicum*. The authors (Garb *et al.*, 1982) postulated that immunoregulation may involve either the direct action of immunoglobulin reacting with SEA and/or anti-idiotypic antibodies, which can regulate lymphoid functions induced by SEA. Evidence for such an idiotypic regulatory affect has recently been provided by studies of Mitchell *et al.* (1983), using monoclonal antibodies. In summary, although the modes of immunomodulation in *S. mansoni* and *S. japonicum* infections may be quite different, both appear to involve specific and nonspecific mechanisms.

IV. SUMMARY

These studies would indicate tremendous variations in the clinical manifestations of parasitic disease, resulting from characteristics of the parasite, the host, and their interaction. They further suggest that the conceptual mechanistic model described in the introduction is highly applicable. Previous evidence to substantiate the validity of such a model in schistosomiasis, a variety of protozoan diseases, and leprosy has already been presented (Phillips and Fox, 1982). This report would appear to lend additional credence to the postulates and suggests that upon scrutiny, the model represents a reasonable explanation for a wide variety of clinical manifestations of a parasitic disease. In addition, it may provide a working hypothesis for the interpretation of the immunopathology found in other diseases such as filariasis. Figure 3 compares and contrasts schistosomiasis and filariasis within the context of this hypothesis. Immunopathology results from the relative balance of host–parasite immuno-

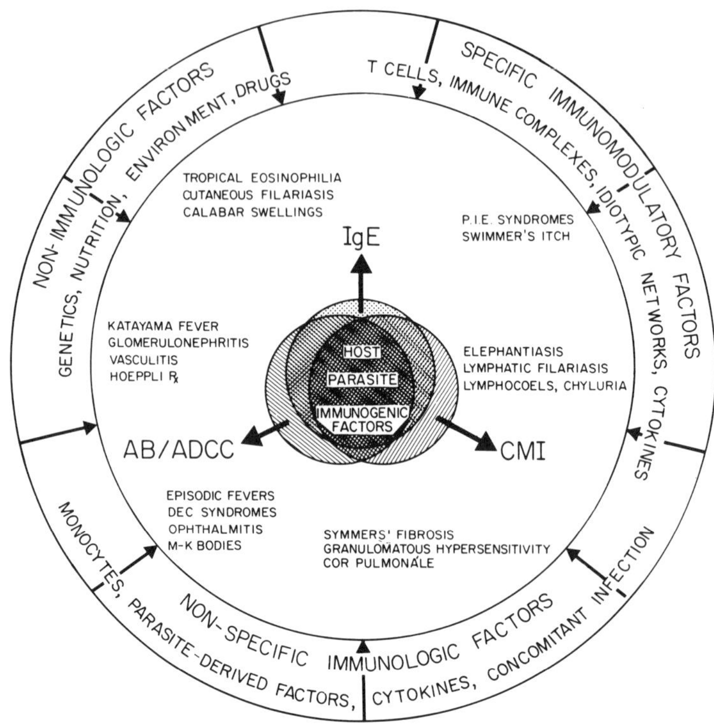

Figure 3. Interplay of immunogenic and immunomodulatory factors in schistosomiasis and filariasis.

genic factors and modulatory specific and nonspecific factors. The resultant immunopathology results from a number of immunologic mechanisms, but for the sake of comparison can be placed in certain analogous groups. Clearly, although a number of experimental questions still exist, *vis-à-vis* these analogies, it would appear that they are reasonable comparisons. It is hoped that such a conceptual approach might provide a useful framework for an understanding of the spectrum of immunopathology resulting from parasitic disease. These concepts might possibly lead to the eventual control of immunopathology.

ACKNOWLEDGMENT

This work was supported by the Edna McConnell Clark Foundation grant 282-0040 and by U.S. NIH grant AI-15193-05.

VI. REFERENCES

Abdel-Hafez, S. K., Phillips, S. M., and Zodda, D. M., 1983, *Schistosoma mansoni*: Detection and characterization of antigens and antigenemia in inhibition enzyme-linked immunosorbent assay (IELISA), *Exp. Parasitol.* **55**:219.

Abdel-Salam, E., Ishaac, S., and Mahmoud, A. A. F., 1979, Histocompatibility-linked susceptibility for hepatosplenomegaly in human schistosomiasis mansoni, *J. Immunol.* **123**:1829.

Ackerman, S. J., Gleich, G. J., Weller, P. F., and Ottesen, E. A., 1981, Eosinophilia and elevated serum levels of eosinophil major basic protein and Charcot-Leyden crystal protein (lysophospholipase) after treatment of patients with Bancroft's filariasis, *J. Immunol.* **127**:1093.

Andrade, Z. A., and Andrade, S. G., 1970, Pathogenesis of schistosomal pulmonary arteritis, *Am. J. Trop. Med. Hyg.* **19**:305.

Andrade, Z. A., and Warren, K. S., 1964, Mild prolonged schistosomiasis in mice: Alterations in host response with time and the development of portal fibrosis, *Trans. R. Soc. Trop. Med. Hyg.* **58**:53.

Askenase, F. W., Hayden, B., and Higashi, G. I., 1976, Cutaneous basophil hypersensitivity and inhibited macrophage migration in guinea pigs with schistosomiasis, *Clin. Exp. Immunol.* **23**:318.

Barrett-Connor, E., 1982, Concise clinical studies: Parasite pulmonary disease, *Am. Rev. Respir. Dis.* **126**:558.

Benjamin, D., and Soulsby, E. J. L., 1976, Homocytotropic and hemagglutinating antibody responses to *Brugia pahangi* infection in the multimmate rat (*Mastomys natalensis*), *Am. J. Trop. Med. Hyg.* **25**:266.

Bentley, A. G., Doughty, B. L., and Phillips, S. M., 1982, Ultrastructural analysis of the cellular response to *Schistosoma mansoni*: III. The *in vitro* granuloma, *Am. J. Trop. Med. Hyg.* **316**:1168.

Billings, F. T., Winkenwerder, W. L., and Huninen, A. V., 1946, Studies of acute schistosomiasis japonica in the Philippine Islands: I. A clinical study of 337 cases with a preliminary report on the results of treatment with Faudin in 110 cases, *Bull. John Hopkins Hosp.* **78**:21.

Bloch, E. H., Abdel-Wahab, M. F., and Warren, K. S., 1972, *In vivo* microscopic observations of the pathogenesis and pathophysiology of hepatosplenic schistosomiasis in the mouse liver, *Am. J. Trop. Med. Hyg.* **21**:546.

Boros, D. L., and Warren, K. S., 1970, Delayed hypersensitivity-type granuloma formation and dermal reaction induced and elicited by a soluble factor isolated from *Schistosoma mansoni* eggs, *J. Exp. Med.* **132**:488.

Bosworth, W., Ewert, A., and Bray, J., 1973, The interaction of *Brugia malayi* and streptococcus in an animal model, *Am. J. Trop. Med. Hyg.* **22**:714.

Bout, D., Santoro, F., Carlier, Y., Bina, J. C., and Capron, A., 1977, Circulating immune complexes in schistosomiasis, *Immunology* **33**:17.

Buchanan, R. D., Fine, D. P., and Colley, D. G., 1973, *Schistosoma mansoni* infection in mice depleted of thymus-dependent lymphocytes. II. Pathology and altered pathogenesis, *Am. J. Pathol.* **71**:207.

Butterworth, A. E., Sturrock, R. S., Houba, V., and Reese, P. H., 1974, Antibody-dependent cell-mediated damage to schistosomula *in vitro*, *Nature* (*Lond.*) **252**:503.

Butterworth, A. E., Sturrock, R. S., Houba, V., Mahmoud, A. A. F., Sher, A., and Reese, P. H., 1975, Eosinophils as mediators of antibody-dependent cell-mediated damage to schistosomula, *Nature* (*Lond.*) **256**:727.

Butterworth, A. E., Wassom, D. L., Gleich, G. J., and Loegering, D. A., and David, J. R., 1979*a*, Damage of schistosomula of *Schistosoma mansoni* induced directly by eosinophil major basic protein, *J. Immunol.* **122**:221.

Butterworth, A. E., Vadas, M. A., Martz, E., and Sher, A., 1979*b*, Cytolytic T lymphocytes recognize alloantigens of schistosomula of *Schistosoma mansoni* but fail to induce damage, *J. Immunol.* **122**:1314.

Byram, J. E., and von Lichtenberg, F., 1981, Altered schistosome granuloma formation in nude mice, *Am. J. Trop. Med. Hyg.* **26**:944.

Canto, A. L., Sesso, A., and de Brito, T., 1977, Human chronic mansonian schistosomiasis–cell proliferation and fibre formation in the hepatic sinusoidal wall: A morphometric, light and electron-microscopy study, *J. Pathol.* **123**:35.

Capron, A. R. G., and Dessaint, J. P., 1982, Schistosome as a potential source of immunological pharmocologic agents, in: *Clinics in Immunology and Allergy*, Vol. 2: *Immunopharmocology* (A. Capron, ed.), No. 3., p. 613, W. B. Saunders, London.

Capron, A., Dessaint, J. P., Capron, M., and Bazin, H., 1975, Specific IgE antibodies in immune adherence of normal macrophages to *Schistosoma mansoni* schistosomules, *Nature (London)*. **253**:474.

Capron, A., Dessaint, J. P., Joseph, M., Torpier, G., Capron, M., Rousseaux, R., Santoro, F., and Bazin, H., 1977, IgE and cells in schistosomiasis, *Am. J. Trop. Med. Hyg.* **26** (Suppl.):39.

Capron, A., Dessaint, J. P., Capron, M., *et al.*, 1980, Role of anaphylactic antibodies in immunity to schistosomes, *Am. J. Trop. Med. Hyg.* **29**:849.

Capron, A., Dessaint, J. P., and Capron, M., 1982, Effector mechanisms and their regulation, *Immunol. Rev.* **61**:41.

Capron, M., Capron, A., and Torpier, G., 1978, Eosinophil-dependent cytotoxicity in rat schistosomiasis. Involvement of IgG2a antibody and role of mast cells, *Eur. J. Immunol.* **8**:127.

Capron, M., Capron, A., Dessaint, J. P., *et al.*, 1981*a*, Fc receptors on human and rat eosinophils, *J. Immunol.* **126**:2087.

Capron, M., Capron, A., Goetzl, E. J., and Austen, K. F., 1981*b*, Tetrapeptides of the eosinophil chemotactic factor of anaphylaxis (ECF-A) enhance eosinophil Fc receptor, *Nature* (*Lond.*) **289**:71.

Carme, B., and Laigret, J., 1979, Longevity of *Wuchereria bancrofti* var. pacifica and mosquito infection acquired from a patient with low level parasitemia, *Am. J. Trop. Med. Hyg.* **28**:53.

Cheever, A. W., 1968, A quantitative post-mortem study of schistosomiasis mansoni in man, *Am. J. Trop. Med. Hyg.* **17**:38.

Cheever, A. W., 1972, Pipe-stem fibrosis of the liver, *Trans. R. Soc. Trop. Med. Hyg.* **66**:946.

Cheever, A. W., and Warren, K. S., 1964, Hepatic blood flow in mice with acute hepatosplenic schistosomiasis mansoni, *Trans. R. Soc. Trop. Med. Hyg.* **58**:406.

Chensue, S. W., and Boros, D. L., 1979*a*, Population dynamics of T- and B-lymphocytes in the lymphoid organs, circulation and granulomas of mice infected with *Schistosoma mansoni*, *Am. J. Trop. Med. Hyg.* **28**:291.

Chensue, S. W., and Boros, D. L., 1979*b*, Modulation of granulomatous hypersensitivity. I. Characterization of T-lymphocytes involved in the adoptive suppression of granuloma formation of in *Schistosoma mansoni*-infected mice, *J. Immunol.* **123**:1409.

Chensue, S. W., Boros, D. L., and David, C. S., 1980, Regulation of granulomatous inflammation in murine schistosomiasis. *In vitro* characterization of T lymphocyte subsets involved in the production and suppression of migration inhibition factor, *J. Exp. Med.* **151**:1398.

Chensue, S. W., Wellhausen, S. R., and Boros, D. L., 1981, Modulation of granulomatous hypersensitivity. II. Participation of Ly-1+ and Ly-2+ T lymphocytes in the suppression of granuloma formation and lymphokine production in *Schistosoma mansoni*-infected mice, *J. Immunol.* **127**:363.

Chensue, S. W., Boros, D. L., and David, C. S., 1983, Regulation of granulomatous inflammation in murine schistomiasis. II. T-suppressor cell-derived, I-C subregion-coded soluble suppressor factor mediates regulation of lymphokine production, *J. Exp. Med.* **157**:219.

Colley, D. G., 1976, Adoptive suppression of granuloma formation, *J. Exp. Med.* **143**:696.

Colley, D. G., 1981, Immune responses and immunoregulation in experimental and clinical schistosomiasis, in: *The Immunology of Parasitic Diseases*, Vol. 1 (J. M. Mansfield, ed.), pp. 1–83, Marcel Dekker, New York.

Colley, D. G., Megalhaes-Filho, Barros-Coelho, R., 1972, Immunopathology of dermal reactions induced by *Schistosoma mansoni* cercariae and cercarial extract, *Am. J. Trop. Med. Hyg.* **21**:558.

Colley, D. G., Savage, A. M., and Lewis, F. A., 1977, Host responses induced and elicited by cercariae, schistosomula and cercarial antigenic preparations, *Am. J. Trop. Med. Hyg.* **26** (Suppl.):88.

Colley, D. G., Lewis, F. A., and Todd, C. W., 1979, Adoptive suppression of granuloma formation by T lymphocytes by lymphoid cells sensitive to cyclophosphamide, *Cell Immunol.* **45**:192.

Coulis, P. A., Lewert, R. M., and Fitch, T. W., 1978, Splenic suppressor cells and cell-mediated immunity in murine schistosomiasis, *J. Immunol.* **120**:1074.

Crandall, R. B., McGreevey, P. B., Connor, D. H., Crandall, C. A., Neilson, J. T., and McCall, J. W., 1982, The ferret (*Mustela putorius furo*) as an experimental host for *Brugia malayi* and *Brugia pahangi*, *Am. J. Trop. Med. Hyg.* **31**:752.

Crowell, W. A., Malone, J. B., Chapman, W. L., and Thompson, P. E., 1973, Amyloidosis in mongolian gerbils infected with *Litomosoides carinii*, *J. Am. Vet. Med. Assoc.* **163**:596.

Danaraj, T. J., 1951, Eosinophilic lung, in: *A study of 150 cases seen in Singapore*, p. 57, Papineau Studios, Singapore.

Date, A., Shastry, J. C. M., and Johny, K. V., 1979, Ultrastructural glomerular changes in filarial chyluria, *J. Trop. Med. Hyg.* **82**:150.

Date, A., Pandandatti, T., Vaska, P. H., and Shastry, J. C. M., 1980, Lymphocytopenia in chyluria, *Trans. R. Soc. Trop. Med. Hyg.* **74**:137.

Date, A., John, T. J., Chandy, K. G., Rajapopalan, M. S., Vaska, P. H., Pundey, A. P., and Shastry, J. C. M., 1981, Abnormalities of the immune system in patients with chyluria, *Br. J. Urol.* **53**:384.

Date, A., Shastry, J. C. M., Vaska, P. H., Pundey, A. P., Sridaran, G., and Koshi, G., 1982, Complement in chyluria, *Trans. R. Soc. Trop. Med. Hyg.* **76**:238.

Davis, B. H., Mahmoud, A. A. F., and Warren, K. S., 1974, Granulomatous hypersensitivity to *Schistosoma mansoni* eggs in thymectomized and bursectomized chickens, *J. Immunol.* **113**:1064.

des Moutis, I., Ouaisse, R., Grzych, J. M., Yarzabal, L., Haque, A., and Capron, A., 1982, *Onchocerca vovulus:* Detection of circulating antigen by monoclonal antibodies in human onchocerciasis, *Am. J. Trop. Med. Hyg.* **32**:533.

Desowitz, R. S., Barnwell, J. W., Palumbo, N. E., Una, S. R., and Perri, S. F., 1978, Rapid increase of precipitating and reagenic antibodies in *Dirofilaria immitis* infected dogs which develop severe adverse reactions following treatment with diethylcarbamazine, *Am. J. Trop. Med. Hyg.* **27**:1148.

Dessaint, J. P., 1982, Anaphylactic antibodies and their significance, in: *Clinics in Immunology and Allergy,* Vol. 2, no. 3: *Immunoparasitology* (A. Capron, ed.), p. 621, W. B. Saunders, London.

Dessaint, J. P., Capron, M., Bout, D., and Capron, A., 1975, Quantitative determination of specific IgE antibodies to schistosome antigens and serum IgE level in patients with schistosomiasis (*S. mansoni* or *S. haematobium*), *Clin. Exp. Immunol.* **20**:427.

Dessaint, J. P., Torpier, G., Capron, M., and Bazin, H., 1979, Cytophilic binding of IgE to the macrophage. I. Binding characteristics of IgE on the surface of macrophages in the rat, *Cell. Immunol.* **46**:12.

Diaz-Rivera, R. S., Ramos-Morales, F., Koppisch, E., Garcia-Palmieri, M. R., Cintron-Rivera, A. A., Marchand, E. J., Gonzalez, O., and Torregrosa, M. V., 1956, Acute Manson's schistosomiasis, *Am. J. Med.* **21**:918.

Doenhoff, M., Musallan, R., Bain, J., and McGregor, A., 1979, *Schistosoma mansoni* infections in T-cell deprived mice, and the ameliorating effect of administering homologous chronic infection serum. I. Pathogenesis, *Am. J. Trop. Med. Hyg.* **28**:260.

Domingo, E. O., and Warren, K. S., 1967, The inhibition of granuloma formation around *Schistosoma mansoni* eggs. II. Thymectomy, *Am. J. Pathol.* **51**:757.

Domingo, E. O., and Warren, K. S., 1968, Endogeneous densitization: Changing host granulomatous response to schistosome eggs at different stages of infection with *Schistosoma mansoni*; *Am. J. Pathol.* **52**:369.

Doughty, B. L., and Phillips, S. M., 1982*a*, Delayed hypersensitivity granuloma formation around *Schistosoma mansoni* eggs *in vitro*. I. Definition of the model, *J. Immunol.* **128**:30.

Doughty, B. L., and Phillips, S. M., 1982*b*, Delayed hypersensitivity granuloma formation around *Schistosoma mansoni* eggs *in vitro*. II. Regulatory T-cell subsets, *J. Immunol.* **128**:37.

Ellner, J. J., Olds, G. R., Kamel, R., Osman, G. S., El Kholy, A., Mahmoud, A. A. F., 1980, Suppressor splenic T lymphocytes in human hepatosplenic schistosomiasis mansoni, *J. Immunol.* **125**:308.

Ellner, J. J., Olds, G. R., Lee, C. W., and Clinhenz, M. E., 1982, Destruction of multicellular parasite *Schistosoma mansoni* by T-lymphocytes, *J. Clin. Invest.* **70**:369.

Epstein, W. L., Fukuyama, K., Danno, K., and Kwan-Wong, E., 1979, Granulomatous hypersensitivity inflammation in normal and athymic mouse infected with *S. mansoni*: An ultrastructural study, *J. Pathol.* **127**:207.

Fox, E. G., 1979, The pathology, immunology and immunogenetics of *Brugia* infection in syngeneic rats, Doctoral dissertation, University of California, Los Angeles.

Fuglsang, H., Anderson, J., and de Marshall, T. F., 1979, Studies on onchocerciasis in the United Cameroon Republic. V. A four year follow-up of 6 rain-forest and 6 Sudan savanna villages. Some changes in skin and lymph nodes, *Trans. R. Soc. Trop. Med. Hyg.* **73**:118.

Galvao-Castro, B., Bina, J., Prata, A., and Lambert, P. H., 1981, Correlation of circulating immune complexes and complement breakdown products with the severity of the disease in human schistosomiasis mansoni, *Am. J. Trop. Med. Hyg.* **30**(6):1238.

Garb, K. S., Stavitsky, A. B., and Mahmoud, A. A. F., 1981, Dynamics of antigen and mitogen-induced responses in schistosomiasis japonica: *In vitro* comparison between hepatic granulomas and splenic cells, *J. Immunol.* **127**:115.

Garb, K. S., Stavitsky, A. B., Olds, G. R., Tracy, J. W., and Mahmoud, A. A. F., 1982, Immune regulation in murine schistosomiasis japonica: Inhibition of *in vivo* antigen- and mitogen-induced cellular responses by splenocyte culture supernatants and by purified fractions from serum of chronically infected mice, *J. Immunol.* **129**(6):2752.

Gazzinelli, G., Ramalho-Pinto, F. J., and Dias da Silva, W., 1969, *Schistosoma mansoni* generation of anaphylatoxin by cercarial extracts, *Exp. Parasitol.* **28**:86.

Goodwin, L. G., Green, D. G., Guy, M. W., and Voller, A., 1972, Immunosuppression during trypanosomiasis, *Br. J. Exp. Pathol.* **53**:40–43.

Green, W. F., and Colley, D. G., 1981, Modulation of *Schistosoma mansoni* egg-induced granuloma formation: I-J restriction of T-cell-mediated suppression in a chronic parasitic infection, *Proc. Natl. Acad. Sci. USA* **78**:1152.

Greenwood, B. M., and Whittle, H. C., 1981, Immunology of medicine, in: *Current Topics in Immunology Series,* Vol. 14 (J. Turk, ed.), Edward Arnold, London.

Greenwood, B. M., Bradley-Moore, A. M., Poalit, A., and Bryceson, A. D. M., 1972, Immunosuppression in children with malaria, *Lancet* **1**:169.

Grove, D. I., and Forbes, I. J., 1979, Immunosuppression in Bancroftian filariasis, *Trans. R. Soc. Trop. Med. Hyg.* **73**:23.

Gutekunst, R. R., Browne, H. G., and Meyers, D. M., 1965, Influence of *Schistosoma mansoni* eggs on growth of monkey heart cells, *Exp. Parasitol.* **17**:194.

Hang, L. M., Warren, K. S., and Boros, D. L., 1974, *Schistosoma mansoni*: Antigenic secretions and the etiology of egg granulomas in mice, *Exp. Parasitol.* **35**:288.

Haque, A., Camus, D., Ogilvie, B. M., Capron, M., Bazin, H., and Capron, A., 1981, *Dipetalonema viteae* infective larvae reach reproductive maturity in rats immunodepressed by prior exposure to *Schistosoma mansoni* or its products and in congenitally athymic rats, *Clin. Exp. Immunol.* **143**:1.

Henson, P. M., Mackenzie, C. D., and Spector, W. G., 1979, Inflammatory reactions in onchocerciasis: A report on current knowledge and recommendations for further study, *Bull. WHO* **57**:667.

Hiatt, R. A., Sotomayor, Z. R., Sanchez, G., Zambrana, M., Knight, W. B., 1979, Factors in the pathogenesis of acute schistosomiasis mansoni, *J. Infect. Dis.* **139**:659.

Hillyer, G. V., 1969, Immunoprecipitins in *Schistosoma mansoni* infections. IV. Human infections, *Exp. Parasitol.* **25**:376.

Hillyer, G. V., 1973, Schistosome deoxyribonucleic acid (DNA), antibodies to DNA in schistosome infections, and their possible role in renal pathology, *Boletin-Asociancion Medica de Puerto Rica,* **65**(Suppl.):1.

Hillyer, G. V., Hernandez-Almenas, N., Knight, W. B., and Cline, B., 1976, Circulating antigens, antibodies and immunoglobulins in the serum of humans with schistosomiasis mansoni, *Biol. Assoc. Med. P. R.* **68**:128.

Houba, F. Sturrock, R. F., and Butterworth, A. E., 1977, Kidney lesions in baboons infected with *Schistosoma mansoni, Clin. Exp. Immunol.* **30**:439.

Houba, V., 1979, Experimental renal disease due to schistosomiasis, *Kidney Int.* **16**:30.

Hsü, H. F., Davis, J. R., Hsü, S. Y. L., and Osborne, J. W., 1963, Histopathology in albino mice and Rhesus monkeys infected with irradiated cercariae of *Schistosoma japonicum, Z. Tropenmed. Parasitol.* **14**:240.

Hsü, S. Y. L., Davis, J. R., and Hsü, H. F., 1965, Histopathology in rhesus monkeys infected four times with the Formosan strain of *Schistosoma japonicum, Z. Tropenmed. Parasitol.* **16**:297.

Hsü, C. K., Hsü, S. H., Whitney, R. A., Jr., and Hansen, C. T., 1976, Immunology of schistosomiasis in athymic mice, *Nature (Lond.)* **262**:397.

Ishizaka, T., Urban, J., Jr., Tatatsu, K., and Ishizaka, K., 1976, Immunoglobulin E synthesis in parasite infection, *J. Allergy Clin. Immunol.* **58**:523.

Jarrett, E. E. E., 1978, Stimuli for the production and control of IgE in rats, *Immunol. Rev.* **41**:52.

Kane, C. A., and Most, H., 1948, Schistosomiasis of the central nervous system: Experiences in World War II and a review of the literature, *Arch. Neurol. Psychol.* **59**:141.

Karavodin, L. M., and Ash, L. R., 1980, Circulating immune complexes in experimental filariasis, *Clin. Exp. Immunol.* **40**:312.

Kasukawa, R., Okada, M., and Igari, S., 1980, Receptors for IgG, Fc and complement in the mammalian aorta, *Int. Arch. Allergy Appl. Immunol.* **61**:175.

Kazura, J. W., Fauning, M. M., Blumer, J. L., and Mahmoud, A. A. F., 1981, Role of cell generated hydrogen peroxide in granulocyte mediated killing of schistosomula of *Schistosoma mansoni in vitro, J. Clin. Invest.* **67**:93.

Klei, T. R., McCall, J. W., and Malone, J. B., 1980, Evidence for increased susceptibility of *Brugia pahangi*-infected jirds (*Meriones unguiculatus*) to subsequent homologous infections, *J. Helminthol.* **54**:161.

Klei, T. R., Enright, F. M., Blanchard, D. P., and Uhl, S. A., 1981, Specific hypo-responsive granulomatous tissue reactions in *Brugia pahangi* infected jirds, *Acta Trop.* **38**:267.

Klei, T. R., Enright, F. M., Blanchard, D. P., and Uhl, S. A., 1982, Effects of presensitization of the development of lymphatic lesions in *Brugia pahangi* infected jirds, *Am. J. Trop. Med. Hyg.* **31**:280.

Knight, D. A., 1979, Pathophysiology of canine dirofilariasis, *Am. Coll. Vet. Inst. Med. Sci. Proc.* p. 41.

Lambert, P. H., 1978, A WHO collaborative study for the evaluation of eighteen methods for detecting immune complexes in serum, *J. Clin. Lab. Immunol.* **1**:1.

Lawley, T. J., Ottesen, R. A., Hiatt, R. A., and Gazze, L. A., 1979, Circulating immune complexes in acute schistosomiasis, *Clin. Exp. Immunol.* **37**:221.

Lehman, J. S., 1973, Urinary schistosomiasis in Egypt: Clinical and parasitological correlation, *Trans. R. Soc. Trop. Med. Hyg.* **67**:385.

Lehman, J. S., Farid, Z., Haxton, S., Jr. Abdel-Wahab, M. F., and Kent, D. C., 1970, Intestinal protein loss in schistosomal polyposis of the colon, *Gastroenterology* **59**:433.

MacFarlane, W. V., 1949, Schistosome dermatitis in New Zealand. Part II. Pathology and immunology of cercarial lesions, *Am. J. Hyg.* **50**:152.

Machado, A. J., Gazzinelli, G., Pellegrino, J., and Dias da Silva, W., 1975, *Schistosoma mansoni*: The role of the complement C3-activating system in the cercaricidal action of normal serum, *Exp. Parasitol.* **38**:20.

McCully, R. M., Barron, C. N., and Cheever, A. W., 1976, Schistosomiasis (bilharziasis), in: *Pathology of Tropical and Extraordinary Diseases*, Vol. 2 (C. H. Binford and D. H. Connor, eds.), p. 482, Armed Forces Institute of Pathology, Washington, D.C.

Magalhaes-Filho, A., 1959, Pulmonary lesions in mice experimentally infected with *Schistosoma mansoni, Am. J. Trop. Med. Hyg.* **8**:527.

Magalhaes-Filho, A., Krupp, I. M., and Malek, E. A., 1965, Localization of antigen and presence of antibody in tissues of mice infected with *Schistosoma mansoni*, as indicated by fluorescent antibody techniques, *Am. J. Trop. Med. Hyg.* **14**:84.

Mahmoud, A. A. F., Peters, P. A., Sevil, R., and Remington, J. S., 1979, *In vitro* killing of schistosomula of *Schistosoma mansoni* by BCG and *C. paruum*-activated macrophages, *J. Immunol.* **122**:1655.

Marcial-Rojas, R. A., and Fial, R. E., 1963, Neurologic complications of the literature and report of two cases of transverse myelitis due to *Schistosoma mansoni, Ann. Intern. Med.* **59**:215.

Manson-Bahr, P. E. C. and Apted, F. I. C., 1982, in: *Manson's Tropical Diseases*, 18th edition, p. 206, Guilliere Tindall, London.

Meyers, F. J., and Kouwenaar, W., 1939, Over hypereosinophile en over en merkwaaridgen vorm van filariasis, *Geneesk. Tidscher. Med. Ind.* **79**:853.

Mitchell, G. F., Garcia, E. G., Cruise, K. M., Tiu, W. U., and Hawking, R. E., 1983, Lung granulomatous hypersensitivity to eggs of *Schistosoma japonicum* in mice analyzed by a radioisotopic assay and effects of hybridoma (idiotype sensitization), *Aust. J. Exp. Biol. Med. Sci.* **60**:401.

Morbidity and Mortality Weekly Report Katayama Syndrome, Ethiopia 1982, *Morbidity Mortality Weekly Rep.* **31**(32):436.

Nash, T. E., Cheever, A. W., Ottesen, E. A., and Cook, J. A., 1982, Schistosome infections in humans: Perspectives and recent findings, *Ann. Intern. Med.* **97**:740.

Nelson, G. S., 1966, The pathology of filarial infections, *Helminthol. Abstr.* **35**:311.

Nelson, G. S., 1979, Current concepts in parasitology: Filariasis, *N. Engl. J. Med.* **300**:1136.

Neva, F. A., and Ottesen, E. A., 1978, Tropical (Filarial) Eosinophilia, *N. Engl. J. Med.* **298**:1129.

Ogilvie, B. M., and Jones, V. E., 1973, Immunity in the parasitic relationship between helminths and hosts, *Prog. Allergy* **17**:93.

Olds, G. R., Ellner, J. J., Kearse, L. A., Jr., Kazura, J. W., and Mahmoud, A. A. F., 1980, Role of agrinase in killing of schistosomula of *Schistosoma mansoni*, *J. Exp. Med.* **151**:1551.

Olds, G. R., Olveda, R., Tracy, J. W., and Mahmoud, A. A. F., 1982, Adoptive transfer of modulation of granuloma formation and hepatosplenic disease in murine schistosomiasis japonica by serum from chronically infected animals, *J. Immunol.* **128**:1393.

Olivier, L., 1949, Schistosome dermatitis: A sensitization phenomenon, *Am. J. Hyg.* **49**: 290.

Olivier, L., and Weinstein, 1953, Experimental schistosome dermatitis in rabbits, *J. Parastiol.* 38:280.

Oliveira, C. A., Bicalho, D. M., Pimenta-Filho, R., Katz, H., Ferreira, H. Bittencourt, D., Dias, R. P., Alvarenga, R. J., and Dias, C. B., 1969, A fase aguda da equistossomose mansoni. Estude laparoscopico disseminacao equistossomoticos, *GEN* **23**:369.

Orange, R. G., Valentine, M. D., and Austen, K. F., 1968, Inhibition of the release of slow-reacting substance of amphylaxis in the rat with diethylcarbamazine, *Proc. Soc. Exp. Biol. Med.* **127**:127–132.

Ottesen, E. A., 1979, Modulation of the host response in human schistosomiasis mansoni. I. Adherent suppressor cells which inhibit blastogenesis, *J. Immunol.* **123**:1639.

Ottesen, E. A., 1980, Immunopathology of lymphatic filariasis in man, *Springer Sem. Immunopathol.* **2**:373.

Ottesen, E. A., and Poindexter, R. W., 1980, Modulation of the host response in human schistosomiasis. II. Humoral factors which inhibit lymphocyte proliferative responses to parasite antigens, *Am. J. Trop. Med. Hyg.* **29**:592.

Ottesen, E. A., and Weller, P. F., 1979, Eosinophilia following treatment of patients with schistosomiasis mansoni and Bancroft's filariasis, *J. Infect. Dis.* **139**:343.

Ottesen, E. A., Weller, P. F. and Heck, L., 1977, Specific cellular unresponsiveness in human filariasis, *Immunology* **33**:413.

Ottesen, E. A., Hiatt, R. A., Cheever, A. W., Sotomayor, Z. R., and Neva, F. A., 1978, The acquisition and loss of antigen-specific cellular immune responsiveness in acute and chronic schistosomiasis in man, *Clin. Exp. Immunol.* **33**:38.

Ottesen, E. A., Neva, F. A., Paranjape, R. S., Tripathy, S. P., Thiruvengadam, V., and Beaver, M. A., 1979, Specific allergic sensitization to filarial antigens in tropical eosinophilia syndrome, *Lancet* **1**:1158.

Ottesen, E. A., Poindexter, R. W., and Hussain, R., 1981*a*, Detection, quantitation and specificity of anti-parasite IgE in schistosomiasis mansoni, *Am. J. Trop. Med. Hyg.* **30**:1225.

Ottesen, E. A., Mandell, N. R., MacQueen, J. M., Weller, P. F., Amos, D. B., and Ward,

F. E., 1981*b*, Familial predisposition to filarial infection–not linked to HLA-A or -B locus specificities, *Acta. Trop.* **38**:205.

Palmieri, J., Purnomo, R., Dennis, D. T., Marwoto, H. A., 1980, Filariid parasites of South Kalimantan (Borneo) Indonesia. *Wuchereria kalimantani* Sp.N. (*Nematoda: Filarioidea*) from the silvered leaf monkey, *Presbytis crisatus* Eshscholtz 1921, *J. Parasitol.* **66**:645.

Palmieri, J. R., Connor, D. H., Purnomo, D. T., Dennis, D. T., and Marwoto, H., 1982, Experimental infection of *Wuchereria bancrofti* in the silvered leaf monkey *Presbytis cristatus* Eschcholz, *J. Helminthol.* **56**:243.

Peck, C. A., Carpenter, M. D., and Mahmoud, A. A. F., 1983, Species-related innate resistance to *Schistosoma mansoni*: Role of mononuclear phagocytes in schistosomula killing *in vitro, J. Clin. Invest.* **71**:66.

Pereira, D. M., and Da, S. M., 1979, Sistemas HLA, ABO e Rh e característicás raciasis em pacientes com hepatoesplenomegalia equistossomótica, Brasilia, Brazil, thesis, p. 152, University of Brasilia.

Phillips, S. M., and Colley, D. G., 1978, Immunologic aspects of host responses to schistosomiasis: Resistance, immunopathology, and eosinophil involvement, *Prog. Allergy* **24**:49.

Phillips, S. M., and Fox, E. G., 1982, The immunopathology of parasitic disease, in: *Clinics of Immunology and Allergy*, Vol. 2, no. 3, *Immunoparasitology* (A. Capron, ed.), p. 667, Fig. 1, W. B. Saunders, London.

Phillips, S. M., and Zodda, D. M., 1983, *Monoclonal Antibodies and Immunoparasitology in Monoclonal Antibodies*, 2nd ed. (R. H. Kennett, T. J. McKearn, and K. B. Bechtol, eds.), Plenum Press, New York (in press).

Phillips, S. M., DiConza, J. J., Gold, J. A., and Reid, W. A., 1977, Schistosomiasis in the congenitally athymic (nude) mouse. I. Thymic dependence of eosinophilia, granuloma formation and host morbidity, *J. Immunol.* **118**:594.

Phillips, S. M., Reid, W. A., Doughty, B. L., and Bentley, A. G., 1980, The immunologic modulation of morbidity in schistosomiasis studies in athymic mice and *in vitro* granuloma formation, *Am. J. Trop. Med. Hyg.* **29**:8.

Piessens, W. F., 1981, Lymphatic filariasis in human: An immunologic maze, *Ann. Intern. Med.* **95**:778.

Piessens, W. F., McGreevy, P. B., Piessens, P. W., McGreevy, M., Kaman, I., Karoso, J. S., and Dennis, D. T., 1980*a*, Immune responses in human infections with *Brugia malayi*. Specific cellular unresponsiveness to filarial antigen, *J. Clin. Invest.* **65**:172.

Piessens, W. F., Ratiwayanto, S., Tuti, S., Palmieri, J. H., Piessens, P. W., Koiman, I., and Dennis, D. T., 1980*b*, Antigen-specific suppressor cells and suppressor factors in human filariasis with *Brugia malayi, N. Engl. J. Med.* **302**:833.

Piessens, W. F., Partono, F., Hoffman, S. L., Ratiwayanto, S., Piessens, P. W., Palmieri, J. R., Koiman, I., Dennis, D. T., and Carney, W. P., 1982, Antigen-specific suppressor T-lymphocyte in human lymphatic filariasis, *N. Engl. J. Med.* **307**:144.

Portaro, J., Britton, S., and Ash, L. R., 1976, *Brugia pahangi:* Depressed mitogen reactivity in filarial infection in the jird, *Meriones unguiculatus, Exp. Parasitol.* **40**:438.

Powell, M. R., and Damian, R. T., 1982, Schistosomiasis mansoni in baboons. VI. Plastic adherent cell suppression of the primary *in vitro* antibody response of peripheral blood mononuclear cells to a heterologous antigen (SRBC) during chronic infection, *Am. J. Trop. Med. Hyg.* **31**:790.

Reis, A. P., Katz, N., and Pellegrino, J., 1970, Immunodiffusion tests in patients with *Schistosoma mansoni* infection, *Rev. Inst. Med. Trop. Sao Paulo* **12**:245.

Rocklin, R. E., Brown, A., Warren, K. S., Pelley, R. P., Houba, V., and Butterworth, A. E., 1977, Immunologic modulation in children with schistosomiasis, *Clin. Res.* **25**:486A.

Rocklin, R. E., Tracy, J. W., and El Kholy, A. E., 1981, Activation of antigen-specific suppressor cells in human schistosomiasis mansoni by fractions of soluble egg antigens nonadherent to Con A sepharose, *J. Immunol.* **127**:2314.

Rogers, R., Denham, D. A., Nelson, G. S., Guy, F., and Ponnudurai, T., 1975, Studies with *Brugia pahangi.* III. Histological changes in the affected lymph nodes of infected cats, *Am. J. Trop. Med. Parasitol.* **69**:77.

Sadun, E. H., 1963, Immunization in schistosomiasis by previous exposure to homologous and heterologous cercariae by inoculation of preparations from schistosomes and by exposure to irradiated cercariae, *Ann. NY Acad. Sci.* **113**:418.

Sadun, E. H., 1972, Homocytophilic antibody response to parasitic infections, in: *Immunity to Animal Parasites* (E. J. L. Soulsby, ed.), p. 97, Academic Press, New York.

Sadun, E. H., von Lichtenberg, F., Cheever, A. W., and Erickson, D. G., 1970, Schistosomiasis mansoni in the chimpanzee. The natural history of chronic infections after single and multiple exposures, *Am. J. Trop. Med. Hyg.* **29**:258.

Sasazuki, T., Ohuta, N., Kanoeoka, R., and Kojima, S., 1980, Association between an HLA haplotype and low responsiveness to schistosomal worm antigen in man, *J. Exp. Med.* **152**:314s.

Scaglia, M., Tinelli, M., and Revoltella, L., 1979, Relationship between serum IgE levels and intestinal parasite load in African populations, *Intl. Arch. Allergy* **59**:465.

Schacher, J. F., and Sayhoun, P. F., 1967, A chronological study of the histopathology of filariid disease in cats caused by *Brugia pahangi* (Buckley and Edeson, 1950), *Trans. R. Soc. Trop. Med. Hyg.* **61**:234.

Schacher, J. F., Edesen, J. F. B., Salahian, A. and Rizk, G., 1973, An 18-month longitudinal lymphographic study of filarial disease in dogs infected with *Brugia pahangi* (Buckley and Edeson, 1956), *Ann. Trop. Med. Parasitol.* **67**:81.

Sher, A., and Scott, P. A., 1982, Genetic factors influencing the interactions of parasites with the immune system, in: *Clinics in Immunology and Allergy,* Vol. 2, no. 21, *Immunoparasitology* (A. Capron, ed.), W. B. Saunders, London p. 489.

Simpson, C. F., and Neilson, J. T. M., 1976, The pathology associated with single and quadruple infections of hamsters with *Dipetalonema viteae, Tropenmed. Parasitol.* **27**:349.

Smith, M. D., 1977, The ultrastructural development of the schistosome egg granuloma in mice, *Parasitology* **75**:119.

Smithers, S. R., and Terry, R. J., 1969, The immunology of schistosomiasis, *Adv. Parasitol.* **7**:41.

Smithers, S. R., and Terry, R. J., 1976, The immunology of schistosomiasis, in: *Advances in Parasitology,* (B. Dawes, ed.), Vol. 14, p. 399, Academic Press, New York.

Spry, C. J. F., 1981, Alterations in blood eosinophil morphology, binding capacity for complexed IgE and kinetitics in patients with tropical (filarial) eosinophilia, *Parasite Immunol.* **3**:1.

Staninuas, R. J., and Hammerberg, B., 1982, Diethylcarbamazine enhanced activation of complement by intact microfilariae of *Dirofilaria immitis* and their *in vitro* products, *J. Parasitol.* **68**:809.

Stenger, R. J., Warren, K. S., and Johnson, E. A., 1967, An ultrastructural study of hepatic granulomas and schistosome egg shells in murine hepatosplenic schistosomiasis mansoni, *Exp. Mol. Pathol.* **8**:116.

Suemura, M., Yodoi, J. Hirashima, M., and Ishizaka, K., 1980, Regulatory role of IgE-binding factors from rat T-lymphocytes. I. Mechanism of enhancement of IgE response by IgE-potentiating factor, *J. Immunol.* **125**:148.

Suzuki, T., and Damian, R. T., 1981, Schistosomiasis mansoni in baboons. IV. The development of antibodies to *Schistosoma mansoni* adult worm, egg, and cercarial antigens during acute and chronic infections, *Am. J. Trop. Med. Hyg.* **30**:825.

Symmers, W., St. C., 1903, Note on a new form of liver cirrhosis due to the presence of the ova of *Bilharzia haematobia, J. Pathol. Bacteriol.* **9**:237.

Tada, T., Kondo, Y., Okumura, K., Sano, M., and Yokogawa, M., 1975, *Schistosoma japonicum*: Immunopathology of nephritis in *Macaca fascicularis, Exp. Parasitol.* **38**:291.

Todd, C. W., Goodgame, R. W., and Colley, D. G., 1979, Immune responses during human schistosomiasis mansoni. V. Suppression of schistosome antigen-specific lymphocyte blastogenesis by adherent phagocytic cells, *J. Immunol.* **122**:1440.

Tsang, V. C. W., Hubbard, W. J., and Damian, R. T., 1977, Coagulation Factor XIIa (Activated Hageman Factor) inhibitor from adult *Schistosoma mansoni, Am. J. Trop. Med. Hyg.* **26**:243.

Tsutsumi, H., and Nakashima, T., 1972, Pathology of liver cirrhosis due to *Schistosoma japonicum*, in: *Research in Filariasis and Schistosomiasis* (M. Yokogawa, ed.), p. 133, University Park Press, Baltimore.

Urban, J. F., Jr., and Ishizaka, K., Jr., 1977, IgE formation in the rat following infection with *Nippostrongylus brasiliensis.* III. Soluble factor for the generation of IgE bearing lymphocytes, *J. Immunol.* **119**:583.

Vadas, M. A., David, J. R., Butterworth, A. E., Pisari, N. T., and Siogok, T. A., 1979, A new method for the purification of human eosinophils and neutrophils, and comparison of the ability of these cells to damage schistosomula of *Schistosoma mansoni, J. Immunol.* **122**:1228.

Verwaede, C., Grzych, J. M., Bazin, H., Capron, M., and Capron, A., 1979, Production d'anticorps monoclonaux anti-*Schistosoma mansoni.* Etude préliminaire de leurs activités biologiques, *Comptes Rendus de l'Academic des Sciences, Paris* (D) **289**:725.

Vincent, A. L., Ash, L. R., Rodrick, G. E., and Sodeman, W. A., 1980, The lymphatic pathology of *Brugia pahangi* in the mongolian jird, *J. Parasitol.* **66**:613.

von Lichtenberg, F., 1964, Studies on granuloma formation. III. Antigen sequestration and destruction in the schistosome pseudotubercle, *Am. J. Pathol.* **45**:75.

von Lichtenberg, F., and Byram, J. C., 1980, Pulmonary cell reactions and resistance to *Schistosoma mansoni*, in: *The Host-Invader Interplay* (H. van Den Bossche, ed.), p. 405, Elsevier-North Holland, Amsterdam.

von Lichtenberg, F., Erickson, D. G., and Sadun, E. H., 1973, Comparative histopathology of schistosome granulomas in the hamster, *Am. J. Pathol.* **72**:149.

von Lichtenberg, F., Bawden, M.P., and Shealey, S. H., 1974, Origin of circulating antigen in the schistosome gut. An immunofluorescent study, *Am. J. Trop. Med. Hyg.* **23**:1088.

von Lichtenberg, F., Sher, A., and MacIntyre, S., 1977, A lung model of schistosome immunity in mice, *Am. J. Pathol.* **87**:105.

Walker, L., Hay, F. C., and Roitt, I. M., 1979, Characteristics of complexes for arming and inhibiting effector cells for antibody-dependent cell-mediated cytotoxicity, *Clin. Exp. Immunol.* **36**:397.

Waltzing, P. and Bloch-Michel, H., 1971, Le phénomène des rosettes filarennes: Intérêt pour le diagnostic des arthrites filarennes, **30**:2061.

Warren, K. S., 1971, Worms, in: *Immunological Diseases*, 2nd ed. (M. Samter, ed.), p. 668, Boston, Little, Brown.

Warren, K. S., 1972*a*, The immunopathogenesis of schistosomiasis: A multidisciplinary approach, *Trans. R. Soc. Trop. Med. Hyg.* **66**:417.

Warren, K. S., 1972*b*, Pipe-stem fibrosis of the liver, *Trans. R. Soc. Trop. Med. Hyg.* **66**:946.

Warren, K. S., 1973, The pathology of schistosome infections, *Helminthol. Abstr. (A)* **42**:591.

Warren, D. S., 1974, Modulation of immunopathology in schistosomiasis, in: *Parasites in the Immunized Host: Mechenisms of Survival,* Ciba Foundation Symposium No. 25, p. 243, Elsevier North-Holland, Amsterdam.

Warren, K. S., 1982, The secret of immunopathogenesis of schistosomiasis *in vivo* models, *Immunol. Rev.* **61**:189.
Warren, K. S., Reboucas, G., and Baptista, A. G., 1965, Ammonia metabolism and hepatic coma in hepatosplenic schistosomiasis. Patients studied before and after portacaval shunt, *Ann. Intern. Med.* **62**:1113.
Warren, K. S., Domingo, E. O., and Cowan, R. B. T., 1967, Granuloma formation around schistosome eggs as a manifestation of delayed hypersensitivity, *Am. J. Pathol.* **51**:735.
Warren, K. S., Boros, D. L., Hang, L. M., and Mahmoud, A. A. F., 1975, *Schistosoma japonicum* egg granuloma, *Am. J. Pathol.* **80**:279.
Warren, K. S.,, Grove, K. I., and Pelley, R. P., 1978; *Schistosoma japonicum* egg granuloma. II. Cellular composition, granuloma size, and immunologic concomitants, *Am. J. Trop. Med. Hyg.* **27**:271.
Webb, J. K. G., Job, C. K., and Gault, E. W., 1960, Tropical eosinophilia: Demonstration of microfilariae in lung, liver, and lymph nodes, *Lancet* **7129**:835.
Webster, E. H., 1946, Filariasis among white immigrants in Samoa, *U.S. Nav. Med. Bull.* **46**:186.
Wei-Hsin, C., 1958, Acute schistosomiasis. Clinical manifestations of 96 cases, *Chinese Med. J.* **76**:1.
Weinstock, J. V., and Boros, D. L., 1981, Heterogeneity of the granulomatous response in the liver, colon, ileum, and ileal Peyer's patches to schistosome eggs in murine schistosomiasis mansoni, *J. Immunol.* **127**:1906.
Weinstock, J. V., Boros, D. L., Gee, J. B., and Ehrinpreis, M. N., 1981, The effect of SQ-14225, an inhibitor of angiotensin I converting enzyme on the granulomatous response to *Schistosoma mansoni* eggs in mice, *J. Clin. Invest.* **67**:931.
Weiss, N., Sturchler, D., and Dietrich, F. M., 1978, Radioallosorbent and indirect fluorescent antibody tests in immunodiagnosis of schistosomiasis, *Lancet* **2**:1231.
Wellhausen, S. R., Chensue, S. W., and Boros, D. L., 1980, Modulation of granulomatous hypersensitivity. Analysis by adoptive transfer of effect and suppressor T lymphocytes involved in granulomatous inflammation in murine schistosomiasis, in: *Basic and Clinical Aspects of Granulomatous Diseases* (D. L. Boros and T. Yoshida, eds.), p. 219, Elsevier North-Holland, Amsterdam.
World Health Organization, 1976, Epidemiology of onchocerceriasis, *Tech. Rep. Ser.* **597**:94.
World Health Organization, 1981, Immunology of filariasis, *Bull. WHO* **59**:1.
Wyler, D. H., and Postlethwaite, A. E., 1983, Fibroblast stimulation in schistosomiasis, **130**:1371.
Wyler, D. H., Wahl, S. M., and Wahl., L. M., 1978, Hepatic fibrosis in schistosomiasis: Egg granulomas secrete fibroblast stimulating factor *in vitro, Science* **202**:438.
Wyler, D. H., Weinbaum, F. I., and Herrod, H. R., 1979, Characterization of in vitro proliferative responses of human lymphocytes to leishmanial antigens, *J. Infect. Dis.* **140**:215.
Zvaifler, N. J., Sadun, E. H., Becker, E. L., and Schoenbechler, M. J., 1967, Demonstration of homologous anaphylactic antibody in rabbits infected with *Schistosoma mansoni, Ex. Parasitol.* **20**:278.

Chapter 12

Cellular Immunity to Malaria and Babesia Parasites: A Personal Viewpoint

Anthony C. Allison

Institute of Biological Sciences
Syntex Research
Palo Alto, California 94304

I. INTRODUCTION

It has long been recognized that falciparum malaria takes a different course in persons living where the disease is endemic, when repeated infection is initiated while they are still infants, and in nonimmune persons. Indeed, Plehn (1902) used the term *relative immunity* to describe the situation in Cameroon Africans. The epidemiology of malaria in Gambians, described by McGregor *et al.* (1956), is representative of the disease in many endemic areas, although the intensity of transmission and the age at which immunity is acquired vary from one location to another. The most severe disease is seen in young children between 6 months and 5 years of age. Relative immunity is first reflected by reduced symptoms, after which parasitemias fall. By school age, children living in endemic areas show only occasional, low parasitemias despite frequent bites by infected mosquitoes. Nevertheless, after as short a period as 6 months abroad, or after elimination of parasites by chemotherapy, severe malaria may again occur (Maegraith, 1974).

The purpose of this chapter is to suggest that the relative immunity of persons living in endemic areas is attributable to two factors: (1) inherited erythrocytic resistance provided by abnormal hemoglobins and glucose-6-phosphate dehydrogenase (G6PD) deficiency, and (2) acquired immunity principally from T-lymphocyte-dependent activation of macrophages and natural killer (NK) cells, which exert oxidant stress on parasitized erythrocytes. Antibodies can collaborate in this process. The effects of acquired immunity are synergistic with inherited erythrocyte abnormalities.

Some of the opinions now put forward have been controversial, and one of the points currently being disputed is the development of ideas about erythrocytic and cellular immunity to malaria and babesia parasites. As a pioneer in both fields of research, I believe it appropriate to review the development of the major concepts, both as a historic exercise and an example of how observations from several different fields can be complementary.

II. RELATIONSHIP OF THE SICKLE CELL TRAIT TO MALARIA

This investigation began in 1949, when I was an undergraduate anthropologist on an Oxford University expedition to Kenya. I studied the distribution of blood groups and other genetic markers in various East African populations (Allison *et al.*, 1952). I found the sickle cell trait to be common in tribes living near the Kenya Coast and Lake Victoria (where falciparum malaria is endemic) but not in tribes living in nonmalarious high or dry areas in between. Since I was aware of the genetics of polymorphism through contact with E. B. Ford in Oxford, I concluded that the sickle cell polymorphism was being maintained because the heterozygotes were relatively resistant to malaria, while the homozygotes were being eliminated through sickle cell anemia.

There was no opportunity to test this hypothesis until I had obtained a medical qualification and returned to East Africa in 1953. I then found that in all three East African countries (Kenya, Uganda, and Tanganyika, later termed Tanzania), high frequencies of sickle cell heterozygotes are present only in areas in which falciparum malaria is endemic (Allison, 1954*a*). This is true for populations with different linguistic affinities and blood groups, suggesting a strong selective effect of a local environmental factor such as malaria. In support of this interpretation, I found in young Ugandan children, 4 months to 4 years of age, lower *P. falciparum* rates and counts in those with HbAS than in those with HbA (Allison, 1954*b*).

The latter publication aroused furious controversy. The statistical analysis used (chi-square test without correction for continuity) was challenged, and Foy and colleagues soon reported that they could find no difference in malaria parasite counts in adult Kenyans with or without HbS. That was no surprise because of the powerful effects of acquired immunity. When Foy and colleagues in Kenya, Raper in Uganda, and Vandepitte in what was then known as Congo Belge (now Zaïre) repeated the investigation on young children, they found exactly the same degree of protection that I had reported. For a review at Cold Spring Harbor (Allison, 1964), I complied all published observations with a statistical analysis by Fisher's exact method, which showed highly significant protective effects against parasites and lethal malaria in children with HbAS and insignificant heterogeneity among observations. Further observations confirming the resistance of HbAS children to malaria were published by Gilles *et al.* (1967) and Martin *et al.* (1979).

With the help of S. M. Smith, an able mathematician, a theoretical analysis of the sickle cell polymorphism was presented (Allison, 1954*c*), which was the basis of all subsequent treatments of the subject. We calculated that the heterozygous advantage could be as much as 25%, implying a higher mortality through direct and indirect effects of malaria than had previously been considered. Out of my work on sickle cells arose the concept that human polymorphisms can be maintained by natural selection through disease, a concept now being considered in relationship to major histocompatibility gene complexes and other polymorphisms.

After my publications on HbS it was obvious that other abnormal Hb polymorphisms might also be maintained by selection through malaria. The most likely example is β-thalassemia, which is common in certain areas in Mediterranean and other countries, but since chlorophenothane (DDT) had eliminated malaria in these countries, direct evidence could not be obtained. Nevertheless, the many observations that have been made on the distribution of Hb S, C, E, and thalassemia (reviewed by Livingstone, 1967) show that high frequencies are confined to areas that were formerly malarious. The resistance of erythrocytes bearing these traits to malaria infection in culture is discussed in Section III.

III. MALARIA AND GLUCOSE 6-PHOSPHATE DEHYDROGENASE DEFICIENCY

The existence of this genetic trait in Americans of African origin was revealed by the correlation of G6PD deficiency with their sensitivity to the antimalarial drug primaquine (Carson *et al.*, 1956). Later it became apparent that there are two polymorphic varieties of G6PD deficiency—the African A− and the Mediterranean B− forms, a nondeficient form polymorphic in Africans (B+) as well as several rare deficient variants. Furthermore, $G6PD^-$ individuals show hemolysis when exposed to several different drugs or dietary constituents, such as fava beans, which exert oxidant stress on erythrocytes.

As soon as an assay became available that could be used under field conditions I studied the distribution of G6PD deficiency in East and West Africa (Allison, 1960; Allison *et al.*, 1961). These were the first analyses of the distribution of the trait on the African continent. The essential findings, which have been repeatedly confirmed, were that G6PD deficiency is common in malarious areas of East Africa (15-25% of males affected) and in West Africa from Nigeria to the Gambia. It is absent from nonmalarious parts of East Africa. I suggested that malaria might again be the major selective factor and, on the basis of my observations in Africa and observations of others on the distribution of the enzyme deficiency in Mediterranean countries, Motulsky (1960) made the same suggestion. Many observations (reviewed by Livingstone, 1967) have supported the view that high frequencies of both the African and Mediterranean varieties of G6PD deficiency occur only where malaria is or has been endemic.

With David Clyde, a respected malariologist in Tanganyika (now Tanzania), I set about testing the hypothesis that young African male $G6PD^-$ hemizygotes have less severe falciparum malaria than do $G6PD^+$ males (Allison and Clyde, 1961). Since our observations subsequently proved controversial, it is worth describing the conditions under which we chose to work, which make the difference between meaningful and uninterpretable research. Clyde had for several years been studying the epidemiology of malaria in a remote valley, where antimalarial drugs were not available. We took blood samples from a high proportion of young boys during the period of maximum transmission of malaria. Clyde counted parasites and I measured G6PD activity; we then exchanged results, so there was no possibility of bias. We found significantly lower *P. falciparum* rates and counts in the $G6PD^-$ children.

Likewise, Gilles *et al.* (1967) observed in male and female Nigerian children with severe malaria (parasite count $\geqslant 100{,}000/mm^3$) significantly fewer $G6PD^-$ cases than in controls. Bienzle *et al.* (1972) reported that Nigerian girls heterozygous for the deficiency (Gd^{A-}/Gd^{A}) and boys hemizygous for another variant (Gd^{B}), but not Gd^{A-} hemizygotes, showed significantly lower *P. falciparum* counts than did other children. There is no obvious reason why females heterozygous for G6PD deficiency should be better protected than male hemizygotes, especially in view of the resistance provided by both in culture (see Roth *et al.*, 1983). But from the point of view of maintaining polymorphism, a selective advantage of Gd^{A-}/Gd^{A} heterozygotes is crucial. In another study from Nigeria (Martin *et al.*, 1979) no protection by G6PD deficiency was found. However, in this small series only six boys and eight girls of all genotypes had severe malaria, so it is difficult to draw any conclusions.

What explanation can there be for such apparent discrepancies? One is the fact that Clyde and I were studying malaria under the natural conditions that had prevailed in Africa for centuries. This was not a population in which some infections had been either attenuated by antimalarial drugs or biased by hospital attendance. Another possibility is that an additional environmental factor made the East African children we were studying more susceptible to oxidant stress and therefore increased the protective effect of G6PD deficiency against malaria.

For example, selenium is a component of glutathione peroxidase, which, together with G6PD, provides the major protection of human erythrocytes against oxidants. Selenium deficiency is common in East Africa. I had long been aware of this, but it was emphasized while I was Director of International Laboratory for Research in Animal Diseases, Nairobi, Kenya (1978–1980). Cattle brought from the field in East Africa were found to have white muscle disease, a manifestation of selenium deficiency. Indeed, Dr. K.-E. Arfors and I have discussed the possibility that endomyocardial fibrosis, which occurs in East Africans, is a manifestation of selenium deficiency analogous to Keshan disease in China.

This opens up a whole area of research in which I hope to participate during the next few years, but it leaves open the question of the conditions under which G6PD deficiency protects against malaria. From observations on erythrocyte cultures, it is likely that the principal effect will be aggravation of oxidant stress by dietary or other factors.

IV. SYNERGISTIC EFFECTS OF INHERITED AND ACQUIRED IMMUNITY TO MALARIA

In the original study (Allison, 1954*b*) and later studies (reviewed by Allison, 1963, 1964) differences in malaria parasite counts between African children with and without HbS were consistently more pronounced in children between 2 and 5 years of age than in infants first exposed to the disease. Moreover, partially immune adult East Africans with HbS infected with *P. falciparum* (Allison, 1954*b*) were more resistant than were nonimmune American blacks, although the latter showed statistically significant protection (Beutler *et al.*, 1955). Allison (1963) concluded that "acquired immunity actually reinforces the protection afforded by the inherited haemoglobin factor during the intermediate phase before the former becomes fully effective." A possible explanation for such synergistic effects is given in Secion V.

V. ROLE OF CELL-MEDIATED IMMUNITY IN MALARIA AND BABESIOSIS

Earlier reports suggesting that immune serum can protect humans against asexual bloodstream forms of malaria parasites were extended by S. Cohen *et al.* (1961), who showed that protection was present in the immunoglobin G (IgG) fraction. Similar protection against *P. berghei* in rats was reported by Diggs and Osler (1969). It was suggested that the principal effect of immune serum was inhibition of the invasion by merozoites of red cells (S. Cohen *et al.*, 1970). However, later studies in Cohen's laboratory showed that resistance of monkeys against *P. knowlesi* is not well correlated with the presence of antibodies blocking reinvasion (Langhorne *et al.*, 1979).

There has also been a long-standing interest in antibodies against sporozoites. The demonstration that monoclonal antibodies against a protein on the surface of *P. berghei* sporozoites can protect mice against mosquito-transmitted infection (Potocnjak *et al.*, 1980) has reinforced this view.

However, the epidemiology of malaria in endemic areas, briefly reviewed in the introduction to this chapter, suggested to me that antibodies against common determinants in sporozoites, or against common components of the apical complexes of merozoites allowing attachment to, and penetration of, erythrocytes, are unlikely to play a major protective role in natural infections. Immunity is slowly acquired as a result of repeated infections and may be rather quickly lost in the absence of parasites.

The epidemiology of malaria is, in fact, reminiscent of the situation termed *premunition* by Sergent (1963). Sergent distinguished between infections of one type in which the microbe kills the host or the host kills all the microbes, and infections of a second type, in which the acute attack ends in a compromise, with the host and parasite both surviving. The host resumes the appearance of good health and the microbes, now rare, live in a latent state in the tissues, blood, and body fluids. Although Sergent did not say so, it seemed to me, when we discussed the situation at a symposium on immunity to protozoa in 1963, that the infections in which he believed premunition occurs (tuberculosis, brucellosis, relapsing fever, rickettsiosis, and herpes simplex virus) are those in which antibodies alone are unable to protect hosts. Cell-mediated immunity might, in fact, be the basis of premunition.

By analogy, the role of cell-mediated immunity in malaria would be worth investigating. That is why, soon after hearing from Jacques Miller that neonatal thymectomy prevents the development in rodents of cell-mediated immunity, I initiated such experiments in malarial infections. We found that in neonatally thymectomized rats *P. berghei* infections are much more severe than in intact rats (I. N. Brown *et al.*, 1968), an observation confirmed by Stechschulte (1969). With the availability of mice congenitally deprived of T lymphocytes (nude or *nu/nu* mice) and strains of malaria (*P. yoelii*) and babesia (*B. microti*) from which intact mice recover, the problem was taken up again (Clark and Allison, 1974). The strain of *P. yoelii* used multiplies in reticulocytes; in nude mice, escalating parasitemias leading invariably to death were observed. *B. microti* multiplies selectively in older erythrocytes, and in the nude mice permanent parasitemias of about 50% were observed; this apparently represents an equilibrium of hemolysis, followed by erythrocyte regeneration, aging, and parasitization. These observations, as well as similar observations from several laboratories that followed, demonstrated that recovery from malarial and babesia infections is thymus dependent. Using mice reconstituted with T cells bearing a chromosome marker, Jayawardena *et al.* (1977) found marked proliferative responses of T cells in *P. berghei* and *P. yoelii* infections. Although their observations are of interest, they did not demonstrate a protective role of T cells, which was evident from our experiments.

The next question was whether T lymphocytes were merely functioning as helpers in the formation of protective antibodies of appropriate isotypes, or whether they were exerting effects through cell-mediated immune responses. Two independent lines of evidence suggest that the latter is true, although antibodies can act synergistically with such responses.

VI. NONSPECIFIC IMMUNITY TO HEMOPROTOZOAN INFECTIONS

Killed *Corynebacterium parvum* (later known as *Propionibacterium acnis*) was found to protect mice against challenge with *P. berghei* sporozoites, but to be much less effective against bloodstream forms of the same parasite (Nussenzweig, 1967). In our laboratory much stronger protection against bloodstream forms of *B. microti*, *B. rodhaini*, and *P. vinckei* was found in mice previously injected with live *Mycobacterium bovis* bacillus Calmette-Guerin (BCG) (Clark *et al.*, 1976) or *P. acnis* (Clark *et al.*, 1977*a*). Clark's experiments with BCG were suggested by J.-L. Virelizier, who was following my work on the production by BCG of type-2 interferon (now known as interferon-γ) and its role in immunosuppression and activation of macrophages to resist virus infections. Mice injected with BCG were found resistant to hemoprotozoan infections even when given immunosuppressive doses of X-radiation, showing that these effects were the result of nonspecific immunity and not of an adjuvant effect accelerating specific immune responses after challenge (Clark *et al.*, 1977*b*).

The major manifestation of immunity to rodent malaria and babesia infections, produced by specific thymus-dependent immune responses or activation of cells nonspecifically by BCG or other agents, was the degeneration of parasites in circulating erythrocytes (Clark *et al.*, 1977*b*). These crisis forms had been previously observed by Taliaferro and Taliaferro (1944) in resolving *P. brasilianum* infections in *Cebus capucinus* monkeys.

VII. ANTIBODY INDEPENDENCE OF IMMUNITY TO HEMOPROTOZOA

The importance of thymus-dependent, antibody-independent mechanisms of immunity to haemoprotozoan parasites has been demonstrated by the use of hosts in which humoral responses have been suppressed. Rank and Weidanz (1976) made chickens agammaglobulinemic by combined chemical bursectomy, i.e., treatment with testosterone and cyclophosphamide. Such chickens injected with *P. gallinaceum* died with fulminating parasitemias. However, when rescued from otherwise fatal infections by chloroquine therapy, the B-cell-deficient chickens resisted challenge infection with the same parasite. Parasites were not completely cleared from the B-cell-deficient mice, which could nevertheless produce a protective immune response.

In mice made B-cell deficient by neonatal injections of antiserum against the μ chain of immunoglobulin, infection with a strain of *P. yoelii* avirulent in intact mice produced lethal disease (Weinbaum *et al.*, 1976). Again, treatment of the B-cell-deficient mice led to nonsterilizing protective immunity (Roberts and Weidanz, 1979). With another parasite, *P. chabaudi adami*, antibodies were not required even for recovery from the first wave of infection (Grun and Weidanz, 1981). The μ-suppressed mice developed a lasting nonsterile immunity and were resistant to homologous challenge as well as to infections with *P. vinckei*, but not to challenge with *P. yoelii* or *P. berghei.* Likewise, μ-suppressed mice recover from infections with *Babesia microti*, although this is usually delayed when compared with intact mice, and the immunity is not sterile (J. I. Grun, W. P. Weidanz, and A. C. Allison, unpublished observations, 1983). This finding suggests that antibodies play at most a minor synergistic role in immunity to murine babesiosis.

Further discussion of the mechanism by which cell-mediated immunity leads to intraerythrocytic death of haemoprotozoa requires background information on the effects of oxidant stress on these parasites.

VIII. MULTIPLICATION OF *P. FALCIPARUM* IN CULTURES OF ERYTHROCYTES WITH ABNORMAL Hbs AND G6PD DEFICIENCY

The next major advance was the development by Trager and Jensen (1976) of a method for continuous propagation of *P. falciparum* in cultures of human erythrocytes. It was then possible to compare under controlled conditions multiplication of the parasite in normal erythrocytes and those with abnormal Hbs and G6PD deficiency. In two laboratories *P. falciparum* was found to grow less well in erythrocytes bearing HbS than in those bearing HbA under low partial pressures of O_2 (Friedman, 1978; Pasvol *et al.*, 1978). Such relatively anoxic conditions would be expected where *P. falciparum* completes its asexual bloodstream cycle of development attached to endothelial cells of postcapillary venules, so the observations nicely support the concept that HbAS heterozygotes would be protected against severe falciparum malaria *in vivo.* One factor limiting parasite growth is loss of K^+ from HbS-bearing cells under low O_2 partial pressures; raising the concentration of K^+ in the extracellular medium allows the parasites to replicate (Friedman *et al.*, 1979).

In contrast, Friedman (1979) found that *P. falciparum* grows less well in G6PD$^-$ and β-thalassemic erythrocytes when they are exposed to oxidant stress by culturing them in 30% O_2 or in the presence of menadione or ascorbic acid. However, Roth *et al.* (1983) have shown that *P. falciparum* grows poorly in erythrocytes from persons hemizygous or heterozygous for the Mediterranean variety of G6PD deficiency even in the presence of 17% O_2; presumably the protection would be even greater under conditions of oxidant stress. Since there is good evidence that erythrocytes containing HbS are sensitive to oxidant stress (Chiu *et al.*, 1982), replication of *P. falciparum* may also be inhibited in these cells under such conditions. *P. falciparum* grows poorly in erythrocytes containing HbC or HbE (Friedman *et al.*, 1979).

IX. SUSCEPTIBILITY OF ERYTHROCYTES TO OXIDANT STRESS

In collaboration with Sydney Brenner, now director of the Molecular Biology Laboratory in Cambridge, I analyzed the mechanisms by which oxidant chemicals, including phenylhydrazine, hydroxylamine, and alloxan, exert oxidant effects on erythrocytes (Brenner and Allison, 1953). We concluded that the principal effects were inhibition of catalase; degradation of Hb to yield denatured globin, bile pigments, and free iron; and lipid peroxidation yielding fluorescent products. Denatured globin and lipid peroxides become associated with the membrane to form inclusions known as Heinz bodies. These conclusions are still valid. There has been much speculation about chemical mechanisms by which they are produced, particularly the role of superoxide, but this field is still controversial.

We showed that vitamin K and its analogues, including menadione, are hemolytic, especially in vitamin E-deficient rats (Allison *et al.*, 1956). Overdoses of a vitamin K analogue in newborn children were found to produce a hemolytic syndrome and brain damage (Allison, 1955). Once the danger was recognized and doses reduced, this syndrome disappeared. During the course of this work, we found that some dogs congenitally lack catalase in erythrocytes and are especially susceptible to hemolysis by oxidant drugs.

We also demonstrated that as erythrocytes age, their enzyme content falls (Allison and Burn, 1955). It had been shown shortly beforehand that normoblasts and reticulocytes have RNA and synthesize proteins, whereas mature erythrocytes do not. We reasoned that since erythrocyte enzymes are unlikely to be acquired from plasma or leukocytes, each human erythrocyte would have to depend for its life span of ~120 days on enzymes synthesized in the precursor. From our observations we calculated that at the end of its life span the activity of several enzymes is about one-tenth that in the young cell. This was shown to be true of G6PD in human erythrocytes (Marks *et al.*, 1958) and later of superoxide dismutase (Bartoz *et al.*, 1978). Hence old erythrocytes are more susceptible to oxidant stress than are young erythrocytes.

As a result of oxidant stress, erythrocytes can be lysed or their surface changed so that they are removed from the circulation by the spleen and liver

(Baehner *et al.*, 1971). Erythrocytes bearing malaria parasites when subjected to oxidant stress can be lysed or modified in such a way that parasite growth is prevented without lysis. Under the latter circumstances, degenerate parasites are observed in intact erythrocytes. A summary of information now available on the biochemical effects of oxidant stress in erythrocytes is relevant to an interpretation of the mechanisms by which abnormal Hbs and G6PD deficiency confer resistance to malaria and by which cell-mediated immune responses are protective.

X. FACTORS THAT INCREASE AND DECREASE SUSCEPTIBILITY OF ERYTHROCYTES TO OXIDANT STRESS

Four components of erythrocytes function in concert to counteract oxidant stress: the erythrocyte membrane, Hb, protective enzymes, and antioxidants such as vitamin E. Defects in Hb or protective enzymes can increase the sensitivity of erythrocytes to oxidants. The first principle of erythrocyte organization is that normal adult Hb (HbA) carries out its oxygen-transporting function in the cytoplasm, separated from the red cell membrane. In the cytoplasm Hb is protected by enzymes from oxidants and, if it is oxidized to metHb, can be reduced again to Hb without damaging the membrane. If Hb is bound to the membrane and oxidized there, it can initiate lipid peroxidation. Thus lipid peroxides are formed in erythrocyte membranes exposed to H_2O_2 only when Hb is present (G. Cohen, 1975).

Recent analysis of the mechanism (Caughey *et al.*, 1983) suggests that deoxy-Hb (Fe^{2+}) is protonated and liganded with anion, followed by a reaction to give O_2 and the Fe^{3+}-anion complex; direct electron transfer from Fe^{2+} to O_2 is precluded. H_2O_2 can form directly from HbO_2 and a one-electron reductant. Changes that open up the ligand-binding site generally enhance reaction rates. Much evidence supports a mechanism of direct electron transfer from reductant to bound O_2, making steric access to the ligand site a critical factor. For this reason, and because changes in Fe coordination can convert heme Fe into a powerful oxidant species, denaturation of Hb, with or without release of heme, can increase its capacity to initiate lipid peroxidation. Denaturation also exposes hydrophobic sites in the Hb molecule, promoting membrane binding, and drugs such as menadione have been shown to liberate heme, which is itself powerfully hemolytic (Fitch, 1983). Associated with lipid peroxidation is the formation of malonaldehyde, which can cross-link proteins and other biologic molecules, and of fluorescent lipid derivatives (Tappel, 1975).

Hb and myoglobin were the first proteins of which the three-dimensional structure was accurately determined by X-ray crystallographic analysis. The two α- and two β-chains of Hb are arranged such that the surface is covered with hydrophilic residues, while inner hydrophobic regions promote assembly into a stable tetramer ($\alpha_2\beta_2$). Free α- or β-chains are unstable, rapidly denature, become bound to the erythrocyte membrane, and form auto-oxidizing systems (Rotilio *et al.*, 1977). Since free heme or α- or β-polypeptide chains are deleterious, the synthesis of heme and globin, as well as of α- and β-chains of globin in erythrocyte precursors, must be carefully balanced. In the case of heme, this is ensured by its special regulatory role: heme binds to and activates a subunit of a factor initiating protein synthesis ($eIF_{2\alpha}$) (Ochoa, 1981). When heme is consumed in the Hb that is formed, protein synthesis is automatically stopped. Approximately equivalent numbers of messengers for α- and β-chains are normally translated, but in the case of α- and β-thalassemia there is a reduction in the number of α- and β-chain messengers, respectively, hence excess β- or α-chains are transcribed (Weatherall and Clegg, 1981). As already stated, these become denatured and bound to the erythrocyte membrane, conferring sensitivity to oxidant stress. Moreover, in β-thalassemia an excess of minor component HbA_2 ($\alpha_2\delta_2$) is formed, and this has a high affinity for the erythrocyte membrane (Klipstein and Ranney, 1960). The same is true of HbS, even when newly synthetized in reticulocytes (De Simone *et al.*, 1977), and of the polymorphic Hbs C, E, O Arab, and D Punjab (Friedman, 1981). All the latter have greater affinity for the membrane than HbA, owing to replacement of negatively charged glutamic acid residues at potential binding sites.

Thus for several reasons thalassemic erythrocytes, as well as those containing HbS, HbC, and HbE, are unusually susceptible to oxidant damage, as reflected by the high content of lipid peroxides in erythrocytes of patients with these conditions (Stocks *et al.*, 1972; Rachmilewitz *et al.*, 1976; Chiu *et al.*, 1982). Observations on the presence of lipid peroxides in erythrocytes of patients with abnormal Hbs were presented at the Ciba Foundation Symposium on Oxygen Free Radicals in 1968, at which I discussed with T. L. Dormandy their implications in terms of increased oxidant stress on the cells, and their consequently decreased capacity to support the replication of malaria parasites.

Mechanisms protecting the erythrocyte against oxidant damage include the cytoplasmic enzyme superoxide dismutase, which catalyzes the conversion of O_2^- to H_2O_2, and catalase, which breaks down H_2O_2 into H_2O and O_2. As shown in Fig. 1, three other enzyme systems function in concert. One is G6PD, which catalyzes the formation of NADPH; this metabolite is used by glutathione reductase to convert GSSG to GSH. In the presence of glutathione peroxidase, GSH reacts not only with H_2O_2, at rates comparable to those of catalase, but also with a wide range of peroxides and hydroperoxides, such as *t*-butylhydroperoxide and lipid hydroperoxides (Flohé, 1978). Thus GSH peroxidase is likely to provide a major protective mechanism against lipid peroxidation in the erythrocyte. The third system linked to G6PD through NADPH and NADH is metHb reductase.

GSH peroxidase contains selenium and is highly specific for GSH but less specific for peroxides (Flohé, 1978). Human erythrocytes deficient in GSH per-

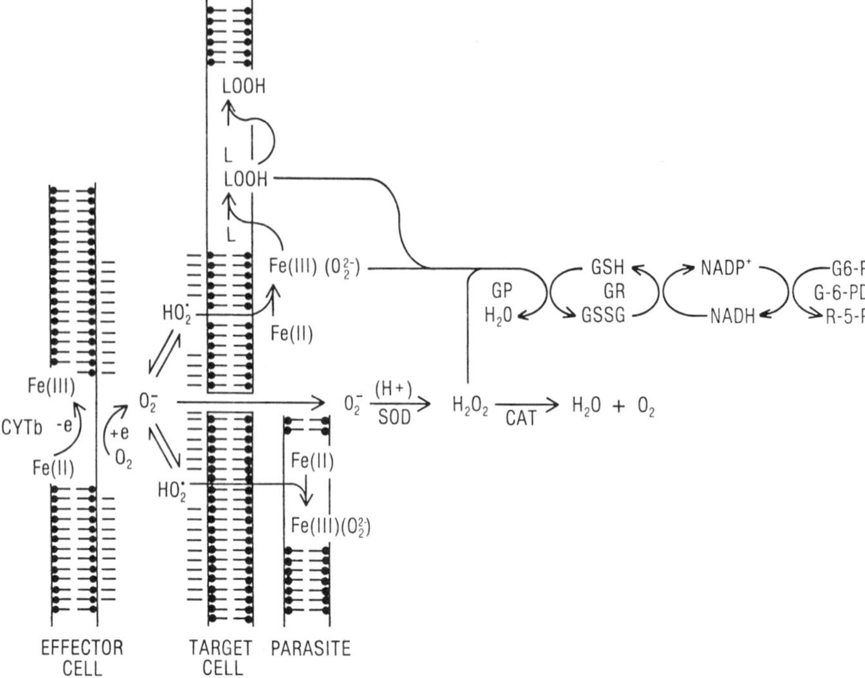

Figure 1. Diagrammatic representation of effector and target cells. Cytochrome b-245 (CYTb) is the terminal oxidase producing O_2^- and $HO_2^{\bullet}$, which oxidizes Fe, e.g., of membrane-bound Hb, which in turn produces lipid hydroperoxide (LOOH). This can also be carried to effector cells by apolipoprotein. O_2 can enter the target cell only through an anion channel. The protective enzymes superoxide dismutase (SOD), catalase (CAT), glutathione peroxidase (GP), and glutathione reductase (GR) are shown. The latter oxidizes glutathione (GSH) to GSSG, reduction of which consumes NADPH. Regeneration of the latter occurs through the hexose monophosphate shunt pathway, converting G6P to ribose-5-phosphate (R5P). Glucose 6-phosphate dehydrogenase (G6PD) activity limits the rate of metabolism by this pathway.

oxidase are very sensitive to oxidant drugs (Necheles, 1974). Erythrocytes of rats made GSH-peroxidase deficient by means of a low selenium diet are likewise susceptible to oxidant hemolysis (Flohé, 1978).

Three mechanisms by which oxidant stress is exerted on erythrocytes have been described, all of which have greater effects on $G6PD^-$ than on $G6PD^+$ cells. The first is a variety of drugs, and the second is dietary constituents. Friedman (1983) recently observed that the sera of some individuals who have consumed fava beans inhibit the growth of *P. falciparum* in $G6PD^-$ but not in $G6PD^+$ erythrocytes. Thus a constituent of the fava bean, or a metabolite produced by some persons, appears to exert oxidant stress on erythrocytes. Third, Baehner *et al.*

(1971) incubated $G6PD^-$ erythrocytes with normal human leukocytes in which the respiratory burst had been triggered with zymosan. The erythrocytes were lysed or, when labeled and injected into human subjects, were rapidly removed in the spleen and liver. When chronic granulomatous disease leukocytes were incubated with $G6PD^-$ erythrocytes, or normal leukocytes were incubated with $G6PD^+$ erythrocytes, no damage was seen. Hence activation of a respiratory burst in leukocytes, with concomitant production of O_2^- and H_2O_2, imposes oxidant stress on erythrocytes. The relevance of this observation to immunity against malaria is discussed in Section XI.

XI. PRODUCTION OF SUPEROXIDE AND HYDROGEN PEROXIDE BY LEUKOCYTES

When appropriately stimulated, phagocytic cells (neutrophils and mononuclear phagocytes) undergo a series of biochemical responses collectively termed the respiratory burst (Badwey and Karnovsky, 1980). Within 1 min of exposure to the stimulus, oxygen consumption is increased 10- to 15-fold. This increase is not inhibited by cyanide, hence is not attributable to mitochondrial oxidative phosphorylation. At the same time, glucose oxidation is stimulated; studies of radioactive CO_2 formation from glucose with carbon selectively labeled in positions 1 and 6 show that the increased metabolism of oxygen occurs through the hexose monophosphate shunt pathway. This generates NADPH, oxidation of which is linked to the reduction of O_2 through a series of enzymes, including a flavoprotein, in the plasma membrane (Badwey and Karnovsky, 1980).

The importance of this system in killing of pathogenic microorganisms is shown by the increased susceptibility to infection of persons with chronic granulomatous disease, a group of inherited defects. In some boys with the typical X-linked form of the disease, we found that a previously undescribed cytochrome b, present in the membranes around phagocytic vacuoles, was either undetectable or showed an abnormal absorption spectrum (Segal *et al.*, 1978).

Our suggestion (Segal and Allison, 1978) that this cytochrome might accept electrons from NADPH oxidase and react with O_2 to form O_2^- provoked strong controversy. However, the original observations on chronic granulomatous disease have been confirmed and extended (Segal *et al.*, 1983), and the enzyme has been purified and characterized by Segal, Jones, and collaborators. It has a redox potential of −245 mV, so it is now termed cytochrome b_{-245}, and it reacts with CO, but not with cyanide. Hence it has the properties we predicted and is becoming widely accepted as the enzyme accepting electrons from NADPH oxidase and catalyzing one-electron reduction of O_2 with the formation of the superoxide anion radical:

$$\mathrm{NADPH} \to \mathrm{NADP}^+ \quad \mathrm{FMN} \rightleftarrows \mathrm{FMNH}\ (\text{Flavoprotein}) \quad \text{Cytochrome b Fe(II)} \underset{+e}{\overset{-e}{\rightleftarrows}} \text{Cytochrome b Fe(III)} \quad O_2 \xrightarrow{+e} O_2^- \tag{1}$$

O_2^- released into the extracellular medium of cells during phagocytosis can be detected by reduction of nitroblue tetrazolium or cytochrome c, the reduction being inhibitable by superoxide dismutase. The respiratory burst in leukocytes also produces chemiluminiscence, amplified in the presence of luminol and other compounds (Johnson, 1982). It is agreed that chemiluminiscence results from the generation of oxygen-reduction products, although the molecular species involved is not identified with certainty.

Two molecules of O_2 can interact to form a two-electron reduction product, the peroxide dianion ($O_2{}^{2-}$). In protic solvents, such as water, protonation and disproportion occur as follows:

$$2O_2^- + 2H \rightarrow H_2O_2 + O_2 \tag{2}$$

This *dismutation* can take place spontaneously, but the reaction is accelerated in the presence of the enzyme superoxide dismutase (SOD), which occurs in the cytoplasm of leukocytes, erythrocytes, and several other cell types (Fridovitch, 1975). H_2O_2 can also be detected in the extracellular medium of leukocytes after phagocytosis, especially when its breakdown by catalase is inhibited. Most likely, O_2^- is an intermediate in the formation of H_2O_2, all of which is thought to be formed in this way in neutrophils.

The chemical interactions of O_2^- and H_2O_2 with biologic molecules are complex and imperfectly understood (Sawyer and Valentine, 1981). In the present discussion, only a few reactions that might be relevant to mechanisms mediating protection against malaria parasites are outlined. O_2^- does not oxidize unsaturated lipids such as linoleic acid; however, this reaction can be catalyzed by iron and by lipid hydroperoxide (Thomas *et al.*, 1982). Thus chain reactions initiated by an iron oxidant system and perpetuated by hydroperoxide can occur. Autooxidation of polyunsaturated fatty acids may be one of the principal reactions by which free radicals damage biologic systems. Some lipid hydroperoxides, e.g., those produced by the lipo-oxygenase and cyclo-oxygenase systems, have potent pharmacologic activity and may regulate biochemical pathways in parasites, as they do in other cell types.

The importance of the conjugate acid of O_2^-, the perhydroxyl radical ($HO_2^\bullet$), is now being appreciated (Bielsky, 1983). In aqueous solutions, $HO_2^\bullet$ is in equilibrium with O_2^- and the corresponding $K_{HO_2^\bullet}$ is of the order of 1.2×10^{-5} *M*:

$$O_2^- \underset{}{\overset{pK_a = 4.8}{\rightleftarrows}} HO_2^\bullet \tag{3}$$

As a consequence of this equilibrium, nearly 1% of any O_2^- formed under physiologic conditions (pH near 7.0) will be present as $HO_2^\bullet$. In acidic microenvironments (e.g., phagocytic vacuoles or where negatively charged effector and target cell membranes are closely apposed), the proportion of $HO_2^\bullet$ will be much greater. In many cases $HO_2^\bullet$ reacts orders of magnitude faster than O_2^- or is the sole reactant with a biologic compound (e.g., vitamin E or polyunsaturated fatty acids).

Since $HO_2^\bullet$ is uncharged, it could readily diffuse into hydrophobic regions of lipid bilayers or through membranes, whereas passage of O_2^- through membranes is confined to anion channels (Lynch and Fridovitch, 1978). Most metal ions and complexes are not oxidized by O_2^- in aqueous solution at rates competitive with its disproportion, even when such oxidations are thermodynamically favored (Sawyer and Valentine, 1981). The probable explanation is that O_2^- must first be coordinated and reduced in the first coordination sphere of the metal. Oxidations by $HO_2^\bullet$ are not subject to the same restriction and readily occur, e.g., in aqueous Fe(II). When Fe(II) is already coordinated, as in heme-containing molecules, $HO_2^\bullet$ will be much more likely than O_2^- to oxidize Fe. The possible role of $HO_2^\bullet$ in initiating lipid peroxidation in membranes is shown diagramatically in Fig. 1.

XII. OXIDANT STRESS EXERTED ON ERYTHROCYTES BY MALARIA PARASITES

Etkin and Eaton (1975) reasoned that since G6PD$^-$ cells are less capable than normal cells of supporting the replication of malaria parasites, and the cardinal characteristic of G6PD$^-$ red cells is a marked sensitivity to oxidant stress, the malaria parasite might itself generate oxidants within the infected red cell. These workers showed that *P. berghei*-infected mouse red cells accumulate metHb *in vivo* and have increased sensitivity to oxidants *in vitro.* In the absence of added oxidants, catalase within infected red cells was readily inactivated by 3-amino-1, 2, 4-triazole, a reaction requiring the presence of H_2O_2, suggesting that H_2O_2 is generated in the parasitized cell.

Friedman (1983) found that catalase in the presence of aminotriazole is also inactivated in cultured human erythrocytes infected with *P. falciparum*, the rate of inactivation depending on the partial pressure of O_2 in the range of 10-30%. These results provide strong evidence that malaria parasites exert oxidant stress on erythrocytes, presumably because of NAD(P)H depletion.

Consistent with the hypothesis that malaria parasites multiply less well in erythrocytes inadequately protected against oxidant stress than in normal erythrocytes is the observation of Eaton *et al.* (1976) that *P. berghei* infections progress slowly in mice on vitamin E-deficient diets. J. W. Eaton reviewed all these findings at the Gordon Research Conference on Parasitology in August 1981, where we discussed their implications in relation to our own observations with polyamine oxidase and the role of oxidant effects in cell-mediated immunity to hemoprotozoa.

XIII. EFFECTS OF OXIDANTS ON MALARIA PARASITES IN ERYTHROCYTES

Brewer and Coan (1969) found that hyperoxia, produced by exposing rats to 3 atm of compressed air, markedly inhibited *P. vinckei* infections. The hyperoxia also had significant inhibitory effects on the early stages of *P. berghei* infections, although this effect was less dramatic than in the case of *P. vinckei.* Preliminary results suggested that hyperoxia also inhibits *P. cynomolgi* in rhesus monkeys. These interesting observations seem to have escaped notice for a decade.

In 1978 we established the existence of cytochrome b_{-245} in human leukocytes and its probable role in superoxide and hydrogen peroxide production, as well as abnormalities in this system in chronic granulomatous disease (Segal and Allison, 1978). Since M. Karnovsky had spent a sabbatical in my laboratory, I was aware of the experiments of his former pupil Robert Baehner, showing that activation of the respiratory burst in normal human leukocytes can exert oxidant stress on G6PD-deficient erythrocytes; leukocytes from patients with chronic granulomatous disease had no such effect (Baehner *et al.*, 1971). It was obvious to draw the conclusion that oxidant effects of leukocytes might limit the proliferation of malaria parasites in G6PD-deficient erythrocytes and that this might be an effector mechanism in cell-mediated immunity.

Because the NADPH-oxidase system in leukocytes is unstable and difficult to study, we used another oxidant enzyme system, polyamine oxidase, that we had shown to be present in activated macrophages, from which it could be released by stimuli such as lipopolysaccharide (Morgan *et al.*, 1980). The macrophage enzyme has properties very similar to those in bovine serum and rat liver, which had been purified and characterized (Morgan, 1980). In the presence of substrate, i.e., spermine or spermidine, polyamine oxidase reacts with molecular oxygen to generate hydrogen peroxide, a well-known oxidant, and aldehydes exhibiting reactivities analogous to those produced by lipid peroxidation (Tappel, 1975). Polyamine oxidase is therefore of interest in its own right as a microbicidal system in macrophages and ruminant serum, as well as being a model oxidant enzyme system.

We found that polyamine oxidase in the presence of substrate eliminated the ineffectivity of *P. yoelii* and greatly reduced the infectivity of *B. rodhaini* and *B. microti* in short-term cultures of mouse erythrocytes (Morgan *et al.*, 1981). These studies were undertaken in 1978; in October 1978 I became director of the International Laboratory for Research on Animal Diseases (ILRAD), Nairobi, Kenya, where we were permitted to work only on African trypanosomes and *Theileria* parasites. An Australian visitor to ILRAD, A. Ferrante, showed that polyamine oxidase also kills trypanosomes (Ferrante *et al.*, 1982). On return to Australia, Ferrante arranged a collaboration with C. Rzepczyk, which showed that polyamine oxidase in the presence of substrate effects the death of *P. falciparum* in cultures of human erythrocytes (Ferrante *et al.*, 1983). Hence the studies with polyamine oxidase were initiated in 1978 and systematically followed through. I was well aware of their general implications. When the work of Friedman (1979) showing the sensitivity of *P. falciparum* to oxidant stress in G6PD-deficient erythrocytes was published, the relationship to our results was evident, as I stated explicitly in a seminar at ILRAD in 1980.

Clark and Hunt (1983) reported that injection of alloxan producted intraerythrocytic degeneration of *P. vinckei* in mice. These workers concluded that this result was attributable to the production of superoxide by the alloxan–dialuric acid redox system and by iron-catalyzed hydroxy radical generation. In support of this interpretation they quoted observations of Fee and Teitelbaum (1972) that superoxide dismutase and catalase protected erythrocytes against dialuric acid hemolysis, from which the suggestion was made that superoxide ion or a product of a reaction between peroxide and superoxide might act as the hemolytic agent. A later paper of Fee *et al.* (1975) reported further analysis showing this simplistic view to be incorrect. These investigators state: "Our results appear to exclude superoxide ion and H_2O_2 (and any substance which might be formed as a result of their combination as candidates for the reactive species in this system." Mezick *et al.* (1970) have shown that dialuric acid-induced hemolysis takes place in 10% ethanol ($\simeq$1.7 *M*), indicating that the hydroxy radical cannot be the active element, since it reacts very rapidly ($k \simeq 2 \times 10^9$ min^{-1} sec^{-1}) with ethanol (Neta and Dorfman, 1968). Fee and colleagues (1975) postulate that dialuric acid and oxygen interact to form a highly reactive substance, perhaps a peroxy derivative or the dialurate radical, which attacks the membrane directly or serves as a precursor of the reactive substance. Such a reactive intermediate could itself attack malaria parasites. Suggestions that iron-catalyzed production of hydroxy radicals is involved in the pathogensis of alloxan diabetes, quoted by Clark and Hunt (1983), are contradicted by observations of Marklund *et al.* (1983) that the iron chelator desferrioxamine aggravates alloxan diabetes. Thus the alloxan experiments of Clark and Hunt provide no evidence that superoxide and hydrogen peroxide, and still less hydroxy radicals, are involved in immunity to malaria.

The same is true of observations of Rigdon *et al.* (1950) that injections of phenylhydrazine protected monkeys from *P. knowlesi* infections. This is quoted by Clark and Hunt (1983) in support of the view that superoxide is involved in immunity, but in erythrocytes phenylhydrazine is converted into highly reactive products, probably including the phenyl radical (Goldberg *et al.*, 1976; Jain and Hochstein, 1979); these could likewise react with malaria parasites. Moreover, K. N. Brown and Hills (1981) have shown that even pretreatment with phenylhydrazine makes rats resistant to malaria infections, perhaps through effects on the immune system. Hence, the mechanism of action of phenylhydrazine on malaria parasites is still not interpretable.

This raises the general point of the role of O_2^- itself in erythrocyte damage and immunity to malaria. As Sawyer and Valentine (1981) and Fee (1982) point out, no direct reaction of aqueous superoxide with a cellular substance has been discovered that can account for the toxicity of oxygen, and the role of hydroxy radicals *in vivo* is questionable. Other reactions, such as conversion of redox-active metals into powerful oxidants, generation of lipid hydroperoxides, and conversion of dialuric acid, hydrazine, or phenylhydrazine to reactive intermediates, are likely to be more important in biologic effects than superoxide, hydrogen peroxide, or hydroxy radicals themselves.

More convincing evidence for an oxidant mechanism comes from observations of Clark *et al.* (1983) and Allison and Eugui (1982) that injections of *t*-butylhydroperoxide reduced the parasitemias of mice infected with *P. vinckei* and *P. yoelii* 17 X L. A lethal infection with the latter could be converted into a nonlethal infection. Parasites were found to degenerate within circulating erythrocytes. If treatment was commenced early enough, the parasitemia recurred, but this was followed by recovery, showing that acquired immune responses are possible under these conditions. The mechanism of action of *t*-butylhydroperoxide is under investigation in our laboratory. Injection of the iron chelator desferrioxamine before *t*-butyhydroperoxide abolished the protection against malaria. In the presence of traces of Fe^{2+}, *t*-butylhydroperoxide oxidizes unsaturated fatty acids; this is prevented by desferrioxamine. Erythrocytes exposed to *t*-butylhydroperoxide show lipid peroxidation (Trotta *et al.*, 1982). The role of lipid oxidation in immunity to malaria and in immunopathology is discussed in Section XIV.

XIV. TUMOR-NECROSIS FACTOR, LIPOPROTEINS AND OXIDIZED LIPIDS

Mycobacterium bovis BCG infection or treatment with killed *Propionibacterium acnis* (formerly known as *Corynebacterium parvum*) induce proliferation of macrophage precursors and activation of macrophages. Animals treated with these agents show increased sensitivity to bacterial endotoxin (LPS). Rabbits or mice treated with these agents and injected intravenously with LPS release substantial amounts of tumor-necrosis factor (TNF) into the plasma within 2 hr

(Carswell *et al.*, 1975). On the basis of studies with chimeric animals constructed from bone marrow-derived cells, believed to be macrophages (Sipe and Rosenstreich, 1981), TNF appears to be a glycoprotein of molecular weight ~60,000 (Mannel *et al.*, 1980).

Rzepczyk and Clark (1981) found that *P. vinckei petteri* taken from mice after challenge with lipopolysaccharide incorporates hypoxanthine poorly in culture. Mouse serum containing TNF inhibits the growth of *P. berghei* and *P. yoelii in vitro* (Taverne *et al.*, 1981). It also effects the degeneration of *P. falciparum* within cultured human erythrocytes (Haidaris *et al.*, 1983).

These effects are analogous to those seen in cells exposed to oxidized lipids (cholesterol or unsaturated fatty acid derivatives) complexed with apolipoproteins. Of special relevance is apoprotein E, which is synthesized by mouse and human macrophages (Basu *et al.*, 1982). Apoproteins, including E, readily form homologous or mixed polymers linked by disulfide bonds (see Weisgraber and Mahley, 1983). Our interpretation is that TNF is a complex of apoprotein E, bearing lipids oxidized before release from macrophages by the respiratory burst triggered by LPS. Another mechanism by which lipoproteins can be oxidized during malarial infections is secondary to hemolysis. Heme becomes attached to albumin (forming methemalbumin) and to other plasma proteins and is a powerful catalyst of their oxidation.

Jensen *et al.* (1982) have found in sera from Southern Sudanese children a factor that inhibits growth of *P. falciparum* in cultured human erythrocytes. The active component appears not to be an immunoglobulin, and the possibility that it is an oxidized lipoprotein is under investigation in a collaborative study with Jensen. A serum component could provide an effector mechanism able to act on *P. falciparum* in erythrocytes adherent to the endothelial cells in postcapillary venules.

Both human low-density lipoproteins and high-density lipoproteins bind to human erythrocytes (Hui *et al.*, 1981) and could carry oxidized cholesterol or unsaturated fatty acids to the cells. Peroxidized lipids can increase the leakage of K^+ from erythrocytes (Chiu *et al.*, 1982), thereby reducing their capacity to support the replication of *P. falciparum* (Friedman *et al.*, 1979). The cytotoxic effects of human high-density lipoprotein on *Trypanosoma brucei* (Rifkin, 1978) may well be caused by the oxidized lipids transferred from lipoprotein bound to the parasite membrane.

Lipoproteins may also be involved in the immunosuppression occuring in malaria and other infections. They are known to have immunosuppressive effects when oxidized (Schuh *et al.*, 1978). Apoprotein E has been shown to suppress phytohemagglutinin-activated phospholipid turnover and DNA synthesis in human peripheral blood mononuclear cells (Avila *et al.*, 1982). Lipoproteins in mice become oxidized during the course of malarial infections. Immunoregulatory effects of murine lipoproteins (which autoxidize during isolation) have been studied by Hsu *et al.* (1982). Marked suppression of splenic and peripheral lymphocyte responses to mitogenic stimulation were observed, with no demonstrable effect on lymph node lymphocytes. This is precisely the pattern of immunosuppression observed in malaria and African trypanosome infections (Weidanz and Rank, 1975; Kar *et al.*, 1981). Injection of low-density lipoproteins from mice infected with *P. chabaudi* was found to reduce immune responses of recipients to tetanus toxoid (Goumard *et al.*, 1982). This raises the possibility that oxidized lipids carried by lipoproteins have an important part in the immunosuppression characteristic of many parasitic infections (Terry and Hudson, 1982).

The role of lipoproteins in immunity and immunoregulatory networks is only just being revealed, and further analysis is obviously required. Nevertheless, it emerges as one of the most interesting contemporary developments.

XV. CELL-MEDIATED IMMUNITY AND PREMUNITION

The observations reviewed in this chapter support my hypothesis that what was formerly termed premunition is a manifestation of cell-mediated immunity. Animals that are unable to make antibodies develop nonsterile immunity to malaria and babesia parasites. Small numbers of parasites persist, but the animals are resistant to challenge with large numbers of homologous and related parasites. Specific T-cell-dependent immunity is elicited as the parasite number increases, thereby triggering the nonspecific effector phase of immunity caused by activation of the NADPH-oxidase system of macrophages and NK cells. When the number of parasites falls below a threshold level, cell-mediated immunity is no longer activated, so parasites may not be completely eliminated.

In some cases (e.g., *P. chabudi adami* in mice), antibodies are not required for recovery from the first wave of parasitemia. Presumably binding of parasitized erythrocytes to effector cells triggers an oxidative burst. In other cases, antibodies facilitate recovery from the first wave of parasitemia, suggesting that they facilitate binding of effector cells to target cells and activation of the former. This is shown by the finding of Makimura *et al.* (1982) that antibodies increase the respiratory burst in macrophages incubated with *P. berghei*-infected erythrocytes. However, if the animals are protected from the first phase of infection by chemotherapy, parasitemia does not recur and they are immune to challenge. Such nonsterile, antibody-independent immunity constitutes *premunition*. Antibodies can produce sterile immunity, perhaps by focusing effector cells on small numbers of parasitized cells.

XVI. SYNERGISM OF CELL-MEDIATED IMMUNITY AND INHERITED ERYTHROCYTIC RESISTANCE

The attachment of *P. falciparum*-parasitized erythrocytes to endothelial cells in postcapillary venules through surface knobs of parasite origin (Udeinya *et al.*, 1981) now has a teleologic explanation. In this microenviroment, where partial pressures of oxygen are low, the parasite can complete the later stages of asexual replication without being subjected to oxidant stress. This adaptation is complicated in two situations. One is when the erythrocyte contains HbS, because under these conditions the cell cannot efficiently support the replication of *P. falciparum*. The second is when the effectors of cell-mediated acquired immunity liberate molecules that impose oxidant stress on erythrocytes. This stress is aggravated when the erythrocytes contain abnormal hemoglobins or are G-6PD deficient. That is why acquired cell-mediated immunity reinforces inherited resistance as a function of erythrocytic traits, as suggested by Allison (1963). These two factors account for the relative resistance to malaria of persons living where the disease is endemic.

XVII. CONTROVERSY AND THE DEVELOPMENT OF SCIENCE

Most of the concepts reviewed in this chapter were controversial when first put forward. My suggestion that sickle cell heterozygotes are favored by selection through relative resistance to malaria was long disputed, but the epidemiologic evidence, supported by observations on growth of *P. falciparum* in cultured erythrocytes, is now overwhelming. The distribution of G6PD deficiency and the poor growth of *P. falciparum* in enzyme-deficient erythrocytes in culture have convinced most investigators that this trait provides protection against malaria. I believe that demonstration of additional factors, such as selenium deficiency, will resolve epidemiologic discrepancies—the only issue still undecided.

For a long time my championship of a role for cell-mediated immunity in hemoprotozoan infections was considered by most malariologists as unlikely to be correct, but that view now prevails. Antibodies can be fitted into this scheme, providing a supplementary mechanism more important in some host–parasite combinations than others. Our postulate (Segal and Allison, 1978) that cytochrome b_{-245} is a terminal oxidase in leukocytes is still disputed, but the balance of evidence is swinging in our favor. My new suggestions about the relationships of effector cells (NK cells as macrophage precursors) and the role of oxidized lipids carried by lipoproteins as effector mechanisms in immunity to hemoprotozoa and of immunosuppression will doubtless be disputed as strongly.

It will be interesting to see whether they stand the test of time as well as my other concepts.

There is no more potent allergen than a new idea: If it is any good, it provokes strong reactions, favorable or unfavorable. But how dull and stagnant science would be if we all thought the same. Controversy can stimulate and entertain; it becomes destructive only when personal allegations are made. The ideas now put forward are as interesting as any of the others reviewed in this chapter, and testing them will be fun, just as following other unconventional thoughts and leads has been enjoyable and more productive than following the crowd.

ACKNOWLEDGMENTS

I am indebted to colleagues who collaborated in various stages of this work: David Clyde, in the field studies of G6PD deficiency; Ivor Brown and Roger Taylor, with whom malaria in thymectomized animals was investigated; Ian Clark, collaborator in experiments on nude mice, nonspecific immunity, and crisis forms, and who has independently studied effects of oxidants on parasites; Elsie Eugui, who analyzed genetic differences in susceptibility to malaria and effector mechanisms; David Morgan, Gillian Christensen, Tony Ferrante, Christine Rzepczyk, Constantine Haidaris, David Haynes, and Monte Meltzer, who collaborated in studies of tumor-necrosis factor and effects of oxidant enzymes on erythrocytes; and James Grun and William Weidanz, with whom the studies of *Babesia* infections in B-cell-deprived mice were performed. Discussions with all these workers, as well as with Edmund Sergent and William H. Taliaferro in 1963, and recently with Milton Friedman, John Eaton, and James Jensen, helped in the evolution of my thinking about cell-mediated immunity to malaria.

XVIII. REFERENCES

Allison, A. C., 1954*a*, The distribution of the sickle-cell trait in East Africa and elsewhere, and its apparent relationship to the incidence of subtertian malaria, *Trans. Soc. Trop. Med. Hyg.* **48**:312.

Allison, A. C., 1954*b*, Protection by the sickle-cell trait against subtertian malarial infection, *Br. Med. J.* **1**:290.

Allison, A. C., 1954*c*, Notes on sickle-cell polymorphism, *Ann. Hum. Genet.* **13**:39.

Allison, A. C., 1955, Danger of vitamin K to newborn, *Lancet* **1**:669.

Allison, A. C., 1960, Glucose-6-phosphate dehydrogenase deficiency in red blood cells of East Africans, *Nature (Lond.)* **185**:531.

Allison, A. C., 1963, Inherited factors in blood conferring resistance to protozoa, in: *Immu-*

nity to Protozoa (P. C. C. Garnham, A. E. Pierce, and I. Roitt, eds.), p. 109, Blackwell Scientific, Oxford.

Allison, A. C., and Burn, G. P., 1955, Enzyme activity as a function of age in the human erythrocyte, *Br. J. Haematol.* **1**:291.

Allison, A. C., and Clyde, D. F., 1961, Malaria in African children with deficient erythrocyte glucose-6-phosphate dehydrogenase, *Br. Med. J.* **1**:1345.

Allison, A. C., and Eugui, E. M., 1982, A radical interpretation of immunity to malaria parasites, *Lancet* **2**:1431.

Allison, A. C., Ikin, E. W., Mourant, A. E., and Raper, A. B., 1952, Blood groups in some East African tribes, *J. R. Anthrop. Inst.* **82**:55.

Allison, A. C., Moore, E. and Sharman, I. M., 1956, Haemolysis and haemoglobinuria in Vitamin E deficient rats after injections of vitamin K substitutes, *Brit. J. Haematol.* **2**:197.

Allison, A. C., Charles, L. S., and McGregor, I. A., 1961, Erythrocyte glucose-6-phosphate dehydrogenase deficiency in West Africa, *Nature (Lond.)* **190**:1198.

Avila, E. M., Holdsworth, G., Sasaki, N., Jackson, R. L., and Harmony. J. A. K., 1982, Apoprotein E suppresses phytohemagglutinin-activated phospholipid turnover in peripheral blood mononuclear cells, *J. Biol. Chem.* **257**:5900.

Badwey, J. A., and Karnovsky, M. L., 1980, Active oxygen species and the function of phagocytic leukocytes, *Annu. Rev. Biochem.* **49**:695;

Baehner, R. L., Nathan, D. G., Castle, W. B., 1971, Oxidant injury of Caucasian glucose-6-phosphate dehydrogenase-deficient red blood cells by phagocytosing leucocytes during infection, *J. Clin. Invest.* **50**:2466.

Bartoz, G., Tannert, C., Fried, R., Leyko, W., 1978, Superoxide dismutase activity decreases during erythrocyte aging, *Experientia* **34**:1464.

Basu, S. K., Ho, Y. K., Brown, M. S., Bilheimer, D. W., Anderson, R. G. W., and Goldstein, J. L., 1982, Biochemical and genetic studies of the apoprotein E secreted by mouse macrophages and human monocytes, *J. Biol. Chem.* **257**:9788.

Beutler, E., Dern, R. J., and Flanagan, C. L., 1955, Effect of sickle-cell trait on resistance to malaria, *Br. Med. J.* **1**:1189.

Bielsky, N., 1983, A comparison of the reactivities of $HO_2^{\cdot}$ and O_2^- with compounds of biological interest, in: *Third International Conference on Superoxide and Superoxide Dismutase* (in press).

Bienzle, V., Lucas, A. O., Ayeni, O., and Luzzatto, L., 1972, Glucose-6-phosphate dehydrogenase deficiency and malaria, *Lancet* **1**:107.

Brenner, S., and Allison, A. C., 1953, Catalase inhibition: A possible mechanism for the production of Heinz bodies in erythrocytes, *Experientia* **9**:381.

Brewer, G. J., and Coan, C. C., 1969, Interaction of red cell ATP levels and malaria, and the treatment of malaria with hyperoxia, *Mil. Med.* **134**:1056.

Brown, I. N., Allison, A. C., and Taylor, R. B., 1968, *Plasmodium berghei* infections in thymectomized rats, *Nature (Lond.)* **219**:292.

Brown, K. N., and Hills, L. A., 1981, Erythrocyte destruction and protective immunity to malaria: enhancement of the immune response by phenylhydrazine treatment, *Tropenmed. Parasitol.* **32**:67.

Carson, P. E., Flanagan, C. L., Ickes, C. E., and Alving, A. S., 1956, Enzymatic deficiency in primaquine-sensitive erythrocytes, *Science* **124**:4814.

Carswell, E. A., Old, L. J., Kassel, R. L., Green, S., Fiore, N. and Williamson, B., 1975, An endotoxin-induced serum factor that causes necrosis of tumors, *Proc. Natl. Acad. Sci. U.S.A.* **72**:3666.

Caughey, W. S., Wallace, W. J., and Kawanishi, S., 1983, Mechanism for superoxide production in autoxidation reactions of hemoglobins and myoglobins, in: *Third International Conference on Superoxide and Superoxide Dismutase* (in press).

Chiu, D., Vichinsky, E., Yee, M., Kletman, K., and Lubin, B., 1982, Peroxidation, vitamin E and sickle-cell anemia, *Ann. N. Y. Acad. Sci.* **393**:323.

Clark, I. A., and Allison, A. C., 1974, *Babesia microti* and *Plasmodium berghei yoelii* infections in nude mice, *Nature (Lond.)* **252**:328.

Clark, I. A., and Hunt, N. H., 1983, Evidence for reactive oxygen intermediates causing hemolysis and parasite death in malaria, *Infect. Immun.* **39**:1.

Clark, I. A., Allison, A. C., and Cox, F. E. G., 1976, Protection of mice against *Babesia* and *Plasmodium* with BCG, *Nature (Lond.)* **259**:309.

Clark, I. A., Wills, E. J., Richmond, J. E., and Allison, A. C., 1977*b*, Suppression of babesiosis in BCG-infected mice and its correlation with tumour inhibition, *Infect. Immun.* **17**:430.

Clark, I. A., Virelizier, J. L., Carswell, E. A., and Wood, P. A., 1981, Possible importance of macrophage-derived mediators in acute malaria, *Infect. Immun.* **32**:1058.

Clark, I. A., Cos, F. E. G. and Allison, A. C., 1977*a*, Protection of mice against *Babesia* spp. and *Plasmodium* spp. with killed *Corynebacterium parvum*, *Parasitology*. **74**:9.

Clark, I. A., Cowden, W. B., and Butcher, G. A., 1983, Free oxygen radical generators as antimalarial drugs, *Lancet* **1**:234.

Cohen, G., 1975, Unusual defense machanisms against H_2O_2 cytotoxicity in erythrocytes deficient in glucose-6-phosphate dehydrogenase or tocopherol, in: *Erythrocyte Structure and Function* (G. S. Brewer, ed.) p. 685, Alan R. Liss, New York.

Cohen, S., McGregor, I. A., and Carrington, S., 1961, Gamma globulin and acquired immunity to human malaria, *Nature, (Lond.)* **192**:733.

Cohen, S., Butcher, G. A. and Crandall, R. B., 1970, Action of antimalarial antibody in vitro, *Nature (Lond.)* **223**:368.

De Simone, J., Adams, J. G., and Shaeffer, J., 1977, Evidence for rapid loss of newly synthesized haemoglobin S molecules in sickle-cell anemia and sickle-cell trait, *Br. J. Haematol.* **35**:373.

Diggs, C. L., and Osler, A. G., 1969, Humoral immunity in rodent malaria. II. Inhibition of parasitemia by serum antibody, *J. Immunol.* **102**:203.

Eaton, J. W., Eckman, J. R., Berger, E., and Jacob, H. S., 1976, Suppression of malaria infection by oxidant-sensitive host erythrocytes, *Nature (Lond.)* **264**:758.

Etkin, N. L., and Eaton, J. W., 1975, Malaria-induced erythrocyte oxidant sensitivity, in: *Erythrocyte Structure and Function* (G. J. Brewer, ed.), p. 219, Alan R. Liss, New York.

Fee, J. A., 1982, Is superoxide important in oxygen poisoning? *Trends Biochem. Sci.* **7**:84.

Fee, J. A., and Teitelbaum, D. D., 1972, Evidence that superoxide dismutase plays a role in protecting red blood cells against peroxidative hemolysis, *Biochem. Biophys. Res. Commune.* **49**:150.

Fee, J. A., Bergamini, R., and Briggs, R. G., 1975, Observations on the mechanism of the oxygen/dialuric acid-induced hemolysis of vitamin E-deficient rat red blood cells and the protective roles of catalase and superoxide dismutase, *Arch. Biochem. Biophys.* **169**: 160.

Ferrante, A., Allison, A. C., and Hirumi, H., 1982, Polyamine oxidase-mediated killing of African trypanosomes, *Parasite Immunol.* **4**:349.

Ferrante, A., Rzepczyk, C. M., and Allison, A. C., 1983, Polyamine oxidase mediates intraerythrocytic death of *Plasmodium falciparum*, *Trans. R. Soc. Trop. Med. Hyg.* (in press).

Fitch, C. S., 1983, Mode of action of antimalarial drugs, in: *Malaria and the Red Cell* (Ciba Foundation Symposium 94), p. 222, Pitman, London.

Flohé, L., 1978, Glutathione peroxidase: Fact and fiction, in: *Oxygen Free Radicals and Tissue Damage* (Ciba Foundation Symposium 65), p. 95, Excerpta Medica, Amsterdam.

Fridovitch, I., 1975, Superoxide dismutases, *Annu. Rev. Biochem.* **44**:147.

Friedman, M. J., 1978, Erythrocytic mechanism of sickle cell resistance to malaria, *Proc. Natl. Acad. Sci. USA,* **75**:1994.

Friedman, M. J., 1979, Oxidant damage mediates variant red cell resistance to malaria, *Nature (Lond.)* **280**:245.

Friedman, M. J., 1981, Hemoglobin and the red cell membrane: Increased binding of polymorphic hemoglobins and measurement of free radicals in the membrane, in: *The Red Cell* (G. J. Brewer, ed.), p. 519, Alan R. Liss, New York.

Friedman, M. J., 1983, Expression of inherited resistance to malaria in culture, in: *Malaria and the Red Cell* (Ciba Foundation Symposium 94), p. 196, Pitman, London.

Friedman, M. J., Roth, G. F., Nagel, R. L., and Trager, W., 1979, The role of hemoglobins C, S and N BALT in the inhibition of malaria parasite development in vitro, *Am. J. Trop. Med. Hyg.* **28**:777.

Gilles, H. M., Fletcher, K. A., Hendrickse, R. G., Lindner, R., Reddy, S., and Allan, N., 1967, Glucose-6-phosphate dehydrogenase deficiency, sickling and malaria in African children in South-Western Nigeria, *Lancet* **1**:138.

Goldberg, B., Stern, A., and Peisach, J., 1976, The mechanism of superoxide anion generation by the interaction of phenylhydrazine with hemoglobin, *J. Biol. Chem.* **251**:3045.

Goumard, P., VuDac, N., Maurois, P., and Camus, D., 1982, Influence of malaria on a preexisting antibody response to heterologous antigens, *Ann. Immunol.* **133D**:313.

Grun, J. I., and Weidanz, W. P., 1981, Immunity to *Plasmodium chabaudi adami* in the B cell deficient mouse, *Nature* (*Lond.*) **290**:143.

Haidaris, C. G., Haynes, J. D., Meltzer, M. S., and Allison, A. C., 1983, Serum containing tumor necrosis factor is cytotoxic for the human malaria parasite, *Plasmodium falciparum, Infect. Immun.* (in press).

Hsu, K. -H. L., Hiramoto, R. N., and Ghanta, V. K., 1982, Immunosuppressive effect of mouse serum lipoproteins. II. In vivo studies, *J. Immunol.* **128**:2107.

Hui, D. Y., Noel, J. G., and Harmony, J. A. K., 1981, Binding of plasma low density lipoproteins to erythrocytes, *Biochem. Biophys. Acta* **664**:53.

Jain, S. K., and Hochstein, P., 1979, Generation of superoxide radicals by hydrazine and its role in phenylhydrazine-induced hemolytic anemia, *Biochim. Biophys. Acta* **586**:128.

Jayawardena, A. N., Targett, G. A. T., Carter, R. L., Leuchars, E., and Davies, A. J. S., The immunological response of CBA mice to *P. yoelii*, 1. Gereral characteristics The effects of T-cell deprivation and reconstitution with thymus grafts, *Immunology* **32**:849.

Jensen, J. B., Boland, M. T., and Akood, M., 1982, Induction of crisis forms in cultured *Plasmodium falciparum* with human immune serum from Sudan, *Science* **216**:1230.

Johnson, R. B., 1982, Enhancement of phagocytosis-associated metabolism as a manifestation of macrophage activation, in *Lymphokines* (E. Pick, ed.), Vol. 3, p. 33, Academic Press, New York.

Kar, S. K., Roelants, G. E., Mayor-Withey, K. S., and Pearson, T. W., 1981, Immunosuppression in trypanosome-infected mice. 6. Comparison of immune responses of defferent lymphoid organs, *Eur. J. Immunol.* **7**:100.

Klipstein, F. A., and Ranney, H. N., 1960, Electrophoretic components of the hemoglobin of red cell membranes, *J. Clin. Invest.* **39**:1984.

Langhorne, J. Butcher, G. A., Mitchell, G. H. and Cohen, S., 1979, Preliminary investigations on the role of the spleen in immunity to *Plasmodium knowlesi* malaria, in: *The Role of the Spleen in the Immunology of Parasitic Diseases*, (G. Torrigiani, ed.) 205, Schwabe, Basel.

Livingstone, F. B., 1967, *Abnormal Hemoglobins in Human Populations*, Aldine, Chicago.

Lynch, R. E. and Fridovitch, I., 1978, Permeation of the erythrocyte stroma by superoxide radical, *J. Biol. Chem.* **253**:4967.

Maegraith, B. G., 1974, Malaria, in: *Medicine in the Tropics* (A. W. Woodruff, ed.), p. 27, Churchill, Livingstone, London.

Makimura, S., Brinkmann, V., Mossmann, H. and Fischer, H. 1982, Chemiluminescence response of peritoneal macrophages to parasitized erythrocytes and lysed erythrocytes from *Plasmodium berghei*-infected mice, *Infect. Immun.* **37**:800.

Mannel, D. N., Meltzer, M. S., and Mergenhagen, S. E., 1980, Gereration and characterization of lipopolysaccharide-induced and serum-derived cytotoxic factor for tumor cells, *Infect. Immun.* **28**:204.

Marklund, S. L., Grankvist, K., and Taljedal, I. B., 1983, Oxy-radicals in the toxicity of cellular toxins, in: *Third International Conference on Superoxide and Superoxide Dismutase* (in press).

Marks, P. A., Johnson, A. B. and Hirschberg, E., 1958, Effect of age on the enzyme activity in erythrocytes, *Proc. Natl. Acad. Sci., U.S.A.* **44**:529.

Martin, S. K., Miller, L. H., Alling, D., Okoye, V. C., Esan, G. J. F., Osunkoya, B. O., and Deane, M., 1979, Severe malaria and glucose-6-phosphate dehydrogenase deficiency: A reappraisal of the malaria G-6-PD hypothesis, *Lancet* **1**:524.

McGregor, I. A., Gilles, H. M., Walters, J. H., Davies, J. H., and Pearson, F. A., 1956, Effect of heave and repeated infections on Gambian infants and children; effects of erythrocyte parasitization, *Br. Med. J.* **2**:686.

Mezick, J. A., Settlemire, C. T., Brierly, G. P., Barefield, K. P., Jensen, W. N., and Cornwell, D. G., 1970, Erythrocyte membrane interactions with menadione and the mechanisms of menadione-induced hemolysis, *Biochim. Biophys. Acta* **219**:361.

Morgan, D. M. L., 1980, Polyamine oxidases, in: *Polyamines in Biomedical Research* (J. M. Gaugas, ed.), p. 285, Wiley, Chichester.

Morgan, D. M. L., Ferluga, J., and Allison, A. C., 1980, Polyamine oxidase and macrophage function, in: *Polyamines in Biomedical Research* (J. M. Gaugas, ed.), p. 303, Wiley, Chichester.

Morgan, D. M. L., Christensen, J. R., and Allison, A. C., 1981, Polyamine oxidase and the killing of intracellular parasites, *Biochem. Soc. Trans.* **9**:563.

Motulsky, A. G., 1960, Metabolic polymorphisms and the role of infectious deseases in human evolution, *Hum. Biol.* **32**:28.

Necheles, T. F., 1974, The clinical spectrum of glutathione peroxidase deficiency, in: *Glutathione* (L. Flohé, H. C. Benohr, H. Sies, H. D. Waller, and A. Wendel, eds.), p. 173, Thieme, Stuttgart.

Neta, P., and Dorfman, L. M. 1968, Pulse radiolysis studies. VIII. Rate constants for the reaction of hydroxyl radicals with aromatic compounds in aqueous solution, *Adv. Chem. Ser.* **81**:225.

Nussenzweig, R. S., 1967, Increased nonspecific resistance to malaria produced by administration of killed *Corynebacterium parvum*, *Exp. Parasitol.* **21**:224.

Ochoa, S., 1981, Molecular mechanisms of control of protein biosynthesis, *Eur. J. Cell Biol.* **26**:212.

Pasvol, G., Weatherall, D. J., and Wilson, R. J. M., 1978, Cellular mechanism for the protective effect of haemoglobin S. against *P. falciparum* malaria, *Nature (Lond.)* **274**:702.

Plehn, A., 1902, *Die Malaria der Afrikanischen Negerbevolkerung, besonders mit Besug auf die Immunitatsfrage*, p. 43, Fischer, Jena.

Potocnjak, P., Yoshida, N., Nussenzweig, R. S., and Nussenzweig, V., 1980. Monovalent fragments (Fab) of monoclonal antibodies to sporozoite surface antigens (Pb 44) protect mice against malarial infection, *J. Exp. Med.* **151**:1504.

Rachmilewitz, E. A., Lubin, B. H., and Shohet, S. B., 1976, Lipid membrane peroxidation in B-thalassemia major, *Blood* **47**:495.

Rank, R. G., Weidanz, W. P., 1976, Nonsterilising immunity in avian malaria: An antibody-independent phenomenon, *Proc. Soc. Exp. Biol. Med.* **151**:257.

Rifkin, M. R., 1978, Identification of trypanocidal factor in normal human serum-high-density lipoprotein, *Proc. Natl. Acad. Sci. USA* **75**:3450.

Rigdon, R. H., Micks, D. W., Breslin, D., 1950, Effect of phenylhydrazine hydrochloride on *Plamodium knowlesi* infection in the monkey, *Am. J. Hyg.* **52**:308.

Roberts, D. W., and Weidanz, W. P., 1979, T-cell immunity in malaria in the β-cell dificient mouse, *Am. J. Trop. Med. Hyg.* **28**:1.

Roth, E. F., Raventos-Suarez, C., Rinaldi, A., and Nagel, R. L., 1983, Glucose-6-phosphate dehydrogenase deficiency inhibits in vivo growth of *Plasmodium falciparum*, *Proc. Natl. Acad. Sci. USA* **80**:298.

Rotilio, G., Fioretti, E., Falcioni, G., and Brunori, M., 1977, Generation of superoxide radical by heme proteins and its biological relevance, in: *Superoxide and Superoxide Dismutases* (A. M. Michelson, J. M. McCord, and I. Fridovitch, eds.), p. 239, Academic Press, New York.

Rzepczyk, C. M., and Clark, I. A., 1981, Demonstration of lipopolysaccharide-induced cytostatic effect on malarial parasites, *Infect. Immune.* **33**:343.

Sawyer, D. T., and Valentine, J. S., 1981, How super is superoxide? *Acc. Chem. Res.* **14**:393.

Schuh, J., Novgorodsky, A., and Haschemeyer, R. H., 1978, Inhibition of lymphocyte mitogenesis by autoxidized low-density lipoprotein, *Biochem. Biophys. Res. Commun.* **84**:763.

Segal, A. W., and Allison, A. C., 1978, Oxygen consumption by stimulated human neutrophils, in: *Oxygen Free Radicals and Tissue Damage* (Ciba Foundation Symposium 65), p. 205, Excerpta Medica, Amsterdam.

Segal, A. W., Jones, V. T. G., Webster, D., and Allison, A. C., 1978, Absence of a newly described cytochrome b from neutrophils of patients with chronic granulomatous disease, *Lancet* **2**:446.

Segal, A. W., Cross, A. R., Garcia, R. C., Borregaard, N., Valerius, N., Soothill, O. F., and Jones, V. T. G., 1983, Absence of cytochrome b-245 in chronic granulomatous disease: A multicenter European evaluation of its incidence and relevance, *N. Engl. J. Med.* **308**:245.

Sergent, E., 1963, Latent infection and premunition. Some definitions of microbiology and immunology, in: *Immunity to Protozoa* (P. C. C. Garnham, A. E. Pierce, and I. Roitt, eds.), p. 39, Blackwell Scientific, Oxford.

Sipe, J. D., and Rosenstreich, D. L., 1981, Serum factors associated with inflammation, in: *Cellular Functions in Immunity and Inflammation* (J. J. Oppenheim, D. L. Rosenstreich, and Potter, eds.), p. 115, Elsevier–North Holland, New York.

Stechschulte, D. J., 1969, Effect of thymectomy in *Plasmodium berghei* infected rats, *Proc. Soc. Exp. Biol. Med.* **131**:748.

Stocks, J., Offerman, G. L., Modell, C. B. and Dormandy, T. L., 1972, The susceptibility to autoxidation of human red cell lipids in health and disease, *Br. J. Haematol.* **23**:713.

Taliaferro, W. H., and Taliaferro, L. G., 1944, The effect of immunity on the asexual reproduction of *Plasmodium brasilianum*, *J. Infect. Dis.* **75**:1.

Tappel, A. L., 1975, Lipid peroxidation and fluorescent molecular damage to membranes, in: *Pathobiology of Cell Membranes* (B. F. Trump and F. Arstila, Eds.), Vol. 1, p. 311, Academic Press, New York.

Taverne, J., Dockrell, H. M., and Playfair, J. H. L., 1981, Endotoxin induced serum factor kills malarial parasites in vitro, *Infect. Immun.* **33**:83.

Terry, R. J., and Hudson, K., 1982, Immunosupression in parasite infections, in: *Immune Reactions to Parasites* (W. Frank, ed.), p. 125, G. Fischer, Stuttgart.

Thomas. M. J., Mehl, K. S.. and Pryor, W. A., 1982, The role of superoxide in xanthine oxidase-induced autoxidation of linoleic acid, *J. Biol. Chem.* **257**:8343.

Trager, W., and Jensen, J. B., 1976, Human malaria parasites in continuous culture, *Science* **193**:673.

Trotta, R. J., Sulluvan, S. G., and Stern, A., 1982, Lipid peroxidation and haemoglobin degradation in red blood cells exposed to *t*-butylhydorperoxide, *Biochem. J.* **204**:405.

Udeinya, I. J., Schmidt, J. A., Aikawa, M., Miller, L. H., and Green, I., 1981, Falciparum malaria-infected erythrocytes specifically bind to cultured human endothelial cells, *Science* **213**:555.

Weatherall, D. S., and Clegg, J. B., 1981, *The Thalassaemia Syndromes* 3rd ed. Blackwell Scientific, Oxford.

Weidanz, W. P., and Rank, R. G., 1975, Regional immunosuppression induced by *Plasmodium berghei yoelii* infection in mice, *Infect. Immun.* **11**:211.

Weinbaum, F. I., Evans, C. B., and Tigelaar, R. E., 1976, Immunity to *Plasmodium berghei yoelii* in mice. 1. The course of infection in T-cell and B-cell-deficient mice, *J. Immunol.* **117**:1999.

Weisgraber, K. H., and Mahley, R. W., 1978, Apoprotein (E-A-11) complex of human plasma lipoproteins. 1. Characterization of this mixed dusulfide and its identification in a high-density lipoprotein in subfraction, *J. Biol. Chem.* **253**:6281.

Index

Abnormal hemoglobins, 463
Acquired immunity, 463
Acute pulmonary schistosomiasis, 438
ADCC (*see* Antibody-dependent cell-mediated cytotoxicity)
Adult worm death and self-cure, 381
African trypanosomiasis, 225, 237, 242, 261
Age, relative to resistance, 10
Aged hosts, 12
A/J mice, 9, 12
A/J strain of mice, 31
AKR mice, 307
Allogeneic thymus grafts, 331
Amino acid sequences, 227
Anopheles mosquitoes, 109
Antibody-dependent lysis, 13
Antibody-dependent cell-mediated cytotoxicity (ADCC), 4, 18, 90, 94, 431, 437
Antibody-mediated clearance of bacteria, 139
Antibody to knob components, 184
Antibody, independence of immunity to hemoprotozoa, 469
Anti-cuticular serum antibodies, 281
Anti-DNA antibodies, 440
Anti-fecundity effects, 383
Antigenic analysis, 114
Antigenic characterization, 113
Antigenic characterization of plasmodia, 110
Antigenic mimicry, 377
Antigenic variation, 145, 163, 166
Antigenic variation of parasites, 162
Antigenically distinct strains of the parasite, 129
Antigens on malaria-infected erythrocytes, 136
Antigens on the surface of larval stages, 280
Antigen–antibody complexes, 142
Antigiardia immune responses, 342
Anti-idiotype (anti-Id) antibodies, 328
Anti-malarial antibody, 186
Anti-trophozoite IgG and IgA antibodies, 343
Ascaridia lineata, 9
Ascaris, 18
Ascaris lumbricoides, 71
Asexual blood stages of plasmodia, 116
Asexual malaria parasites, 128
Asexual parasitemia, 161
Avian schistosomes, 359

B cells, 131, 232
Babesia microti, 5, 22
Babesia bovis, 9, 170
Babesia bigemina, 9
Bacillus Calmette-Guerin (BCG), 29, 30, 56, 234, 334
Balb/c, 10, 17, 206
Balb/c mice, 26, 202, 237, 329, 331, 342
Balb/c strain, 20
BALT (*see* Bronchus-associated lymphoid tissue)
BCG (*see* Bacillus Calmette-Guerin; *Mycobacterium bovis*)
Biozzi low responder mice, 344
Biozzi strains of mice, 9
Breinlia booliati, 277
Bronchus-associated lymphoid tissue (BALT) 74, 75
Brugia, 228
Brugia malayi, 288, 424
Brugia pahangi, 228

C57BL/6 mice, 241, 341, 393
C57BL/6 strain, 9, 18, 307
C3H mice, 20, 29
CBA mice, 22, 26, 29, 205, 206, 237
CBA/CaJ mice, 23, 239
CBA/H mice, 18, 21, 240, 301
CBA/J mice, 393
CBA/N, 21, 22, 23, 237, 301
CBA strain of mice, 10, 20, 23
Candida albicans, 19
Cattle, 170, 231, 246, 249
Cell-mediated immunity, 235, 482
Cell-mediated reactions, 437
Cell-mediated resistance to, 4
Cellular immunity to *Babesia* parasites, 463
Cellular immunity to malaria, 463
Central nervous system schistosomiasis, 442
Cercarial dose size, 386
Cercarial-egg antigenic cross-reactions, 398
Chronic schistosomiasis, 440
Chickens, 9
Children, 467
Chimpanzee, 276
Circulating immune complexes, 432
Circulating parasite antigens, 429
Clearance of *D. vitaea* microfilariae, 303
Clearance of microfilariae, 305
Clinical and histologic pathology, 425
Clinical spectrum of parasitic disease, 423
Cloned parasites, 145
Cloned T cells specific for *Schistosoma* antigens, 219
Cobra venom factor (CVF), 15
Complement (C), 14, 15, 38, 90, 94, 234, 236, 299, 438
Complement-mediated clearance, 141
Complement-mediated lysis, 142
Concomitant immunity, 367
Concomitant immunity hypothesis, 374
Corynebacterium parvum, 29, 469
CVF (*see* Cobra venom factor)
Cyclophosphamide, 27
Cytotoxicity, 28

Defined antigen, 325
Delayed-type hypersensitivity (DTH), 27, 202, 211
Dermatologic reactions, 435
Digenetic trematodes, 359
Dipetalonema viteae, 9, 277
Dose effects using dead vaccines, 389
Dose effects using irradiated larvae, 389
DTH (*see* Delayed-type hypersensitivity)

Effective antigenic load, 391
Egg-induced hypersensitivity, 397
Endothelial cells, 482
Entamoeba histolytica, 9
Enteroendocrine cells, 80
Eosinophils, 92, 93, 296
Epidemiology of malaria in endemic areas, 468
Epithelial cells, 73
Epithelium, 79, 86
Experimental filariasis, 282
Experimental filariasis in resistant rodents, 291

Fasciola hepatica (murine fascioliasis), 347
Filariasis, 275, 422
Filarial antigens, 287

γ-Globulin, 232
G6PD cells, 477
GALT (*see* Gut-associated lymphoid tissue)
Gastrointestinal infection, 71
Gastrointestinal schistosomiasis, 442
Genetic analysis, 258
Genetic control, 4, 28
Genetic control of natural resistance, 38
Genetic markers in various East African populations, 464
Genetics of murine susceptibility to microfilariae, 306
Genetically based resistance in young mice of some strains (e.g., C57BL/7 and Balb/c), 326
Genetically based variations in susceptibility to disease caused *by L. tropica*, 331
Genes controlling resistance, 8
Genomic arrangements of VSG genes, 229
Giardia lamblia, 28
Giardia muris (murine giardiasis), 342
Giardiasis, 81
Glucose 6-phosphate dehydrogenase deficiency, 465
Goblet cells, 80, 87

Granuloma formation, 340, 443
Granuloma modulation, 394, 443, 444
Granuloma-associated nonspecific modulatory phenomena, 399
Granulomatous response to eggs of *S. mansoni*, 371
Gut-associated lymphoid tissue (GALT), 73, 74, 75, 76

H_2O_2, 63, 477
H_2O_2 generation, 57
H_2O_2 release, 61
Hamsters, 290, 297, 300
HDL (*see* High-density lipoprotein)
Helper activity of *L.tropica*-primed LN cells, 204
Helper T lympocytes, 436
Hepatic schistosomiasis, 441
High-responder line, 10
High-density lipoprotein (HDL), 16, 245
Host-antigen acquisition, 378
Host parasite relationships, 3
Host-protective (i.e., parasite-inhibitory) immune responses, 326, 346
Host range, 1
Host specificity, 364
Human filariae, 294
Human filarial infections, 297
Human monocytes, 161
Human mononuclear phagocytes, 58
Human peripheral blood cells, 32
Human peripheral blood monocytes, 57
Human polymorphonuclear leukocytes, 65
Human schistosomiasis, 394
Humans, 296, 433
Hybridoma techniques, 348
Hyperergic response to suboptimal immune reactivity, 422
Hyperergic state, 422
Hydrogen peroxide, 475
Hydrocortisone, 59
Hypergammaglobulinemia, 238
Hypersensitivity reaction, 438
Hypothymic mice, 17, 20, 33

Idiotype (Id) sensitization, 341
IgA, 75, 78, 82, 86, 88, 89, 90, 95, 344
IgE, 75, 82, 89, 95, 296, 343, 428, 435, 436
IgE antibody, 91, 92
IgG, 76, 95, 121, 132, 232, 233, 260, 296, 298, 347 (*see also* Immunoglobulin G)
IgG_1, 302
IgG_2, 23
IgG2a, 435, 436
IgM, 23, 76, 121, 232, 233, 260, 298, 301, 302, 303, 347
Immune complexes, 380, 429, 440
Immune evasion, 73, 166, 347, 377
Immune unresponsiveness induced by *Brugia* infections in Jirds, 285
Immune unresponsiveness to filarial antigens, 289
Immunity in schistosomiasis, 364
Immunity to malaria parasites, 130
Immunodeficient mice, 17
Immunogenetic systems, 393
Immunogenic role of eggs, 370
Immunogenic role of larval stages, 369
Immunoglobulin G (IgG), 467
Immunologic crossreactivity among different developmental stages, 309
Immunopathologic host changes, 397
Immunopathology, 421
Immunopathology of filariasis, 428
Immunoprecipitation, 117
Immunoprecipitation techniques, 326
Immunoprotective phenomenon, 422
Immunoregulatory molecules on the parasite surface, 378
Immunosuppression, 168, 432
Immunosuppressive mechanisms, 225
Immunosuppressor T cell, 239
Inbred mice, 202, 301
Inbred strains of mice, 11, 244, 256
Inflammation, 71
Inhibitors of rat and human lymphocyte proliferation, 380
Innate immunity to primary *S. mansoni* infections in the rat, 366
Innate resistance, 38
Interferon, 32
Intestinal lumen, 85
Intracellular parasites, 24
Intraepithelial lymphocytes, 88
In vivo vaccination, 310
Itr^r gene, 10
Ity^s, 23, 24

Jird-*Brugia* model, 283

Knob protrusions on infected erythrocytes, 178

Laboratory rodents, 231, 294
Lactoperoxidase, 90
Lamina propria, 82, 89
Latent filariasis, 295
Latent filariasis in rodents, 299
Leishmania, 54, 64
Leishmania antigens, 203, 206
Leishmania-specific T-cell lines, 208
Leishmania-specific T-cell populations, 209
Leishmania donovani, 5, 8, 65
Leishmania tropica, 27, 65, 201
Leishmania tropica (murine cutaneous leishmaniasis), 329
Lewis rat model, 288
Life cycle of malaria parasites, 127
Lipoproteins, 38, 481
Listeria monocytogenes, 5, 8, 19, 39, 73
Litomosoides carinii, 277
Loa loa, 424
Lps, 27
Lr, 28
Lsh, 8, 25
Lymphatic filariae, 288
Lymphokines, 57, 63
Lymph nodes, 75
Lysozyme, 94

Macrophages, 10, 24, 25, 26, 27, 28, 37, 38, 56, 61, 63, 64, 65, 75, 93, 211, 225, 234, 241, 242, 243, 303, 331, 334, 447, 463, 478
Macrophage activating factor (MAF), 203
Macrophage-mediated *in vitro* killing of microfilariae, 296
Macrophage precursors and activation of macrophages, 480
MAF (*see* Macrophage activating factor)
Major exoantigen of *Giardia*, 343
Major histocompatibility complex (MHC), 5, 8, 39
Major histocompatibility complex (MHC),-associated antigen recognition, 333
Malaria, 109
Malaria infected red blood cells, 127
Malaria vaccines, 109
Malarial antigens, 121, 131
MALT (*see* Mucosa-associated lymphoid tissue)
Manifestations of schistosomiasis, 434
Mast cells, 77, 81, 84, 91
Melanoma cells, 179
Merozoite, 136
MHC (*see* Major histocompatibility complex)
Mice, 292, 304, 394, 438
Mice of the C57BL/6 strain, 20
Microfilariae of the genus *Onchocerca* in inbred mice, 307
Microfilarial killing *in vitro*, 304
Migration inhibitory factor (MIF), 56
MLC (Mixed lymphocyte cytotoxicity), 447
Monanema globulosa, 277
Monoclonal antibody, 72, 112, 116, 118, 121, 140, 163, 325, 340
Mononuclear phagocytes, 53, 54
Mosquito vectors, 424
Mouse strain variations in susceptibility to *S. japonicum* infection, 337
Mucosa-associate lymphoid tissue (MALT), 74, 75, 76, 77
Mucosal immune responses, 71
Murine cutaneous leishmaniasis (*see Leishmania tropica*)
Murine cysticercosis (*see Taenia taeniaformis*)
Murine fascioliasis (*see Fasciola hepatica*)
Murine giardiasis (*see Giardia muris*)
Murine nematodiasis (*see Nematospiroides dubius*)
Murine schistosomiasis japonica (*see Schistosoma japonicum*)
Mycobacterium bovis (BCG), 5, 8, 39, 480

Natural cytotoxicity, 29
Natural cytotoxic cells, 30
Natural killer (NK), 13, 30, 31, 32, 33, 34, 37, 89, 463
Natural resistance, 1, 2
Natural resistance in invertebrates, 4
Nematospiroides dubius, 85, 344
Nematospiroides dubius (murine nematodiasis), 344
Neonatal thymectomy, 468

Neutrophils, 93
New antigens, 138
New antigens expressed on malaria-infected erythrocytes, 137
New antigens on infected cells, 142
New components on *P. knowlesi* schizont-infected cells, 159
NK (*see* Natural killer)
NK and NC cells, 38
NK-like, 437
NIH-white Swiss nudes, 22
Nippostrongylus brasiliensis, 72, 84, 87, 88, 344
Nonpermissiveness, 34, 364
Nonspecific immunity to hemoprotozoan infections, 469
Nude mice, 18, 20, 344

Occult filarial syndrome, 427
Onchocerca volvulus, 424
Oncospheral antigens, 328
Opsonization, 140
Opsonizing antibody, 167
Oxidant hemolysis, 474
Oxidant stress exerted on erythrocytes by malaria parasites, 477
Oxidant stress on G6PD-deficient erythrocytes, 478
Oxidative mechanisms, 66
Oxidative mechanisms of mononuclear phagocytes, 53

Paneth cells, 80
Parasite antigens, 72, 155, 203
Parasitic evasion of immunity, 160
Parasiticidal mechanisms, 139
Parasitized erythrocytes, 189
Passive transfer of immunity, 163
Pathogenesis of acute schistosomiasis mansoni, 439
Peroxidases, 94
Peyer's patches (PPs), 73, 76
Phagocytic cells, 475
Phagocytosis, 141
Plasma cell, 76, 82
Plasmodial antigens, 116
Plasmodial life cycle, 110
Plasmodium, 109, 111, 129
Plasmodium chaubaudi, 5
Plasmodium cynomolgi, 170
Plasmodium falciparum, 128
Plasmodium falciparum antigens, 120
Plasmodium knowlesi, 155
Plasmodium yoelii, 13, 22
Platelets, 38
Poly I: poly C (poly I:C), 29
Polyamine oxidase, 478
Polyclonal activation, 238
Polyclonal antibodies, 116
Polymorphonuclear leukocytes, 299
PPs (*see* Peyer's patches)
Premunition, 468
Primaquine, 465
Primates, 438
Primate infections of *P. falciparum*, 175
Protective immunity, 167
Proteinases of *S. mansoni* cercariae, 373
Protozoan parasites, 24
Pulmonary schistosomiasis, 441

Radioiodinated antigens, 151
Rats, 289, 291, 300, 365, 467
Rat trypanosome *T. lewisi*, 247
Reactive oxygen, 64
Reactive oxygen intermediates, 63
Recombinant DNA, 325
Relative immunity, 463
Resistance of mice, 26, 28, 37
Respiratory burst, 54
Rhesus monkeys, 169, 170, 364
Rhesus monkeys infected with *S. mansoni*, 368
Ric, 8
Rickettsia tsutsugamushi, 8
Rodent models of filariasis, 276
Rodent trypanosomes, 2

Salmonella enteritidis, 73
Salmonella typhimurium, 5, 8, 10
Schistosoma antigens, 217, 218
Schistosoma haematobium, 360, 433
Schistosoma japonicum, 336, 360, 433
Schistosoma japonicum, immunopathogenesis, 448
Schistosoma japonicum infections in rabbits, 382

Schistosoma mansoni, 22, 33, 85, 201, 216, 360, 433, 443
Schistosome-specific T-cell lines, 219
Schistosomiasis, 433
Schizogony, 133
Schizont-infected erythrocytes, 149, 151
Schizont polypeptides, 118
SDS-PAGE, 112, 114, 117, 347
Secondary infection schistosomes, 366
Sequestration in the *falciparum*-type malarias, 180
Sequestration of erythrocytes infected with rodent malaria parasites, 175
Serum factors involved in the killing of microfilariae, 295
Sex, relative to resistance, 10
SICA [-] phenotype, 147
SICA-antigen, 145, 154, 173, 183
SICA [+] parasites, 169, 171
SICA [-] parasites, 169, 171
SICA [-] phenotype, 147
SICA-variant antigens, 148, 152, 155
Soluble blocking antigen, 165
Specific cell-mediated immune responses, 431
Specific immunosuppression, 447
Specific T lymphocytes, 204
Splenectomy, 9
Stage-common antigens, 372
Stage-specific antigens, 372
Strain A mice, 31
Superoxide, 475, 479
Suppressor cells, 27
Suppressor T cells, 225
Suppressor T-cell (T_S)-mediated inhibition of anti-egg responses, 339
Surface antigens, 127
Surface antigens of microfilariae, 306
Surface labeling, 116
Surface protein of *P. knowlesi*, 112
Surgical transfer experiments, 388
Susceptible strains of mice, 26
Susceptibility of erythrocytes to oxidant stress, 471
Systemic schistosomiasis, 439

T cells, 77, 217, 243, 437, 447
T-cell clones, 214
T-cell-dependent immunity, 482
T-cell lines, 201, 212
T-cell responses, 83, 201, 205
T lymphocyte, 80, 93, 202, 468
T-lymphocyte lineage, 27
T-lymphocyte–macrophage interactions, 25
T-lymphocyte–dependent activation, 463
T-suppressor cells, 241
Taenia taeniaformis, 5
Taenia taeniaformis (murine cysticercosis), 326
Target antigens, 132
Thalassemic erythrocytes, 473
Thy-1, 89
Thymectomy, 26
Thymus, 19
Thymus dependence of the suppression, 286
Toxoplasma gondii, 5, 12, 13, 31, 54, 63
Transmissable gastroenteritis, 71
Trichinella spiralis, 5, 10, 12, 72, 84
Trichomonas vaginalis, 28
Trypanosoma brucei brucei, 16, 238, 240
Trypanosoma brucei rhodesiense, 241, 242
Trypanosoma congolense, 5, 238, 243, 244, 252, 253, 256
Trypanosoma cruzi, 5, 10, 56, 61
Trypanosoma lewisi, 11
Trypanosoma musculi, 5, 12, 13
Trypanosoma rhodesiense, 5, 23
Trypanosoma vivax, 34, 253
Trypanosomes, 11
Trypanosome antigens, 226
Trypanosome clearance, 248
Trypanosome elimination, 233
Trypanosome-infected mice, 239
Trypanosome-induced immunodepression, 233
Trypanosome-specific T cells, 236
Tumor-necrosis factor, 480

Unembryonated eggs of adult worms, 346
Urinary tract schistosomiasis, 442

Vaccines against parasites, 323
Vaccination trials in nonhuman primates, 385
Variant antigen, 157
Variant antigen phenotypes, 145

Variants of *P. knowlesi*, 165
Variant-surface glycoproteins (VSG), 226, 227
Variant-surface glycoproteins (VSG) of the African trypanosomes, 174
Vascular lesions in the pulmonary arteries of dogs, 429
VSG (*see* Variant-surface glycoproteins)

Worm proteinases, 380
Wuchereria bancrofti, 424

Xid, 23
Xid mutant mice, 22
Xid mutation, 21
X-linked, B-lymphocyte defect, 21